Techniques, Tools and Methodologies Applied to Quality Assurance in Manufacturing

Jorge Luis García Alcaraz ·
Cuauhtémoc Sánchez-Ramírez ·
Alfonso Jesús Gil López
Editors

Techniques, Tools and Methodologies Applied to Quality Assurance in Manufacturing

Springer

Editors
Jorge Luis García Alcaraz
Department of Industrial Engineering
and Manufacturing
Universidad Autónoma de Ciudad Juárez
Ciudad Juárez, Chihuahua, Mexico

Division of Research and postgraduate
Studies
Instituto Tecnológico de Ciudad Juárez
Ciudad Juárez, Chihuahua, Mexico

Alfonso Jesús Gil López
Department of Economics and Business
University of La Rioja
Logroño, La Rioja, Spain

Cuauhtémoc Sánchez-Ramírez
Division of Research and Postgraduate
Studies
Instituto Tecnológico de Orizaba
Tecnológico Nacional de México
Orizaba, Mexico

ISBN 978-3-030-69316-9 ISBN 978-3-030-69314-5 (eBook)
https://doi.org/10.1007/978-3-030-69314-5

This Springer imprint is published by the registered company Springer Nature Switzerland AG
The registered company address is: Gewerbestrasse 11, 6330 Cham, Switzerland

Preface

The product quality and price are two characteristics that are widely evaluated by the final consumer; therefore, managers are focused on obtaining adequate and balanced levels for both sides. Although, traditionally, it is considered that these two characteristics affect each other, because the presence of one characteristic may restraint the other, because a high quality may refer to a high cost, however, this perspective is not always presented, since it is possible to offer high-quality products and low prices at the same time.

This book presents a collection of real cases from industrial practices that production system and quality managers implement to ensure high quality as well as a low cost in products. This book is divided into sections that are focused on:

- The quality and lean philosophies implemented to production systems, starting from the product design as well as from the supply system.
- The principal statistical techniques applied to the quality assurance (statistical quality control, analysis of tests and failure, quality function deployment, accelerated life tests, among others), the process of gathering information, its validation, its reliability process, and techniques for data analysis.
- The techniques applied to the integration of human resources in the process of quality assurance, such as managers and operators' participation, education, and training processes.

The specific objectives of the book can be summarized as follows:

- Generate a dissemination venue from both the academia and the industry in the topics studied in the book, presenting cases of new approaches, applications, methods and techniques in quality and manufacturing.
- Generate a collection of theoretical and practical research works in the field of quality in manufacturing.
- Establish the state of the art in the field of quality.

This book is composed of a set of chapters, each of the kind of regular research papers. These works have been edited according to the norms and guidelines of Springer Verlag Editorial. Several calls for chapters were distributed among the main mailing lists of the field for researchers to submit their works to this issue. Twenty-eight expressions of interest in the form of abstracts were received in total, which were subject to a screening process to ensure their clarity, authenticity, and relevancy to this book. These proposals came from several countries such as Colombia, Mexico, Spain, Peru, Ecuador, Turkey, and Jordan.

After the abstract reviewing process, 25 proposals were accepted and asked to submit full versions. These versions were reviewed by at least two pairs in order to ensure the relevance and quality of the documents. After this process, 21 chapters were finally accepted for their publication once the corrections requested by the pairs and the editors were completed.

The book content is structured into three parts: (1) Quality Improvement Applications, (2) Optimization in Quality Development, and (3) Lean Techniques in Quality. The chapters in every of these parts are as follows.

Quality Improvement Applications: This part contains seven chapters.

The first chapter, entitled "An Integrated Quality Tools Approach for New Product Development", indicates that new product development (NPD) process requires a combined approach. However, the NPD team may not have experience with all stages of NPD processes. This chapter proposes an integrated tool that includes quality function deployment (QFD), failure modes and effects analysis (FMEA), Pareto analysis, and poka-yoke techniques by generating some recommendations in NPD processes to improve results.

The second chapter, entitled "Recent Optical Approaches for Quality Control Monitoring in Manufacturing Processes", identifies that optical methods have proved to be suitable for performing quality control processes in manufacturing since they present several advantages over traditional approaches such as non-destructive, non-contact, and high-speed techniques, among others. In this chapter, recent optical methods involving laser-based metrology systems and multi-parameter fiber optic sensors for quality control monitoring in areas such as materials, pharmaceutical, and chemical industry are presented.

In the third chapter, entitled "Reduction of the Scrap KPI in the Cutting Area of an Automotive Electrical Harness Company Using the Six Sigma DMAIC Methodology", the authors present the implementation of the define, measure, analyze, improve, and control (DMAIC) methodology as a tool for the reduction of scrap generated in the lines of the Ks area (cutting) of the company Electrics Plant located in Lagos de Moreno, Jalisco. The purpose sought with the implementation is to reduce the KPI of scrap in the area during the cutting and crimp process.

In the fourth chapter, entitled "Empirical Bayes Monitoring for Univariate and Multivariate Processes and Other Techniques", the authors used the basic concepts given in the celebrated Kalman filter, which can be derived using a Bayesian approach. Such an approach is implemented through Bayesian empirical monitoring for process control. The application analyzed data taken from a molding process of a critical quality characteristic of an automotive sensor. For the multivariate case, measures of the characteristics to be controlled of a molded part were taken.

The fifth chapter, entitled "Augmented Reality as an Innovative and Efficient Technology to Increase Quality in Manufacturing Processes", proposed the application of augmented reality (AR) in the clothing industry which allows, through the visualization of patterns in real time, manufacturing processes to be achieved with greater efficiency. AR provides cognitive ergonomics that lead to important benefits for the company such as optimization of manufacturing times, fast delivery of products, minimization of manufacturing costs, and greater economic profits.

In the sixth chapter entitled "Towards an Analysis of the Relationship Between Quality Management and Project Management", the authors indicate that modern quality management complements project management. This chapter presents the characteristics and models of project management that are related to quality management. Besides, a comparative analysis of the literature on quality management and project management is carried out. From research on quality, knowledge of project management should be deepened.

In seventh chapter, entitled "A Quality Management and Excellence Philosophy from an Islamic Standpoint", the author's studies adopt the inductive and descriptive approach to follow the most prominent principles of quality management in Islam's primary sources strengthened with real application cases and the deductive approach to extract the most prominent principles of quality management derived from them when compared to the modern quality philosophies.

Optimization in Quality Development: This part contains seven chapters.

In Chapter Eight, entitled "Multi-process Assessment Considering the Error of Measurement Systems Within the Process Capacity Indices", the purpose of this chapter is to make available to young researchers and process engineers the importance of integrating measurement systems with multi-process capability analysis systems using Z-values (short and long term) for continuous data considering the error of measurement systems, as well as the modification of process capacity indicators with the inclusion of the measurement error, determining its significance through ANOVA analysis.

In the ninth chapter, entitled "Experimentation and Multi-Objective Optimization in Manufacturing of Rubber for Shoe Sole", the authors indicate that improving quality and designing new products is an important research activity in industrial engineering. In this chapter, we present the study that was carried out on shoe soles through an experimental design. To find the optimal conditions of the process, the main characteristics of the sole were modeled. In that direction, the classical and generic multi-response optimization methods were compared, in this case both gave similar results. This experience will be useful to professionals working in the industry as a guide for doing research.

The tenth chapter, entitled "A Multi-agent System for the Inventory and Routing Assignment", presents a multi-agent system for solving the joint inventory and routing assignment problem, which integrates the individual and autonomously capacity and demand decisions of every network's members in a collaboration-based process that allows reducing the global distribution costs, demonstrating the impact of agent-based system on improving logistics processes.

The eleventh Chapter, entitled "Multi-objective Product Allocation Model in Warehouses", presents a multi-objective optimization model and a genetic algorithm procedure for its solutions, with the aim of simultaneously minimizing product handling costs and the required time to fulfill orders in an industrial warehouse, which allows defining the positions to locate the products in the warehouse' shelves, considering both critical objectives in warehouses.

In the twelfth chapter, entitled "Model Design of Material Requirement Planning (MRP) Applied to a Surgical Sutures Company", the authors implement the material requirements planning (MRP) to improve productivity in a company dedicated to the manufacturing of supplies for the health area. The use of material requirements planning improved the estimation of the number of raw materials required and optimized the scheduled deliveries within the company.

The thirteenth chapter, entitled "An EWMA Chart with Varying Sample Interval to Monitor Calibration Processes", is proposing an exponentially weighted moving average (EWMA) control chart with a varying sample interval based on Croarkin and Varner's statistical control chart, which was adopted by NIST. The original Croarkin and Varner chart has lower performance than some charts proposed in recent years; nevertheless, the performance of the modified chart that is proposed is comparable to that of better charts, given its detection ability in the face of changes to the calibration process.

In the fourteenth chapter, entitled "Application of Constraint Theory (TOC) on Information and Communication Technologies in Quality and Its Impact on the Circular Economy", the authors applied the TOC in the entire business value chain of a bank financial services company, from the highest level (strategy) to the primary activities of the company. The TOC model is proposed considering an analysis of the problem (design thinking), analysis related to the business processes of the payment line using LEAN-VSM (virtual stream mapping). It is applied in TOC's Drum Buffer Rope (DBR) method to define time and capacity buffers on the pay line.

Lean Techniques in Quality: This part contains seven chapters.

In the fifteenth chapter, entitled "Application of Lean Techniques and Queuing Theory in Food Services", the authors indicate that food services, specifically cafeterias, undesirable factors such as low servers' capacity, poor layout, and demand's variability may impact negatively by creating queues. This scenario is worsened during the rush time, generating the loss of customers and revenue decrease. Therefore, solutions such as lean principles jointly with queuing theory can address such problems and improve service performance. This chapter presents how the simultaneous application of such methodologies can boost the performance of a Mexican cafeteria and, thus, reduce the queues.

The sixteenth chapter, entitled "Dynamic Study of Soil Improvement for Sugarcane Cultivation in Colombia", was proposed to evaluate through the system dynamics methodology and the possible long-term impacts that this crop could generate in the Valle del Cauca's soils. The simulation's model was applied using Vensim DSS software, and it explored soil recovery scenarios using compost, which is produced from sugarcane residues composting (cachaça and bagasse). It was evident that the utilization of this by-product can represent an important contribution in the soil's loss and degradation reduction, plus economic and environmental benefits. Although the proposed model has been applied to the specific case of sugarcane, it can be replicated in other types of crop, thus becoming a valuable tool for the decision-making process involved in crop planning.

The seventeenth chapter, entitled "Lean Manufacturing Implementation in Management of Residues from Automotive Industry—Case Study", indicates that different types of waste can appear in companies' processes and in the recycling industry is not an exemption. Therefore, the adoption of lean manufacturing techniques helps to tackle these problems by diminishing or eliminating them. This chapter presents the case of an Ecuadorian company dedicated to the management of residues from the automotive industry, which experienced an improvement due to the successful implementation of lean tools, showing an enhancement in their CTQs and the employees' culture.

In the eighteenth chapter, entitled "Case Study of Lean Manufacturing Application in a New Process Introduction into a Rail Company", the authors indicate that productivity is a determining factor in the development of organizations. The objective of this chapter is introducing a new process in the remanufacturing of compressors by documenting, analyzing, and designing the process. Using define, measure, analyze, improve, and control (DMAIC) methodology and lean manufacturing techniques eliminated the waste and was the level of quality required by the client ensured.

The nineteenth chapter, entitled "Personnel Training as a Tool for Quality Assurance: Case of Study at Plastic Injection Enterprise", described how an inappropriate training to new personnel may severely affect the performance of a factory and the quality of its final products. A case of study of an automotive business shows how the production was often affected by delays due to errors made by the quality auditors' team, damage in equipment, and high staff turnover. The lack of a proper training for new personnel was a common issue to these problems.

In the twentieth chapter, entitled "Operational Risk Management in the Supply Chain of Blood Products", the blood must be a scarce resource in the world. Blood transfusions help to save and improve the quality of life of thousands of people around the world all the time. The rate of donation is currently considered low in Colombia. However, this is not the only problem for institutions responsible for the reception, storage, and distribution of blood products. Other factors within the biological and logistical control of the chain directly affect the safety and availability of blood. Due to the above, in this chapter, the authors proposed the identification, prioritization, and definition of actions aimed at mitigating or eliminating the main risks in the chain.

Finally, in the twenty-first chapter, entitled "Taguchi's Loss Function in the Weight Quality of Products: Case Study of Cheese Making", the authors applied the Taguchi methodology in a cheese company that belongs to the dairy industry to quantify the financial loss for both the consumer and the company by deviating a quality characteristic from its nominal value (product weight), determining the economic loss for five presentations, of product, and later establishing a mathematical equation and then applying an experimental design to improve the quality of the product in five different presentations.

Once a summary of chapters has been provided, the editors would like to express their gratitude to reviewers who kindly accepted to contribute to the chapters' evaluation at all stages of the editing process.

Ciudad Juárez, Mexico — Jorge Luis García Alcaraz
Orizaba, Mexico — Cuauhtémoc Sánchez-Ramírez
Logroño, Spain — Alfonso Jesús Gil López

Acknowledgements

Editors would like to thank all the authors of this book since, despite the vicissitudes caused by the pandemic period due to COVID-19, they continued with the editorial project. We glad for your confidence.

In addition, the editors wish to thank all colleagues who served as anonymous reviewers. Due to their timely support, this book has notably increased its quality. Our sincere gratitude is for that effort, especially in these difficult times, our gratitude for authors and co-authors who promptly responded to each recommendations and suggestions for improvement.

Finally, the editors wish to thank our families, for their support and for allowing us to dedicate time to this project, time that sometimes we should have spent with them. We appreciate their understanding, and we hope that this book is truly a contribution that justifies the time we have not spent with them in harmony.

Ciudad Juárez, Mexico — Jorge Luis García Alcaraz
Orizaba, Mexico — Cuauhtémoc Sánchez-Ramírez
Logroño, Spain — Alfonso Jesús Gil López

Contents

Editors and Contributors

About the Editors

Jorge Luis García Alcaraz is a full-time professor with adscription to the Department of Industrial Engineering at the Autonomous University of Ciudad Juárez. He received a bachelor degree and a M.Sc. in Industrial Engineering from the Instituto Tecnológico de Colima (Mexico), a Ph.D. in Industrial Engineering from Instituto Tecnológico de Ciudad Juárez (Mexico), a Ph.D. in Innovation in Product Engineering and Industrial Process from University of La Rioja (Spain), a Ph.D. in Engineering and Industrial Technologies, and a Postdoc in Manufacturing Process from University of La Rioja (Spain). His main research areas are related to multi-criteria decision making applied to lean manufacturing, production process modeling, and statistical inference. He is a founding member of the Mexican Society of Operation Research and an active member in the Mexican Academy of Industrial Engineering. Currently, he is National Researcher level III recognized by the National Council of Science and Technology of Mexico (CONACYT) and is working with research groups from Colombia, Spain, Chile, and Dominican Republic. He is author/co-author of around 150 papers published in journals indexed in the Journal Citation Reports, more than 200 international conferences and congress around the world. He is author of 12 books published by international publishers as Springer and IGI Global, all then related to lean manufacturing, the main tools, and techniques.

Scopus Author ID: 55616966800

Cuauhtémoc Sánchez-Ramírez is a full-time researcher from the Division of Research and Postgraduate Studies of the Orizaba Technology Institute. He received a Ph.D. in Industrial Engineering from COMIMSA, Center of Research of National Council of Science Technology of Mexico (CONACYT). His research projects have been recognized by CONACYT and PRODEP. He is a founding member in the Industrial Process Optimization Network (ROPRIN) as well as a member of the National Researcher System by CONACYT level 1. He has won the National Logistics Prize in

2017 that was given by the Secretariat of Economy, The Secretariat of Communications and Transportation of Mexico, and Soy Logístico Association in the Academic Category.

Similarly, he has supervised six bachelor theses, nine master theses, and three Ph.D. theses, and currently he is working on four master theses and one Ph.D. thesis that are in progress. He has been the editor of the following books: *New Perspectives on Applied Industrial Tools and Techniques*, published by Springer Verlag; *Handbook of Research on Managerial Strategies for Achieving Optimal Performance in Industrial Processes*, published by IGI Global Publishing. He has given six magistral conferences and workshops on industrial engineering, mainly focused on supply chain performance; he has two international patents.

Moreover, his research interests are modeling and simulation of logistics process as well as supply chain from a system dynamics approach. He is the author/co-author in around 50 journal and conference papers in logistics and supply chain management.

Scopus Author ID: 50361872500

Alfonso Jesús Gil López is an associate professor attached to the Department of Economics and Business at the University of La Rioja in Spain; he obtained a degree in Economics and Business from the National University of Distance Education UNED in Madrid, Spain, a doctorate in Sciences of the Education, and a doctorate in Economics and Business at UNED. Besides, he is a collaborating professor at the UNED Faculty, a tutor professor at the UNED Faculty of Economics and Business, and a visiting professor at the Universidade Estadual do Oeste do Paraná in Brazil. His main research areas are related to learning in organizations and its consequences in organizational development and knowledge management; he is currently working with research groups of renowned prestige in Spain and Mexico. He is the author of more than 25 articles published in high-impact journals. He has published numerous book chapters by international publishers as Springer and IGI Global, and he presented more than 50 conferences at international conferences.

Scopus Author ID: 55935671100

Contributors

Nataly Agudelo-Ibarguen Escuela de Ingeniería Industrial, Universidad del Valle, Valle del Cauca, Colombia

Hesham Al-Momani Department of Industrial Engineering, The Hashemite University, Zarqa, Jordan

G. Angelina López Pérez Ingeniería en Tecnologías de Manufactura, Universidad Politécnica de Durango, Durango, México

Martín Darío Arango-Serna Universidad Nacional de Colombia, Medellín, Colombia

Eddie N. Armendáriz-Mireles Department of Manufacturing Technologies Engineering, Polytechnic University of Victoria, Cd. Victoria, Tamaulipas, Mexico

Jonnatan Fernando Avilés-González Departamento de Ingeniería de Producción y Operaciones, Universidad del Azuay, Cuenca, Azuay, Ecuador

Sonia Valeria Avilés-Sacoto Departamento de Ingeniería Industrial, Colegio de Ciencias e Ingenierías, Universidad San Francisco de Quito, Quito, Pichincha, Ecuador

R. María Blanca Becerra Tecnológico Nacional de México: Campus Instituto Tecnológico de Querétaro, Santiago de Querétaro, Qro, Mexico

Carlos A. Calles-Arriaga Department of Mechatronics Engineering, Polytechnic University of Victoria, Cd. Victoria, Tamaulipas, Mexico

Elías Carrum-Siller COMIMSA-CONACYT, Saltillo, Coah, Mexico

José A. Castillo-Robles Department of Automotive Systems Engineering, Polytechnic University of Victoria, Cd. Victoria, Tamaulipas, Mexico

Ufuk Cebeci Department of Industrial Engineering, İstanbul Technical University, İstanbul, Turkey

Fiorella Cerna Universidad del Pacífico, Lima, Peru

Mario Chong Universidad del Pacífico, Lima, Peru

María Merced Cruz Rentería Department of Industrial Engineering, Instituto Tecnológico de San Luis Potosí, Soledad de Graciano Sánchez, México

Lidilia Cruz-Rivero Subdirección de Posgrado e Investigación, Tecnológico Nacional de México, Tantoyuca, Veracruz, México

Ma. Dolores Delgado Celis Department of CEA, Tecnológico Nacional de México, Soledad de Graciano Sánchez, México

Onur Dogan Department of Industrial Engineering, Izmir Bakircay University, Izmir, Turkey

Jorge Domínguez Domínguez Centro de Investigación en Matemáticas, Unidad Aguascalientes, Aguascalientes, Aguascalientes. C.P, México

Claudia Leticia Dávila Martínez Department of Industrial Engineering, Instituto Tecnológico de San Luis Potosí, Soledad de Graciano Sánchez, México

Beatriz Adriana Esparza-Ramírez Department of Industrial Engineering, Instituto Tecnológico José Mario Molina Pasquel y Henríquez Campus Lagos de Moreno, Jalisco. Adscrita al Posgrado de CIATEQ, A.C., Lagos de Moreno, Jalisco, Mexico

Rosa Hilda Félix Jácquez Department of Industrial Engineering, Instituto Tecnológico de San Luis Potosí, Soledad de Graciano Sánchez, México

Jorge Luis García Alcaraz Department of Industrial Engineering and Manufacturing, Universidad Autónoma de Ciudad Juárez, Juárez, Chihuahua, Mexico; Division of Research and postgraduate Studies, Instituto Tecnológico de Ciudad Juárez, Ciudad Juárez, Chihuahua, Mexico

Alfonso J. Gil Departamento de Economía y Empresa, Universidad de La Rioja, La Rioja, Spain

Cristian Giovanny Gómez-Marín Instituto Tecnológico Metropolitano, Medellín, Colombia

Catherine Gómez Universidad del Pacífico, Lima, Peru

Arturo Hernández-Hernández Universidad Politécnica de Querétaro, Santiago de Querétaro, Qro, Mexico

Marisol Hernández Hernández Autonomous Mexico State University, Mexico State, Mexico

Manuel Horna Universidad del Pacífico, Lima, Perú

Alfonso Alejandro Lara Negrete Department of Industrial Engineering, Instituto Tecnológico de San Luis Potosí, Soledad de Graciano Sánchez, México

Ana Luna Universidad del Pacífico, Lima, Peru

Juan López-Hernández Department of Automotive Systems Engineering, Polytechnic University of Victoria, Cd. Victoria, Tamaulipas, Mexico

Armando Mares Castro Tecnológico Nacional de México/ITS de Purísima del Rincón, Guardarrayas, Purísima del Rincón, Guanajuato. C.P., México

Mara Mataveli Departamento de Economía y Empresa, Universidad de La Rioja, La Rioja, Spain

Nicolás Francisco Mateo-Díaz Departamento de Ciencias Básicas E Ingenierías-Ingeniería Industrial, Universidad del Caribe, Cancún, Quintana Roo, México

Luis Adolfo Meraz-Rivera Subdirección de Posgrado e Investigación, Tecnológico Nacional de México, Tantoyuca, Veracruz, México

Galo Eduardo Mosquera-Recalde Departamento de Ingeniería Industrial, Colegio de Ciencias e Ingenierías, Universidad San Francisco de Quito, Quito, Pichincha, Ecuador

Mayerli Daniela Naranjo-Sánchez Escuela de Ingeniería Industrial, Universidad del Valle, Valle del Cauca, Colombia

Jesús Ortiz-Martínez Dirección de posgrados, Universidad Euro-Hispanoamericana, Xalapa, Veracruz, México

Luisa Fernanda Ortíz-Vasquez Universidad de San Buenaventura, Medellín, Colombia

Juan Carlos Osorio-Gómez Faculty of Engineering, Logistic and Production Research Group, Juan Carlos Osorio Gomez, Universidad del Valle, Cali, Colombia; Escuela de Ingeniería Industrial, Universidad del Valle, Valle del Cauca, Colombia

Wilian J. Pech-Rodríguez Department of Mechatronics Engineering, Polytechnic University of Victoria, Cd. Victoria, Tamaulipas, Mexico

Eduardo Rafael Poblano-Ojinaga Division of Postgraduate Studies and Research, TecNM/I.T. de Ciudad Juárez, Juárez Chihuahua, Mexico

Mercedes Estefanía Pérez-Naranjo Departamento de Ingeniería Industrial, Colegio de Ciencias e Ingenierías, Universidad San Francisco de Quito, Quito, Pichincha, Ecuador

Fernando Ricárdez-Rueda Department of Industrial Engineering, Instituto Tecnológico José Mario Molina Pasquel y Henríquez Campus Lagos de Moreno, Jalisco. Adscrita al Posgrado de CIATEQ, A.C., Lagos de Moreno, Jalisco, Mexico

Enrique Rocha-Rangel Department of Manufacturing Technologies Engineering, Polytechnic University of Victoria, Cd. Victoria, Tamaulipas, Mexico

Manuel Arnoldo Rodríguez Medina Division of Postgraduate Studies and Research, TecNM/I.T. de Ciudad Juárez, Juárez Chihuahua, Mexico

Manuel Alonso Rodríguez Morachis Division of Postgraduate Studies and Research, TecNM/I.T. de Ciudad Juárez, Juárez Chihuahua, Mexico

José A. Rodríguez-García Department of Manufacturing Technologies Engineering, Polytechnic University of Victoria, Cd. Victoria, Tamaulipas, Mexico

Manuel Iván Rodríguez Division of Postgraduate Studies and Research, TecNM/I.T. de Ciudad Juárez, Juárez Chihuahua, Mexico

Vicente Onofre Rojo-Romo Department of Industrial Engineering, Instituto Tecnológico José Mario Molina Pasquel y Henríquez Campus Lagos de Moreno, Jalisco. Adscrita al Posgrado de CIATEQ, A.C., Lagos de Moreno, Jalisco, Mexico

Gerardo Reyes Ruiz Autonomous Mexico State University, Mexico State, Mexico; Naval Higher Studies Center, Mexico City, Mexico

María Guadalupe Russell-Noriega Facultad de Ciencias Físico-Matemáticas, Universidad Autónoma de Sinaloa, Ciudad Universitaria, Culiacán, Sinaloa, Mexico

Abril Salaices Medrano Ingeniería en Tecnologías de Manufactura, Universidad Politécnica de Durango, Durango, México

Conrado Augusto Serna-Urán Instituto Tecnológico Metropolitano, Medellín, Colombia

Cuauhtémoc Sánchez-Ramírez Tecnológico Nacional de México: Campus Instituto Tecnológico de Orizaba, Veracruz, Mexico

Nadia Sánchez Universidad del Pacífico, Lima, Peru

Luz Elena Terrazas Mata Division of Postgraduate Studies and Research, TecNM/I.T. de Ciudad Juárez, Juárez Chihuahua, Mexico

Daniel Torres Ramírez Department of Industrial Engineering, Instituto Tecnológico de San Luis Potosí, Soledad de Graciano Sánchez, México

Luz Karime Torres-Lozada Faculty of Natural Sciences and Engineering, Luz Karime Torres Lozada, Fundación Universitaria Unisangil, Yopal, Colombia

Patricia Torres-Lozada Faculty of Engineering, Study and Control of Environmental Pollution –ECCA Research Group, Patricia Torres Lozada, Universidad del Valle, Cali, Colombia

Gerardo Alonso Torres-Ávalos Department of Industrial Engineering, Instituto Tecnológico José Mario Molina Pasquel y Henríquez Campus Lagos de Moreno, Jalisco. Adscrita al Posgrado de CIATEQ, A.C., Lagos de Moreno, Jalisco, Mexico

Dayhanna Stephanía Vargas-Mesa Faculty of Engineering, Logistic and Production Research Group, Dayhanna Stephania Vargas Mesa, Universidad del Valle, Cali, Colombia

Enrique Villa-Diharce Department of Probability and Statistics, Centro de Investigación en Matemáticas, Guanajuato, Mexico

Darwin Young-Vazquez CILSS Consulting, Saltillo, Coah, Mexico

Julián Andrés Zapata-Cortes Institución Universitaria CEIPA, Sabaneta, Colombia

List of Figures

An Integrated Quality Tools Approach for New Product Development

Recent Optical Approaches for Quality Control Monitoring in Manufacturing Processes

Reduction of the Scrap KPI in the Cutting Area of an Automotive Electrical Harness Company Using the Six Sigma DMAIC Methodology

Empirical Bayes Monitoring for Univariate and Multivariate Processes and Other Techniques

Augmented Reality as an Innovative and Efficient Technology to Increase Quality in Manufacturing Processes

Towards an Analysis of the Relationship Between Quality Management and Project Management

Multi-process Assessment Considering the Error of Measurement Systems Within the Process Capacity Indices

Experimentation and Multi-Objective Optimization in Manufacturing of Rubber for Shoe Sole

A Multi-agent System for the Inventory and Routing Assignment

Multi-objective Product Allocation Model in Warehouses

Model Design of Material Requirement Planning (MRP) Applied to a Surgical Sutures Company

An EWMA Chart with Varying Sample Interval to Monitor Calibration Processes

Application of Constraint Theory (TOC) on Information and Communication Technologies in Quality and Its Impact on the Circular Economy

Application of Lean Techniques and Queuing Theory in Food Services

Dynamic Study of Soil Improvement for Sugarcane Cultivation in Colombia

Lean Manufacturing Implementation in Management of Residues from Automotive Industry—Case Study

Case Study of Lean Manufacturing Application in a New Process Introduction into a Rail Company

Personnel Training as a Tool for Quality Assurance: Case of Study at Plastic Injection Enterprise

Operational Risk Management in the Supply Chain of Blood Products

Taguchi's Loss Function in the Weight Quality of Products: Case Study of Cheese Making

List of Tables

Augmented Reality as an Innovative and Efficient Technology to Increase Quality in Manufacturing Processes

Towards an Analysis of the Relationship Between Quality Management and Project Management

Multi-process Assessment Considering the Error of Measurement Systems Within the Process Capacity Indices

Experimentation and Multi-Objective Optimization in Manufacturing of Rubber for Shoe Sole

A Multi-agent System for the Inventory and Routing Assignment

Multi-objective Product Allocation Model in Warehouses

Model Design of Material Requirement Planning (MRP) Applied to a Surgical Sutures Company

An EWMA Chart with Varying Sample Interval to Monitor Calibration Processes

Application of Constraint Theory (TOC) on Information and Communication Technologies in Quality and Its Impact on the Circular Economy

Application of Lean Techniques and Queuing Theory in Food Services

Dynamic Study of Soil Improvement for Sugarcane Cultivation in Colombia

Lean Manufacturing Implementation in Management of Residues from Automotive Industry—Case Study

An Integrated Quality Tools Approach for New Product Development

Onur Dogan and Ufuk Cebeci

Abstract New product development (NPD) is critical in competitive environment, because of short product lifecycles, the rapidly changing technology and customers' new requirements. When developing new products, the NPD team needs an integrated approach, because they should go back to the necessary stages according to the feedback and outputs gathered in NPD process. Sometime, team members do not know when and how to go to the necessary stages. Therefore, an integrated quality tools framework for NPD is proposed. Quality function deployment (QFD), failure modes and effects analysis (FMEA), Pareto analysis and poka-yoke techniques are used mainly. Pareto analysis is used to start corrective actions for risk priority numbers (RPN) values in FMEA stage, instead of threshold value to decrease subjectivity. The proposed methodology also gives some recommendations about teamwork and NPD documents such as quality control plan and product manual to increase the efficiency of NPD process and design quality. The proposed integrated framework is applied in an industrial production company as a real case study, and the results are satisfactory.

Keywords New product development · Quality function deployment · Failure modes and effects analysis · Poka-Yoke · Total quality management tools

1 Introduction

In increasing competition environment, new product development (NPD) has an important role. The competitive business environment is characterized by the following factors (Bellary and Murthy 1999):

- Rapidly changing technologies have made the product life cycles shorter.
- Dismantling of trade barriers has resulted in globalization of the marketplace.

O. Dogan (✉)
Department of Industrial Engineering, Izmir Bakircay University, 35665 Izmir, Turkey
e-mail: onur.dogan@bakircay.edu.tr

U. Cebeci
Department of Industrial Engineering, İstanbul Technical University, 34367 İstanbul, Turkey

J. L. García Alcaraz et al. (eds.), *Techniques, Tools and Methodologies Applied to Quality Assurance in Manufacturing*,
https://doi.org/10.1007/978-3-030-69314-5_1

- Customer expectations are ever increasing.

As the product life cycles become shorter, continuous improvement in the new product development is getting critical for manufacturing firms. Because of shortened product life cycles and the desire of firms to reduce the new product development (NPD) cycle time, NPD speed has become an important research issue (Ozer and Cebeci 2009). They studied the role of globalization in NPD. New product development consists of the activities of the firm that lead to a stream of new or changed product market offerings over time. This includes the generation of opportunities, their selection and transformation into artifacts (manufactured products) and activities (services) offered to customers and the institutionalization of improvements in the NPD activities themselves (Loch and Kavadias 2008). NPD process is a long and dynamic period which includes creating a new idea and converting it to a commercialization product. Kotler and Keller (2011) classify eight NPD phases that are idea generation, idea screening, concept development and testing, market strategy development, business analysis, product development, market testing and commercialization.

When the product risks could be defined before manufacturing, product can be launched to market as soon as possible at a high-quality level. One of the different standard approaches is failure modes and effects analysis (FMEA). FMEA is a technique that aims to define failures and then remove or decrease their effects before they exist about system, design, process or service. In this method, the failures are sorted with respect to risk priority number (RPN). Not only quality function deployment (QFD) but also FMEA has been implemented for different problems such as sorting failures according to RPN based on a fuzzy rule-based Bayesian reasoning (Yang et al. 2008), reliability of a wind turbine (Arabian-Hoseynabadi et al. 2010) or decreasing negative impacts in harsh operating conditions of floating offshore wind turbine (Zammori and Gabbrielli 2012).

An integrated quality tools methodology including quality function development, failure modes and effects analysis, poka-yoke and Pareto analysis facilitates and improves efficiency for new product development processes. QFD is a quality tool for developing a design quality aimed at satisfying the consumer and then translating the consumer's demand into design targets and major quality assurance points to be used throughout the production phase (Akao 2004; Dogan and Cebeci 2016). QFD includes different subjects such as developing customer satisfaction and gaining competitive advantages at the initial design stage of the product (Cohen 1995), optimizing the product development (Delice and Güngör 2009), identifying service quality in a library (Chen and Chou 2011) or converting customer needs to service requirement (Na et al. 2012).

NPD team members sometimes do not know when and how to go to the necessary stages. Therefore, an integrated framework for NPD is developed to support their decisions. Classical FMEA technique starts a corrective action if any RPN is greater than 100. Pareto analysis is used to start corrective actions for RPN values in FMEA stage of proposed framework, instead of threshold value to decrease subjectivity. Because if a threshold is fixed, there is a tendency of some people manipulating data to keep values within threshold so that no corrective action is needed. Another advantage of Pareto analysis is that FMEA team can use smaller values for severity, probability and detectability. Therefore, they can obtain smaller RPN (<100), and they consider that no action is needed.

Documentation recommendation should be recorded as quality records, because the company can use the results for new NPD projects. The rest of this paper has been organized as follows:

- The related works are given to show the literature gap about the study scope.
- Methodologies: QFD, FMEA, poka-yoke and Pareto analysis quality tools are explained.
- The developed integrated methodology is explained.
- A real case study is applied in a manufacturing company.
- Conclusion and further research of the study are argued.

The remainder of this study is organized as follows. Firstly, related works are reviewed to show literature gaps in Sect. 2. The techniques used in the study are presented in Sect. 3. Section 4 introduces the developed integrated methodology. The results of the applied methodology are given in Sect. 5. Finally, the discussion and conclusion section is shown in Sect. 6.

2 Literature Review

QFD and FMEA are two methods that can be used together for various purposes. Ginn et al. (1998), Tan (2003), Braglia et al. (200)7, Almannai et al. (2008), Korayem and Iravani (2008), Chen and Ko (2009), Hassan et al. (2010) used both QFD and FMEA in their studies. Braglia et al. (2007) integrated the house of quality (QFD/HoQ) concepts to FMEA and house of reliability (HoR). Doshi and Desai (2017) applied FMEA to four automotive suppliers for continuous improvement and reduced quality rejections. Nevertheless, none of them is about NPD. Hunt (2005b), Moldovan (2014) and Zaim et al. (2014) focused on product development and design only using only QFD.

Some researchers focused on development of the integration of QFD and FMEA in the field of NPD (Chen and Ko 2009; Ko 2013; Lin et al. 2015). Chen and Ko (2009) applied fuzzy linear modeling to QFD. FMEA was treated as the constraint in the models. Ko (2013) first used FMEA to evaluate potential failure modes based on the functions of the parts of the new product. Then by using QFD, FMEA parameters which are occurrence probability, detectability and severity were decreased. Lin et al. (2015) used fuzzy logic to evaluate QFD and FMEA value, and the parameters were used as inputs to NPD. Ma et al. (2019) proposed a new index to calculate the importance of components by integrating the internal and external failure effects. They developed a fuzzy permanent function for the internal failure effects and the external failure effect to measure the external failure effects.

Table 1 compares some previous works with the proposed methodology. Some works applied QFD and FMEA together (Chen and Ko 2009; Ko 2013; Lin et al. 2015; Shaker et al. 2019). Gu et al. (2019) studied QFD to fully consider the interaction between various failure modes and the customer satisfaction degree with product performance, economy and service, and a certain diesel engine fuel system for FMEA is used as a case study.

Pun et al. (2019) studied to minimize the risks of new product development and shorten time-to-market, particularly for high-tech enterprise where the complexity of

Table 1 Comparison of similar approaches

	QFD	FMEA	Recommendation	Quality documentation	Pareto	Poka-yoke	Real case
Hunt (2005b)	+	–	–	–	–	–	+
Chen and Ko (2009)	+	+	–	–	–	–	+
Wang and Chen (2012)	+	–	–	–	–	–	+
Ko (2013)	+	+	–	–	–	–	–
Moldovan (2014)	+	–	–	–	–	–	+
Zaim et al. (2014)	+	–	–	–	–	–	+
Ionica and Leba (2015)	+	–	–	–	–	–	+
Lin et al. (2015)	+	+	–	–	–	–	+
Chen et al. (2017)	+	–	–	–	–	–	+
Goda et al. (2017)	+	+	–	–	–	–	+
Ng et al. (2017)	+	+	–	–	+	+	+
Tonchia (2018)	+	+	–	+	+	–	+
Paprocki (2018)	+	+	–	–	–	–	+
Pun et al. (2019)	–	+	–	–	–	–	+
Shaker et al. (2019)	+	+	–	–	–	–	+
Baskar et al. (2019)	–	+	+	–	–	–	+
Proposed methodology	+	+	+	–	+	+	+

the product generates vast amount of failure mode and applied in flexible electronics industry. Tonchia (2018) highlighted ICT tools such as CAD/CAM, rapid prototyping, product data management (PDM) and PM software to manage NPD. Variety reduction program, design for manufacturability/assembly, concurrent/simultaneous engineering, DOE, FMEA, QFD, etc., are involved and reported the design by platform. Goda et al. (2017) studied how the failure mode and effect analysis (FMEA), the rapid prototype modeling and the quality function deployment (QFD) are extended among the agricultural machinery manufacturers for NPD in Hungary by applying a survey. Ng et al. (2017) reviewed the integration of FMEA with other problem-solving techniques such as QFD, root cause analysis (RCA) and seven basic tools of quality. Paprocki (2018) presented the techniques aiding ecological design of the product development, such as DFMA, DFE, FMEA, QFD and LCA. Baskar et al. (2019) proposed fuzzy VIKOR-based fuzzy failure mode and effect analysis (FMEA). Based on the results of fuzzy FMEA, the developed prototype model is redesigned and then finally recommended for commercialization and applied for sesame seed separator development.

The proposed integrated methodology uses also additional quality tools including poka-yoke and Pareto analysis, and according to the results, the team can go back to the necessary stages. The integrity of the methodology increases NPD process efficiency because it gives some useful recommendations about teamwork, quality control documentation and other stages. A real industrial case study was applied to test the proposed methodology. As it is visible, most of the related works on the NPD, especially the integrative strategy, used only QFD and FMEA techniques. The common aspect of the proposed methodology and previous studies is that QFD is first applied, and then, the results were used into FMEA for additional examination. However, for successful NPD implementations, some additional quality tools and recommendations based on executed activities in the process should be considered. The contribution of the study is about this significant point.

3 Quality Management Tools

The study includes four tools for new product development: quality function deployment, failure modes and effect analysis, poka-yoke and Pareto analysis. The general overview of the study is shown in Fig. 1.

3.1 Quality Function Deployment (QFD)

The literature offers a number of definitions for QFD. It is a method for understanding customer outcomes and developing comprehensive product specifications (Hunt et al. 2005a). QFD is a concept that provides translating customer requirements into technical characteristics for product development. It uses house of quality

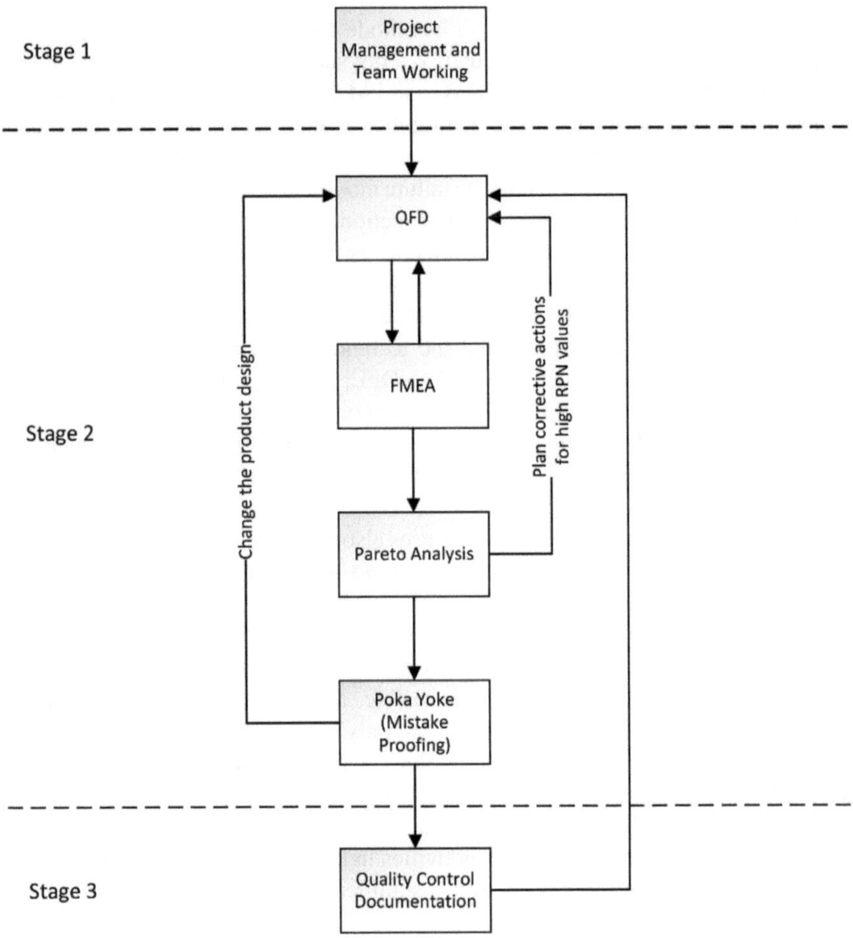

Fig. 1 Overview of the integrated framework

(HOQ) to development product. HOQ, such as the one shown in Fig. 2, is a matrix that compares customer requirements and quality characteristics. It is the most important phase of QFD since it is in this phase that the customer requirements for the product are identified, and then technical measures and their priorities are determined.

QFD can be thought in concept development and testing phases according to Kotler's NPD (2011) process phases. Steps of the QFD are explained below briefly.

Step 1: Customer Requirements—"Voice of the Customer": The first step in a QFD project is to identify who the customers are. Then, information is gathered from customers for the product. These requirements are rated the importance of each requirement on a scale from 1 to 5. This scale will be used later in the relationship matrix.

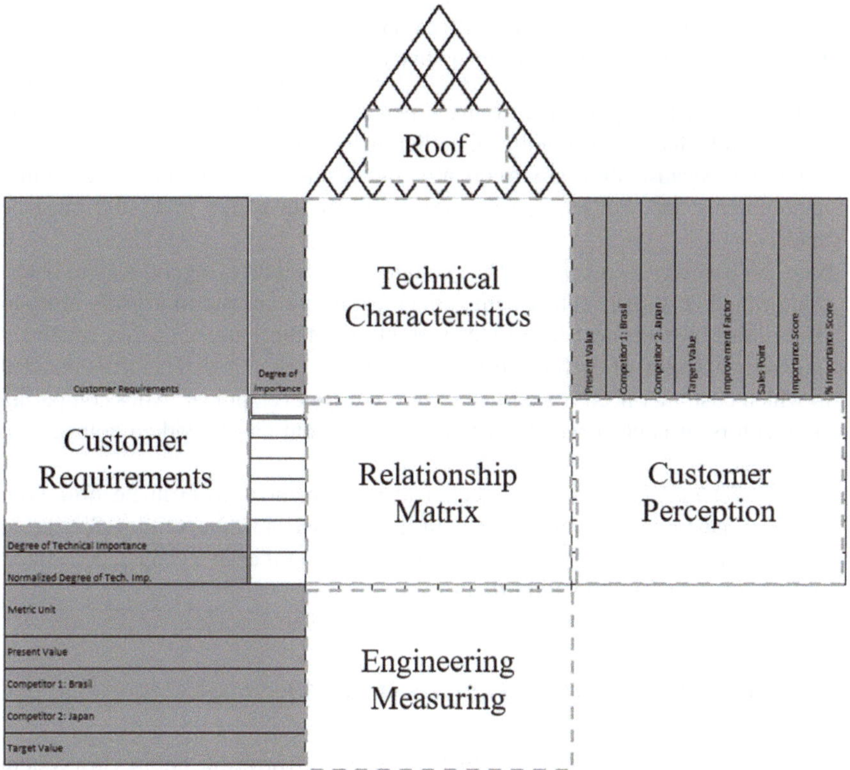

Fig. 2 House of Quality (HOQ)

Step 2: Technical Characteristics—"Voice of the Engineer": There are some technical requirements which the customers cannot identify. These requirements will be either technical or regulatory requirements. Regulatory requirements are things such as government legislation, safety requirements, quality and product standard requirements and classification requirements. Technical requirements occur because the company may have some specific plans for the new product.
Step 3: Relationship Matrix: The relationship matrix is where it determined the relationship between customer needs and the company's ability to meet those needs. It asked the question, "What is the strength of the relationship between the technical characteristics and the customer needs?". Relationships can be "weak, moderate or strong" and carry a numeric value of "1, 3 or 9". We demonstrate them as symbols, respectively, ▲, ■ and ●.
Step 4: Interrelationship Matrix between Technical Characteristics: The roof of the house of quality, called the correlation matrix, is used to identify any interrelationships between each of the technical characteristics. Symbols are used to describe the strength of the interrelationships, for example,

- A plus sign (+) represents a positive relationship.
- A minus sign (−) represents a negative relationship.

Correlation matrix allows identifying which technical characteristics support one another and which are in conflict. Conflicting technical descriptors are extremely important because they are frequently the result of conflicting of customer requirements and, consequently, represent points at which trade-offs must be made.

Step 5: Customer Assessment: The customer assessment is a good way to determine which customer requirement is the most important and to identify areas to concentrate on in the next design. At the right side of the house of quality, relatively importance scores of each of customer needs are calculated. Then it is decided to which customer need is significant for product development. After comparing competitors, it is clear which customer requirement can provide a competitive advantage.

Step 6: Technical Assessment: Technical assessment is to evaluate how your company performs in comparison with its most serious competitors. Similar to the customer assessment, the products are evaluated for each technical descriptor, and weights are assigned all of them.

3.2 Failure Mode Effects Analysis (FMEA)

The FMEA is an analysis procedure which documents all probable failures in a system within specified ground rules, determines by failure mode analysis the effect of each failure on system operation, identifies single failure points and ranks each failure according to a severity classification of failure effect (US Department of Defense 1980).

One of the best features of FMEA is its action manner rather than reaction in dealing with failure. In other words, this is an action before failure rather than after; because usually a lot of money will be spent to resolve problems and damage caused (Bahrami et al. 2012). Traditionally, applying FMEA has been done by developing risk priority number (RPN). RPN is the value obtained by the product of three components: the occurrence probability of a failure mode (P), the severity of the failure mode (S) and the detectability of the failure mode (D). Higher the value of the RPN, higher is the risk associated with the corresponding failure mode. The purpose of RPN is to prioritize the failure modes of a product or system, so that the available resources can be effectively allocated (Mandal and Maiti 2014). Mathematically, the RPN is formulated as

$$\text{RPN} = P \times S \times D \tag{1.1}$$

Table 2 Example of a point scale of linguistic expressions (Lange et al. 2001)

Probability of failure	Number of occurrence	Ranking
Very high: failure is almost inevitable	Larger than 1 in 2	10
	1 in 3	9
High: repeated failures	1 in 8	8
	1 in 20	7
Moderate: occasional failures	1 in 80	6
	1 in 400	5
	1 in 2000	4
Low: relatively few failures	1 in 15,000	3
	1 in 150,000	2
Remote: failure is unlikely	Less than 1 in 1,500,000	1

where risk parameters, *P*, *S* and *D*, are measured in the point scale from 1 to 10. In addition, some linguistic expressions can be used corresponding to numerical value in the point scale (Table 2).

3.3 Poka-Yoke

The Japanese concept of poka-yoke, also called as mistake-proofing, is a quality improvement methodology. It aims to detecting and correcting problems as close to the source as possible to minimize the negative consequences (Krajewski et al. 2013). Some approaches use data from process outputs to suggest how best to manage the process. In these approaches, two functions are required: (1) The defect or its cause must be detected, and (2) corrective action must be taken (Grout and Toussaint 2010). Therefore, poka-yoke solutions are developed on the basis of the recorded or forecasted mistakes that can be made in a process.

There are six mistake-proofing principles as follows:

- Elimination principle eliminates possibility of errors by redesigning the product or process. For example, product simplification avoids a part defect or assembly error in the first place. In addition, it is a principle related to FMEA.
- Replacement principle is used to improve reliability by simply substituting an unpredictable process with a more reliable process. For example, using robotics or automation prevents a manual assembly error.
- Prevention principle indicates that design engineers should design the product or process so that it is impossible to make a mistake at all. For example, part features only allow assembly the correct way or unique connectors to avoid misconnecting wire harnesses or cables.

- Facilitation principle makes the assembly process easier to perform by utilizing specific methods and grouping steps. For examples, visual controls include color coding, marking or labeling of parts.
- Detection principle is used to detect process or product errors before they move to the next processing step so that the user can quickly correct the problem. For example, sensors are in the production process to identify when parts are incorrectly assembled. In addition, it is one of the principles related to FMEA.
- Mitigation principle is an attempt to decrease the effects of errors. For examples, products are designed with low-cost, simple rework procedures when an error is discovered.

Poka-yoke principles can be prioritized by performing FMEA. Elimination, detection and mitigation principles are related to FMEA parameters *P*, *D* and *S*, respectively.

3.4 Pareto Analysis

Pareto analysis is a problem-solving technique used to distinguish important causes of a problem from relatively less important causes. In the Pareto analysis, there is a rule known as 80/20. Accordingly, in many cases, 80% of the results are due to 20% of the causes. Juran (1975) recognized the principles of the "vital few" and "trivial many" where "80% of the overall impact of errors in any industrial scenario is due to a small number of error types, termed the 'vital few' and 20% of the impact due to other error types, called the 'trivial many'" (Sarkar et al. 2013).

The Pareto diagram can be drawn by following the steps below.

- The criterion to be examined is determined.
- All reasons related to these criteria are listed.
- The frequency with which the observations are observed is measured.
- The most frequent observations, starting from the beginning, are noted.
- The cumulative frequency value is calculated and plotted.

4 Developed Integrated Methodology

The framework consists of mainly three stages. The first stage is about project team and project management. For reliable solutions, qualified and informed team members are an important part to apply the integrated framework. The second stage includes integrated form of the QFD and FMEA techniques. Developing new product starts with QFD. Customer requirements are gathered, and technical characteristics are determined. Then by using FMEA, risks are determined in the design stage of the new product. In last stage of the methodology, some documents such as quality control plan and working instructions to apply the methodology are prepared.

4.1 Stage 1: Project Management and Team Building Recommendations

Figure 3 shows the first main stage of the integrated framework. The first step of the suggested methodology is to form a multidisciplinary team. The approach gives some recommendations as follows:

- Team should be multidisciplinary, and members should be related to expertise fields such as quality, manufacturing, research and development and material.

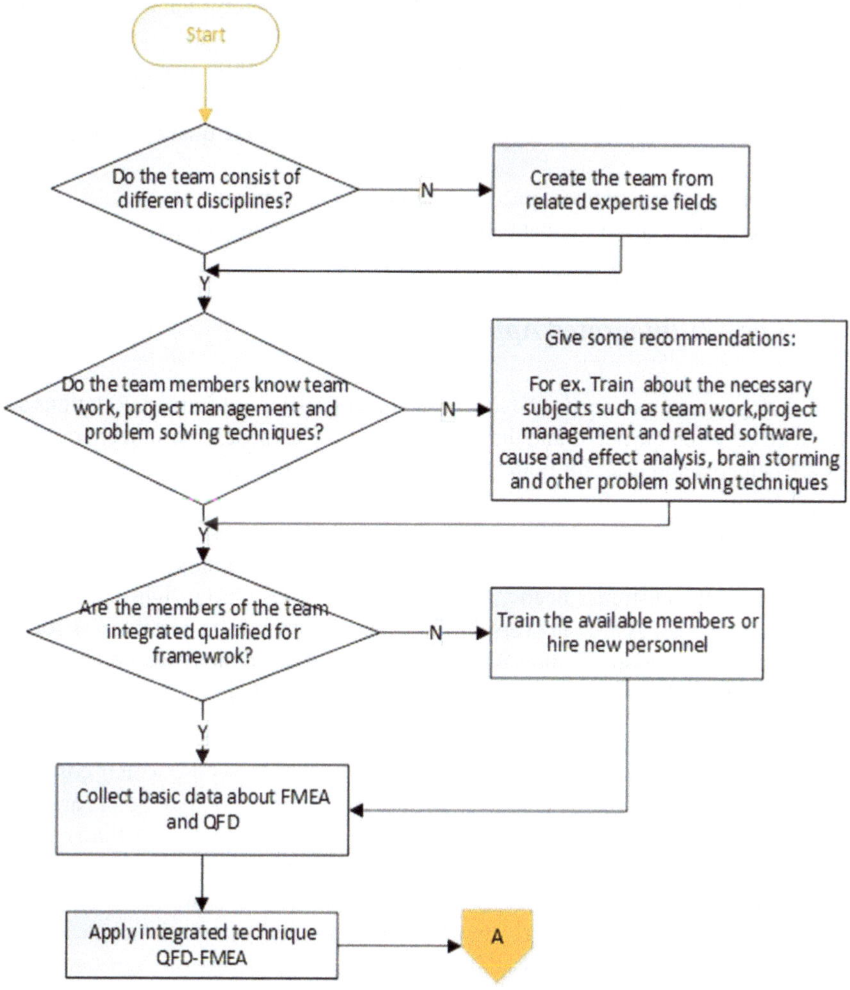

Fig. 3 Stage 1 flowchart: project management and team building

- Various training lessons for the team members should be given such as FMEA, QFD, project management and teamwork necessary. If the members do not have any basic and specific knowledge, they should train about the necessary subjects such as teamwork, project management and related software, cause and effect analysis, brainstorming and other problem-solving techniques.
- It is also important that the team member should be qualified for the framework. There are two possible solutions: either training the available members or hiring new personnel.
- If the sector is automotive, train about ISO 16949 automotive quality standard. In this sector, FMEA table is different, use standard's own FMEA table.
- If the sector is food, train about ISO 22000 food quality safety management system standard and HACCP chapter (hazard analysis of critical control points).
- If the sector is related to information, train about ISO 27001 information security management which helps organizations keep information assets secure.
- If the sector is aerospace industry, train about AS9100 quality management systems.
- Requirements for aviation, space and defense organizations.
- After building team, basic data about QFD and FMEA are gathered.

4.2 *Stage 2: Integrated Approach*

In QFD stage, customer needs and technical characteristics are defined. Relations are evaluated between the needs and the characteristics. The most important characteristics are detected and used for both QFD and FMEA. A risk assessment is conducted using FMEA technique for the characteristics.

In the developed methodology, QFD answers the questions "What are the most important customer needs?" and "What are the most important technical characteristics to meet the customer needs?". So, the firm focuses on the right option in its current system. FMEA also answers the questions "What may be possible risk for the determined characteristic?" and "What activities may avoid the potential failures?". The flowchart of the integrated framework shown in Fig. 4 is briefly explained.

- The new idea can belong to a new product or an available product to develop it. Result of the QFD study is that to meet customer needs, which customer requirement and technical characteristic are more important than others are.
- According to the defined new idea, it is determined that what customer needs should be and to meet the needs what technical characteristics should be.
- For a new product, conditions of the rivals are important. In order to compare with rivals, some data of the rivals are entered into the system to evaluate results. Data about rivals can be guessed by experts.
- Consideration of gathered data entered house of quality.
- Relation matrix, correlation matrix (roof of the house) and competitors' data are filled into the house of quality.

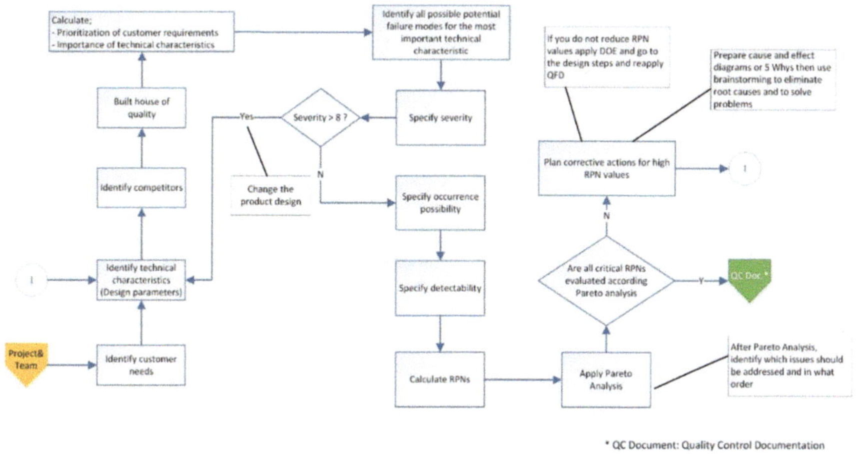

Fig. 4 Stage 2 flowchart: integrated approach

- Prioritization of customer requirements (customer perception) and importance of technical characteristics (engineering measures) are calculated. QFD stage ends here.
- The first step of the FMEA is to identify possible potential failure modes of the product for the technical characteristic which has the biggest importance.
- Specify for all possible potential failure modes possibility of occurrence, severity and detectability, respectively. Risk priority number is calculated. In addition, it is controlled whether it is larger than 100 or not. For failure modes which has their RPNs larger than 100, Pareto analysis is applied to determine root causes. Then, some corrective activities are planned to avoid the cause.

4.3 *Stage 3: Quality Control Documentation*

The last stage is about some documentations related to implementation of the methodology (Fig. 5). As a result of the second stage, preparing some documents is useful in application. Final documentations should be prepared and published such as quality control plan and work instructions. These help to organize the product development process. In addition, safety instructions are prepared, which is related to FMEA results.

Standard operation procedure and other related documents are prepared. When the product needs Conformity Europe (CE) certification, safety risk analysis is added by using FMEA. Finally, list of raw materials and semi-finished necessary to produce the product are prepared, and the methodology ends.

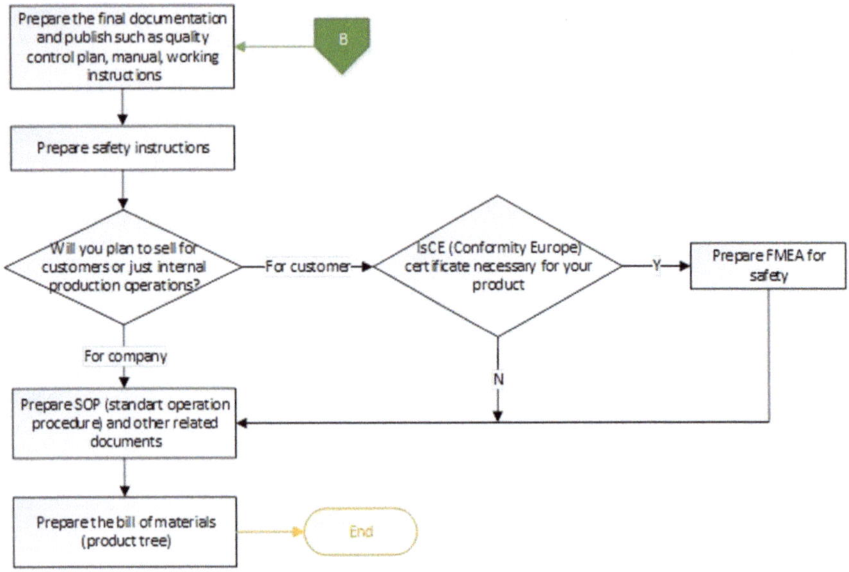

Fig. 5 Stage 3 flowchart: quality control documentation

5 Industrial Application

The developed methodology is applied in a metal plating firm which is established in 1986. The firm has been growing by investing new equipment and having complete customer satisfaction. The firm wants to be a worldwide company by enhancing its technological infrastructure. Therefore, the firm intends to increase customer satisfaction and production capacity.

5.1 Stage 1: Project Management and Team

A team consisting of five members was formed. Two are from quality department, one is from marketing department, and two are from manufacturing department. A meeting was organized to train the members. Teamwork, project management, cause and effect analysis, brainstorming and problem-solving techniques are taught. Some recommendations were made:

- Since the team consists of related expertise fields, train about the necessary subjects such as teamwork, project management and related software, cause and effect analysis, brainstorming and other problem-solving techniques.
- The company works metals for the automotive industry, so ISO 16949 automotive quality standards must be conformed.
- Training the available members is more suitable for the company conditions.

5.2 Stage 2: Integrated Approach

Customer needs were determined by considering all demands, complaints and satisfactions from customer by e-mail, phone and fax in last 5 years. Generally, they are classified into five classes: price, rust, quick delivery, plating quality and homogeneous quality. After determining the customer requirements, technical characteristics were also determined. Voice of the engineer was determined corresponding to each customer need. Generally, they are classified into eight classes: number of baths, mix of chemicals, bath parameters, quality of pure water, test equipment, degree of automation, processing time of oven and transportation equipment. For benchmarking, two competitors, one of them from Brazil and the other is from Japan, are determined. Some targets are set according to customers' level in customer assessment and technical assessment because becoming leader in sector, one of the aims of the firm, depends on this.

Relationship between customer needs and the company's ability to meet those needs was determined together with the team members (Relationship matrix). Then relations between the technical characteristics are evaluated (Correlation matrix). Then some calculations are made (Fig. 6).

As a result of the QFD, it can be seen easily that the most important customer need is "plating quality" with 22.59% scores. According to outcomes of QFD, "processing time of oven" is the most important technical characteristic, and at the same time, it is the input for FMEA. Failure modes and recommended activities are determined

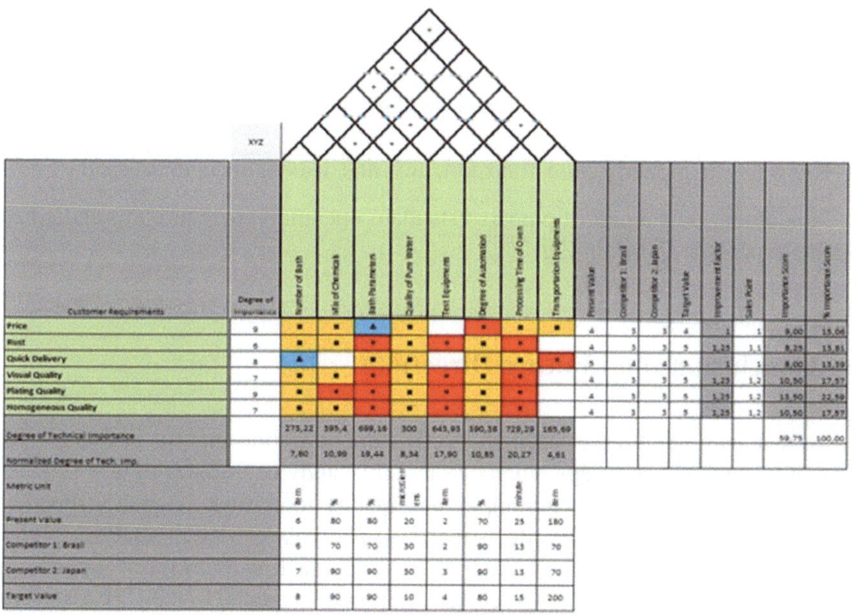

Fig. 6 House of quality created

by experts and company officials. At the end of the FMEA, for processing time of oven, the critical failure mode was found low corrosion. To prevent it, regular maintenance and control with litmus paper are advised for recommended activity. Therefore, designing the product in parallel with customers' need is less risky. Table 3 shows only a part of the study.

5.3 *Stage 3: Documentation*

All testing results were recorded on the daily quality control report. Any concerns or deviations from the required material specifications and the actions taken to correct the problems were noted on the report. Information recorded on the daily quality control report includes

- Number of personnel
- Types and numbers of tests performed
- Nature of defects or cause for rejection
- Proposed corrective action(s), corrective actions applied and closing date
- Delays encountered
- Directions received from the engineer and actions taken
- Health and safety issues or deficiencies and how they were resolved
- Deficiencies (rework and scrap).

Some evaluations and recommendations are made for general guidance only. A part of prepared safety instruction and lessons learned can be shown below:

- You must get training in chemical safety and proper work procedures.
- Read the Material Safety Data Sheet (MSDS) to understand the hazards and safe use of the chemicals.
- Know how to properly store, transport, mix, and dispose of chemicals and wastes.

In addition, only a part of prepared standard operation procedure to standardize plating process is shown in Fig. 7.

6 Conclusions and Future Directions

Firms that analyze customer needs correctly can convert the analysis results for the market to catch an advantage among rivals. This study contributes to new product development domain by integrating several quality tools such as quality function deployment, failure modes and effects analysis, poke-yoke and Pareto analysis. Moreover, it includes a systematic and comprehensive approach by offering documentation and recommendations for NPD processes. Proposed integrated framework asserts to form a multidisciplinary team and train them in Stage 1. When considering the first

Table 3 Part of FMEA for processing time of oven

Potential failure modes	Reasons of the failure	P	S	D	RPN	Recommended activity	P′	S′	D′	RPN′
Low coating thickness	Long processing time	6	2	3	36	Find appropriate time with design of experiment	4	2	3	24
Low corrosion	Breakdown of resistance	7	5	6	210	Regular maintenance	5	5	4	100
	Less than 6 value of pH	6	5	4	120	Control with litmus paper	5	5	2	50
Poor coating	Less than 18 concentration level	5	5	3	75	Control before putting into the oven	4	5	3	60

A. Purpose

1. To describe the startup, operation, and conclusion processes required to use the plating process.
2. The process is designed to be used by researchers in the metal coating and other innovative and under study research for metal bonding.

B. Scope

1. This SOP is intended for staff individuals.

C. Prerequisites

1. Training requires prior approval from the lab manager

D. Responsibilities

1. The equipment coordinator and his appointees are responsible for conducting the training.

Emergency Contact:

- Call ***
- Call EHS & Risk Management at ***
- Call Lab Technician **
- Call Head of Quality Department ***

Fig. 7 Prepared standard operation procedure

step of the converting process, QFD is a strong method. Defining potential failures before design is also important in terms of cost, time and customer satisfaction. FMEA is a useful technique to determine failure modes. Therefore, in Stage 2, an integrated methodology combining QFD and FMEA is developed and applied in a firm plating metals. While QFD is applied, customer needs are determined considering voice of customers for meeting the customer requirements and technical characteristics are revealed. At the first step of QFD, customer requirements are determined by experts using complaint and offers coming from customers with email, fax or telephone. Then, to meet the requirements, technical characteristics are determined. Finally, for comparing, two rivals are decided and their data are entered in the system. In duration of developing with QFD application of available process, consumer needs and technical characteristics are analyzed. The outcome of QFD used the input for FMEA, and for one technical characteristic, possible failure modes and their potential effects are determined. To avoid risk, some activities are recommended. Stage 3 is about some documentations such as safety instruction and standard operation procedure. During the integrated process, quality tools such as Pareto analysis, cause and effect analysis and brainstorming are used to enhance process.

Future studies can expand our research by using different techniques like fuzzy approach while giving scale from 1 to 5 for present and target values and 1–10 for customer needs. Design of experiment (DOE) can be added to the FMEA in recommended activities part. Our developed methodology can be improved for integration of QFD, FMEA and design of experiment. According to collected data, we may apply a knowledge-based machine learning system to improve the methodology.

References

Akao Y (2004) Quality function deployment: integrating customer requirements into product design. SteinerBooks

Almannai B, Greenough R, Kay J (2008) A decision support tool based on QFD and FMEA for the selection of manufacturing automation technologies. Robot Comput Integr Manuf 24(4):501–507

Arabian-Hoseynabadi H, Oraee H, Tavner P (2010) Failure modes and effects analysis (FMEA) for wind turbines. Int J Electr Power Energy Syst 32(7):817–824

Bahrami M, Bazzaz DH, Sajjadi SM (2012) Innovation and improvements in project implementation and management; using FMEA technique. Procedia Soc Behav Sci 41:418–425

Baskar C, Parameshwaran R, Nithyavathy N (2019) Implementation of fuzzy-based integrated framework for sesame seed separator development. Soft Comput 1–20

Bellary A, Murthy D (1999) New product development process and total quality management. In: PICMET'99: Portland international conference on management of engineering and technology. Proceedings book of summaries (IEEE Cat. No. 99CH36310), vol 1. IEEE, p 329

Braglia M, Fantoni G, Frosolini M (2007) The house of reliability. Int J Qual Reliab Manag 24(4):420–440

Chen LH, Ko WC (2009) Fuzzy linear programming models for new product design using QFD with FMEA. Appl Math Model 33(2):633–647

Chen LH, Ko WC, Yeh FT (2017) Approach based on fuzzy goal programing and quality function deployment for new product planning. Eur J Oper Res 259(2):654–663

Chen YT, Chou TY (2011) Applying GRA and QFD to improve library service quality. J Acad Librarianship 37(3):237–245

Cohen L (1995) Quality function deployment: how to make QFD work for you. Prentice Hall, Upper Saddle River

Delice EK, Güngör Z (2009) A new mixed integer linear programming model for product development using quality function deployment. Comput Ind Eng 57(3):906–912

Dogan O, Cebeci U (2016) A methodology for new product development by using QFD, FMEA and its application in metal plating industry. In: 16th production research symposium

Doshi J, Desai D (2017) Application of failure mode and effect analysis (FMEA) for continuous quality improvement-multiple case studies in automobile SMES. Int J Qual Res 11(2)

Ginn D, Jones D, Rahnejat H, Zairi M (1998) The "QFD/FMEA interface". Eur J Innov Manage

Goda A, Medina V, Zsidai L (2017) Examination of the hungarian agricultural machinery manufacturers' product planning, quality management techniques and production coordination. Hung Agric Eng 32:16–21

Grout JR, Toussaint JS (2010) Mistake-proofing healthcare: why stopping processes may be a good start. Bus Horiz 53(2):149–156

Gu YK, Zx Cheng, Gq Qiu (2019) An improved FMEA analysis method basedon QFD and topsis theory. Int J Interactive Des Manufact (IJIDeM) 13(2):617–626

Hassan A, Siadat A, Dantan JY, Martin P (2010) Conceptual process planning–an improvement approach using QFD, FMEA, and ABC methods. Robot Comput Integr Manuf 26(4):392–401

Hunt RA, Killen MCP, Killen CP, Walker M et al (2005a) Strategic planning using GFD. Int J Qual Reliab Manag

Hunt RA, Killen MCP, Miguel PAC et al (2005b) Evidence of QFD best practices for product development: a multiple case study. Int J Qual Reliab Manag

Ionica AC, Leba M (2015) QFD integrated in new product development-biometric identification system case study. Procedia Econ Finan 23:986–991

Juran JM (1975) The non-pareto principle; mea culpa. Qual Prog 8(5):8–9

Ko WC (2013) Exploiting 2-tuple linguistic representational model for constructing HOQ-based failure modes and effects analysis. Comput Ind Eng 64(3):858–865

Korayem M, Iravani A (2008) Improvement of 3p and 6r mechanical robots reliability and quality applying FMEA and GFD approaches. Robot Comput Integr Manufact 24(3):472–487

Kotler P, Keller K (2011) Marketing management, 14th edn. Prentice Hall, Upper Saddle River

Krajewski LJ, Ritzman LP, Malhotra MK (2013) Operations management. Pearson Education, UK
Lange K, Leggett S, Baker B (2001) Potential failure mode and effects analysis (FMEA) reference manual. AIAG, Southfield, Michigan
Lin CY, Lee AH, Kang HY (2015) An integrated new product development framework–an application on green and low-carbon products. Int J Syst Sci 46(4):733–753
Loch C, Kavadias S (2008) Handbook of new product development management. Routledge, Milton Park
Ma H, Chu X, Xue D, Chen D (2019) Identification of to-be-improved components for redesign of complex products and systems based on fuzzy QFD and FMEA. J Intell Manuf 30(2):623–639
Mandal S, Maiti J (2014) Risk analysis using FMEA: fuzzy similarity value and possibility theory based approach. Expert Syst Appl 41(7):3527–3537
Moldovan L (2014) QFD employment for a new product design in a mineral water company. Procedia Technol 12(2014):462–468
Na L, Xiaofei S, Yang W, Ming Z (2012) Decision making model based on QFD method for power utility service improvement. Syst Eng Procedia 4:243–251
Ng W, Teh S, Low H, Teoh P (2017) The integration of FMEA with other problem solving tools: a review of enhancement opportunities. J Phys Conf Ser 890: 012139 (IOP Publishing)
Ozer M, Cebeci U (2009) The role of globalization in new product development. IEEE Trans Eng Manage 57(2):168–180
Paprocki M (2018) The use of methods and computer aided systems in the ecological design of product development. Mechanik 91(1):73–75
Pun KP, Rotanson J, Cheung CW, Chan AH (2019) Application of fuzzy integrated FMEA with product lifetime consideration for new product development in flexible electronics industry. J Ind Eng Manag 12(1):176–200
Sarkar A, Mukhopadhyay AR, Ghosh SK (2013) Issues in pareto analysis and their resolution. Total Qual Manag Bus Excellence 24(5–6):641–651
Shaker F, Shahin A, Jahanyan S (2019) Developing a two-phase QFD for improving FMEA: an integrative approach. Int J Qual Reliab Manag
Tan CM (2003) Customer-focused build-in reliability: a case study. Int J Qual Reliab Manag
Tonchia S (2018) Product design. In: Industrial project management. Springer, pp 49–61
US Department of Defense (1980) Procedures for performing a failure mode, effects and criticality analysis. Department of Defense, Washington, DC (Standard No MIL-STD-1629A)
Wang CH, Chen JN (2012) Using quality function deployment for collaborative product design and optimal selection of module mix. Comput Ind Eng 63(4):1030–1037
Yang Z, Bonsall S, Wang J (2008) Fuzzy rule-based bayesian reasoning approach for prioritization of failures in FMEA. IEEE Trans Reliab 57(3):517–528
Zaim S, Sevkli M, Camgöz-Akdağ H, Demirel OF, Yayla AY, Delen D (2014) Use of ANP weighted crisp and fuzzy QFD for product development. Expert Syst Appl 41(9):4464–4474
Zammori F, Gabbrielli R (2012) ANP/RPN: a multi criteria evaluation of the risk priority number. Qual Reliab Eng Int 28(1):85–104

Recent Optical Approaches for Quality Control Monitoring in Manufacturing Processes

Carlos A. Calles-Arriaga, Eddie N. Armendáriz-Mireles, Enrique Rocha-Rangel, Wilian J. Pech-Rodríguez, José A. Rodríguez-García, Juan López-Hernández, and José A. Castillo-Robles

Abstract Quality assurance in manufacturing engineering requires precise methods in order to implement quantitative techniques for processes control. Optical methods have proved to be suitable in manufacturing since they present several advantages over traditional approaches such as non-destructive, non-contact, and high speed, among others. For example, laser technologies provide a fundamental tool in optical metrology, and fiber optics offers versatility for multiparameter measurements. Optical methods offer improved technology for quality control monitoring in areas such as materials, pharmaceutical, and chemical.

Keywords Quality control · Manufacturing · Optical sensor

1 Introduction

Optical spectra comprise the ultraviolet (UV), visible (Vis), and infrared (IR) regions (see Fig. 1). UV light is the more energetic region of the optical spectra according to the formula $E = h\nu$, where E is the energy, h represents the Planck's constant, and ν is the frequency. Since wavelength is inversely proportional to frequency, shorter wavelengths (as UV) contain more energy than regions with larger wavelengths, for

C. A. Calles-Arriaga (✉) · W. J. Pech-Rodríguez
Department of Mechatronics Engineering, Polytechnic University of Victoria, Av. Nuevas Tecnologías 5902, Parque Científico y Tecnológico de Tamaulipas, Cd. Victoria, Tamaulipas, Mexico
e-mail: ccallesa@upv.edu.mx

E. N. Armendáriz-Mireles · E. Rocha-Rangel · J. A. Rodríguez-García
Department of Manufacturing Technologies Engineering, Polytechnic University of Victoria, Av. Nuevas Tecnologías 5902, Parque Científico y Tecnológico de Tamaulipas, Cd. Victoria, Tamaulipas, Mexico

J. López-Hernández · J. A. Castillo-Robles
Department of Automotive Systems Engineering, Polytechnic University of Victoria, Av. Nuevas Tecnologías 5902, Parque Científico y Tecnológico de Tamaulipas, Cd. Victoria, Tamaulipas, Mexico

J. L. García Alcaraz et al. (eds.), *Techniques, Tools and Methodologies Applied to Quality Assurance in Manufacturing*,
https://doi.org/10.1007/978-3-030-69314-5_2

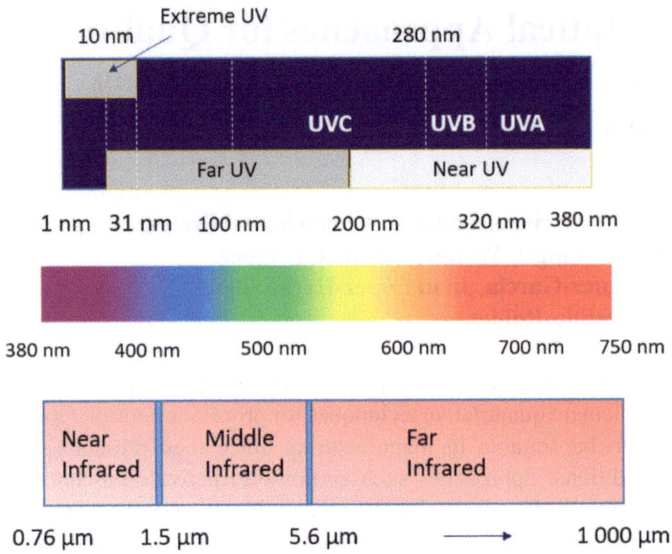

Fig. 1 Optical spectrum regions

example, IR. Interaction of UV light with matter can cause some optical phenomena such as fluorescence which can provide relevant information of a study sample.

Visible light is widely used in optical metrology since its versatility, for example, it can use in low-cost UV-Vis spectrometers to analyze absorbance, reflectance, or transmittance. This can be useful in quality inspection, not only in the optics industry but also in the diverse applications, for example, in the elaboration of food products or the manufacturing process of solar cells. Lasers in visible and near IR regions are currently used for chemical identification by the Raman scattering method. Nowadays the pharmaceutical industry performs quality control procedures based on Raman scattering. Infrared devices are frequently used in optoelectronics for transmission/reception systems. However, there are plenty of techniques for materials characterization by IR light. For instance, Fourier transform infrared (FTIR) spectroscopy typically operates at wavenumbers ranging from 500 to 5000 cm^{-1} (2–20 μm), i.e., middle and far IR. This method is suitable for the identification of functional groups in organic and inorganic substances.

Optical spectroscopy is based on the interaction of light-matter which can provide useful information for quality purposes in manufacturing processes. Figure 2 shows an example of equipment used for characterization by FTIR method. A list of industries and optical methods examples is shown in Table 1.

This chapter aims to study optical techniques, based on non-contact approaches and fiber sensors technology that has been implemented in the last years as a viable alternative to conventional methods for quality monitoring in industrial processes. This work contributes to the analysis of non-contact and high-resolution optical techniques developed for quality control in manufacturing.

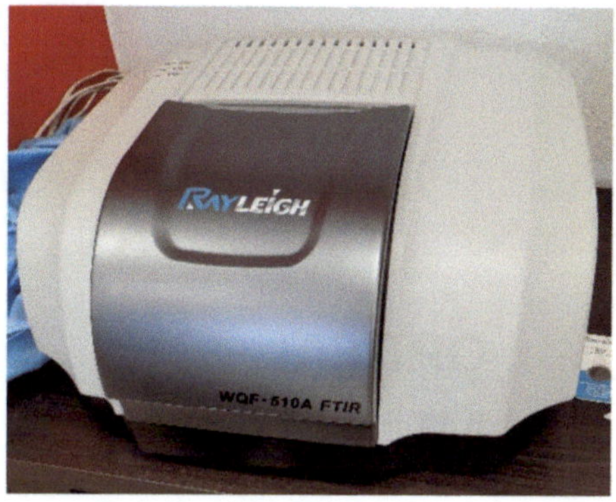

Fig. 2 FTIR commercial equipment

Table 1 Optical industry and methods

Industry	Methods
Materials	Diffuse reflectance
Biopharmaceutical	Colorimetry
Food products	Absorbance
Industry 4.0	Photoluminescence
	Raman
	UV-Vis
	FTIR

2 Non-contact Optical Methods for Quality Monitoring

One of the main advantages of optical approaches for monitoring manufacturing processes is the ability to take measurements without physical contact with the corresponding sample. These techniques are often based on lasers or incoherent light sources with a wide spectra emission. Among the optical inspection methods are UV-Vis, FTIR and computer vision as shown in Fig. 3. The main advantages and disadvantages of each technique are shown in Table 2. In this section, the main non-contact free-space optical methods and tools for quality control are reviewed.

2.1 Laser in Manufacturing

Laser is the acronym of light amplification by stimulated emission of radiation; this means that basically it is amplified light. However, it has unique properties, e.g., it is monochromatic, it is highly directional, and all waves that form the laser are in the same phase which is called coherence. These distinctive characteristics make the laser a highly versatile tool, being called a solution in search of problems. Its areas in manufacturing could be divided into two kinds of applications: materials processing

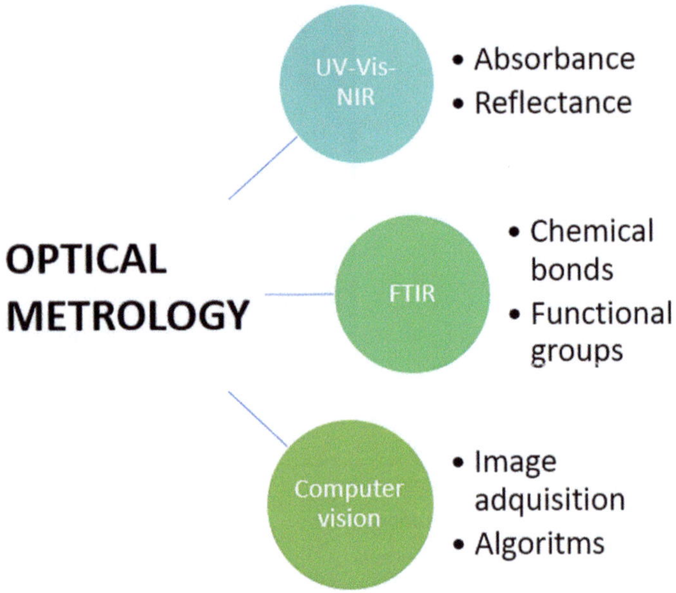

Fig. 3 Optical methods for quality inspection

Table 2 Advantage and disadvantages of optical metrology techniques

Technique	Characteristics
Raman	Pros: It is an accurate method Cons: depends on a weak signal and requires a laser and a spectrometer
Absorbance	Pros: it is a fast and low-cost method Cons: it cannot perform in situ measurements
Reflectance	Pros: it can perform in situ measurements Cons: it requires additional accessory in spectrometer (reflectance probes)
FTIR	Pros: it is appropriate for detecting functional groups present in materials and chemical products Cons: it requires

involving cutting, marking, drilling, welding, and other specialized processes such as surface treatment; and optical metrology which includes technologies such as 3D laser scanners, fiber optics sensors, and free-space optical detectors, among others. In this case, we will be focus on the optical metrology area.

Automated measurement is linked to automated manufacturing processes. Optical technology allows non-contact dimensional measurements with high-quality standards such that required by industries such as aerospace or automotive. A particular technology that has result successful is the 3D laser scanner. There are many advantages of this optical device such as high speed, a resolution in the order of several microns, and ease of use. Although the main areas of application of laser scanners are injection molded plastics, stampings, and castings (turbine blades), and there are emerging fields such as orthopedic implants or rapid prototyping.

2.2 *Optical Monitoring of Profiles*

Industries related to biomaterial products, renewable energy, or electronics, among others, require accurate methods to measure the surface topography. Although there are well-established technologies such as contact profilometers and laser-based scanning systems, some manufacturing processes require more precise systems. Recently, novel methodologies have been proposed to improve quality control by optical means. For example, in low emittance (low-e) glass manufacturing, optical profile measurements by a novel approach have been carried out (Wu et al. 2018). Low-e glass contains a material such as metal oxide coating transparent to visible wavelengths but highly reflective to infrared radiation. Hence, a quality inspection process of the optical profile is required. In this case, a new method called piecewise polynomial random coefficient (PRC) model was proposed. In this approach, data obtained by laser scanning is analyzed with PRC and multivariate control charts using MATLAB. Through this methodology, shape and intrinsic variations were of optical profiles were accurately measured.

Aerodynamic profiles are fundamental in wind turbine blades. Some of the traditional methods for turbine blades surface measuring include coordinate machine measurements, laser radar technique, and photogrammetry. Although these techniques have shown high accuracy, in general, are complex, slow, and high-cost systems. In a recent work, a novel optical profilometer was developed for quality inspection in turbine blades (Moreno-Oliva et al. 2019). The system is based on the laser triangulation technique (LTT). In this case, a laser diode module (633 nm) and a CCD camera (752 × 480 px) were installed in an XYZ linear stage to perform the surface scanning (Fig. 4). The camera was connected to a computer and images

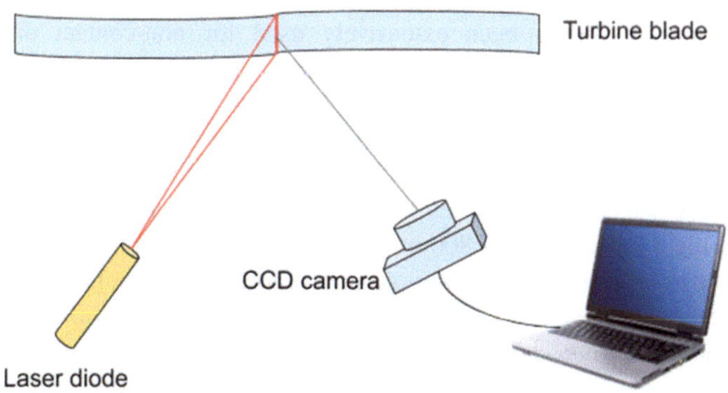

Fig. 4 Scheme for profile measurements of wind turbines (based on Moreno-Oliva et al. 2019)

were analyzed by the Xfoil free software. Results compared with a reference profile showed a high-precision system with a resolution in the order of 0.1 mm. Moreover, this is a low-cost experimental setup that could be adapted to the renewable energy industry.

Optical measurements of surfaces can be difficult in special cases, for example, when a transparent layer is added to the sample. Although several optical techniques could lead to acceptable results, in complicated surfaces, it is appropriated to perform evaluations with different methods. Feng et al. measured surface topographies consisting of a metallic substrate with a transparent coating (Feng et al. 2019). In order to compare with different methods, data were first obtained from scanning electron microscopy (SEM) and optical microscope imaging with focus stacking (FS). In a second stage, areal topography was measured by focus variation microscopy (FVM), point autofocus instruments (PAI), imaging confocal microscopy (ICM), atomic force microscopy (AFM), and coherence scanning interferometry (CSI). In comparison with AFM results, it was found that the CSI method showed better accuracy.

High-precision mechanical parts are demanded in fields such as robotics and mechatronics among others. In particular, gears' quality is critical for the construction of machinery. Coordinate-measuring machines (CMMs) are commonly used for tooth gear inspections. Although this is a highly accurate method, it requires a contact probe which requires a considerable time for measurements. Optical methods are a good alternative since they are non-contact techniques that can perform high-speed measurements. Some optical methods used for 3D reconstruction of gears are based in interferometry, laser triangulation, and Moiré systems. Cheng and Chen developed an optical inspection system to reconstruct a gear tooth surface (Chen and Chen 2019). The system consisted of a halogen lamp as the illumination source, lenses and gratings to produce Moiré fringes and a CCD camera directed to the gear. The approach is based on five-step phase-shifting captured by the camera. Images are processed by MATLAB to carry out a 3D reconstruction of the gear. The authors reported that the difference of the mean values of this Moiré system compared to the CCM method is smaller than 3 ums for three variables: the involute profile, lead profile, and 3D topology.

Optical metrology has been extensively used for non-contact dimensional measurements. Nowadays, optical devices have a resolution comparable with the wavelength of its light source, for example, 633 nm for a He–Ne laser. Recently, Yuan and Zheludev proposed an optical ruler with a resolution of a small fraction of the wavelength (Yuan and Zheludev 2019). The system is based on Pancharatnam-Berry metasurface which creates a diffraction pattern with peaks on the order of subwavelengths. Using a light source of 800 nm, the resolution is about 1 nm. In principle, it is possible to have resolving power of lambda/4000 which could measure atoms. Among the possible applications of this high-resolution optical ruler is the monitoring of microelectromechanical systems (MEMS), measurements of mechanical properties (e.g., the Young modulus), or the inspection of high-precision lenses.

2.3 Materials

Materials have a significant role in manufacture and engineering since electrical, mechanical, or thermal properties among others are key factors in a product's quality. Cement is a high-used product in civil engineering; their physical properties such as strength, consistency, density, or hydration depend on its fabrication process. However, many fabrication processes are empirical, and hence, there is a lack of information in the fabrication method which could vary the results. Optical methods can be used to characterize the physical properties of types of cement, for example, diffuse reflection was used to evaluate the hydration of cement paste (De León Martínez et al. 2015). During the setting process of the cement pass (water added to cement), there is a generation of hydration heat that presents variations related to chemical reactions. Changes in temperature are associated with the stages of the Powers-Brunauers model. It was found that using a He–Ne laser with emission at 633 nm the diffuse reflectance pattern is similar to the temperature pattern (Fig. 5), hence providing important information about the cement hydration process.

Mechanical properties of asphalt determine the quality of pavements. Complex elasticity module has a fundamental importance in asphalt durability. Currently, there are standard methods to determine complex elasticity, and however, conventional approaches require contact devices, e.g., accelerometers or piezo-electric devices, and are limited to low frequencies (30 Hz). Hasheminejad et al. proposed an optical technique to obtain the complex modulus (Hasheminejad et al. 2019). The system consists of a scanning laser Doppler vibrometer (SLDV) and an identification technique based on Timoshenko's beam theory which is associated with the mechanical

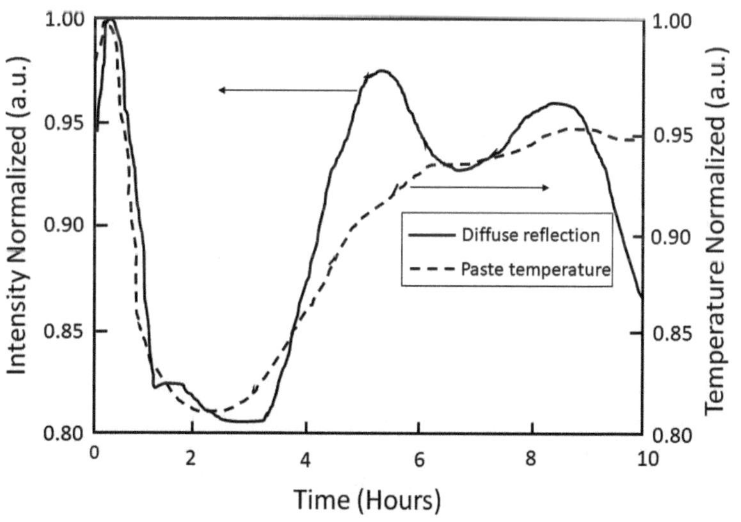

Fig. 5 Temperature versus reflectance on the setting of cement paste. Adapted from De León Martínez et al. (2015)

behavior of a structure under bending. This non-contact method has the capacity of taking measurements at high frequencies (15 kHz). Compared with conventional stiffness tests the SLDV technique reduces the system cost by 20%.

Fibrous porous materials have several applications such as clothing, filters, and medicine among others. Their properties are highly dependent on the orientation of fibers, and hence, automatized systems to evaluate this parameter are required. Computer vision is a suitable method for non-contact optical inspection. In general, this method consists of four stages: image acquisition, image enhancement, image segmentation, and image representation and description. Tunák and Antoch proposed an effective system to inspect the orientation of fibers from image segmentation (Tunák and Antoch 2018). The method is based on the transformation from spatial to frequency domain using the 2D discrete Fourier transform. The Fourier spectrum is transformed into a binary image which direction is related with the real spatial orientation of fibers. To analyze a great sample, the image is divided in small parts that are individually analyzed by the same method, and at the end the total results are represented by a histogram that shows the dominants fiber orientations.

2.4 Automobile Optical Inspection (AOI)

Several industrial manufacturing processes such as automotive parts rely on visual inspection carried out by humans or automated systems. Personal with the required training is often used to perform quality inspection. However, human inspection is limited to low speeds and personal rotation is commonly required. A system for automatic optical inspection was recently proposed for optical inspection of head-up displays (HUD) used in the automotive industry (Ferreira et al. 2017). The system has no moving parts and is intended to be used for regular patterns. This method is based on Fourier filtering and operates below the Shannon–Nyquist criterion which is related to the minimum sampling rate that will not distort the signal information. This system has the advantages of not requiring a high-resolution camera and neither a scanning system.

Quality control can be performed by vibration measurements, for example, in the automotive industry, acoustic waves caused by vibrations are used to evaluate operating conditions of internal combustion engines. However, there are some limitations to implement this method, for instance, a sound isolated space to avoid interference is required, consequently, the quality inspection could affect productivity. Optical vibration measurement is a reliable option for acoustic methods since is acoustic interference-free, and hence, inspections can be carried out in situ. The physical principle of optical vibration measurements is the laser Doppler effect. This effect consists of changes in light wave frequencies corresponding to changes in positions (vibration) as shown in Fig. 6. This method has been used on motors production as inline and end-line process (Polytec).

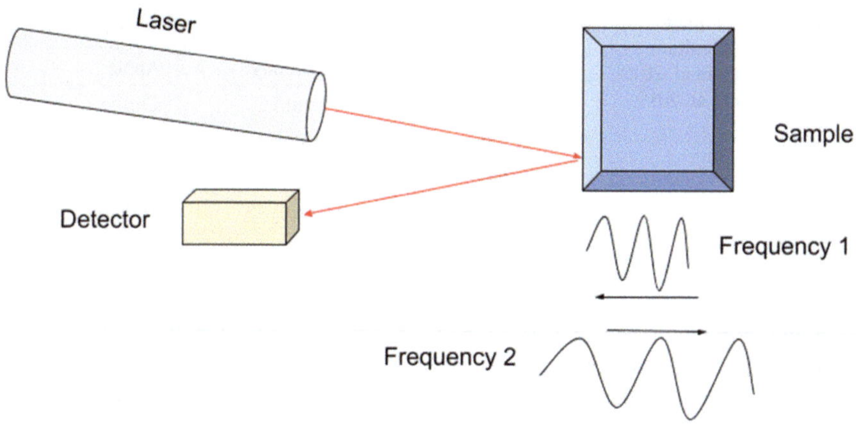

Fig. 6 Laser doppler effect

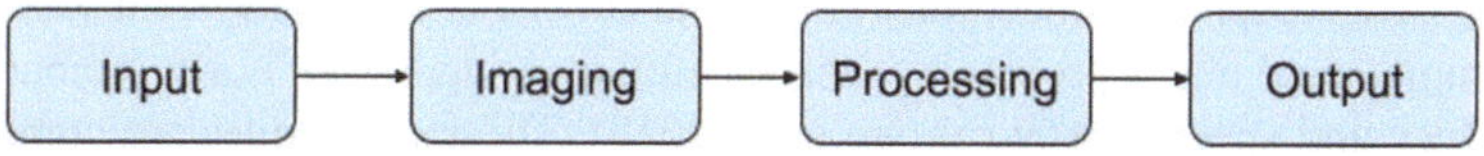

Fig. 7 Machine vision system process

2.5 *Advanced Manufacturing*

Advanced manufacturing has become a fundamental tool to increase industrial productivity. It is characterized by a high level of innovation and technology involved in the production process. Some of the main areas are virtual manufacturing, assembly technologies, tests and measurements, non-conventional procedures, and monitoring and control of production processes. The advanced manufacturing technologies are robotics and automation, big data, augmented reality, and additive manufacturing.

Some products based on additive manufacturing require high-precision equipment to analyze its surface. Hence, companies such as ZYGO are developing advanced optical metrology equipment, for example, for 3D imaging, a laser Fizeau interferometer-based profilometer is capable of taking 250 million surface topography points per second (Zygo). Other companies such as Keyence offer commercial solutions to quality inspection for example by advanced vision systems (Keyence). Figure 7 shows a scheme of an industrial vision system.

2.6 *Food Industry 4.0*

The foodstuff industry requires to accomplish high-quality standards to satisfy international import/export norms. Non-contact optical methods traditionally used on

Table 3 Optical quality inspection in agriculture (based on Yeong et al. 2019 and El-Mesery et al. 2019)

Wavelength (nm)	Parameter	Product
400–800	Chlorophyll	Apple
450–1000	Vitamin C	Chillie
600–2200	Moisture	Mushroom
1000–2400	Anthocyanin	Jambu
1200–2200	Maturity	Mango

other industries such as pharmaceutical and chemical manufacturing products have been incorporated in the last years to the food industry. Optical methods can be used to inspect the quality, for example, to measure foodstuff firmness, size, shape, color, substance concentrations, and fat levels, among others. The optical spectrum region often used is between visible and near-infrared (NIR), ranging from 400 to 2600 nm. Some of the optical approaches recently used in quality monitoring of foodstuff include spectroscopy which is based on absorbance, transmittance, reflectance, fluorescence, or phosphorescence; imagology, that consist of a vision system based on a CCD camera and image processing; and spectral imaging, which is a combination of imagology and spectroscopy (Yeong et al. 2019). These techniques are commonly utilized with other tools such as statistical methods or special algorithms to obtain the desired characteristics.

Food product components such as proteins, carotenoids, or chlorophyll, among others, absorb or reflect wavelengths, mainly in the visible-NIR region. Table 3 shows some examples of quality measurements by optical spectroscopy:

Reflectance spectra is a simple and practical method to detect and quantify substances in food products. A general scheme of optical spectroscopy based on reflectance is shown in Fig. 8.

Wang et al. determined the vitamin C concentration in chilies by means of diffuse reflectance spectroscopy (Wang et al. 2011). The studied optical spectrum ranged from 400 to 1800 nm, i.e., visible to near-infrared region. Reflectance data were compared with vitamin C contents obtained with standard methods. In order to obtain a more accurate model, mathematical methods were applied. It was found that the first derivative preprocessing method allows a better correlation between reflectance data and vitamin C concentration.

Optical reflectance has also been implemented to study apple contents such as anthocyanin and chlorophyll (Merzlyak et al. 2003). Pale-green, yellow, and red apples were inspected by visible light ranging from 400 to 800 nm. From this study, it was found that in green and yellow apples, there is a high reflectance correlation at 550 and 700 nm, while in red apples this correlation disappears due to the anthocyanin content. On the other hand, reflectance at 678 nm showed an inverse proportional relation with respect to chlorophyll a and b concentration.

In a recent work regarding as external as internal apple's quality a new approach was proposed (Li et al. 2020). In order to analyze the external appearance, an inline camera installed directly in the production line was used (Fig. 9). The external quality method is based on image processing utilizing the isohypse line extraction

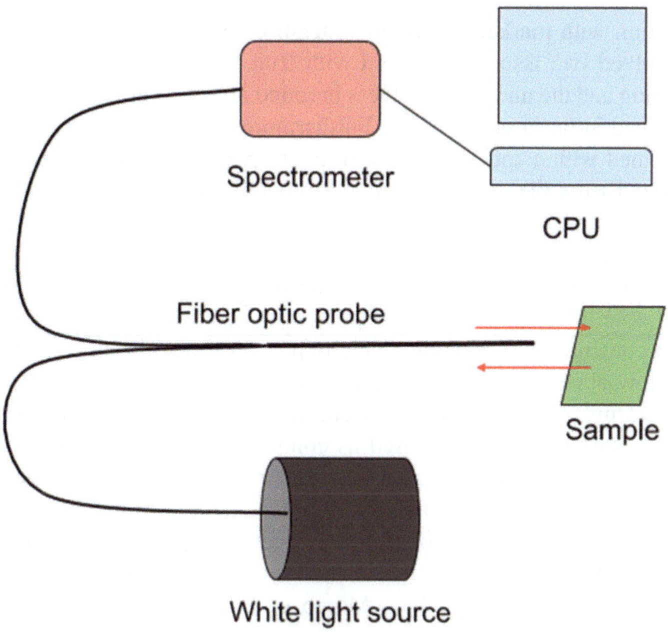

Fig. 8 Optical methods for quality inspection

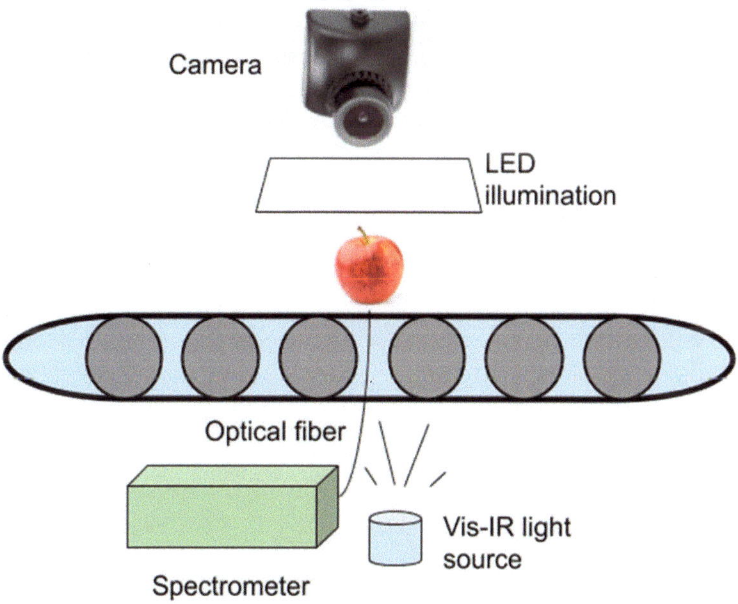

Fig. 9 Inline apple's quality inspection system (adapted from Li et al. 2020)

in combination with marker constraint watershed segmentation (ILE-WSM). This method resolved two issues associated with fruit analysis by vision systems: the light scattering and the uneven brightness in round objects. The detection of external defects showed an accuracy of 97%. Furthermore, internal quality measurements were performed with a spectrometer with fiber optic output. The optical spectrum studied ranged from 600 to 1100 nm. Additionally, the normalized spectral ratio (NSR) method was utilized in order to optimize results. The advantages of this system are compactness and high speed.

Chlorophyll is one of the most studied substances in food products. Chlorophyll plays a key role in the photosynthesis process to convert light in oxygen. Species with a high content of chlorophyll generally present a greenish appearance. This is due to the low absorption at wavelengths around 500 nm, hence the reflection of green light occurs. Besides plants, there are food products that contain chlorophyll. For example, spirulina algae are used as complement food. Figure 10 shows the absorbance spectra of spirulina algae. As can be seen, there are peaks that represents high absorption at 440, 620, and 680 nm.

Traditionally, fermentation evolution is associated with pH levels, hence, contact pH meters are often used. However, test probes are required to take measurements which may contaminate the product. Moreover, the equipment has to be cleaned after each use. Dairy product quality can also be inspected by optical spectroscopy. Arango et al. proposed IR scattering to measure the fermentation process in yogurt based on a fiber optic sensor (Arango et al. 2020). Since IR light scattering showed a correlation with pH, a mathematical model was developed to determine ph levels from 5.2 to 4.6. This non-destructive method is suitable for inline monitoring and could be adapted to similar products.

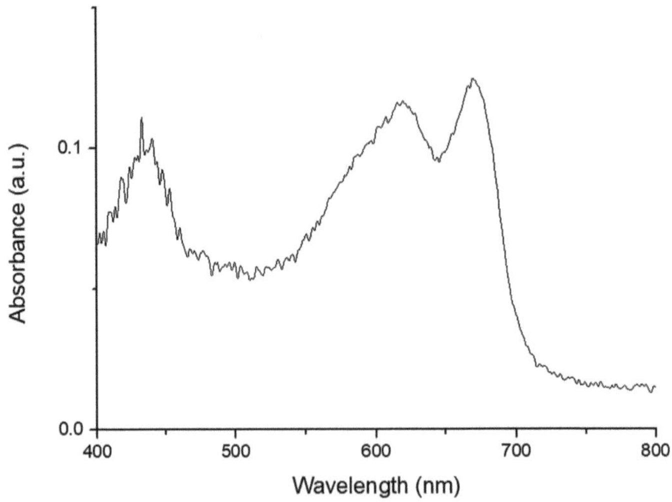

Fig. 10 Spiruline algae absorbance spectra

2.7 *Pharmaceutical Industry*

One of the manufacturing industries that demands the highest quality standards is pharmaceutical. Therefore, innovative approaches are demanded in order to offer better products. One of the key processes is pharmaceutical cleaning validation. This is important since stainless steel plates, polymer surfaces or instruments, and materials in general, intended to manufacture a pharmaceutical product are required to be free of any substances before being reused. Traditional methods to perform the cleaning inspection task include analytical approaches, for example, the total organic carbon method. Although analytical methods have shown high accuracy, their main drawback is that they are time-consuming since it is required to take a sample to be analyzed in a laboratory which could take several hours. Optical methods such as Raman scattering, phosphorescence, or fluorescence can provide a solution to reduce the time analysis. As shown in Fig. 11, fluorescence occurs when energy is absorbed by matter. In this case, atoms move to ground state to higher energy levels. Atoms tend to return to the ground state which evokes non-radiative and radiative transitions. Energy from radiative transitions consists of photons with a fixed wavelength that depends on the matter energetic levels.

A recent work proposed a laser-induced fluorescence analysis to validate pharmaceutical cleaning (Chullipalliyalil et al. 2020). This method consists of a laser with emission at the deep UV region which produces fluorescence in active pharmaceutical ingredients (API), for instance, Paroxetine, an antidepressant with fluorescent emission at 353 nm. This approach allows for taking in situ measurements to reduce considerably the inspection time. This method allows detecting traces in the order as low as 0.2 $\mu g/cm^2$.

The biopharmaceutical industry requires accurate and high-speed methods for their manufacturing processes. Conventional processes, e.g., mass spectroscopy (M.S.) has some disadvantages that affect productivity. This method requires sample preparation, hence complexity is increased, besides it is a time-consuming method. M.S. may be combined with liquid chromatography to obtain highly accurate data, however, information from this method could be difficult to interpret. Optical spectroscopy is a non-contact analytical method suitable for the biopharma industry. Optical spectroscopy can be divided into electronic (e.g., UV-Vis) and vibrational-based techniques (Raman).

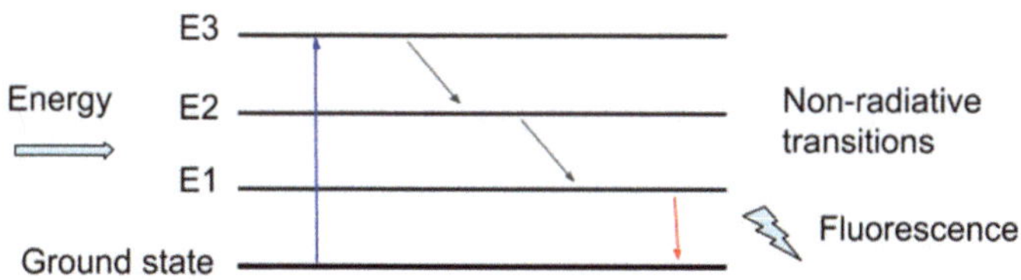

Fig. 11 Fluorescence mechanism

Raman spectroscopy is based on inelastic scattering. These phenomena occur when light interacts with matter and produces scattering. Most of this scattering is elastic and it is associated with Rayleigh scattering. However, a small amount of scattered photons changes its wavelength due to inelastic scattering, which is called Raman scattering. Although Raman spectroscopy has been widely used in forensic and semiconductors, and similar processes can be used in biopharma, it has some limitations. One of them is associated with the weak signal produced as an inherent result of inelastic scattering. Another problem associated with this method is that the emission produced can be difficult to measure when fluorescence is present. Nonetheless, there are alternatives to diminish these problems such as pulsed emissions or enhanced Raman scattering (ERS) that make use of materials such as gold or silver to increase the sensibility.

3 Fiber Sensors

Fiber optics are currently used in telecommunications for Internet transmission; medicine, (endoscopy), or industrial sensing applications. Conventional fiber optics consist of a core surrounded by a cladding (Fig. 12) and are made by silica (SiO_2) although there are also plastic fibers made by polymethylmethacrylate (PMMA) and other polymers. On the other hand, special fibers include double-clad fiber, thin-core fiber, and photonic crystal fiber, among others.

Fiber sensors typically use structural changes or materials coating to produce changes in light traveling inside the fiber due to external variables. For example, long period gratings (LPG) inscribed in fiber enhance its sensitivity to temperature changes. LPG can be fabricated by CO_2 laser (Fig. 13), electric arc, or UV interference patterns. The principle of operation is that periodic perturbations in fiber refractive index, cause rejection bands at punctual wavelengths. This wavelength is affected by external variables.

Among the fiber modifications for sensors are the fiber tapers (Fig. 14). Commonly, fiber is thinned by a method consisting of heating and pulling. Popular heat sources include mini-flames and electric discharge. In fiber tapers, light from the core has close interaction with external media. Fiber sensors based on tapers are

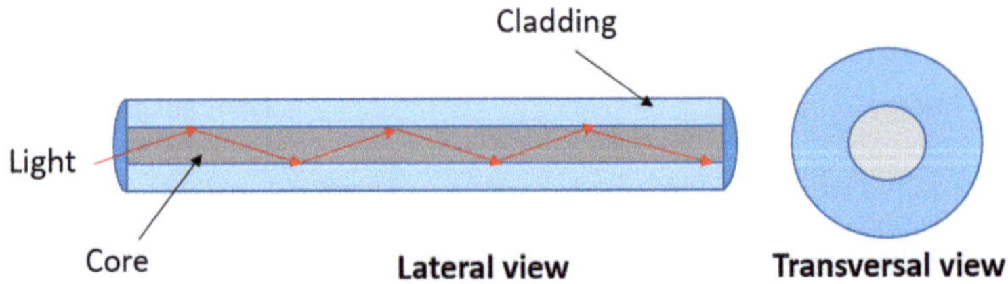

Fig. 12 Fiber optic structure

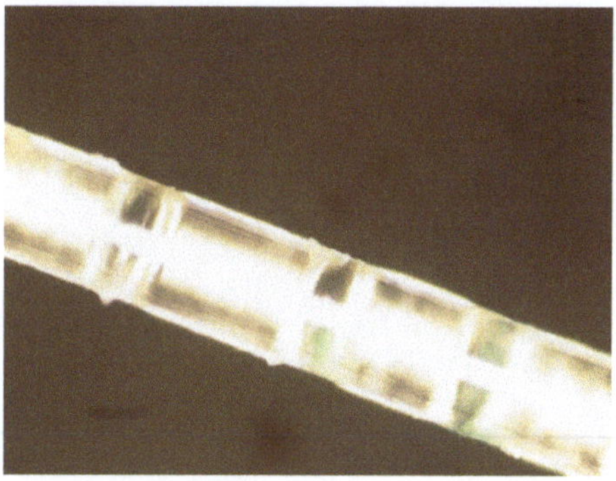

Fig. 13 Long period grating fiber manufactured with laser

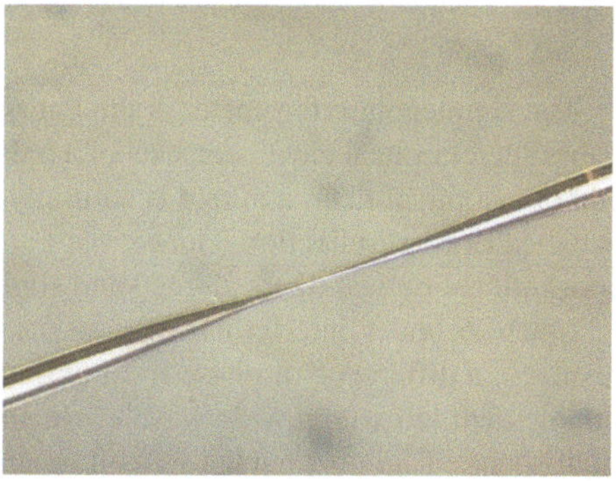

Fig. 14 Fiber taper fabricated by electric arc

used for refractive index or organic/inorganic components. Fiber tapers have also been used for the development of fiber interferometers for sensing applications.

3.1 Fiber Interferometers

Interferometry occurs when two or more signals are out of phase. This phenomenon can be employed in an optical fiber to create sensors. The response of this kind of sensor consists of an interference pattern formed by peaks and valleys (Fig. 15) that can be visualized in an optical spectrum analyzer (OSA). This pattern can be sensible to external variables (pressure, temperature or refractive index, among others) with a corresponding change on frequency or amplitude.

In general, there are four different types of fiber interferometers (Lee et al. 2012):

1. Fabry–Perot.
2. Mach–Zehnder.
3. Michelson.
4. Sagnac.

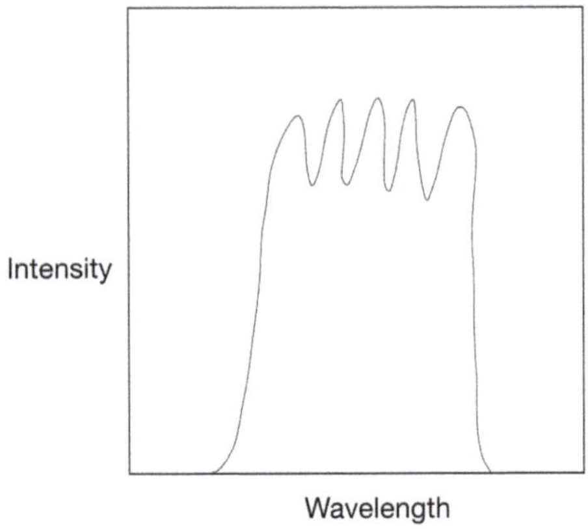

Fig. 15 Interference pattern response of fiber interferometers

The simplest interferometer is the Fabry–Perot configuration. Conventionally, a Fabry–Perot optical cavity consists of a pair of mirrors aligned along an axis. At the end of an optical fiber, a mirror is naturally formed by the interface silica-air which is called Fresnel reflection. Hence, the Fabry–Perot implemented simply with two conventional optical fibers aligned and slightly separated (Fig. 16).

Mach–Zehnder interferometer consists of a difference of optical paths which results in a difference of phase to form the interference pattern. Typically, a super-luminescent led with a wide wavelength spectrum of about 100 nm is utilized as a light source. Light is coupled into an optical fiber connected to a coupler with one input and two outputs, hence, light from on fiber is divided and travels along two different fibers, one of them acting as the sensing arm. Among the elements utilized as sensing fibers are long period gratings or fiber tapers. Finally, light from the two fibers is matched into a second coupler and the output is observed in an OSA. The experimental setup is shown in Fig. 17.

Michelson interferometers-based fiber sensors are often used for temperature or refractive index measurements. In principle, this configuration is similar to the Mach–Zehnder interferometer but it is simpler. The optical arrangement also requires a light source and just one coupler, but in this case, a 2 × 2 type is needed (2 inputs and 2 outputs). Coupler outputs are connected with optical fibers of different length to induce the phase difference. The end of each fiber has a reflecting element that may

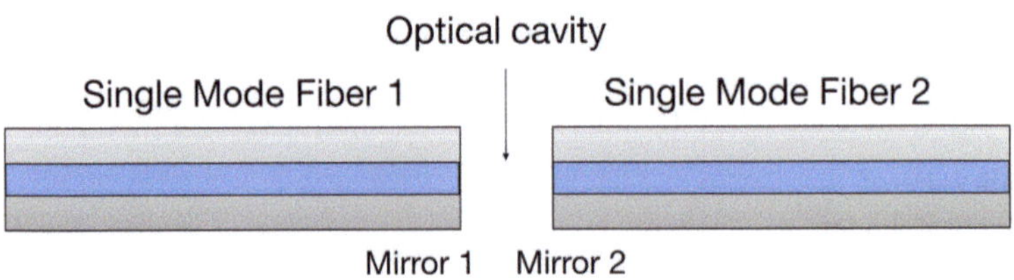

Fig. 16 Fabry–Perot fiber interferometer

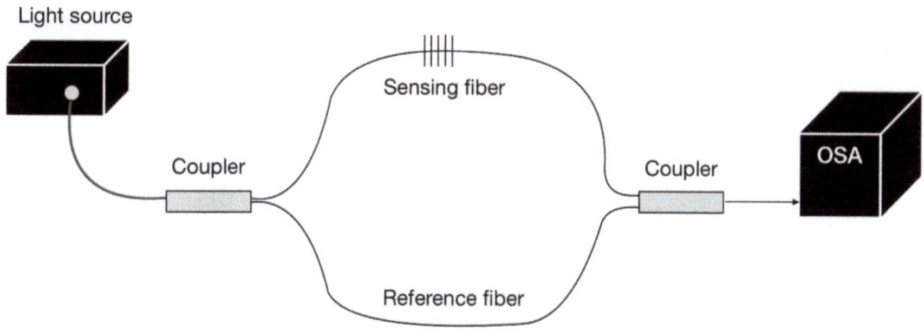

Fig. 17 Mach–Zehnder interferometer

be the Fresnel reflection, a thin film or a FBG among others. A resulting signal returns to the coupler and is visualized in one of the inputs connected to an OSA as shown in Fig. 18.

Sagnac interferometer-based fiber sensor is one of the most practical and easy-to-implement schemes since it is based on a fiber loop. Besides the utilization of a light source and an OSA, this configuration only requires three elements: an optical fiber, a coupler, and a polarization controller (PC) as shown in Fig. 19. The input passes through the coupler and is divided in a 50/50 proportion. Each of the divided signals follows opposite paths and again passes through the coupler to the output. This interferometer usually uses birefringent fibers since it is a polarization-dependent device and can be used as temperature sensors. However, sensing of other variables such as strain or bending, use photonic crystal fibers due to thermal stability.

Although conventional fiber sensors interferometers offer good performance in terms of resolution, robustness, and compactness, most of the schemes need optical couplers which may add an extra space to the sensors besides being expensive. Recently, in-line fiber interferometers had been designed and constructed (Zhu et al. 2012). These types of sensors do not require external elements since the optical schemes are all-fiber. The simplicity and compactness of this kind of fiber sensors make them suitable for industrial applications. Figure 20 shows the principle of

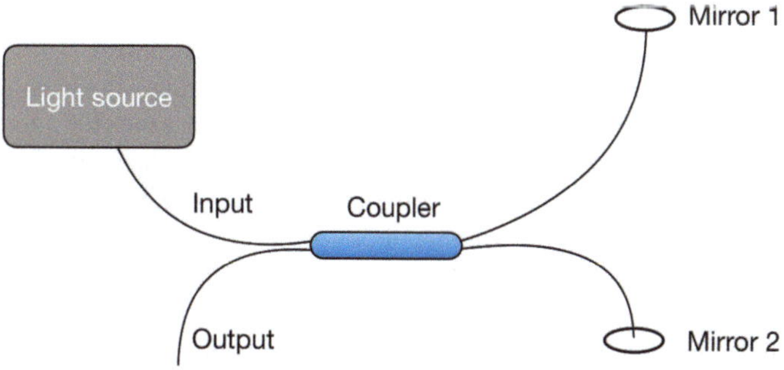

Fig. 18 Michelson interferometer

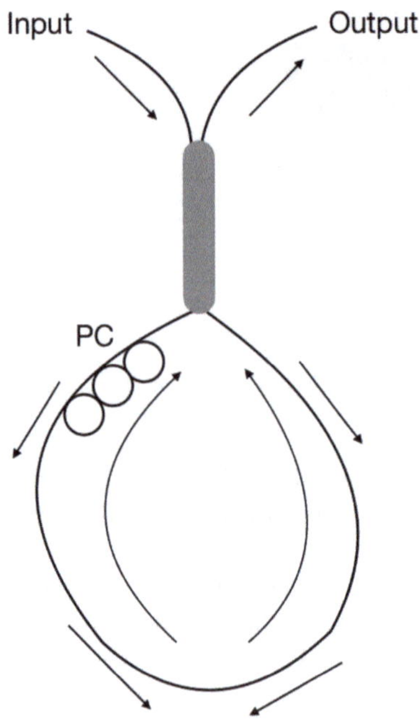

Fig. 19 Sagnac interferometer

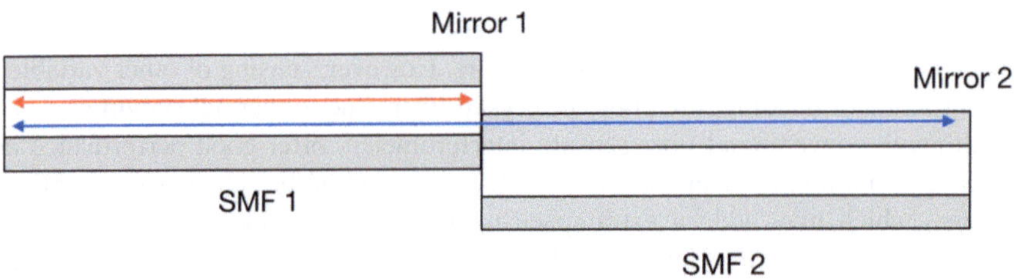

Fig. 20 Inline Fabry–Perot interferometer

operation of an all-fiber Fabry–Perot interferometer. As can be seen, light traveling along the core of an optical fiber (SMF 1) will be reflected at two different distances (represented by red and blue lines), hence interference is produced.

Conventional Mach–Zehnder fiber interferometer is perhaps the least easy setup since utilizing two couplers. However, in-line Mach–Zehnder interferometers can be fabricated from optical fibers spliced with a minimal offset as shown in Fig. 21. Given that light travels along materials with different refractive index (core and cladding) in the middle fiber, both signals become out of phase.

Michelson inline fiber interferometer can be made from just one fiber with a taper and a reflective thin film as shown in Fig. 22. Light is partially coupled from core to cladding due to the fiber taper, hence follows a different optical path and consequently

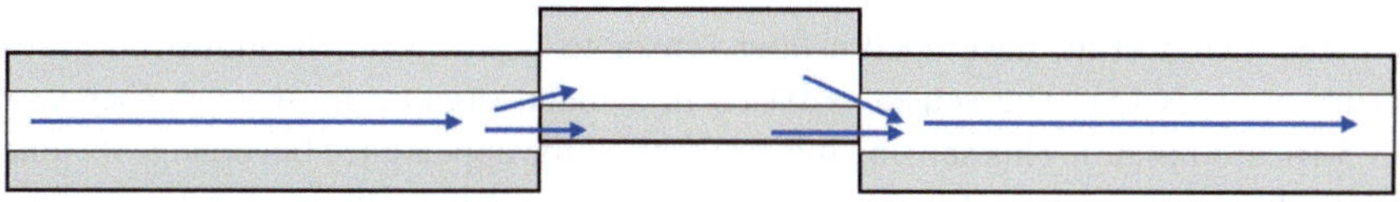

Fig. 21 Inline Mach–Zehnder interferometer

Fig. 22 Inline Michelson interferometer

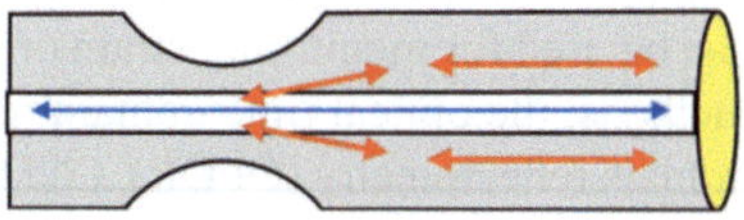

causes interference. Gold or other highly reflective material is deposited on the fiber end to act as a mirror.

3.2 *Fiber Sensors in Quality Control*

Fiber sensors have been used in quality control monitoring for several applications. Ghahrizjani et al. developed a low-cost fiber sensor for analysis of engine oil quality (Taheri Ghahrizjani et al. 2016). The sensor consists of a fiber taper, hence light traveling through the core interacts with parameters of the oil such as particle size or contaminants. The detection system uses optical power as interrogation method.

Noiseux et al. proposed a system of two fiber sensors for quality inspection of wine (Noiseux et al. 2004) Monitoring process consisted of a micromachined V-bend fiber to measure refractive index and an absorption sensor based on an air-gap design. The combination of the two fiber sensors allowed to measured sugar contents and color density.

Gases production process requires high standards of quality monitoring. Fiber sensors, in combination with materials (Fig. 23), can be used to detect compounds

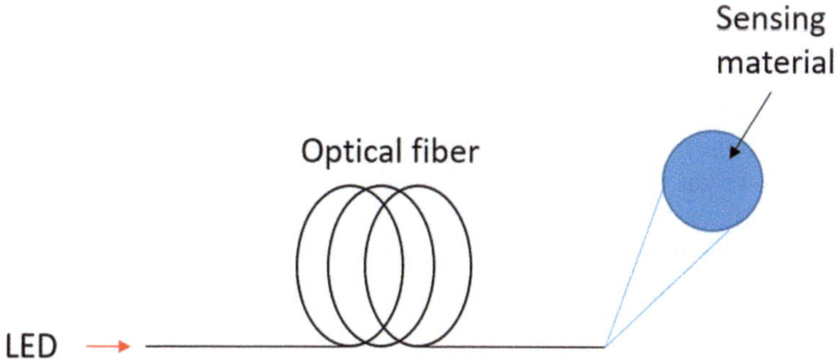

Fig. 23 Fiber sensor based on materials

in gases. Ohira et al. developed a fiber sensor to detect water in industrial gases (Ohira et al. 2015). The sensor is based in a metal organic framework (benzene 1,3,5-tricarboxilate). This sensing material has a blue appearance but change its tone to light blue in presence of wet gases. Sensor response was studied for N_2, Ar, and He with relative sensitivities of 1.00, 0.96, and 1.02, respectively.

Fiber sensors may operate based on different approaches such as interferometers, gratings or Rayleigh backscattering among others. The Rayleigh distributed sensors can be used for monitoring strain or temperature in manufacturing processes by two methods: the optical time domain reflectometry (OTDR) and the optical frequency domain reflectometry (OFDR). OTDR use a similar method that is utilized by radars, i.e., they measure the time a signal travels to a certain point and returns to its origin. Although this is a popular technique, its main drawback is a low spatial resolution. On the other hand, OFDR, although requires an additional step involving the Fourier transform, external variables can be accurately quantified by changes in frequency.

Recently, a fiber sensor based on the OFDR technique was proposed to test printed circuit boards (PCB) strain as part of the quality control in electronics industry (Gomes et al. 2018). In PCB manufacturing, strain excess could cause a future failure in the product even working under normal conditions. Traditionally, PCB strain is measured with foil strain gauges. However, the demand for more components in circuit boards requires more compact sensors. OFDR fiber sensors, besides being compact, are immune to electromagnetic interference which represents an advantage in electronic circuits inspection and can have several sensing points in one optical fiber. The fiber sensor proposed consisted of 390 sensing points distributed along a 1 m fiber. Results showed a similar response compared with foil staring gauges with only a difference of 3.5%. Fiber sensors based on OFDR method could create a two-dimensional map for strain monitoring of PCB with multiple electronic components.

Quality control of fuels is fundamental, especially for companies that produce or distribute gasoline with additions of alcohol (commonly ethanol). This mixture called gasohol is widely used in several countries, and hence, monitoring of the right mixture proportion is required. Although there are chemical methods to analyze gasohol quality, these kinds of tests depend on laboratory studies which delay the results. Moreover, electronic sensing approaches are not suitable for fuel studies for security reasons, i.e., risk of fire. In a fiber, sensor was studied for monitoring of gasohol quality (Rodriguez et al. 2014). The sensor was based on the phenomenon called multimode interference (MMI). This effect is produced when a multimode fiber is spliced to a single-mode fiber in each tip. The original single-mode signal (input) passes through the multimode fiber, which produces interference with periodic focal points as shown on Fig. 24. If the multimode fiber is uncladded, the interference pattern depends on the surrounding media, for example gasohol. Changes in the concentration of a substance produce a variation on the signal wavelength. An output single-mode fiber transmits the signal to an optical spectrum analyzer (OSA) to monitor the substance quality.

Although spectral responses obtained from OSA can provide accurate information in fiber sensors, the main drawback in their implementation is the high cost of this equipment. Photo-detectors convert the amplitude of light signal to a voltage which

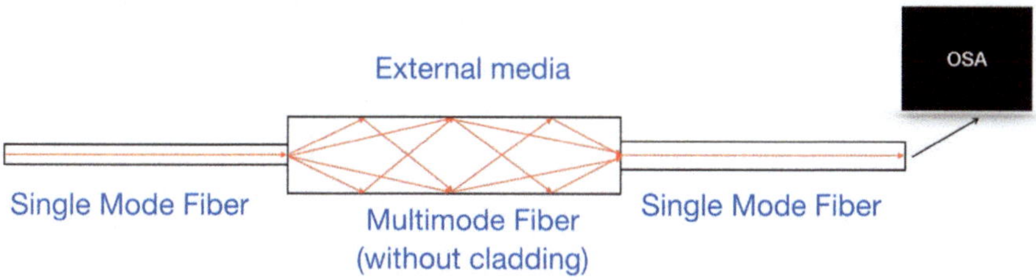

Fig. 24 MMI fiber sensor

can be used as a low-cost alternative in sensors. In a quality amplitude-based fiber, sensor was proposed to determine concentrations of gasoline-ethanol (Aristilde et al. 2019). The device consists of fiber Bragg gratings (FBG) to reflect a wavelength range employed for sensing. The fiber sensor is based on a tilted FBG sensitive to external changes of refractive index that occur with variations in the gasoline–ethanol proportion. These changes are measured by two photodiodes in which output voltages are directly proportional to the percentage of ethanol in gasoline with a resolution of 1.5%. Moreover, this sensor can register temperature variations with a resolution of 0.5 ºC.

A composite is a material made up of more than one substance. Composites are created to improve mechanical properties, for example hardness, strength, ductility, or toughness with respect to conventional materials. These properties make composites suitable for a wide range of applications, mainly in vehicles, aeronautics, and aerospace industries. However, a current challenge is the quality monitoring in the manufacturing process. A conventional technique to fabricate composite parts is called resin transfer molding (RTM). In this method, the composite in the form of resin is emptied into a mold where the material is gradually solidified. Although RTM is apparently a simple process, it requires the monitoring of a uniform distribution of the resin to avoid what is called dry spots which are regions with lack of material. A possible solution is to utilize transparent materials at the top or bottom of the mold to make a visual inspection. However, most composite parts are produced from opaque materials, and hence, sensors are required to monitor the manufacturing process quality. In a recent work (Keulen et al. 2011), a fiber sensor-based experimental setup was proposed to monitor the RTM production process. The monitoring process consists of a fiber optic sensor formed by etched fiber sensors (EFS) and FBG. EFS are formed by uncladder regions of optical fiber that change the intensity of the light traveling through the core fiber depending on the surrounding material. In this way, it can detect the presence of a substance such as a composite resin when it is in contact with the fiber. On the other hand, FBG are used to quantify the strain level (with corresponding shifts on Bragg wavelength monitored in an OSA) that is directly proportional to the material above the sensor. To evaluate the results, Bragg wavelength shifts were studied against strain measured with conventional strain gauge. Results showed a linear response with an R^2 value of 0.999 in a range from 0 to 1700 micro strain units with a resolution of 0.001 nm per microstrain.

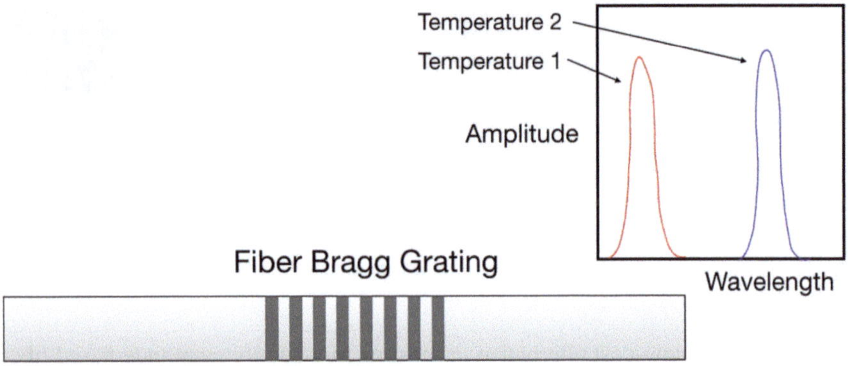

Fig. 25 Bragg wavelength at different temperatures

Metal-mechanic industry is a huge market worldwide. Demand from this industry includes auto parts, oil and aircraft companies. Metal parts manufacturing requires a well-controlled thermal process, since mechanical properties that determine the quality depend on fabrication temperature. Although an option for temperature monitoring is the use of thermal cameras, this technology is relatively expensive and only provides data from the surfaces. Fiber Bragg gratings (FBG) is based on the thermal-optical property of SiO_2 (fiber optic material) to sense temperature. FBG consist of periodic changes in refractive index along a length of optical fiber which cause light reflection of mainly a wavelength called Bragg wavelength. Usually, the light source in a FBG experimental setup is a superluminescent source (SLS). Under temperature variations, Bragg wavelength increase or decrease according the temperature as shown in Fig. 25.

In Alemohammad and Toyserkani (2011), a temperature fiber sensor based on FBG was embedded in a metallic piece for in situ monitoring. To develop this sensor, an optical fiber with FBG was coated with silver nanoparticles to create a conductive thin film. The coated FBG was embedded in a sample part of nickel and steel (with similar mechanical and thermal properties) since area of interest is tools manufacturing.

A layer of tungsten carbide and cobalt was deposited on the steel piece since this material increases the hardness of machining tools. Results showed a linear relation of Bragg wavelength with respect to temperature with a sensitivity of 25.8 pm/ºC.

Embedded fiber sensors have also been implemented to study the manufacturing process of plastics composites. This material is used in aircraft structures, however, its use is limited since more control instruments are needed in their manufacturing process to achieve a uniform quality. In Takeda et al. (2017), FBG were embedded in a thermoplastic composite to monitor the strain and temperature during the fabrication of an L-shaped part. In this case, two FBG were utilized in the positions shown transversely in Fig. 26. A heating monitoring from room temperature to 300 ºC was carried out with a wavelength spectra ranging from 1549 to 1554 nm. Press and demolding processes were also monitored as strain changes reflecting in Bragg

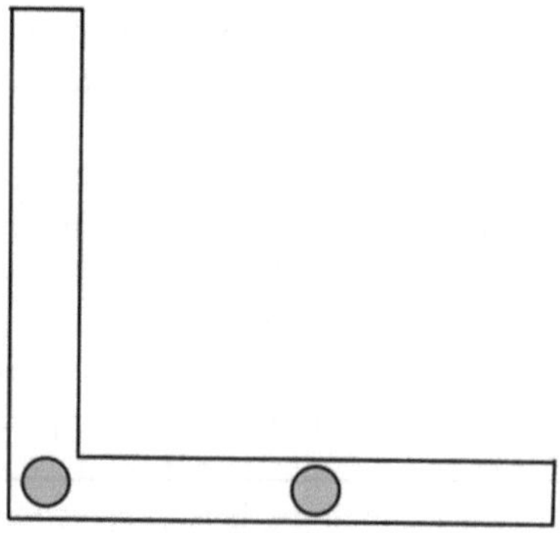

Fig. 26 Embedded fiber sensors based on FBG

wavelength shift. Although, variations in the spectra were minimal (less than one nanometer) which makes it difficult to sense the deformations.

Metal additive manufacturing represents an option to create complex products in a short time. However, there are problems associated with this method such as cracking and delamination which are caused by excessive stress in certain zones. Although there are well established methods to measure the quality of manufactured products, they are designed to evaluate the finished piece and not the manufacturing process. In order to overcome this issue, a smart build-plate based on fiber optic sensors was proposed (Hehr et al. 2020). In this device, the fiber was embedded in the build-plate which is allowed to measure the strain distribution in the manufactured piece. Data from fiber sensing was collected and processed in MATLAB and CAD model data. The prototype was effectively used to monitor quality in a laser-powder bed fusion process.

4 Conclusions

Optical technologies are suitable for quality monitoring in different industrial processes due their characteristics of versatility, compactness, high speed, and high resolution, among others. In general, optical approaches for manufacturing processes can be divided in two areas: free-space measurements involving lasers and contact methods based on fiber optics. Laser-based non-contact methods have been successfully used for quality control of aerodynamic profiles in turbine blades (Moreno-Oliva et al. 2019), 3D reconstruction of gears using interference patterns (Chen and Chen 2019), and in dimensional metrology with an outstanding resolution of 1 nm (Yuan and Zheludev 2019). Moreover, non-contact optical approaches have been used in conjunction with computer vision systems and image processing techniques for inline inspection quality in the food industry 4.0 (Li et al. 2020). Raman spectroscopy and reflectance have been also utilized for quality inspection of food and pharmaceutical products (Chullipalliyalil et al. 2020). On the other hand, fiber sensors are an emerging technology with high potential in monitoring manufacturing process with particular advantages as non-electromagnetic interference issues, distributing sensing, i.e., several sensing points along the same optical fiber and micrometric sizes. Fiber sensors could be based on fiber devices as fiber Bragg gratings, long period gratings, fiber tapers or in fiber interferometers configurations (Fabry–Perot, Mach–Zehnder, Michelson or Sagnac) (Lee et al. 2012). Fiber sensors have been used for quality inspection of oil (Taheri Ghahrizjani et al. 2016), wine (Noiseux et al. 2004), and gases production (Ohira et al. 2015), as well as embedded sensor for inspection of manufacturing processes of metal (Alemohammad and Toyserkani 2011) and composite parts for vehicles and aircraft industries. Based on current optical technology, it is expected an increasing demand in the areas of industrial application for quality monitoring.

References

Alemohammad H, Toyserkani E (2011) Metal embedded optical fiber sensors: laser-based layered manufacturing procedures 133(1):31015-1–31015-12. http://manufacturingscience.asmedigitalcollection.asme.org

Aristilde S, Cordeiro CM, Osorio JH (2019) Gasoline quality sensor based on tilted fiber Bragg gratings. Photonics 6(2):1–8. https://doi.org/10.3390/photonics6020051

Arango O, Trujillo AJ, Castillo M (2020) Inline control of yoghurt fermentation process using a near infrared light backscatter sensor. J Food Eng 277: https://doi.org/10.1016/j.jfoodeng.2019.109885

Chen YC, Chen JY (2019) Optical inspection system for gear tooth surfaces using a projection moiré method. Sensors (Switzerland) 19(6):E450. https://doi.org/10.3390/s19061450

Chullipalliyalil K, Lewis L, McAuliffe MAP (2020) Deep UV laser-induced fluorescence for pharmaceutical cleaning validation. Anal Chem 92(1):1447–1454. https://doi.org/10.1021/acs.analchem.9b04658

De León Martínez HA, Soto Bernal JJ, González Mota R, Rosales-Candelas I (2015) Optical evaluation on the setting of cement paste. J Phys Conf Ser 582(1):

El-Mesery HS, Mao H, Abomohra AEF (2019) Applications of non-destructive technologies for agricultural and food products quality inspection. Sensors 19:1–23. https://doi.org/10.3390/s19040846

Feng X, Senin N, Su R et al (2019) Optical measurement of surface topographies with transparent coatings. Opt Lasers Eng 121:261–270. https://doi.org/10.1016/j.optlaseng.2019.04.018

Ferreira FP, Forte PMF, Felgueiras PER et al (2017) Automatic optical inspection of regular grid patterns with an inspection camera used below the Shannon-Nyquist criterion for optical resolution. Photonic Instrum Eng IV 10110:101101Q. https://doi.org/10.1117/12.2267193

Gomes M, Cruz S, Lopes S, Arcipreste B, Magalhaes R, da Silva AF, Viana JC (2018) Distributed optical fiber sensors for PCB-strain analysis. IEEE Trans Ind Electron 66(10):8181–8188. https://doi.org/10.1109/TIE.2018.2885686

Hasheminejad N, Vuye C, Margaritis A et al (2019) Characterizing the complex modulus of asphalt concrete using a scanning laser doppler vibrometer. Materials (Basel) 12(21):3542. https://doi.org/10.3390/ma12213542

Hehr A, Norfolk M, Kominsky D et al (2020) Smart build-plate for metal additive manufacturing processes. Sensors 20:360. https://doi.org/10.3390/s20020360

Keulen C, Rocha B, Yildiz M, Suleman A (2011) Embedded fiber optic sensors for monitoring processing, quality and structural health of resin transfer molded components. In: 9th International conference on damage assessment of structures, 11–13 July 2011, University of Oxford

Lee BH, Kim YH, Park KS, Eom JB, Kim MJ, Rho BS, Choi HY (2012) Interferometric fiber optic sensors. Sensors 12(3):2467–2486. https://dio.org/10.3390/s120302467

Li L, Peng Y, Yang C, Li Y (2020) Optical sensing system for detection of the internal and external quality attributes of apples. Postharvest Biol Technol 162: https://doi.org/10.1016/j.postharvbio.2019.111101

Merzlyak MN, Solovchenko AE, Gitelson AA (2003) Reflectance spectral features and non-destructive estimation of chlorophyll, carotenoid and anthocyanin content in apple fruit. Postharvest Biol Technol 27:197–211. https://doi.org/10.1016/S0925-5214(02)00066-2

Moreno-Oliva VI, Román-Hernández E, Torres-Moreno E et al (2019) Measurement of quality test of aerodynamic profiles in wind turbine blades using laser triangulation technique. Energy Sci Eng 7:2180–2192. https://doi.org/10.1002/ese3.423

Noiseux I, Long W, Cournoyer A, Vernon M (2004) Simple fiber-optic-based sensors for process monitoring: an application in wine quality control monitoring. Appl Spectrosc 58(8):1010–1019. https://doi.org/10.1366/0003702041655476

Ohira SI, Miki Y, Matsuzaki T et al (2015) A fiber optic sensor with a metal organic framework as a sensing material for trace levels of water in industrial gases. Anal Chim Acta 886:188–193. https://doi.org/10.1016/j.aca.2015.05.045

Rodriguez AJ, Baldovino-Pantaleon O, Cruz RF, Zamarreño CR, Matías IR, May-Arrioja DA (2014) Gasohol quality control for real time applications by means of a multimode interference fiber sensor. Sensors 14(9):17817–17828. https://doi.org/10.3390/s140917817

Taheri Ghahrizjani R, Sadeghi H, Mazaheri A (2016) A novel method for onLine monitoring engine oil quality based on tapered optical fiber sensor. IEEE Sens J 16(10):3551–3555. https://doi.org/10.1109/JSEN.2016.2523805

Takeda S, Tsukada T, Minakuchi S, Takeda N, Iwahori Y (2017) Fiber-optic sensing for press forming of L-shaped thermoplastic composites. Procedia Eng 188(1):348–353. https://doi.org/10.1016/j.proeng.2017.04.494

Tunák M, Antoch J (2018) Monitoring homogeneity of textile fiber orientation. Text Res J 88(11):1226–1243. https://doi.org/10.1177/0040517517698983

Wang X, Xue L, He X, Liu M (2011) Vitamin C content estimation of chilies using Vis/NIR spectroscopy. In: International conference on electric information and control engineering, Wuhan, 2011, pp 1894–1897. https://doi.org/10.1109/ICEICE.2011.5777721

Wu Q, Zeng L, Zhou Q (2018) Phase I monitoring of optical profiles with application in low-emittance glass manufacturing. J Qual Technol 50(3):262–278. https://doi.org/10.1080/00224065.2018.1474687

Yeong TJ, Jern KP, Yao LK et al (2019) Applications of photonics in agriculture sector: a review. Molecules 24. https://doi.org/10.3390/molecules24102025

Yuan GH, Zheludev XI (2019) Detecting nanometric displacements with optical ruler metrology. Science 364(6442):771–775. https://doi.org/10.1126/science.aaw7840

Zhu T, Wu D, Liu M, Duan DW (2012) In-line fiber optic sensors in single-mode fibers. Sensors 12(8):10430–10449. https://doi.org/10.3390/s120810430

Reduction of the Scrap KPI in the Cutting Area of an Automotive Electrical Harness Company Using the Six Sigma DMAIC Methodology

Beatriz Adriana Esparza-Ramírez, Fernando Ricárdez-Rueda, Vicente Onofre Rojo-Romo, and Gerardo Alonso Torres-Ávalos

Abstract The following research project presents the implementation of the define, measure, analyze, improve and control (DMAIC) methodology as a tool for the reduction of scrap generated in the lines of the Ks area (cutting) of the company Electrics Plant located in Lagos de Moreno, Jalisco. It is recognized as a company that supplies parts or systems directly to automotive manufacturers tier 1 company. That the degree of demand in their products is robust; the purpose sought with the implementation is to reduce the KPI of scrap in the area during the cutting and crimp process which will help to increase the degree of satisfaction of the client and the company looking for a goal of 6.9 g/min compared with the established until the month of June of 8 g/min. Within the DMAIC stages used, SIPOC and CTQ graphics are used for the systematization of the process, Ishikawa graphics for the determination of the root cause of the problem, as well as control charts, AMEF, Gage R&R, and techniques. Lean Manufacturing which results in a sample average of the KPI with a value of 6.89 g/min falling within the satisfactory range agreed by management and showing a process that satisfices the specific requirements of the client and a downward trend in the associated indicator to the generation of scrap.

Keywords Tier 1 · DMAIC · KPI · Lean manufacturing and scrap

1 Introduction

This research work shows through the scientific method the application of define, measure, analyze, improve and control (DMAIC) methodology as a tool for the

B. A. Esparza-Ramírez (✉) · F. Ricárdez-Rueda · V. O. Rojo-Romo · G. A. Torres-Ávalos
Department of Industrial Engineering, Instituto Tecnológico José Mario Molina Pasquel y Henríquez Campus Lagos de Moreno, Jalisco. Adscrita al Posgrado de CIATEQ, A.C., Libramiento Tecnológico 5000, Colonia Portugalejo de los Romanes, Lagos de Moreno, Jalisco, Mexico
e-mail: beatriz.esparza@lagos.tecmm.edu.mx

J. L. García Alcaraz et al. (eds.), *Techniques, Tools and Methodologies Applied to Quality Assurance in Manufacturing*,
https://doi.org/10.1007/978-3-030-69314-5_3

reduction of scrap generated in the cutting and crimp production lines for automotive harnesses. The company served as an experimental unit for the application of the research, and this has noticed non-conformities in compliance with the key performance indicator (KPI) scrap in the cutting area Ks previously identified by the management area, reaching these numbers above 11 g/min in periods prior to the month of June.

Therefore, it has been decided to study the impact of the implementation of the DMAIC methodology for the solution of this problem by identifying, reducing and eliminating defects of any product. The problem was defined with support of critical to quality (CTQ), supplier, process, output and costumer (SIPOC) and Ishikawa diagrams, thus defining the activities that add value to the process and characterized the problem and its possible factors, thus collected process data and were quantified in formats and diagrams.

The measurement system for significant characteristics was verified and the KPI's indicators to be worked on during the project were defined, culminating in this stage, the collected data were interpreted to identify significant differences through the statistical inference and the NPR indexes (index yielded by AMEF tool) of potential failure modes and effects were calculated using the AMEF methodology (potential mode analysis and failure effect) (Corporation et al. 2001).

The drafting and structuring of the ideas of improvement for the process were reached, performing the corresponding simulation to complete by recording the process management sheets (X-R Charts) and evaluating the plots of dispersion and behavior of KPI and documented the process clearly to identify areas for improvement (Chrysler Corporation et al. 2005).

2 Literature Review and Hypotheses Statement

2.1 The Six Sigma and the DMAIC Methodology

We must differentiate the meaning of 6σ as we are referring to improvement programs or to the measure of the quality level of a process. From the point of view of improvement programs, and whose implementation we will see in this project, we can define 6σ as a methodological application philosophy whose objective is the elimination of waste, such as rework, improving the level of quality, costs and process time. From a capacity point of view, 6σ is a measure that relates process variability to specification limits, allowing it to be defined as: "A 6σ process means that it is capable of producing with a maximum of 3.4 ppm defects." In order to give a definition as concise as possible, let us say: "Six Sigma is a business system to achieve and maintain success through customer orientation, process management and process improvement, as well as analytical use of facts and data." (Ponsa 2015).

Improving the quality of a manufacturing or service system requires a formal approach to system performance analysis and finding ways to improve system performance. The define, measure, analyze, improve and control (DMAIC) is the process improvement methodology used by Six Sigma and is an iterative method that follows a structured and disciplined format based on the approach of a hypothesis, the conduct of experiments and its subsequent evaluation to confirm or reject the previously raised hypothesis. In recent years, Six Sigma concepts together with their DMAIC methodology have become the standard way to solve operational and design problems in both manufacturing and service systems. However, this methodology is sour of the use of simulation and optimization tools that take into account the complexity associated with non-normal statistical distributions, random failures, etc. (Ocampo and Pavón 2012).

Six Sigma projects use an improvement cycle called DMAIC that includes the following phases:

- Define. Define the problem or select the project, describing the effect caused by an adverse situation, or the improvement project that you want to perform, in order to be able to understand the starting situation and define the objectives. In this phase, the team is configured, which must be of a multidisciplinary type.
- Measure. Define and describe the process with focus on its metrics, determining its elements, its phases, inputs, outputs and characteristics. Evaluate measurement systems, analyzing their capacity and stability through studies of repeatability, reproducibility, linearity, accuracy and stability.
- Analyze. Determine significant variables. The process variables defined in the "measure" phase should be analyzed by means of experiment design and/or multivaried studies, to measure the contribution of these factors in process variability. The hypothesis and confidence interval tests will also be applied at this stage. Evaluate the stability and capacity of the process. Determine the process ability to produce products within specifications, using short- and long-term capacity studies, while evaluating defective fractions.
- Improve. Optimize and strengthen the process. If the process is not capable, it will have to be optimized to reduce its variation. This phase will use quality tools such as experiment design, regression analysis and ability to validate improvement.
- Control. Control and track the process, keeping it under statistical control. Continuously improve. Once the process is capable, better operating conditions, materials, procedures, etc. should be sought, leading to robust process performance less sensitive to variations in input metrics (Ponsa 2015).

Six Sigma has been involving from its application as a quality tool to being considered a key concept for world-class companies. In the Lean Six Sigma, while the Lean Manufacturing acts on waste in production activities in a rapid way the Six Sigma faces the analysis of the causes to avoid its repetition. Lean Six Sigma is more than just a conventional improvement program.

2.2 VSM Value Mapping

Value is a concept of perception of a product or service. It is everything that makes the functionalities expected by the customer, with an expected level of quality, at an expected cost and in an expected time frame and for which the customer is willing to pay. Anything that is not value or does not help to increase it directly and comes at a cost to the company is called waste (Lobato 2012).

The value chain is a sequence of activities or steps (with or without input of value) developed to get a certain product or service through the three typical tasks of managing a business:

- Troubleshooting Tasks: From design to launch of a product.
- Information management tasks: From order receipt to shipping planning.
- Physical transformation tasks: Transformation from raw materials to finished product.

The VSM is a tool used to analyze the value chain globally, beyond the analysis of a single process and collecting only certain general data of the different operations that are performed (Lobato 2012).

From the information collected, you must establish what is the target situation with the future map of the value chain. Finally, an action plan is established that specifies the changes that need to be made and those responsible for them.

The goal may be to obtain a view of the "door-to-door" production flow in a plant, including shipping the product to the plant's customer and delivering parts and material; or, in the case of large companies, the study of the value chain of a product goes through several facilities (Lobato 2012).

2.3 Repeatability and Reproducibility Study Gage R&R

Measurement repeatability and reproducibility studies determine that some of the variation observed in the process is due to the measurement system used. Repeatability can be expressed quantitatively in terms of the characteristic dispersion of the results. It is defined, according to the International Metrology Vocabulary (VIM), as the proximity of concordance between the results of successive measurements of the same measuring under the same measurement conditions, where:

1. These conditions are called repeatability conditions.
2. Repeatability conditions include: the same measurement procedure, the same observer, the same measuring instrument, used under the same conditions, the same place and repetition in a short period of time.

Reproducibility is defined as the concordance proximity between the results of successive measurements of the same by measuring under changing measurement conditions.

2.4 Hypotheses

It is possible to reduce the Scrap KPI to 6.9 g/min of the Ks area of the LME power plant using the DMAIC methodology as a tool.

3 Methodology

3.1 Definition of the Problem

This activity refers to the DMAIC—Define stage, which technically breaks down the problem studied by industrial engineering tools. For the correct definition of the problem, it is necessary to know the process with the input and output variables involved in it, by means of observations and information collection in the plant the SIPOC diagram tool was used for represent. The systemic approach to the interpretation of production processes serves as the basis for the improvement of synergy at the different stages of the flow; it is necessary to contemplate concepts such as entropy and the feedback when performing a complete analysis of the system Soft.

- Suppliers:
  Tool room refers to the area of the company that supplies Ks machines with tools necessary for crimping, sealing and stamping functions.
  Receipts refer to the department responsible for receiving and inspecting shipments of materials, taking care of the arrivals of raw materials for the cutting process such as cable, seal, terminals and indirect raw materials for the packaging of the cuts; this area is at the helm of the company's logistics staff and is in constant communication with the different areas for making accurate forecasts for the purchase of materials.
  External suppliers: refer to all suppliers who supply raw materials to the Ks area; these can be consulted in the receipt area in company warehouses.
- Inputs
  Automotive cable: Electrical conductors used for current transmission in vehicles. For cable identification, the company operates a coding system.
  Terminal: They have the function of establishing the connection of two or more cables with an electronic part. The surface of the base body and the surface of the terminal contact area can be pluggable and screwable.
  Seal: Fulfills the function of avoiding moisture and oxidation of crimps.
  As regards indirect inputs such as the protective cone, polytech tape and label ink, they are installed in each of the machines and are controlled by the equipment offers and relays that are on the lookout for the maternal inputs necessary ales.
- Process
  The process corresponding to the area represents the first step in the creation of the automotive harness, in which the cutting, waxing, crimping, sealing, stamping,

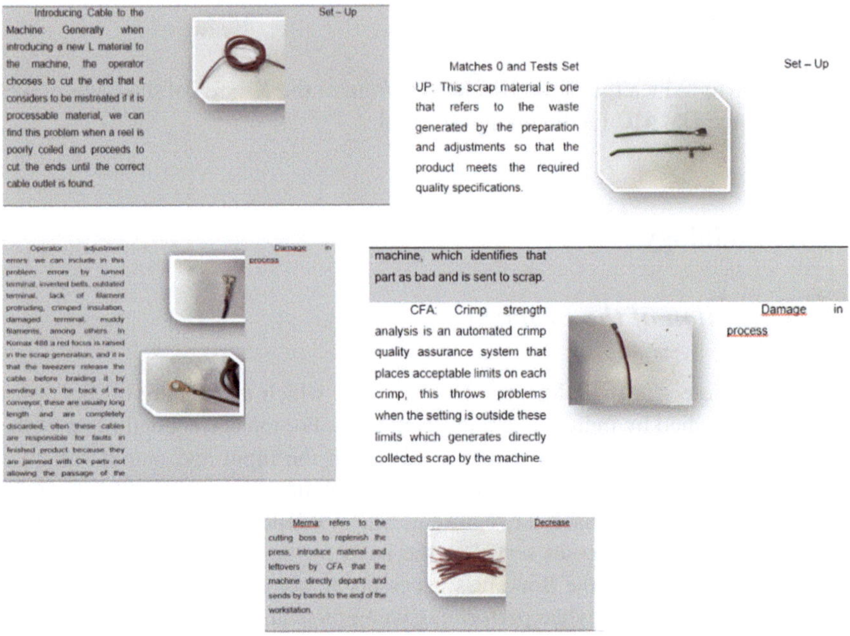

Fig. 1 Types of scrap-causing problems

braiding and labeling of the cables are carried out and then assembled in modules and they will turn into a harness.

- Outputs
  Cable cut, waxed, crimped, sealed, stamped and packed.
- Costumers
  The main customers are the brands AUDI, BMW, VW and MINI; in the same way, the internal customer of the Ks area is sub-assembly and modules.
  Identification of the main problems causing scrap. Figure 1 presents the identification table for the types of scrap problems in the area.

3.2 Determining Process Metrics

By representing the scrap (Gómez Giraldo 2009) (Gutiérrez Pulido and De la Vara Salazar 2009) damage in process 70% of the accumulated total, this is taken as the critical concept to study and apply efforts in its reduction and analyze the different causes or problems that lead to it being generated; it is done through a cell matrix taking the following criteria:

3: Frequent occurrence, routine failure.
2: Occasional failure, occurs suddenly.

```
         Total
Variable Count  N  N*  CumN  Percent   Mean  SE Mean  StDev  Variance  Minimum
Q1
Kpi         12 12   0    12      100  7.175    0.229  0.792     0.627    6.400
6.500
Variable Median     Q3  Maximum  Range
Kpi       6.950  8.100    8.500  2.100
```

Fig. 2 Descriptive statistics: KPI

1: Spontaneous failure, not very common but if occurring.

The descriptive statistics tools are applied in the Minitab software to interpret the information collected from the indicator at week 41 as illustrated in Figs. 2, 3 and 4.

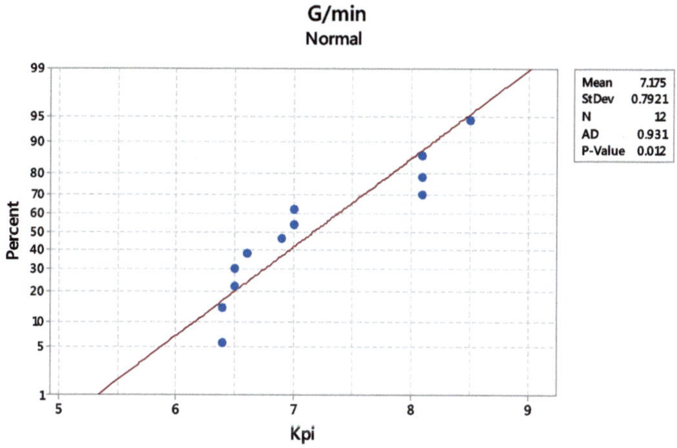

Fig. 3 Normally test: KPI

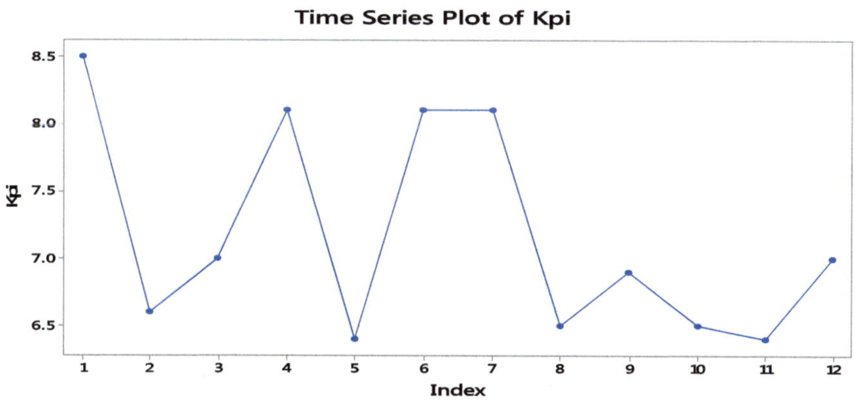

Fig. 4 Time series plot of KPI

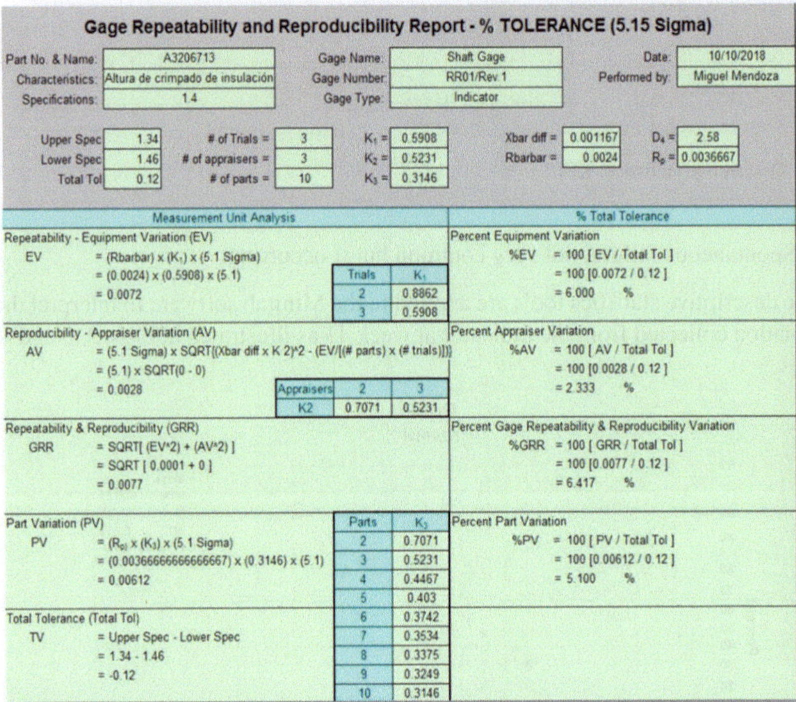

Gage Repeatability and Reproducibility Report - % TOLERANCE (5.15 Sigma)

Part No. & Name:	A3206713	Gage Name:	Shaft Gage	Date:	10/10/2018
Characteristics:	Altura de crimpado de insulación	Gage Number:	RR01/Rev 1	Performed by:	Miguel Mendoza
Specifications:	1.4	Gage Type:	Indicator		

Upper Spec	1.34	# of Trials =	3	K_1 =	0.5908	Xbar diff =	0.001167	D_4 =	2.58
Lower Spec	1.46	# of appraisers =	3	K_2 =	0.5231	Rbarbar =	0.0024	R_p =	0.0036667
Total Tol	0.12	# of parts =	10	K_3 =	0.3146				

Measurement Unit Analysis | % Total Tolerance

Repeatability - Equipment Variation (EV)
EV = (Rbarbar) x (K_1) x (5.1 Sigma)
= (0.0024) x (0.5908) x (5.1)
= 0.0072

Trials	K_1
2	0.8862
3	0.5908

Percent Equipment Variation
%EV = 100 [EV / Total Tol]
= 100 [0.0072 / 0.12]
= 6.000 %

Reproducibility - Appraiser Variation (AV)
AV = (5.1 Sigma) x SQRT[(Xbar diff x K 2)^2 - (EV/[(# parts) x (# trials)])]
= (5.1) x SQRT(0 - 0)
= 0.0028

Appraisers	2	3
K2	0.7071	0.5231

Percent Appraiser Variation
%AV = 100 [AV / Total Tol]
= 100 [0.0028 / 0.12]
= 2.333 %

Repeatability & Reproducibility (GRR)
GRR = SQRT[(EV^2) + (AV^2)]
= SQRT [0.0001 + 0]
= 0.0077

Percent Gage Repeatability & Reproducibility Variation
%GRR = 100 [GRR / Total Tol]
= 100 [0.0077 / 0.12]
= 6.417 %

Part Variation (PV)
PV = (R_p x (K_3) x (5.1 Sigma)
= (0.00366666666666667) x (0.3146) x (5.1)
= 0.00612

Parts	K_3
2	0.7071
3	0.5231
4	0.4467
5	0.403
6	0.3742
7	0.3534
8	0.3375
9	0.3249
10	0.3146

Percent Part Variation
%PV = 100 [PV / Total Tol]
= 100 [0.00612 / 0.12]
= 5.100 %

Total Tolerance (Total Tol)
TV = Upper Spec - Lower Spec
= 1.34 - 1.46
= -0.12

Fig. 5 Gage R&R

In order to validate is the current measurement system of the special quality characteristics, MSA analysis is used using the Gage R&R method (repeatability and reproducibility of gages) Fig. 5 taking as reference the crimp height characteristic of discharge measured with vernier. This was selected because it is a critical feature that is measured without a digital measuring instrument connected to the machine.

3.3 *Process Analysis*

For the phase corresponding to DMAIC analysis, the data collected from the previous phase was taken to process the information using statistical tools.

Week 41 reported incidents of high weights of kilograms of scrap reported on machine 894 hypothesized that the labor factor was the cause of the increase in the machine's scrap kilograms, because shift B has an operator r new entry. To statistically verify this assumption in Fig. 6, the data is subjected to a one-factor ANOVA, 3 levels, significance coefficient 0.05 two-way Tukey comparison.

One-way ANOVA: A, B, C

Method

Null hypothesis	All means are equal
Alternative hypothesis	At least one mean is different
Significance level	$\alpha = 0.05$

Equal variances were not assumed for the analysis.

Factor Information

Factor	Levels	Values
Factor	3	A, B, C

Welch's Test

Source	DF Num	DF Den	F-Value	P-Value
Factor	2	6.85169	1.43	0.303

Model Summary

R-sq	R-sq(adj)	R-sq(pred)
9.80%	0.00%	0.00%

Means

Factor	N	Mean	StDev	95% CI
A	5	4.080	2.077	(1.502, 6.658)
B	5	6.56	6.82	(-1.90, 15.02)
C	5	7.26	3.67	(2.70, 11.82)

Fig. 6 One-factor ANOVA, 3 levels, significance coefficient 0.05 two-way Tukey comparison

Since *P*-value 0.303 greater than the significance level 0.05, the null hypothesis (all means are the same according to ANOVA) is accepted; therefore, the existence of significant differences between the operators of the 894 machine cannot be considered.

3.4 Improve the Process

For the application of DOE in the plant, it sought to attack a latent problem in scrap generation, a 2*k* factorial model was raised to find the optimal treatment of the CFA+ sensitivity zones for the terminal A3606748 crimped with the applicator 1892.

Each factor is studied at only two levels and the experiments cover all combinations of each level of one factor with all levels of the other factors.

- Objectives of the experiment.

  I: Treatment factors and their levels.
  II: Experimental unit: Adjustment 1892-748.
  III: Noise factors. The factors selected for the experiment Fig. 7 are the sensitivity zones in the CFA+ specifications for standard parameters corresponding to the terminal to be examined.

The following Fig. 8 shows the experiment matrix assigned for a factorial DOE 2^3.

Figure 9 to illustrate the use of a 2^3 factorial design, consider a CFA reaction. You want to check what effect the 3 sensitivity zones S1, S2 and S3 on the increase of scrap problems by CFA+.

To know the effect of a factor, it is sufficient to make it vary between two values. The most suitable are the extremes in their experimental domain, and in addition to

Factor	Name	Type	Low	High
A	S1	Numeric	0.1	A1
B	S2	Numeric	1	A2
C	S3	Numeric	0.6	A3

Fig. 7 Coded notation

Fig. 8 Experiment matrix for 2*k* factorial

Design Table (randomized)

Run	A	B	C
1	-	-	-
2	-	+	-
3	-	-	+
4	+	+	-
5	+	-	+
6	+	-	-
7	+	+	+
8	-	+	+

C1	C2	C3	C4	C5	C6	C7	C8-T
StdOrder	RunOrder	CenterPt	Blocks	S1	S2	S3	Response
1	1	1	1	0,1	1	0,6	y1
3	2	1	1	0,1	2	0,6	y2
5	3	1	1	0,1	1	3,0	y3
4	4	1	1	1,0	2	0,6	y4
6	5	1	1	1,0	1	3,0	y5
2	6	1	1	1,0	1	0,6	y6
8	7	1	1	1,0	2	3,0	y7
7	8	1	1	0,1	2	3,0	y8

Fig. 9 Sensitivity zones

this, variation must be performed for each possible combination of the values of the other factors. This will allow you to discover whether the effect depends on what value the other factors take. All these combinations are contemplated in the factorial design; in total, eight experiments corresponding to two levels of S1 by two levels of S2 by two levels of S3.

3.5 *Process Control*

As a process, a Poka Yoke applied to the Pull Tester measurement tool was designed, because it reflects the special stretch strength feature considered critical to customers. The manufacturing practices by operators to this tool are not carried out in the proper way giving operators this test with the force of their hands without introducing the sample to run the test causing it not to be guaranteed if the part actually complies with the specification given by the customer thus causing non-conformities.

The function of the Poka Yoke falls on the inclusion of a second button on the left side, in order to program the test so that it can only be executed by pressing both buttons continuously until the test is complete, keeping the hands busy so as not to give them an opportunity to use in applying force manually.

Once a better understanding of the process is achieved, the process itself must be maintained at an appropriate skill level. The processes are dynamic and change; the performance of the process should be monitored so that effective measures can be taken to prevent unwanted changes. The control graphs tool was used for efficient monitoring. For effective management of variations during production, there must be effective means of detecting special causes, and the elements considered for the elaboration of the control charts are as follows:

(a) Appropriate scale: The natural variation of the process is respected and can be easily visualized.

(b) UCL, LCL: Specification limits for determining process skill based on a sample distribution.
(c) Centerline: Based on sample distribution to allow the determination of non-random patterns that signal special causes.

For the documentation of the process, a weekly registration format was created for the control of scrap damage in process registered by machine; this document shows us indicators of percentage contribution, top 5 of scrap generating machines, average scrap per machine and the corresponding scatter plot.

4 Results

4.1 Game Rules for the Team

The following rules were defined for the group:

1. Meetings. A weekly meeting will be held by the control group, as well as a weekly meeting with each assigned development group.
2. Communication. The official communication will be made through the email created for this purpose, which must be sent with a copy to all members of the group.
3. Decision-making mechanism: Some decisions are made among all, through a group consensus. If a member is not at the time of a meeting where there is consensus, the decision taken will be approved, he/she must accept and comply with the decisions that have been made there. In the event that the decision to be made depends largely on the absent person, the group leader shall have the authority if he considers it to defer the decision to the next meeting.
4. Decision-making mechanism: Some decisions are made among all, through a group consensus. If a member is not at the time of a meeting where there is consensus, the decision taken will be approved, he/she must accept and comply with the decisions that have been made there. In the event that the decision to be made depends largely on the absent person, the group leader shall have the authority if he considers it to defer the decision to the next meeting.
5. Task report. Commitment to have the progress of each member's tasks on the Mondays before the meeting. On Monday at the meeting, the project leader will follow up respectively. Once the assigned task is complete, each member must upload the task to the group repository and send the notification to the planning leader, to perform the compliance registration, and the latter will inform the quality leader to perform quality control and with the notification was sent to the leader of the group to report the tasks to the dot project and these to be reviewed by the monitor and the teacher of the subject. If any group member does not submit the completed task and with the quality approval, they will be deemed unfulfilled of the assigned task. Therefore, a record of non-compliance

should be carried out in the assignments measured in hours/days as appropriate. For all deliverables, delivery date and time would be defined.

6. Document quality: All generated documents must be approved by the quality leader and then reviewed by the group leader before being delivered or published.
7. Conflict resolution: If there are conflicts between two or more people, they can escalate to the team leader who can personally try to resolve it, or you can set it as a point on the weekly meeting agenda.

4.2 Simulation

The respective simulation of the process after the implementation of the DMAIC in ProModel is performed, and the data was updated according to the improvements obtaining the following result in Fig. 10.

In Fig. 11, the cutting minutes feature is taken as the output variable; the function of maximizing the feature is sought.

In Fig. 12, the total output result yields a figure of 16,500 min of average outage as system capacity.

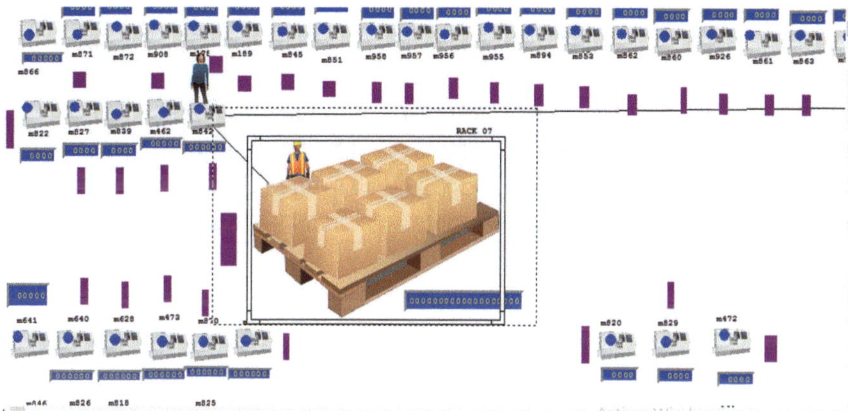

Fig. 10 Layout simulation after DMAIC

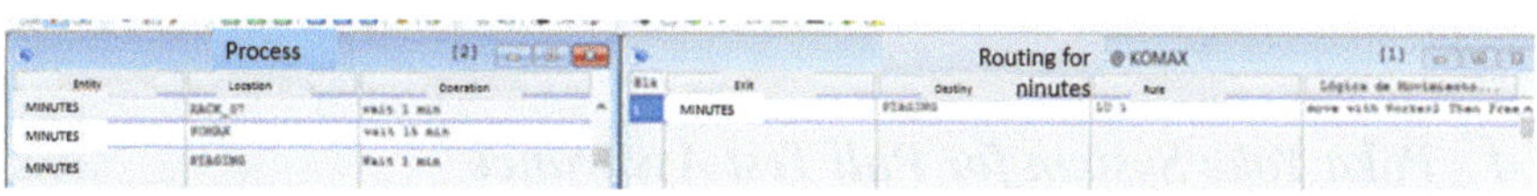

Fig. 11 Simulation process after DMAIC

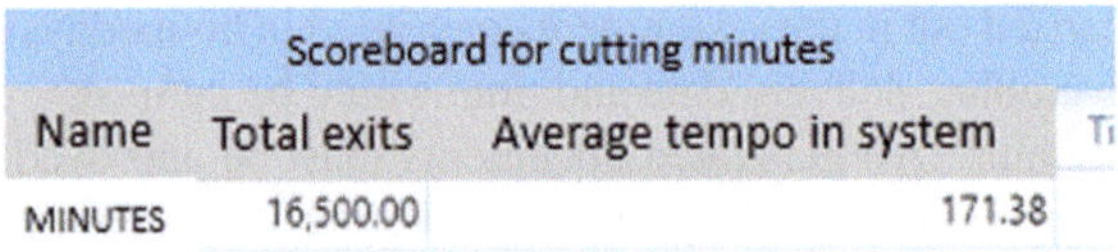

Scoreboard for cutting minutes			
Name	Total exits	Average tempo in system	Ti
MINUTES	16,500.00	171.38	

Fig. 12 Scoreboard for cutting minutes

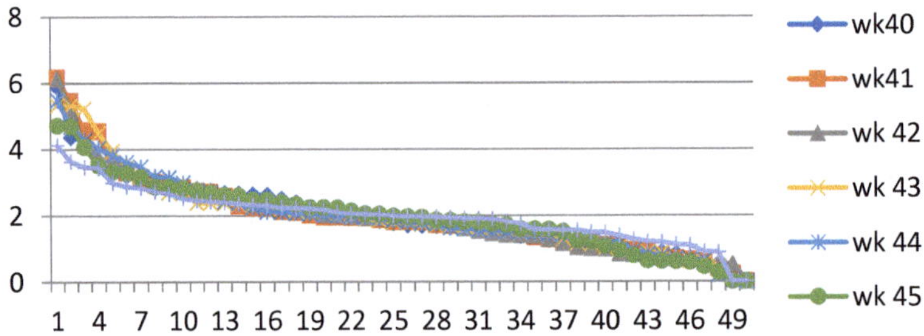

Fig. 13 Linearity chart for average damage in process

Table 1 Top 5 average scrap damage in process

wk40	wk41	wk42	wk43	wk44	wk45	wk46
5.876871	6.175852	6.141197	5.328789	5.477396	4.714761	4.122319
4.361061	5.453492	5.190349	5.313299	4.689602	4.686157	3.625467
4.280301	4.574946	4.13941	5.2281	4.280841	4.066209	3.454674
3.634218	4.548915	3.7605	4.500039	4.043017	3.508871	3.446911
3.590731	3.429583	3.610366	3.95012	3.812624	3.326159	2.988875

4.3 KPI Reduction

In Fig. 13 and Table 1, results are displayed in scrap kilograms in-process damage from week 40 to 46 by recording values per machine and interpreting in time series charts. In Fig. 14, the first object of study was the average scrap variable per machine by analyzing its behavior and obtaining the following results.

4.4 Poka Yoke System for Pull Test Assurance

A Poka Yoke device for the Ks area was planned to ensure compliance with the significative characteristics (SC) specifically the Pull Test stretch strength test.

The problem lies in the operator applying the Pull Test with the force of his hands by pulling the clamping part of the terminal to the clamping part of the cable;

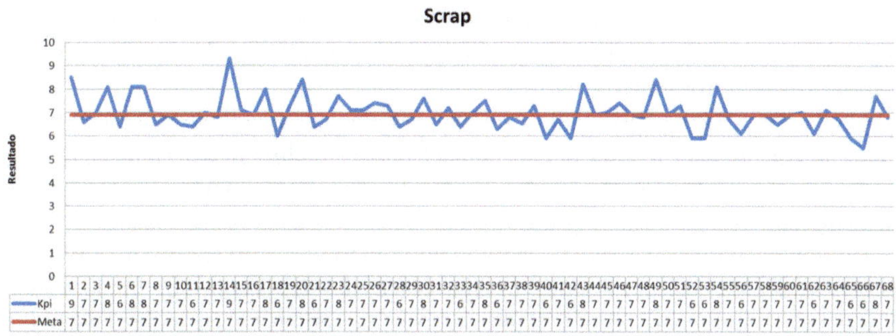

Fig. 14 Scrap KPI variation chart

Fig. 15 Left side view of the Poka Yoke button measuring instrument

this creates uncertainty in the finished parts because it cannot be guaranteed if the product products they actually meet the quality requirements because the test is not performed with the sample.

The proposal for the Poka Yoke device is proposed to install in the Pull Tester measuring tool is a right button at the bottom left so that the measurement is done only by pressing the original and secondary button at the same time throughout the test, thus forcing the operator to keep his hands busy and there is no other way to perform the test.

Figure 15 shows a CAD-illustrated prototype of the Poka Yoke system.

5 Conclusions and Industrial Implications

In conclusion, it is emphasized that the application of scientific research in this case focused on the DMAIC methodology provides an effective continuous improvement tool through the use of statistical tools that are applied in the improvement of processes, customer satisfaction and resource optimization. It is important to visualize expenditures on these projects as short- and medium-term investments and that systems are only efficient with monitoring based on statistical tools for live data control that are used to analyze the position and trend of the production system with respect to its control limits.

The research applied in the improvement and innovation of industrial processes is a tool that facilitates very precisely the objectives to be achieved as a team, analyzing the technical feasibility of the project, the necessary estimation of resources and the costs thereof.

By implementing the DMAIC methodology in the Ks workspace, the KPI indicator was decreased from 8 g/min to an average of 6862 g/min. The implemented X-R control chart record sheet will allow corrective actions to be executed when the production system has a trend that leads it to produce non-compliant parts.

The best way to deal with the improvement of a process is to think about how to transform it into a new process more efficient; in short, the actions of improvement are the basis of the solution or persistence of the problem. As recommendations, it is proposed to share the experience of the multidisciplinary project team with other teams as well as provide support to other production lines for the DMAIC methodology.

The modern automotive industry is a precursor to the stiffest and most innovative production systems in the global market, so the demand in quality systems and indicators turns out to be challenging and extremely robust.

References

Chrysler Corporation D, Ford MC, General Motors C (2005) Control estadístico de los procesos (SPC). Lahser Road, Southfield, Michigan, pp 69–86

Corporation DC, Company FM, Corporation GM (2001) Análisis de modos y efectos de fallas potenciales. AIAG, Estados Unidos, pp 31–51

Gómez Giraldo H (2009) Estadística. Ciudad de México, pp 20–64

Gutiérrez Pulido H, De la Vara Salazar R (2009) Control estadístico de calidad y Seis Sigma. Mc Graw Hill, Ciudad de México, pp 96–292

Lobato MB (2012) Análisis y mejora de los procedimientos. Ingeniería Organizacional Sevilla, Sevilla, pp 26–117

Ocampo JR, Pavón AE (2012) Integrando la Metodologia DMAIC de Seis Sigma con la con la Simulación de eventos discretos. In: Tenth LACCEI Latin American and caribbean conference for engineering and technology, pp 21–34

Ponsa XP (2015) Aplicación de la metodología "dmaic" en la resolución de problemas de calidad. Na, UVIC, pp 55–109

Empirical Bayes Monitoring for Univariate and Multivariate Processes and Other Techniques

Manuel Arnoldo Rodríguez Medina, Manuel Alonso Rodríguez Morachis, Luz Elena Terrazas Mata, Manuel Iván Rodríguez, and Eduardo Rafael Poblano-Ojinaga

Abstract The purpose of this document is to use the basic concepts given in the celebrated Kalman filter, which can be derived using a Bayesian approach. Such an approach is implemented through Bayesian empirical monitoring for process control. The purpose of the Kalman filter is to obtain real-time estimates of the results of critical process variables, subject to process variations and noise conditions, such as environmental and measurement conditions. This document analyzes both the univariate and multivariate case of Bayesian empirical monitoring. The first application is analyzed to data taken from a molding process of a critical quality characteristic of an automotive sensor. For the multivariate case, measures of the characteristics to be controlled of a molded part were taken. The three-dimensional behavior was analyzed, first, by means of Bayesian empirical monitoring and then illustrating tests of multivariate normality and the management of Hotelling's square T.

Keywords Kalman's filter · Bayes · Empirical Bayes monitoring · Multivariate normality analysis

1 Introduction

The improvements of companies in quality and productivity, considering products and processes, have been strongly based on the techniques of statistical process control. The control techniques have been established according to the type of variables to be controlled, the required sensitivity of the graph, and under concepts of inferential statistics under the frequentist approach. It is not common to find process monitoring jobs based on Bayesian inference.

Colosimo and Del Castillo (2007) and Box and Tiao (1973) mention that Bayesian inference combines previous experience or knowledge about the parameters of the

M. A. Rodríguez Medina · M. A. Rodríguez Morachis · L. E. Terrazas Mata · M. I. Rodríguez · E. R. Poblano-Ojinaga (✉)
Division of Postgraduate Studies and Research, TecNM/I.T. de Ciudad Juárez, Av. Tecnológico 1340, Fraccionamiento el Crucero, Juárez Chihuahua, Mexico

J. L. García Alcaraz et al. (eds.), *Techniques, Tools and Methodologies Applied to Quality Assurance in Manufacturing*,
https://doi.org/10.1007/978-3-030-69314-5_4

model with evidence of the data through the management of Bayes theorem. In this approach, there is a subjective approach to probability, unlike the frequentist approach where the probability of occurrence of an event is seen as the limit of the ratio of the frequencies of events.

The Bayes Theorem

Bayes theorem (Box and Tiao 1973) for probability density functions, where X represents the data and θ a parameter or parameter vector of a population of interest, is mathematically defined as:

$$f(\theta|X) = \frac{f(X|\theta)f(\theta)}{f(X)} \tag{1}$$

where $f(\theta|X)$ is the posterior probability density function,

$f(X|\theta)$ is the likelihood function,

$f(\theta)$ is the a priori probability density function and

$f(X) = \int_{\forall\theta} f(\theta)f(X|\theta)d\theta$ represents the marginal probability.

This can be used to make inferences or decisions about the parameter, with respect to future measurements or observations, conditionally on past events and under the consideration that the parameters behave randomly.

In this section, the basic concepts of Bayesian inference will be used to generate monitoring systems for the population mean and variance of a univariate process. Subsequently, concepts for empirical monitoring of multivariate processes will be generalized.

2 Empirical Bayes Monitoring of Molded Parts Parameters

2.1 The Kalman Filter

Castañeda Cárdenas et al. (2013) define the Kalman filter as an algorithm based on the state space of a system to estimate the future state and future output through optimal filtering of the output signal and with adequate sampling, it will fulfill a function of estimates of parameters Kim and Bang (2019) agree that the Kalman filter is an algorithm that generates estimates of unknown variables given the measurements observed over time; They also mention that the Kalman filter has a relatively simple form and requires little computational power.

Kalman (1960) describes the type of problems to study (or solve) as follows: The signal $x_1(t)$ and the noise $x_2(t)$ are given. Only the sum $y(t) = x_1(t) + x_2(t)$ can be observed. Assuming that the values of $y(t_0), \ldots, y(t))$ are observed and known exactly. What can be inferred from the knowledge about the (not observable) value of the signal $t = t_1$ where t_1 may be: If $t_1 < t$, the problem is interpolation, If $t_1 = t$, the problem is called filtering and if $t_1 > t$, the problem is prediction or, also called estimation. Among the problems to be solved, Wiener's problem is considered, which is stated as follows:

Consider the dynamic model

$$\boldsymbol{X}(t+1) = \boldsymbol{\Phi}(t+1; t)\boldsymbol{X}(t) + \boldsymbol{u}(t) \tag{2}$$

$$\boldsymbol{y}(t) = \boldsymbol{M}(t)\boldsymbol{X}(t) \tag{3}$$

where $\boldsymbol{u}(t)$ it is an independent Gaussian random process of n vectors with zero mean, $\boldsymbol{X(t)}$ it has an n-vector, $\boldsymbol{y(t)}$, it has a p-vector $(p \leq n)$,$\boldsymbol{\Phi}(t+1; t)$, $\boldsymbol{M}(t)$, they are matrices of $n \times n$ and $p \times n$ whose elements are non-random functions of time. Given the observed values of $y(t_0), \ldots, y(t)$, the problem is finding an estimate $X^*(t_1|t)$ of $X(t_1)$ which minimizes the expected loss. Munuera (2018) mentions that the Kalman filter will generate an estimator that will minimize the mean square error, which will be observed when evaluating the Kalman gain.

Youngjoo and Hyochoong (2018) mention that the Kalman filter, as seen in the processes, cited above (such as the Wiener process), is used to estimate process states based on dynamic linear systems in state space format, where the model defines the evolution of the state of time $k - 1$ to time k as:

$$x_k = Fx_{k-1} + Bu_{k-1} + w_{k-1} \tag{4}$$

where

F: is the transition state matrix applied to the previous state vector x_{k-1}

B: is the input control matrix applied to the control vector u_{k-1}

u_{k-1}: it is the noise vector of the process that is Gaussian with zero mean with covariance Q, $w_{k-1} \sim N(0, Q)$.

The relationship between the process model and the measurement model is given by

$$z_k - Hx_k + v_k \tag{5}$$

where z_k is the vector of measurements, H is the measurement matrix and, v_k is the measurement noise vector which is assumed Gaussian $v_k \sim N(0, R)$, where R is the covariance.

The purpose of the Kalman filter is to generate estimates of x_k in the time k setting an initial value of x_0, the set of measurements $z_1, z_2, \ldots, z_k$, and the information matrices of the system F, B, H, Q, and R, which are assumed without variation in time.

2.2 The Classic EWMA Control Chart

Abbas et al. (2014) cite the work of Roberts (1959) on the EWMA statistic for monitoring the process average by

$$Z_i = \gamma \bar{X}_i + (1-\gamma)Z_{i-1} \tag{6}$$

where i is the sample number, γ is the smoothing constant and should be chosen as $0 < \gamma \leq 1$ and $\bar{X}_i$ is the average of the ith sample, while Z_{i-1} represents the past information and its initial value Z_0 is assumed equal to its target value or the average of the initial samples.

The value of γ determines the reason at which the information passed into the calculation of the EWMA statistic. The large values of γ assign more weight to current information. When $\gamma = 1$, the EWMA chart matches the charts X of Shewhart.

The mean in control and the variance of the EWMA statistic are:

$$E(Z_i) = \mu_0,\ \mathrm{Var}(Z_i) = \sigma^2_{\bar{X}}\left(\frac{\gamma}{2-\gamma}\left(1-(1-\gamma)^2\right)\right) \tag{7}$$

where μ_0 represents the target value of the average of X and $\sigma^2_{\bar{X}}$ represents the variance of X. When either or both μ_0 and $\sigma^2_{\bar{X}}$ are unknown, they can be estimated from the first sample values. Thus, the upper and lower control limits and the center line of the EWMA chart are calculated as:

$$\begin{aligned} \mathrm{LCL}_i &= \mu_0 - K\sigma_{\bar{X}}\sqrt{\frac{\gamma}{2-\gamma}\left(1-(1-\gamma)^{2i}\right)} \\ \mathrm{LC} &= \mu_0 \\ \mathrm{UCL}_i &= \mu_0 + K\sigma_{\bar{X}}\sqrt{\frac{\gamma}{2-\gamma}\left(1-(1-\gamma)^{2i}\right)} \end{aligned} \tag{8}$$

where K determines the width of the control limits, which at the same time determines the ARL in control of the EWMA chart.

2.3 Empirical Monitoring Techniques

Shiau and Feltz (2007) show empirical process monitoring techniques and consider the following aspects as basic:

- The likelihood function may be used as a sampling distribution and analyze the variability of the current process due to the measurement and the form of selection of the elements.
- Empirically estimate the distribution a priori and make inferences about the process average and its variation over time.
- The estimation of the subsequent distribution may be used to create an estimated distribution for the variation of the parameter of interest of the process.
- Weight the observations differently to create short and long-term estimates of the process parameters. The weights performed act in a similar way to the exponentially weighted moving averages (EWMA).
- Use recursive equations to estimate parameters based on sufficient statistics.

Using the basic concepts given for Bayes empirical monitoring, provided by Shiau and Feltz (2007), we will analyze the height (critical characteristic) of the parts produced in an over molding operation represented by X_t (measured with vernier) which assumes normal behavior with mean μ_t and variance due to sampling and measurement σ^2 at time t. The conditional probability density (p.d.f.) function of the given characteristic X_t/μ_t can be written.

$$g_1(X_t|\mu_t) = \frac{1}{\sqrt{2\pi\sigma^2}}\exp\left[\frac{-(X_t-\mu_t)^2}{2\sigma^2}\right] \tag{9}$$

The parameter of interest will be the average of the quality characteristic μ_t. The average of the process may change over time due to manufacturing imperfections and changes in ambient temperature and humidity; the latter due to the moisture absorption of the molding materials. The "a priori" distribution in the Bayesian analysis is established based on the prior knowledge of the process managers and here it can be assumed that the average quality characteristic behaves $\mu_t \sim \text{NID}(\mu, \gamma^2)$ and its probability density function will be:

$$g_2(\mu_t) = \frac{1}{\sqrt{(2\pi\gamma^2)}}\exp\left[\frac{-(\mu_t-\mu)^2}{2\gamma^2}\right] \tag{10}$$

where μ and γ^2 are called hyperparameters, which are unknown constants and must be estimated. Here, γ^2 represents the process variance, or variance due to the change in the behavior of the process parameters. Consider again the Bayes theorem in Eq. 3, where $f(\mu_t|X_t)$ is the posterior distribution, $g_1(X_t|\mu_t)$ the likelihood function, and $g_2(\mu_t)$ is the a priori distribution for μ_t. The posterior distribution will be used to obtain the arithmetic mean of the quality characteristic X_t

$$f(\mu_t|X_t) = \frac{g_1(X_t|\mu_t)g_2(\mu_t)}{\int g_1(X_t|\mu_t)g_2(\mu_t)\mathrm{d}\mu_t} \tag{11}$$

The posterior distribution will be useful to make inferences in relation to the behavior of the process, by successive estimation of the behavior of the parameters, that is, μ_t, σ_t^2, Var_t and γ^2.

The posterior distribution of μ_t/X_t is $\mathrm{NID}(\mu_{pt}, \sigma_p^2)$, where

$$\mu_{pt} = w\mu + (1-w)X_t \tag{12}$$

and

$$\sigma_p^2 = \gamma^2 w \tag{13}$$

where $w = \frac{\sigma^2}{\sigma^2+\gamma^2}$. The value of X_t is the value measured at time t and is used to empirically estimate the values of, σ^2 and γ^2, and also estimate the posterior value of μ_t, that is, where the process is at time t.

2.4 Parameter Estimation

A basic assumption for estimation is that the process parameter μ_t is relatively constant in small intervals of time, and from here, the X_t is independent and identically distributed with the function g_1 as a function of probability density. The X_t can be considered unbiased estimators of μ, whereby they can be used to calculate an estimator $\widehat{\mu}$ of μ:

$$\widehat{\mu} = \frac{\sum_{t=1}^{T} X_t}{T} \tag{14}$$

To estimate the variance of the sampling, it should be assumed that the adjacent observations do not present sudden changes and could be considered of the same distribution. So, the variance

$$\mathrm{Var}\,(X_t - X_{t-1})^2 = 2\sigma^2 \tag{15}$$

for identically distributed random variables, it is estimated σ^2 averaging square mobile ranges

$$\widehat{\sigma}^2 = \frac{1}{T-1}\sum_{t=2}^{T}(X_t - X_{t-1})^2/2 \tag{16}$$

The estimator of the total variance will be

$$\widehat{V} = \frac{1}{T}\sum_{t=1}^{T}(X_t - \widehat{\mu})^2 \tag{17}$$

Remembering that $V = \sigma^2 + \gamma^2$, we can write, $\widehat{\gamma}^2 = \widehat{V} - \widehat{\sigma}^2$, $\widehat{\gamma}^2$ is positive. If this does not happen let us do $\widehat{\gamma}^2 = k\widehat{V}$, where k is a very small arbitrary number, $0 < k < 1$.

Considering now that current data have more weight than previous data, Sturm et al. (1991) proposed the following way of obtaining exponentially weighted estimators for μ, σ, and V. Weighted estimators are obtained as follows:

$$\widehat{\mu}_T = \frac{\sum_{t=1}^{T}\lambda^{T-t}X_t}{\sum_{t=1}^{T}\lambda^{T-t}} \tag{18}$$

$$\widehat{\sigma}_T^2 = \frac{\sum_{t=2}^{T}\lambda^{T-t}(X_t - X_{t-1})^2/2}{\sum_{t=1}^{T}\lambda^{T-t}} \tag{19}$$

$$\widehat{V}_T = \frac{\sum_{t=1}^{T}\lambda^{T-t}(X_t - X_{t-1})^2}{\sum_{t=1}^{T}\lambda^{T-t}} \tag{20}$$

where T represents the most current observation and λ is a positive number less than 1 (usually $0.8 < \lambda < 1$). The incorporation of the weights in the Bayes empirical approach helps to maintain the distribution structure and obtain the separation of the variations within the sampling and the process and also allows to identify the changes in the process average.

The recursive equations above can also be obtained by:

1. Calculate the weighting in period t

$$\theta_t = \frac{1}{\sum_{k=1}^{t}\lambda^{t-k}} \tag{21}$$

where the values of θ_t can be obtained through the series

$$\theta_m = \frac{1}{1+\sum_{n=2}^{m}\lambda^n} = \frac{1-\lambda^{n+1}}{1-\lambda} - \lambda;\quad \text{to } n > 0 \tag{22}$$

With the values of $\lambda = 0.946$ and $\lambda = 0.99$ for short term and long term, respectively, weighting coefficients are calculated. Table 1 shows the coefficient values for the fifty observations taken from the process. Note that in both cases the coefficient θ_1 is equal to 1.

Table 1 Values of weighting coefficients

λ = 0.99		λ = 0.946		Obs
θ_i	$1-\theta_i$	θ_i	$1-\theta_i$	
1.00000	0.000000	1.000000	0.000000	1.345
0.50500	0.494900	0.527730	0.472270	1.348
0.33890	0.661060	0.364760	0.635240	1.346
0.25570	0.744310	0.282300	0.717700	1.349
0.20570	0.794320	0.232560	0.767440	1.353
0.17230	0.827690	0.199333	0.800667	1.352
0.14850	0.851530	0.175600	0.824399	1.356
0.13057	0.869420	0.157820	0.842170	1.354
0.11670	0.883330	0.144032	0.855960	1.356
0.10550	0.894470	0.133033	0.866960	1.353
0.09642	0.903580	0.124070	0.875930	1.354
0.08883	0.911170	0.116637	0.883360	1.358
0.08240	0.917597	0.110381	0.889610	1.355
0.07689	0.923102	0.105050	0.894940	1.350
0.07220	0.927880	0.100460	0.899540	1.359
0.06795	0.932050	0.096470	0.903530	1.346
0.06427	0.935730	0.089905	0.907010	1.347
0.06100	0.938990	0.087175	0.910090	1.345
0.05807	0.941930	0.084740	0.912820	1.346
0.05544	0.944590	0.082560	0.915260	1.345
0.05306	0.946940	0.080599	0.917439	1.345
0.05089	0.949110	0.078830	0.919400	1.347
0.04892	0.951080	0.077220	0.921170	1.349
0.04711	0.952890	0.075760	0.922770	1.345
0.04544	0.954557	0.074430	0.924239	1.348
0.04391	0.956093	0.073211	0.925572	1.345
0.04248	0.957516	0.072100	0.926788	1.345
0.04116	0.958835	0.071070	0.927904	1.345
0.03994	0.960063	0.070129	0.928930	1.348
0.03879	0.961210	0.069260	0.929870	1.347
0.03772	0.962280	0.068460	0.930740	1.346
0.03672	0.963290	0.067720	0.931542	1.344
0.03577	0.964230	0.067030	0.932284	1.345
0.03488	0.965115	0.066390	0.932972	1.346
0.03410	0.965950	0.065790	0.933610	1.348
0.03326	0.966740	0.065250	0.934200	1.345

(continued)

Table 1 (continued)

λ = 0.99		λ = 0.946		Obs
θ_i	$1-\theta_i$	θ_i	$1-\theta_i$	
0.03252	0.967480	0.064730	0.934750	1.346
0.03181	0.968190	0.064260	0.935265	1.346
0.03114	0.968859	0.063810	0.935714	1.347
0.03050	0.969495	0.063397	0.936187	1.345
0.02990	0.970099	0.063010	0.936602	1.342
0.02933	0.970670	0.062647	0.936990	1.353

2. Obtain the value of the arithmetic mean for the period t

$$\widehat{\mu}_0 = 0, \widehat{\mu}_t = \theta_t X_t + (1-\theta_t)\widehat{\mu}_{t-1}, t = 1, \ldots, T \tag{23}$$

The following R program calculates the values of the short- and long-term average giving priority to the new data by entering the weighting coefficients:
> prom<-vector (mode = "integer", length = length(n)).
> n<-50.
> for (i in 1: n) {
+ prom[i + 1]<-prod[i] + THETA2[i]*prom[i]}.
Figure 1 shows the behavior of the data taken in real time and the predicted data with the algorithm. It can be seen that they tend to behave normally.

3. Obtain the error variance of period t, making the initial value, $\widehat{\sigma}_1^2 = 0$

$$\widehat{\sigma}_t^2 = \theta_t (X_t - X_{t-1})^2/2 + (1-\theta_t)\widehat{\sigma}_{t-1}^2, t = 2, \ldots, T \tag{24}$$

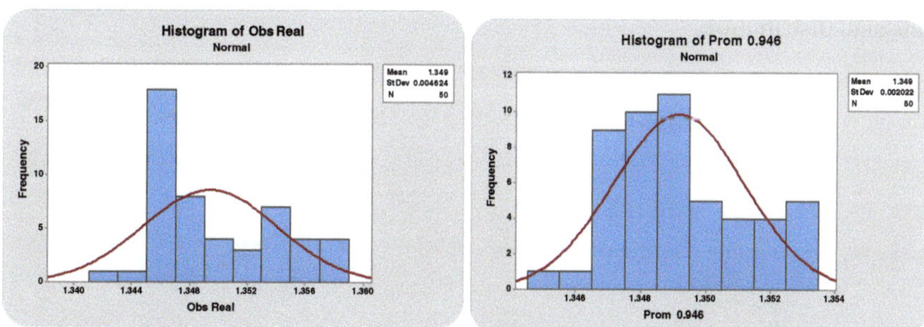

Fig. 1 Real-time data behavior and short-term predicted values

The program for calculating error variance values is:

```
> sigma2<-vector (mode = "integer", length = length(n)).
> n<-50.
> for (i in 1: n) {
+ sigma2[i + 1] < THETA1[i + 1]*(0.5*(Lecture[i + 1]Lecture[i])^2)*sigma2[i].
```

4. Obtain the total variance by doing, $\widehat{V}_0 = 0$

$$\widehat{V}_t = \theta_t\left(X_t - \widehat{\mu}_t\right)^2 + (1 - \theta_t)\left(\widehat{V}_{t-1} + \left(\widehat{\mu}_t - \widehat{\mu}_{t-1}\right)^2\right), t = 1, \ldots, T \qquad (25)$$

and the value of the total variance is calculated by:

```
>VarT<-vectorv(mode="integer", length=length(n))
>n<-50
>for (i in 1: n) {
>VarT<-THETA1[i+1](X[i+1]-prom[i+1])^2(VarT[i]+(prom[i+1]-prom[i])^2
```

The process variance γ^2 is calculated by $\gamma^2 = \mathrm{Var}T - \mathrm{Sigma2}$.

Figure 2a shows a comparison between the values obtained in real time and the values calculated with the recursive procedure. It can be seen that there is not a large difference between the predicted values in the short term and long term (a). It can be assumed that the process remains in control in relation to the behavior of the population arithmetic mean.

Table 3 shows the results of the calculations of prom $= \widehat{\mu}_{pt}$, sigma2 $= \widehat{\sigma}^2_{pt}$, Var$T = \widehat{V}_t$, and the result of the difference $\gamma^2 = \mathrm{Var}T - \mathrm{Sigma2}$. The results in Table 2 show the values of the subsequent distribution for the population mean, the variance of the measurement error and influence of the environment, the total variance, and the process variance. It is necessary to emphasize that the a priori distribution for the behavior of the population's arithmetic mean was a normal or Gaussian distribution.

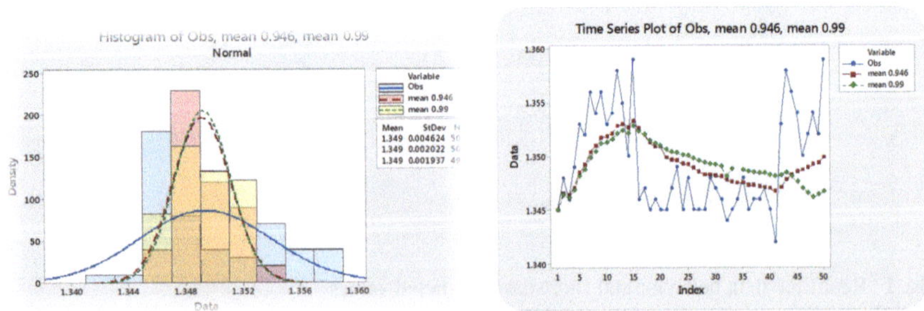

(a) (b)

Fig. 2 Histogram (**a**) and comparative series (**b**) of the behavior of the variable and the short-term and long-term predictions

Table 2 Analysis of variances for the comparison of population means

Source	DF	Analysis of variance			
		Adj SS	Adj MS	F-value	P-value Fac
Factor	2	0.000001	0.000001	0.06	0.942
Error	146	0.001428	0.000010		
Total	148	0.001430			

Figure 3 compares the variances of the error, the process, and the total variation, considering short- and long-term variations. It is possible to observe that the variance of the long-term process tends to increase. See Fig. 2b.

To verify the equality of arithmetic means between the data obtained in real time and the short-term and long-term predictions, an ANOVA was carried out, and the p-value = 0.942 tested the hypothesis of equal population means (see Table 3).

However, when observing the graphs of the variances, it was considered necessary to prove the assumption of equal variances, making a comparison between them, finding that there really is a difference (see Table 4).

The variances comparison test showed that there were differences in variances, which is why a nonparametric test was performed, specifically the Kruskal Wallis test, again demonstrating the equality between the means (see Table 5).

3 Bayes Empirical Monitoring Process for a Multivariate Continuous Variable

Jain et al. (1993) cited by Shiau and Feltz (2005) [cited in Colosimo and Del Castillo (2007)] present the following multivariable analysis scheme: Assuming that it is desired to monitor p quality characteristics of a process simultaneously. The quality characteristic observed at time t can be represented by the following vector of

$$p \times 1.\boldsymbol{X}_t = \begin{bmatrix} x_1 \\ \vdots \\ x_p \end{bmatrix}$$

Since the multivariate observation $\boldsymbol{X}_t$ (that has sampling errors and measurement errors), it is considered to be normally distributed with mean vector $\boldsymbol{\mu}_t$ and covariance matrix Σ and its probability density function (p.d.f) is

$$g_1(\boldsymbol{X}_t|\mu_t) = \frac{1}{(2\pi)^{p/2}|\boldsymbol{\Sigma}|}\exp\left[-\frac{1}{2}(\boldsymbol{X}_t - \boldsymbol{\mu}_t)'\boldsymbol{\Sigma}^{-1}(\boldsymbol{X}_t - \boldsymbol{\mu}_t)\right] \tag{26}$$

Generally, in univariate analysis, it is assumed that the arithmetic mean of the process may undergo slight changes, but not the variance. This assumption can be extended to multivariate analysis, considering that the covariance matrix Σ remains

Table 3 Values of the short-term and long-term process parameters

Obs	Short-term dist ($\lambda = 0.946 of\, \mu_t/X_t$)				Long-term dist ($\lambda = 0.99 of\, \mu_t/X_t$)			
	$\widehat{\mu}_{pt}$	$\widehat{\sigma}^2_{pt}$	$\widehat{V}_t$	$\widehat{\gamma}^2$	$\widehat{\mu}_{pt}$	$\widehat{\sigma}^2_{pt}$	$\widehat{V}_t$	$\widehat{\gamma}^2$
1.345	1.3450	0.00000e + 00	0.00000e + 00	0.0000000	1.3450	0.000000e + 00	0.000000e + 00	0.000000
1.348	1.3466	2.374785e−06	2.243079e−06	0.0000002	1.3464	2.272500e−06	2.267810e−06	0.00000023
1.346	1.3464	2.238078e−06	1.503650e−06	0.0000001	1.3462	2.180059e−06	1.534341e−06	0.00000015
1.349	1.3471	2.876619e−06	2.480580e−06	0.0000002	1.3469	2.773290e−06	2.636436e−06	0.00000026
1.353	1.3485	4.068112e−06	8.089093e−06	0.0000040	1.3482	3.848479e−06	8.118899e−06	4.2704E−06
1.352	1.3492	3.356870e−06	8.451923e−06	0.0000051	1.3488	3.271498e−06	8.775119e−06	5.5036E−06
1.356	1.3504	4.172200e−06	1.369518e−05	0.0000095	1.3499	3.973779e−06	1.394912e−05	9.9753E−06
1.354	1.3510	3.829342e−06	1.327656e−05	0.0000094	1.3505	3.716023e−06	1.399261e−05	1.0277E−05
1.356	1.3517	3.565827e−06	1.452465e−05	0.0000110	1.3512	3.515874e−06	1.552207e−05	1.2006E−05
1.353	1.3519	3.690078e−06	1.280102e−05	0.0000091	1.3513	3.619594e−06	1.420870e−05	1.0589E−05
1.354	1.3521	3.294285e−06	1.172690e−05	0.0000084	1.3516	3.318803e−06	1.347195e−05	1.0153E−05
1.358	1.3528	3.843136e−06	1.395175e−05	0.0000101	1.3521	3.734634e−06	1.562791e−05	1.1893E−05
1.355	1.3530	3.915607e−06	1.289616e−05	0.0000090	1.3524	3.797689e−06	1.496033e−05	1.1163E−05
1.350	1.3527	4.817358e−06	1.239438e−05	0.0000076	1.3522	4.466779e−06	1.420803e−05	9.7413E−06
1.359	1.3533	8.402036e−06	1.475647e−05	0.0000064	1.3528	7.068735e−06	1.631540e−05	9.2467E−06
1.346	1.3526	1.574321e−05	1.799951e−05	0.0000023	1.3523	1.233019e−05	1.811464e−05	5.7845E−06
1.347	1.3520	1.432569e−05	1.899571e−05	0.0000047	1.3520	1.156986e−05	1.864994e−05	7.0801E−06
1.345	1.3513	1.321748e−05	2.125031e−05	0.0000080	1.3515	1.098599e−05	2.029815e−05	9.3122E−06
1.346	1.3509	1.210877e−05	2.165501e−05	0.0000095	1.3512	1.037706e−05	2.079517e−05	1.0418E−05
1.345	1.3509	1.112504e−05	2.247877e−05	0.0000114	1.3510	9.829791e−06	2.166662e−05	1.1837E−05

(continued)

Table 3 (continued)

Obs	Short-term dist ($\lambda = 0.946 of\, \mu_t / X_t$)				Long-term dist ($\lambda = 0.99 of\, \mu_t / X_t$)			
	$\widehat{\mu}_{pt}$	$\widehat{\sigma}^2_{pt}$	$\widehat{V}_t$	$\widehat{\gamma}^2$	$\widehat{\mu}_{pt}$	$\widehat{\sigma}^2_{pt}$	$\widehat{V}_t$	$\widehat{\gamma}^2$
1.345	1.3499	1.020655e−05	2.279818e−05	0.0000126	1.3506	9.308223e−06	2.227194e−05	1.2964E−05
1.347	1.3497	9.545096e−06	2.159032e−05	0.0000120	1.3504	8.936307e−06	2.176310e−05	1.2827E−05
1.349	1.3496	8.950316e−06	1.992183e−05	0.0000110	1.3503	8.596983e−06	2.079135e−05	1.2194E−05
1.345	1.3493	8.876843e−06	1.990767e−05	0.0000110	1.3501	8.568859e−e−06	2.109387e−05	1.2525E−05
1.348	1.3492	8.545245e−06	1.850973e−05	0.0000100	1.3500	8.383944e−06	2.032514e−05	1.2266E−05
1.345	1.3489	8.244174e−06	1.832314e−05	0.0000101	1.3498	8.213426e−06	2.047936e−05	1.2673E−05
1.345	1.3486	7.640602e−06	1.798844e−05	0.0000103	1.3496	7.864486e−06	2.053793e−05	1.2976E−05
1.345	1.3483	7.089745e−06	1.754327e−05	0.0000105	1.3494	7.540745e−06	2.051677e−05	1.2351E−05
1.348	1.3483	6.905692e−06	1.630306e−05	0.0000094	1.3493	7.419320e−06	1.976991e−05	1.2053E−05
1.347	1.3482	6.456460e−06	1.526892e−05	0.0000088	1.3492	7.150920e−06	1.920441e−05	1.196E−05
1.346	1.3481	6.043916e−06	1.452397e−05	0.0000085	1.3491	6.900047e−06	1.885964e−05	1.2372E−05
1.344	1.3478	5.767081e−06	1.457522e−05	0.0000088	1.3481	6.720186e−e−06	1.909182e−05	1.2446E−05
1.345	1.3476	5.410418e−06	1.407444e−05	0.0000087	1.3488	6.497690e−06	1.894351e−05	1.2258E−05
1.346	1.3475	5.081283e−06	1.328956e−05	0.0000082	1,3487	6.288458e−06	1.854607e−05	1.1792E−05
1.348	1.3475	4.876717e−06	1.242352e−05	0.0000075	1.3487	6.142536e−06	1.793492e−05	1.1699E−05
1.345	1.3473	4.851884e−06	1.199715e−05	0.0000071	1.3486	6.087906e−06	1.778720e−05	1.1517E−05
1.346	1.3473	4.567923e−06	1.132434e−05	0.0000068	1.3485	5.906187e−06	1.742341e−05	1.1348E−05
1.346	1.3472	4.272219e−06	1.068666e−05	0.0000064	1.3484	5.718311e−06	1.706584e−05	1.1042E−05
1.347	1.3471	4.029705e−06	1.000258e−05	0.0000060	1.3484	5.555807e−06	1.659757e−05	1.0986E−05

(continued)

Table 3 (continued)

Obs	Short-term dist ($\lambda = 0.946of\,\mu_t/X_t$)				Long-term dist ($\lambda = 0.99of\,\mu_t/X_t$)			
	$\widehat{\mu}_{pt}$	$\widehat{\sigma}^2_{pt}$	$\widehat{V}_t$	$\widehat{\gamma}^2$	$\widehat{\mu}_{pt}$	$\widehat{\sigma}^2_{pt}$	$\widehat{V}_t$	$\widehat{\gamma}^2$
1.345	1.4370	3.900177e−06	9.633036e−06	0.0000057	1.3483	5.447327e−e−06	1.643313e−05	1.167E−05
1.342	1.3467	3.938200e−06	1.049610e−05	0.0000066	1.3481	5.418997e−06	1.708936e−05	1.0237E−05
1.353	1.3471	7.502159e−06	1.220414e−05	0.0000047	1.3482	7.034523e−06	1.727169e−05	1.2243E−05
1.358	1.3478	7.815259e−06	1.846255e−05	0.0000106	1.3485	7.191833e−06	1.943508e−05	1.3795E−05
1.356	1.3483	7.452934e−06	2.128990e−05	0.0000138	1.3481	7.044148e−06	2.083947e−05	1.4703E−05
1.354	1.3486	7.114886e−06	2.188117e−05	0.0000148	1.3476	6.903190e−06	2.160570e−05	1.4624E−05
1.350	1.3487	7.169468e−06	2.064327e−05	0.0000135	1.3470	6.929431e−06	2.155326e−05	1.517E−05
1.352	1.3489	6.851989e−06	2.000664e−05	0.0000132	1.3466	6.796316e−06	2.196660e−05	1.643E−05
1.354	1.3492	6.555242e−06	2.028229e−05	0.0000137	1.3462	6.668929e−06	2.309895e−05	1.6803E−05
1.352	1.3494	6.277782e−06	1.949759e−05	0.0000132	1.3464	6.547690e−06	2.335066e−05	1.9733E−05
1.359	1.3499	7.383262e−06	2.360625e−05	0.0000162	1.3467	7.006757e−06	2.673958e−05	1.1941E−05

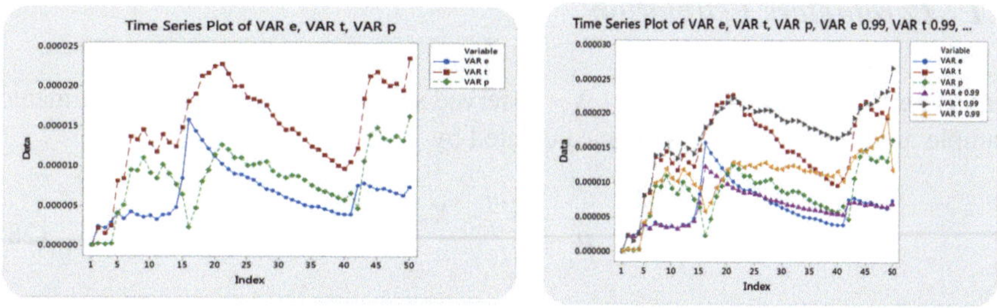

(a) (b)

Fig. 3 Comparative graph of error variance, process, and total short- and long-term variation (**b**)

Table 4 Test for variance comparison

Method	Test	
	Statistic	*p*-value
Multiple comparisons		0.000
Levene	21.75	0.000

Table 5 Kruskal Wallis test for comparison of means

PRONOSTICO	*N*	Median	Ave rank	*Z*
Obs real	50	1.348	68.7–1.27	
Prom 0.946	50	1.349	77.2–0.44	
Prom 0.99	49	1.349	79.2–0.84	
Overall 149			75.0	

$H = 1.66$, DF $= 2$, $P = 0.435$

unchanged. It can also be assumed that the mean of the process $\boldsymbol{\mu}_t$ also has a normal multivariate distribution with vector of means μ and covariance matrix $\boldsymbol{g}$. The p.d.f in this case is

$$g_2(\boldsymbol{\mu}_t) = \frac{1}{(2\pi)^{p/2}|\boldsymbol{g}|} exp\left[-\frac{1}{2}(\boldsymbol{\mu}_t - \boldsymbol{\mu})'\boldsymbol{g}^{-1}(\boldsymbol{\mu}_t - \boldsymbol{\mu})\right] \tag{27}$$

The model given in (1) assumes the existence of two sources of variation, the first being the variability due to sampling when the quality characteristic is considered centered in μ_t, and the variability due to the variation of the process (changes in the mean) in relation to time.

3.1 Parameters Estimation

Let us get T data vectors $\boldsymbol{X}_1, \ldots, \boldsymbol{X}_T$ observed sequentially in time. The estimated sample arithmetic mean $\widehat{\boldsymbol{\mu}}$ will be calculated by

$$\widehat{\boldsymbol{\mu}} = \frac{\sum_{t=1}^{T} \boldsymbol{X}_t}{T} \tag{28}$$

and the covariance matrix of the sample is given by

$$\widehat{\boldsymbol{V}} = \frac{\sum_{t=1}^{T} (\boldsymbol{X}_t - \widehat{\boldsymbol{\mu}})(\boldsymbol{X}_t - \widehat{\boldsymbol{\mu}})'}{T} \tag{29}$$

To estimate the variability of sampling, it is necessary to have multiple independent observations at time t given that $\boldsymbol{\mu}_t$ shows variation in time. Obtaining observations in industrial processes is not feasible, which is why the assumption is established that $\boldsymbol{\mu}_t' s$ remain constant in relatively small time intervals. Under this assumption, the estimate of Σ will be given by

$$\widehat{\boldsymbol{\Sigma}} = \frac{\sum_{t=2}^{T} (\boldsymbol{X}_t - \boldsymbol{X}_{t-1})(\boldsymbol{X}_t - \boldsymbol{X}_{t-1})'/2}{T - 1} \tag{30}$$

The covariance matrix $\boldsymbol{\mathcal{g}}$ of the distribution of the mean $\boldsymbol{\mu}$, can be calculated with

$$\boldsymbol{\mathcal{g}} = \boldsymbol{\Sigma} - \boldsymbol{V} \tag{31}$$

It is necessary to mention that if $\boldsymbol{\mathcal{g}}$ is not a definite positive matrix, it indicates that the process variation is small compared to the variation due to sampling, and we can make $\boldsymbol{\mathcal{g}} = \boldsymbol{kV}$, where k is given as $0 < k < 0.1$, depending on the application.

The incorporation of the weighting is achieved by the following equations, Jain et al. (1993):

$$\widehat{\boldsymbol{\mu}}_t = \frac{\sum_{t=1}^{T} \lambda^{T-1} \boldsymbol{X}_t}{\sum_{t=1}^{T} \lambda^{T-1}} \tag{32}$$

$$\widehat{\boldsymbol{\Sigma}}_T = \frac{\sum_{t=2}^{T} \lambda^{T-1} (\boldsymbol{X}_t - \boldsymbol{X}_{t-1})(\boldsymbol{X}_t - \boldsymbol{X}_{t-1})'/2}{\sum_{t=1}^{T} \lambda^{T-1}} \tag{33}$$

$$\widehat{\boldsymbol{V}}_t = \frac{\sum_{t=2}^{T} \lambda^{T-1} (\boldsymbol{X}_t - \widehat{\boldsymbol{\mu}})(\boldsymbol{X}_t - \boldsymbol{\mu})'}{\sum_{t=1}^{T} \lambda^{T-1}} \tag{34}$$

The value of λ is taken in the range $0.8 < \lambda < 1$ and depends on the application (Sturm et al. 1991). If m observations are taken, the value of λ can be chosen as

$\epsilon 1/m$ where ϵ is a very small number, of the order of 0.0001. Sturm et al. (1991) recommend values of $0.95 < \lambda < 0.99$.

The calculations of the values in Eqs. (7)–(9) can be obtained by recursive Eqs. (10)–(12), in which the starting values are established, $\widehat{\boldsymbol{\mu}}_0 = 0, \widehat{\boldsymbol{\Sigma}}_1 = 0$ and $\widehat{\boldsymbol{V}}_0 = 0$,

$$\widehat{\boldsymbol{\mu}}_t = \theta_t \boldsymbol{X}_t + (1-\theta_t)\widehat{\boldsymbol{\mu}}_{t-1}, \text{ to } t = 1, \ldots, T \quad (35)$$

$$\widehat{\boldsymbol{\Sigma}}_T = \theta_t(\boldsymbol{X}_t - \boldsymbol{X}_{t-1})(\boldsymbol{X}_t - \boldsymbol{X}_{t-1})'/2 + (1-\theta_t)\widehat{\boldsymbol{\Sigma}}_{T-1}, \text{ to } t = 2, \ldots, T \quad (36)$$

$$\widehat{\boldsymbol{V}}_t = \theta_t\left(\boldsymbol{X}_t - \widehat{\boldsymbol{\mu}}_t\right)\left(\boldsymbol{X}_t - \widehat{\boldsymbol{\mu}}_t\right)' + (1-\theta_t)\left(\widehat{\boldsymbol{V}}_{t-1} + \left(\widehat{\boldsymbol{\mu}}_t - \widehat{\boldsymbol{\mu}}_{t-1}\right)\left(\widehat{\boldsymbol{\mu}}_t - \widehat{\boldsymbol{\mu}}_{t-1}\right)'\right), \quad t = 1, \ldots T \quad (37)$$

where $\theta_t = 1/\sum_{t=1}^{T} \lambda^{t-k}$. The values of weighting coefficients used are shown in Table 6.

Table 6 Values of weighting coefficients used

$\lambda = 0.99$		$\lambda = 0.946$	
θ_i	$1-\theta_i$	θ_i	$1-\theta_i$
1.00000	0.000000	1.000000	0.000000
0.50500	0.494900	0.527730	0.472270
0.33890	0.661060	0.364760	0.635240
0.25570	0.744310	0.282300	0.717700
0.20570	0.794320	0.232560	0.767440
0.17230	0.827690	0.199333	0.800667
0.14850	0.851530	0.175600	0.824399
0.13057	0.869420	0.157820	0.842170
0.11670	0.883330	0.144032	0.855960
0.10550	0.894470	0.133033	0.866960
0.09642	0.903580	0.124070	0.875930
0.08883	0.911170	0.116637	0.883360
0.08240	0.917597	0.110381	0.889610
0.07689	0.923102	0.105050	0.894940
0.07220	0.927880	0.100460	0.899540
0.06795	0.932050	0.096470	0.903530

3.2 Using Recursive Equations

First, we will define the values of the weighting coefficients, θ_i, defined above as $\theta_t = \frac{1}{\sum_{k=1}^{t} \lambda^{t-k}}$ and calculated by $\theta_m = \frac{1}{1+\sum_{n=2}^{m} \lambda^n} = \frac{1-\lambda^{n+1}}{1-\lambda} - \lambda$; para $n > 0$

Let the vectors be sequentially obtained in samples of size three (measured pieces). For the calculation of the parameters, only one piece will be taken (the vector).

$$X_1 = \begin{bmatrix} 8.95 & 16.54 & 18.54 \\ 8.85 & 15.98 & 18.05 \\ 8.78 & 15.43 & 18.40 \end{bmatrix} X_2 = \begin{bmatrix} 8.98 & 15.45 & 18.41 \\ 8.98 & 14.83 & 18.45 \\ 9.00 & 15.91 & 18.32 \end{bmatrix} X_3 = \begin{bmatrix} 9.09 & 15.36 & 18.37 \\ 9.08 & 15.45 & 18.47 \\ 9.07 & 15.22 & 18.39 \end{bmatrix}$$

$$X_4 = \begin{bmatrix} 9.11 & 15.06 & 18.35 \\ 9.01 & 15.33 & 18.46 \\ 9.17 & 15.26 & 18.65 \end{bmatrix} X_5 = \begin{bmatrix} 8.97 & 15.22 & 18.51 \\ 9.05 & 15.04 & 18.94 \\ 8.74 & 14.76 & 18.41 \end{bmatrix} X_6 = \begin{bmatrix} 9.04 & 15.40 & 18.42 \\ 8.98 & 16.02 & 18.39 \\ 9.02 & 15.87 & 18.46 \end{bmatrix}$$

Only the first sixteen read for the monitoring were taken.

Let $\widehat{\boldsymbol{\mu}}_0 = 0 = \begin{bmatrix} 0 \\ 0 \\ 0 \end{bmatrix}, \widehat{\boldsymbol{\Sigma}}_1 = 0 y \widehat{\boldsymbol{V}}_0 = 0$

$$\widehat{\boldsymbol{\mu}}_1 = \theta_1 \boldsymbol{X}_1 + (1-\theta_1)\widehat{\boldsymbol{\mu}}_0 = 1.000 \begin{bmatrix} 8.95 \\ 16.54 \\ 18.54 \end{bmatrix} + 0 = \begin{bmatrix} 8.95 \\ 16.54 \\ 18.54 \end{bmatrix}$$

$$\widehat{\boldsymbol{\mu}}_2 = \theta_2 \boldsymbol{X}_2 + (1-\theta_2)\widehat{\boldsymbol{\mu}}_1 = 0.505 \begin{bmatrix} 8.85 \\ 15.88 \\ 18.05 \end{bmatrix} + 0.495 \begin{bmatrix} 8.95 \\ 16.54 \\ 18.54 \end{bmatrix} = \begin{bmatrix} 8.8995 \\ 16.2067 \\ 18.2925 \end{bmatrix}$$

$$\widehat{\boldsymbol{\mu}}_3 = \theta_3 \boldsymbol{X}_3 + (1-\theta_3)\widehat{\boldsymbol{\mu}}_2 = 0.3389 \begin{bmatrix} 8.78 \\ 15.93 \\ 18.4 \end{bmatrix} + 0.6611 \begin{bmatrix} 8.8995 \\ 16.2067 \\ 18.2925 \end{bmatrix} = \begin{bmatrix} 8.8546 \\ 15.9358 \\ 18.3198 \end{bmatrix}$$

and so on. Some results are analyzed below to $\lambda = 0.946$

$$\widehat{\boldsymbol{\mu}}_1 = \theta_1 \boldsymbol{X}_1 + (1-\theta_1)\widehat{\boldsymbol{\mu}}_0 \tag{38}$$

$$\widehat{\mu}_1 = \begin{pmatrix} 8.95 \\ 16.54 \\ 18.54 \end{pmatrix}$$

$$\widehat{\mu}_2 = \theta_2 \boldsymbol{X}_2 + (1-\theta_2)\widehat{\mu}_1 = 0.527730 \begin{pmatrix} 8.85 \\ 15.98 \\ 18.05 \end{pmatrix} + 0.47227 \begin{pmatrix} 8.95 \\ 16.54 \\ 18.54 \end{pmatrix} = \begin{pmatrix} 8.8972 \\ 16.2445 \\ 18.2814 \end{pmatrix}$$

$$\widehat{\mu}_3 = 0.36476\begin{pmatrix} 8.78 \\ 15.43 \\ 18.4 \end{pmatrix} + 0.63524\begin{pmatrix} 8.8972 \\ 16.2445 \\ 18.2814 \end{pmatrix} = \begin{pmatrix} 8.8545 \\ 15.9474 \\ 18.3247 \end{pmatrix}$$

$$\widehat{\mu}_4 = 0.28230\begin{pmatrix} 8.98 \\ 15.45 \\ 18.41 \end{pmatrix} + 0.7177\begin{pmatrix} 8.8545 \\ 15.9474 \\ 18.3247 \end{pmatrix} = \begin{pmatrix} 8.8899 \\ 15.807 \\ 18.3488 \end{pmatrix}$$

$$\widehat{\mu}_5 = 0.23256\begin{pmatrix} 8.98 \\ 14.83 \\ 18.45 \end{pmatrix} + 0.76744\begin{pmatrix} 8.8899 \\ 15.807 \\ 18.3488 \end{pmatrix} = \begin{pmatrix} 8.91 \\ 15.5798 \\ 18.3723 \end{pmatrix}$$

$$\widehat{\mu}_6 = 0.199333\begin{pmatrix} 9.00 \\ 15.91 \\ 18.32 \end{pmatrix} + 0.800667\begin{pmatrix} 8.91 \\ 15.5798 \\ 18.3723 \end{pmatrix} = \begin{pmatrix} 8.9286 \\ 15.6456 \\ 18.3619 \end{pmatrix}$$

$$\widehat{\mu}_7 = 0.1756\begin{pmatrix} 9.09 \\ 15.36 \\ 18.37 \end{pmatrix} + 0.824399\begin{pmatrix} 8.9286 \\ 15.6456 \\ 18.3619 \end{pmatrix} = \begin{pmatrix} 8.957 \\ 15.5954 \\ 18.3633 \end{pmatrix}$$

And the calculations for the 16 vectors are carried out in the same way and are shown in Table 7.

Table 7 shows 16 results for lambda 0.99 and lambda 0.946.

Figure 4 shows the behavior of the three dimensions and the values through the Bayesian monitoring scheme. It is observed that in the three dimensions there is an evident downward trend. It should be noted that the calculated values are for long-term forecasting with lambda equal to 0.99.

The first six values of $\widehat{\boldsymbol{V}}_t$ (considering $\widehat{\boldsymbol{V}}_0 = 0$) are:

$$\widehat{\boldsymbol{V}}_1 = 0$$

$$\widehat{\boldsymbol{V}}_2 = \begin{bmatrix} 0.00245 & 0.01617 & 0.0120025 \\ 0.001617 & 0.106722 & 0.0792168 \\ 0.012005 & 0.0792168 & 0.0588004 \end{bmatrix}$$

$$\widehat{\boldsymbol{V}}_3 = \begin{bmatrix} 0.0091120 & 0.002715 & 0.0146006 \\ 0.002715 & 0.241937 & 0.0605804 \\ 0.0146006 & 0.06058 & 0.0614729 \end{bmatrix}$$

$$\widehat{\boldsymbol{V}}_4 = \begin{bmatrix} 0.0029928 & -0.0263852 & 0.0031070 \\ -0.0263852 & 0.257858 & -0.0290328 \\ 0.0031070 & -0.0290328 & 0.0033322 \end{bmatrix}$$

Table 7 Parameter values $\widehat{\mu}_t$, $\widehat{\Sigma}_t$, and $\widehat{V}_t$ for $\lambda = 0.99$ and $\lambda = 0.946$

		$\lambda = 0.99$	$\lambda = 0.946$
	X_t	$\widehat{\mu}_t$	$\widehat{\mu}_T$
1	8.95,16.54,18.54	[8.95,16.54,18.54]	[8.95,16.54,18.54]
2	8.85, 15.88, 18.05	[8.8995,16.2067,18.2925]	(8.8972,16.2445,18.2814)
3	8.78,15.43,18.40	[8.8546,15.9358,18.3198]	8.8545, 15.9474, 18.3247
4	8.98, 15.45, 18.41	8.8867, 15.8116, 18.3428	8.8899, 15.807 18.3488
5	8.98, 14.83, 18.45	[8.9059, 15.6096, 18.3649]	8.91, 15.5798, 18.3723
6	9.00, 15.91, 18.32	[8.9221,15.6614,18.3571]	8.9286, 15.6456, 18.3619
7	9.09, 15.36, 18.37	[8.947,15.6166,18.3591]	8.957, 15.5954, 18.3633
8	9.08. 15.45, 18.47	[7.7787,13.5774,15.9617]	8.9765, 15.5723, 18.38
9	9.07, 15.22, 18.39	[7.9686,13.8374,16.3254]	8.9897, 15.5215, 18.3813
10	9.11, 15.06, 18.35	[8.098,13.9664,16.534]	9.1058, 15.6256, 18.5788
11	9.01, 15.33, 18.46	[8.041,13.8821,16.4391]	9.094, 15.5889, 18.5641
12	9.17, 15.26, 18.65	[8.1405,14.0025,16.6355]	9.1028, 15.5505, 18.574
13	8.97, 15.22, 18.54	[8.2088,14.1046,16.7924]	9.0881, 15.5139, 18.5668
14	9.05, 15.04, 18.94	[7.5776,13.02,15.5011]	9.084, 15.464, 18.6058
15	8.74, 14.70, 18.41	[7.6621,13.1423,15.7124]	9.0076, 15.3221, 18.5006
16	9.04, 15.40, 18.42	[7.6681,13.1542,15.7220]	9.0108, 15.3296, 18.4928

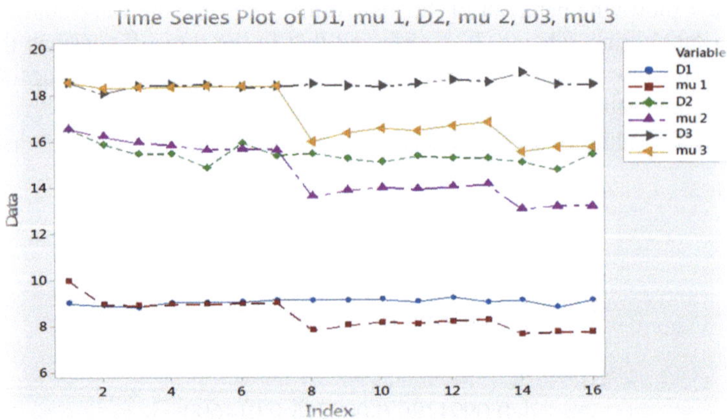

Fig. 4 Behavior of the real-time values of the three dimensions and the calculated values

$$\widehat{V}_5 = \begin{bmatrix} 0.258602 & 0.094808 & -0.0114593 \\ 0.094808 & 0.243547 & -0.0294696 \\ -0.0114593 & -0.0294696 & 0.0034995 \end{bmatrix}$$

$$\widehat{V}_6 = \begin{bmatrix} 0.219421 & 0.06882 & -0.00907 \\ 0.06882 & 0.243547 & -0.0251368 \\ -0.00907 & -0.0251368 & 0.0028872 \end{bmatrix}$$

Considered the initial value of $\widehat{\mathbf{\Sigma}}_1 = 0$, was calculated the first six values of $\widehat{\mathbf{\Sigma}}_t$ using

$$\widehat{\mathbf{\Sigma}}_t = \theta_t(\mathbf{X}_t - \mathbf{X}_{t-1})(\mathbf{X}_t - \mathbf{X}_{t-1})'/2 + (1-\theta_t)\widehat{\mathbf{\Sigma}}_{T-1} \tag{40}$$

$$\widehat{\mathbf{\Sigma}}_t = \theta_t(\mathbf{X}_t - \mathbf{X}_{t-1})(\mathbf{X}_t - \mathbf{X}_{t-1})'/2 + (1-\theta_t)\widehat{\mathbf{\Sigma}}_{T-1} \tag{41}$$

and posing the recursive equation for the following calculated values, there is

$$\widehat{\mathbf{\Sigma}}_2 = \theta_2(\mathbf{X}_2 - \mathbf{X}_1)(\mathbf{X}_2 - \mathbf{X}_1)'/2 + (1-\theta_1)\widehat{\mathbf{\Sigma}}_1 \tag{42}$$

$$\widehat{\mathbf{\Sigma}}_2 = 0.505\left[\begin{pmatrix} 8.85 \\ 15.98 \\ 18.05 \end{pmatrix} - \begin{pmatrix} 8.95 \\ 16.54 \\ 18.54 \end{pmatrix}\right]\left[\begin{pmatrix} 8.85 \\ 15.88 \\ 18.05 \end{pmatrix} - \begin{pmatrix} 8.95 \\ 16.54 \\ 18.54 \end{pmatrix}\right]'/2 + 0$$

$$= \begin{bmatrix} 0.00252 & 0.016665 & 0.0123725 \\ 0.016665 & 0.109989 & 0.0816585 \\ 0.0123725 & 0.0816585 & 0.0606252 \end{bmatrix}$$

$$\widehat{\Sigma}_3 = \frac{0.3389\left[\begin{pmatrix} 8.78 \\ 15.43 \\ 18.40 \end{pmatrix} - \begin{pmatrix} 8.85 \\ 15.98 \\ 18.05 \end{pmatrix}\right]\left[\begin{pmatrix} 8.78 \\ 15.43 \\ 18.40 \end{pmatrix} - \begin{pmatrix} 8.85 \\ 15.98 \\ 18.05 \end{pmatrix}\right]'}{2}$$

$$+ 0.661\begin{bmatrix} 0.00252 & 0.016665 & 0.0123725 \\ 0.016665 & 0.109989 & 0.0816585 \\ 0.0123725 & 0.0816585 & 0.0606252 \end{bmatrix}$$

$$\widehat{\Sigma}_3 = \begin{bmatrix} 0.004275 & 0.017539 & 0.0040267 \\ 0.0175377 & 0.123961 & 0.0213571 \\ 0.0040267 & 0.021357 & 0.0608309 \end{bmatrix}$$

$$\widehat{\Sigma}_4 = 0.2557\left[\begin{pmatrix} 8.98 \\ 15.45 \\ 18.41 \end{pmatrix} - \begin{pmatrix} 8.78 \\ 15.43 \\ 18.40 \end{pmatrix}\right]\left[0.2\ 0.02\ 0.01\right](0.5)$$

$$+ 0.7443\begin{bmatrix} 0.004275 & 0.017539 & 0.0040267 \\ 0.0175377 & 0.123961 & 0.0213571 \\ 0.0040267 & 0.021357 & 0.0608309 \end{bmatrix}$$

$$\widehat{\Sigma}_4 = \begin{bmatrix} 0.009381 & 0.01805 & 0.0042824 \\ 0.01805 & 0.124012 & 0.0213827 \\ 0.0042824 & 0.021383 & 0.0608437 \end{bmatrix}$$

$$\widehat{\Sigma}_5 = \begin{bmatrix} 0.0074517 & 0,014338 & 0.0034015 \\ 0.014338 & 0.138039 & 0.0144336 \\ 0.0034015 & 0.0144336 & 0.0484927 \end{bmatrix}$$

$$\widehat{\Sigma}_6 = \begin{bmatrix} 0.0066866 & 0.01466 & 0.0028125 \\ 0.01466 & 0.223712 & 0.0007894 \\ 0.0028125 & 0.000789 & 0.0447454 \end{bmatrix}$$

Figure 4 shows the behavior of the measured values in real time and the predicted values with the Bayes empirical monitoring.

3.3 *The Hotelling T2 Statistic*

Johnson and Wicherin (2002) they mention that Student's t variable is given by

$$t = \frac{\overline{X} - \mu}{s/\sqrt{n}} \tag{43}$$

Which is used to test the hypothesis

$$H_0 : \mu = \mu_0 y \ H_1 : \mu \neq \mu_0$$

and $H0$ is rejected if the observed value $|t|$ exceeds a specified percentage of a student's t-distribution with $n - 1$ degrees of freedom. Squaring the variable t

$$t^2 = \frac{\left(\overline{X} - \mu_0\right)^2}{s^2/n} = n\left(\overline{X} - \mu_0\right)\left(s^2\right)^{-1}\left(\overline{X} - \mu_0\right) \tag{44}$$

and, rejecting H_0 when $|t|$ is large, it is equivalent to rejecting it if its square is large, so, at the level of significance of α, if

$$n\left(\overline{X} - \mu_0\right)\left(s^2\right)^{-1}\left(\overline{X} - \mu_0\right) > t_{\alpha/2,n-1} \tag{45}$$

where $t_{\frac{\alpha}{2},n-1}$ is the percentile $100(\alpha/2)$ of the student's t-distribution with $n - 1$ degrees of freedom.

To generalize toward a multivariable analog inference, consider a vector $\boldsymbol{\mu}_0$ of order $p \times 1$, and analyze if this is a plausible value for the average of a multivariate distribution. Analogically, for a multivariate square distance (Johnson and Wichering 2002),

$$T^2 = \left(\bar{\boldsymbol{X}} - \boldsymbol{\mu}_0\right)' \left(\frac{1}{n}\boldsymbol{S}\right)^{-1} \left(\bar{\boldsymbol{X}} - \boldsymbol{\mu}_0\right) = n\left(\bar{\boldsymbol{X}} - \boldsymbol{\mu}_0\right)' \boldsymbol{S}^{-1}\left(\bar{\boldsymbol{X}} - \boldsymbol{\mu}_0\right) \tag{46}$$

where

$$\bar{\boldsymbol{X}} = \frac{1}{n}\sum_{j=1}^{n} \boldsymbol{X}_j,\ \boldsymbol{S} = \frac{1}{n-1}\sum_{j=1}^{n}\left(\boldsymbol{X}_j - \bar{\boldsymbol{X}}\right)\left(\boldsymbol{X}_j - \bar{\boldsymbol{X}}\right)',\ \text{and}\ \boldsymbol{\mu}_0 = \begin{bmatrix} \mu_{10} \\ \mu_{20} \\ \vdots \\ \mu_{p0} \end{bmatrix} \tag{47}$$

The vectors $\bar{\boldsymbol{X}}$ and $\boldsymbol{\mu}_0$ are of the order $(p \times 1)$ and $\boldsymbol{S}$ is an array of order $(p \times p)$.

The T^2 statistic is called Hfotelling's T^2 in honor of Harold Hotelling, who first obtained its sampling distribution. If the distance between $\bar{\boldsymbol{X}}$ and $\boldsymbol{\mu}_0$ is very big, the hipothesis $H_0{:}\boldsymbol{\mu} = \boldsymbol{\mu}_0$ is rejected (Ghute and Shirke 2009). There are no special tables for percentage points of the T^2 statistic for hypothesis testing, which is possible because T^2 is distributed as $\frac{(n-1)}{(n-p)}F_{p,n-p}$.

Where $F_{p,n-p}$ is a random variable with distribution F with p and n-p degrees of freedom.

When handling multivariate data, it should be an important requirement to verify whether the data is a normal multivariate distribution. Next, in addition to developing Q-Q graphs, the Mardia test and Henze-Zirkler test.

3.4 *Henze-Zirkler Test*

It is based on the non-negative functional distance, which measures the distance between two distribution functions. When the data follow a normal multivariate distribution, the test is distributed as a lognormal, and the mean, variance, and smoothing parameters are calculated, then they are log normalized and the p-value is calculated (Porras Cerron 2016).

$$T = \frac{1}{n}\sum_{i=1}^{n}\sum_{j=1}^{n} e^{-\frac{\beta^2}{2}D_{ij}} - 2\left(1+\beta^2\right)^{-p/2} \times \sum_{i=1}^{n} e^{-\frac{\beta^2}{2\left(1+\beta^2\right)}D_i} + n\left(1+2\beta^2\right)^{-\frac{p}{2}} \tag{48}$$

where

p: Number of variables

$$\beta = \frac{1}{\sqrt{2}}\left(\frac{n(2p+1)}{4}\right)^{\frac{1}{p+4}}$$

$$D_{ij} = \left(\boldsymbol{x}_i - \boldsymbol{x}_j\right)' \boldsymbol{S}^{-1}\left(\boldsymbol{x}_i - \boldsymbol{x}_j\right)$$

$$D_i = \left(\boldsymbol{x}_i - \bar{\boldsymbol{x}}\right)' \boldsymbol{S}^{-1}\left(\boldsymbol{x}_i - \bar{\boldsymbol{x}}\right) = m_{ii}$$

where D_i is the square of the distance of Mahalanobis of the ith observation to the centroid and D_{ij} is the distance of Mahalanobis between the ith and jth observation. The statistic HZ for normal multivariate data is approximately lognormal with mean μ and variance σ^2

$$\mu = 1 - \frac{a^{\frac{p}{2}}\left(1 + p\beta^{\frac{1}{a}} + \left(p(p+2)\beta^4\right)\right)}{2a^2} \tag{49}$$

$$\sigma^2 = 2\left(1 + 4\beta^2\right)^{\frac{p}{2}} + \frac{2a^{-p}\left(1 + 2p\beta^4\right)}{a^2} - 4\omega_\beta^{-\frac{p}{2}}\left(1 + \frac{3p\beta^4}{2\varpi_\beta} + \frac{p(p+2)\beta^8}{2\varpi_\beta^2}\right) \tag{50}$$

where

$$a = 1 + 2\beta^2 \text{and } \varpi_\beta = \left(1 + \beta^2\right)\left(1 + 3\beta^2\right)$$

The mean and lognormalized variance of the HZ statistic can be defined as

$$\log(\mu) = \log\left(\sqrt{\frac{\mu^4}{\sigma^2 + \mu^2}}\right) \tag{51}$$

$$\log\left(\sigma^2\right) = \log\left(\sqrt{\frac{\sigma^2 + \mu^2}{\sigma^2}}\right) \tag{52}$$

which can be used using the lognormal distribution to prove the significance of multivariate normality.

The Wald test (named in honor of the statistician Abraham Wald) can be used to check normal multivariate behavior. Using the standardized Z of the normal distribution, the Wald test for the multivariate normal distribution will be given by

$$Z = \frac{\log(HZ) - \log(\mu)}{\log(\sigma)} \tag{53}$$

Table 8 Measurements of three different dimensions of a molded part

x1	x2	x3
8.95	16.54	18.54
8.85	15.88	18.05
8.78	15.43	18.40
8.98	15.45	18.41
8.98	14.83	18.45
9.00	15.91	18.32
9.09	15.36	18.37
9.08	15.45	18.47
9.07	15.22	18.39
9.11	15.06	18.35
9.01	15.33	18.46
9.17	15.26	18.65
8.97	15.22	18.59
9.05	15.04	18.94
8.74	14.70	18.41
9.04	15.40	18.42
8.98	16.02	18.39
9.02	15.87	18.46

Example 1 Obtaining the Henzel-Zirkler Statistic.

Table 8 shows the measurements of three different dimensions of a molded part. The measurements were made with a vernier with the appropriate resolution and recently calibrated.

The calculations were made with the R language version 3.6.2 (2019-12-12)—“Dark and Stormy Night”. The basic packages required were: MVN, MSQC, mvtnorm. The results provided by running the Henzel–Zinkler test are the significance and power for the given sample size.

The R program developed is shown below.

```
> data<-data.frame(x1, x2, x3).
> data # shows the 18 × 3 data matrix.
> data<-as.matrix(data).
> library(MSQC).
> HZ.test(data).
[1] 0.02536657 0.87422317.
```

The first value given shows the significance of the test, while the second is the power of the test for a sample size of $n = 18$. The p-value of significance equal to 0.02536657 checks the multivariate normal behavior and the value of 0.87422317 shows a good test power.

3.5 Mardia's Test

Porras Cerron (2016), they quote Mardia (1970) who proposed a multivariate normality test, based on the extent of bias ($\widehat{\gamma}_{1.p}$) (asymmetry) and the kurtosis $\left(\widehat{\gamma}_{2.p}\right)$, where

$$\widehat{\gamma}_{1.p} = \frac{1}{n^2}\sum_{i=1}^{n}\sum_{j=1}^{n} m_{ij}^2 \tag{54}$$

$$\widehat{\gamma}_{2.p} = \frac{1}{n}\sum_{i=1}^{n} m_{ii}^2 \tag{55}$$

here $m_{ij} = \left(x_i - \bar{x}\right)' S^{-1}\left(x_i - \bar{x}\right)$ represents the square distance of Mahalanobis and p is the number of variables. The statistical test for asymmetry $n/6\gamma_{1.p}$ is distributed approximately as a distribution χ^2 with $p(p+1)(p+2)/6$ degrees of freedom. Similarly, the statistical test for kurtosis $\gamma_{2.p}$ is distributed.

$\sim N(p(p+2), 8p(p+2)/n)$.

The following R program shows results for data bias and kurtosis.

```
> data < -data.frame(×1, × 2, × 3)
> library(MSQC)
> MardiaTest(data)
$skewness
[1] 3.722144
$p.value
[1] 0.3939576
$kurtosis
[1] 15.11844
$p.value
[1] 0.4822217
```

Example 2 Obtaining the Hotelling square T statistic.

The data given in Table 9 are the variables in a welding process and will serve to illustrate the obtaining of the Hotelling square T (Johnson and Wichern 2002).

Arithmetic stockings vector is:

$$\bar{X} = \begin{bmatrix} 22.68 \\ 276.8 \\ 288.86 \end{bmatrix}$$

Table 9 Data of a welding process

Voltage (x1)	Current (x2)	Feed (x3)
22.9	278	288.0
22.5	277	290.0
22.4	275	287.0
22.6	274	287.5
22.8	276	289.0
22.9	280	290.0
22.7	284	289.0
22.6	277	289.4
22.5	272	290.1
22.9	275	288.6
$\bar{x}_1 = 22.68$	$\bar{x}_2 = 276.8$	$\bar{x}_3 = 288.86$

3.6 Normality Verification. Construction of a Q-Q Chart

A Quantile–Quantile graph allows us to analyze how close the distribution of a data set to some ideal distribution or compare the distribution of two datasets. The Q-Q chart (see Fig. 5) was constructed using the R language. The program for its construction is shown below.

R program to build a Q-Q chart

```
> qqplot(×1, x2,cex.lab=2,cex.axis=1.5)
> qqplot(×1, x3,cex.lab=2,cex.axis=1.5)
> qqplot(×2, x3,cex.lab=2,cex.axis=1.5)
```

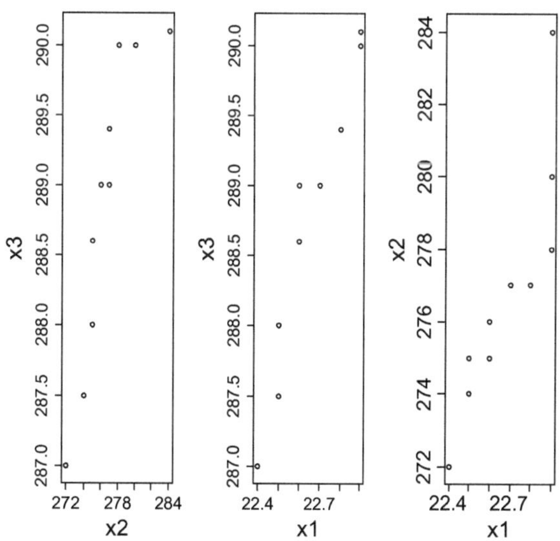

Fig. 5 Q-Q chart

Tables 10 and 11 show the matrix of covariance and matrix correlation, respectively.

Objective value: $\mu_0 = \begin{bmatrix} 23 \\ 277 \\ 289 \end{bmatrix}$

Inverse matrix of the correlation matrix is as follow:

$$\begin{matrix} 1.17179 & -0.43952 & -0.04263 \\ -0.43952 & 1.19402 & -0.15725 \\ -0.04263 & -0.15725 & 1.03071 \end{matrix}$$

The value of the T^2 statistic is obtained by:

$$T^2 = n\left(\bar{X} - \mu_0\right)' S^{-1}\left(\bar{X} - \mu_0\right) \tag{56}$$

where S is estimated from a sample of n multivariate observations obtained (MacGregor and Kourti 1995)

$$T^2 = 3\left(-0.32\ -0.20\ -0.14\right)\begin{bmatrix} 1.17179 & -0.43952 & -0.04263 \\ -0.43952 & 1.19402 & -0.15725 \\ -0.04263 & -0.15725 & 1.03071 \end{bmatrix}\begin{pmatrix} -0.32 \\ -0.20 \\ -0.14 \end{pmatrix}$$

$$= 0.3573$$

The hypothesis $H_0\colon \mu = \mu_0$ versus $H_1\colon \mu \neq \mu_0$ can be rejected at the level of significance α if

$$T^2 = \frac{(n-1)p}{(n-p)} F_{\alpha,p,n-p} \tag{57}$$

Table 10 Matrix of covariance

Variable	Voltage	Current	Feed
Voltage	0.035111	0.2400	0.02022
Current	0.240000	11.2889	0.61333
Feed	0.020222	0.6133	1.17600

Table 11 Matrix of correlation

	Matrix of correlation		
	Voltage	Current	Feed
Voltage	1.00000	0.38121	0.09952
Current	0.38121	1.00000	0.16833
Feed	0.09952	0.16833	1.00000

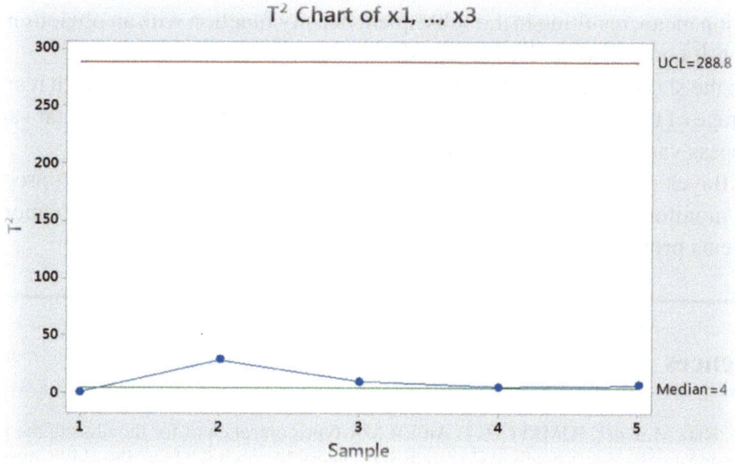

Fig. 6 Hotelling square T chart

And for three variables with $\alpha = 0.05$, $F_{0.05,2,7} = 4.74$ and

$$\text{LCS} = \frac{2(3)}{7} 4.74 = 4.062$$

Figure 6 shows the Hotelling square T chart for the data in Table 8. The points plotted are the values of the T^2 of Hotelling. Figure 6 is graph of Hotelling square T value.

4 Conclusion

Inspectors generally carry out manual control of the variables, using instruments such as verniers, manometers, optical comparators, manual scales, and voltmeters, a control that, until then, was sufficient; the automation of industrial processes, the complexity of the products, the increase in customer requirements in relation to volumes and delivery times, has generated the need to establish local and/or remote monitoring systems.

Monitoring aims to be aware of the state of a system, to observe a situation of changes that may occur over time, on the control of variables. The control of the continuous and discrete variables consists of verifying trends in the population arithmetic mean or in the desired target value (or nominal value), as well as in the population variance.

In the process of Bayes empirical monitoring of the characteristic, the normal distribution was used as a likelihood function and the a priori distribution for the population mean, resulting in the subsequent density function with an obvious normal behavior. It could also be demonstrated.

With the short- and long-term mean that did not show a difference with respect to the average of the data observed in real time, it was also shown that the total variance and process variance varied markedly.

The Bayes empirical monitoring process is a strongly recommended procedure for the monitoring of automatic processes, of course by choosing the smoothing coefficients properly.

References

Abbas N, Riaz M, Does, RJMM (2014) An EWMA-type control chart for monitoring the process mean using auxiliary information. Commun Stat Theory Methods 43(16):3486–3498. https://doi.org/1080/03610926.2012.700368

Box GEP, Tiao GC (1973) Bayesian inference in statistical analysis. Wiley Classics Library, New York. https://doi.org/10.1002/9781118033197

Castañeda Cárdenas JA, Nieto Arias MA, Ortíz Bravo VA (2013) Análisis y aplicación del filtro de kalman a una señal con ruido aleatorio. Sci Tech 18(1):267–274. https://www.redalyc.org/pdf/849/84927487039.pdf

Colosimo BM, Del Castillo E (2007) An introduction to Bayesian inference in process monitoring, control and optimization. Bayesian process monitoring, control and optimization, Taylor & Francis Group, New York. https://doi.org/10.1201/9781420010701

Ghute VB, Shirke DT (2008) A multivariate synthetic control chart for monitoring process mean vector. Commun Stat Theory Methods 37(13):2136–2148. https://dx.doi.org/10.1080/03610920701824265

Johnson RA, Wichern DW (2002) Applied multivariate statistical analysis, 5th edn. Prentice Hall, Upper Saddle River, NJ, pp 210–221

Kalman RE (1960) A new approach to linear filtering and prediction problems. Trans ASME J Basic Eng 82:35–45. https://www.unitedthc.com/DSP/Kalman1960.pdf

Kim Y, Bang H (2019) Introduction to Kalman filter and its applications. Introduction and implementations of the Kalman filter. https://doi.org/10.5772/intechopen.80600

Mardia KV (1970) Measures of multivariate skewness and kurtosis with applications. Biometrika 57(3):519–530

Munuera MC (2018) Filtro de Kalman y sus aplicaciones. Documento de Trabajo Final de Grado. Facultad de Matemáticas e Informática, Universidad de Barcelona, Barcelona

MacGregor JF, Kourti T (1995) Statistical process control of multivariate processes. Control Eng Pract 3(3):403–414

Porras Cerron JC (2016) Comparación de pruebas de Normalidad Multivariada. An Cient 77(2):141–146. https://doi.org/10.21704/ac.v77i2.483

Roberts SW (1959) Control chart based on geometric moving average. Techometrics 1(3):239–250. https://doi.org/10.1080/00401706.1959.10489860

Shiau JJH, Feltz CJ (2007) Empirical Bayes process monitoring techniques. In: Colosimo BM, del Castillo E (eds) Bayesian process monitoring control and optimization, pp 109–138. https://doi.org/10.1111/j.1751-5823.2007.00030_4.x

Sturm GW, Feltz CJ, Yoursi MA (1991) An Empirical Bayes Strategy for analyzing manufacturing data in real time. Qual Reliab Eng Int 7:159–167. https://doi.org/10.1201/9781420010701

Youngjoo K, Hyochoong B (2018) Introduction to Kalman filter and its applications. Intechopen, London

Augmented Reality as an Innovative and Efficient Technology to Increase Quality in Manufacturing Processes

Marisol Hernández Hernández and Gerardo Reyes Ruiz

Abstract Manufacturing processes are divided into stages whose purpose is to achieve the desired object: the initial phase of these stages is the design of the product, continuing with the cutting of the fabric and the assembly of the pieces. Thus, in the design stage, the prototypes of the product are made, which are distributed to the manufacturing microcompanies so that their assemblers first visualize them according to their instant perception and, later, they keep them in their memory. This process is carried out according to the memories of what people perceived, which can lead to assembly errors and which, in turn, will be translated into losses of time and economic resources. This is where the use of augmented reality makes sense, because to make manufacturing processes more efficient it is important to use digital technologies with tools that simulate the physical structures of products, that is, to allow users to use their cognitive ergonomics in the production process. In this research, a system based on augmented reality is developed that shows a different way of building patterns and, of course, allows multiple benefits in manufacturing processes. The system was focused on the garment industry because of the complexity and variety of garment patterns. The results indicate that augmented reality is an efficient and extremely important technology for the manufacturing area and, consequently, allows the optimization of delivery times, costs, and material waste.

Keywords Augmented reality · Manufacturing · Clothing industry · New technologies

M. H. Hernández (✉) · G. R. Ruiz
Autonomous Mexico State University, Mexico State, Mexico
e-mail: mhernandezh@uaemex.mx

G. R. Ruiz
Naval Higher Studies Center, Mexico City, Mexico

J. L. García Alcaraz et al. (eds.), *Techniques, Tools and Methodologies Applied to Quality Assurance in Manufacturing*,
https://doi.org/10.1007/978-3-030-69314-5_5

1 Introduction

The manufacturing areas are key elements in the manufacture of products of any kind (Moghaddam et al. 2018). Each of these areas is of elementary importance, both individually and jointly, for the creation of a product, but they also require new methods to help optimize their processes (Grubor and Milovanov 2017; Guo et al. 2018). For their part, industry leaders have recognized the need and importance of developing approaches to improve competitive advantage in the development of new products. However, the process of product innovation is complex and involves the effective management of different actions (Awazu 2006; Ning et al. 2020). Fortunately, the cost of technologies is falling exponentially over time, which leads to a drop in the costs of their implementation and, consequently, it is intended that more innovative technologies are available to SMEs (Abulrub et al. 2012). In this way, avant-garde actions and methods are currently emerging that help to complement the production processes, such is the case of additive manufacturing or also called 3D printing (Bowman et al., 2005), which is revolutionizing manufacturing with small-scale 3D printing and, later, will serve to create large-scale parts (Roschli et al. 2019). Furthermore, immersive 3D visualization (Mazikowski and Lebiedź 2014; Lebiedź et al. 2010) is becoming the innovative tool for new product development (NPD) in several sectors, including manufacturing. This latest technology can be integrated into NPD processes, or facilitate the understanding of complex information, as well as used in simulation, planning and training. These technologies have the potential to provide a distinct advantage for companies to continue or become leaders in their field.

Other inventions that include intelligent technologies of various types, and that come together as important pieces in the production process, are computer-based models, manufacturing systems and sensor technologies, virtual/augmented reality, the Internet of things and technological trends; all correspond to global patterns, which individually and jointly reveal a range of multiple sector applications (Turovets et al. 2019). These tools must be contextualized, especially in research that directs them toward points of support and thus build intelligent and avant-garde factories, according to the technological era in which we live today (Zhong et al. 2019; Strozzi et al. 2017; Hozdić 2015). Of course, these tools must be able to improve and facilitate both manufacturing processes and derived products and be useful to people (Rao 2011). Therefore, manufacturing areas are key elements in the manufacture of products of any kind and require new methods to help optimize those processes (Šatanová et al. 2015; Afteni and Frumuşanu 2017).

The production phase begins with the design of a new product, which will serve as a prototype and must meet certain parameters related to measurements, materials, and specifications (Shao et al. 2018; Katrijn 2012). The pattern of a product is made geometrically following the indications of the model created by the designer, who is a fully trained person in this profession. Once the new product has been designed, the base model is translated into assembly terms by the manufacturing area staff, who give it their own construction interpretation, especially in the elementary details of the garment; this interpretation can sometimes be wrong or different from the base model proposal. It is clear that if this interpretation is totally correct then a quality product will be obtained. Otherwise, a product with defects could be obtained and, as a consequence, it could not be sold or its value and/or quality could be diminished. These drawbacks show the need to introduce cutting-edge technological tools that help optimize manufacturing processes through the visualization of patterns, which must be coupled or adapted to each person who interacts with the base model. On this type of technologies adjacent to factories, the research trend is focusing on advancing processes through a digital world, which is present in everyday life and is essential for any type of activity (Mekni and Lemieux 2014; Alkhamisi and Monowar 2013).

Companies are currently applying innovative technologies to reduce production costs, introduce product and service innovations, promote their patents, facilitate growth, and raise barriers to entry (Nie et al. 2018; Bravo et al. 2015). Among these innovative technologies is augmented reality (AR), which can help and catapult a company to the extent of gaining a competitive advantage in a market segment (Bulearca and Tamarjan 2010; Baratali et al. 2016).

Despite its enormous potential to streamline business innovation processes, AR adoption is not without risks in traditional areas such as human resources and expected performance levels (Martinetti et al. 2019). On the other hand, the risks of using this new technology can be mediated through careful planning and management, i.e., successful adoption of a technology can push companies even ahead of their competitors (Ong et al. 2008; Nee and Ong 2013; Bloching et al. 2015). A tool that lately converges toward these requirements in the production processes is the AR, which contains, in turn, several technological tools with different purposes for the construction of objects (Bottani and Vignali 2019). The elements that integrate the AR, such as videos, 3D images and animations, open a new world of opportunities from the perspective of the supply chain (Rejeb 2019; Koul 2019; Cirulis and Ginters 2013). In other words, this technology not only makes it possible to increase the efficiency and flexibility of manufacturing and distribution processes, but also modifies, with a special emphasis on the consumer, the relationship between the different steps of the supply chain (Merlino and Sproġe 2017; Condino et al. 2018).

AR goes beyond the use of tricks in games, entertainment and the use of technology, and it is precisely in this context that the benefits of AR in manufacturing production should be highlighted (Etonam et al. 2019). In these production processes, where technologies are not used in a taxing way for users to do new things, AR allows for increased creativity and simultaneously persuades workers to perform their daily activities with other alternatives that provide them with better experiences when performing their activities (Siriborvornratanakul 2018). In this context, it is time to use these new technologies to provide benefits different from what people are used to. The AR has facilitated the creation of new building techniques and even, depending on the type of manufacture to be built, systems based entirely on the AR have been codified and applied in different ways and in multiple contexts.

The AR presents the product design, but not only as a sample of what an object could be or how it would be visualized at a given time; it visualizes it as a useful way to represent both objects and their constructions more deeply, and thus to understand its implications in other areas, for example, the maintenance of a building. To exemplify this case, Khalek et al. (2019) investigated the design of a building, where decisions made at the beginning of that phase had a significant impact, measured by different means of visualization, which allowed inexperienced people to identify, through the implementation of an AR-based system, maintenance problems in a design model.

Nowadays, there are tutorials that allow to complete the construction processes, even some of them are not limited to the use of 2D images and are based on more innovative technologies such as a software that helps to build frames of wooden structures, which uses a software based on AR with a full-scale visualization and audio. In the USA, this technology allows builders to follow extremely specific and exact measurements and thus facilitates the adaptation of frames in wooden houses (Cuperschmid et al. 2016). Without a doubt the success of the AR, applied to this case, is due to the fact that the frames are of a standard type.

Another company that manufactures high-precision instruments and tooling equipment for machining parts in industrial companies is the German company Emuge-Franken (see https://www.emuge-franken-group.com/de/en/). This industry has implemented virtual technical manuals for the tasks related to the creation of tools in thread cutting and milling technology, which constitute a basis for their business. This company has also adopted the animated models of the AR-based HF20 threaded spindle assembly. The experience of this company has shown that the use of the latter product is necessary for companies in the Russian market, as they deal with metal processing by drilling and generating threads (Gren et al. 2018). Thus, the use of these new technologies has been considered, as the key tools for companies to improve their competitiveness and develop their relations with the nodes of the upstream and downstream supply chain. Consequently, the applications of these technologies for supply chain management have become a fruitful area of research mainly due to their clear and strong management implications.

In this way, the AR has been assessed for inclusion in future manufacturing processes. The main purpose of this is to allow workers to perform multiple tasks, which will allow the change from mass production to customized production. There is definitely much to be done to fulfill these promises in an industrial scenario or scale. However, in terms of the development of AR-based tutorials for assembling objects, we can see a greater performance in the tasks with the highest degree of difficulty (Uva et al. 2018). This background is the basis for the implementation of future intelligent factories, which are waiting for the best techniques to facilitate manufacturing processes (Rabah et al. 2018). Therefore, this research is built by envisioning the creation of these new manufacturing companies totally focused on improving the quality of their processes and not only on design forms, that is, these new companies will move from the discernment of the abstract to superimpose it as AR (Segovia et al. 2015).

This research focuses on the area of the clothing industry where the design must have an adaptation that leads to the quality of the garment, which will be achieved by following exactly the geometric characteristics of a prototype, created before starting with the manufacture of the garment and function as a guide for the transformation of raw material into clothing. The workshops are the spaces where the garments are manufactured, that is, where a physical sample is prepared and whose design be memorized by the "constructors". They must memorize even the hidden, but necessary, details so that the garment is visualized and functions as in the original model. A disadvantage in this process is that in many cases the ergonometric or operating details are complex and, consequently, will be difficult to memorize or specify so that, not having the garment that serves as a pattern in sight, the assembly can be done in a wrong way. With this, the manufacturing industry loses material resources, time in assembly and, of course, work done. In addition, when corrections are made to errors in manufacturing, delivery times are delayed and, consequently, the allocation and obtaining of profits, which ultimately result in a significant deficit for the company.

As mentioned above, the AR has been adopted in various ways in factories of some specific sectors, for example, through tutorials that allow the repair of machinery and where inventories are required that require optimization of space (Henderson and Feiner 2007). The AR also works with instructions for building tools or other types of objects; however, this proposal focuses on complementing the manufacturing processes, helping from assembly to quality control. All this with the purpose of obtaining better-elaborated products that led to optimize the cost-benefit equation of a manufacturing company (Sabarinathan et al. 2018). In other words, and as will be detailed in later sections, this research is intended to help ensure that the manufacture of garments is done in such a way that anyone who has this technology, usually with a mobile device, is involved in the production chain in a more efficient and reliable way. This with a view to providing the optimization of manufacturing times and, consequently, obtaining greater profits.

1.1 Objective

Create an AR-based system with previously designed resources that support workers during the assembly process in clothing. Implementing a system of this nature in a factory where garments are made aims to improve and speed up the manufacturing process with the support of technological tools that help optimize delivery times, resources used, and quality of garments. Analyzing the performance of a system with these characteristics will serve to generate recommendations.

2 Methodology

The XP methodology, also called Extreme Programming, is an agile framework where the proximity of the client to the development team is important for the correct design of a small project. The fundamental principles of this framework are based on a quick feedback, its simplicity, the acceptance of incremental changes, and high-quality software. The main stages that were considered for the development of this application were: planning, design, coding, testing, and release, which will be described below.

2.1 Planning

In the garment industry, garment production is done in stages. Each of these stages is a process that at first sight might seem easy and autonomous, however, each of these stages has its own degree of difficulty and, simultaneously, each is different. On the other hand, there are different levels of difficulty in the assembly, where it is essential to take into account several aspects such as the union of pieces that are different, the details that are hidden and the combination of different colors, just to mention some of them. One of the main problems that the manufacturing process faces is, precisely, when the assembly is not done properly. That is to say, when the assembly of the pieces is not done according to what is established by the prototype of the garment. This problem leads to large production errors that, consequently, generate considerable losses in time, money, and effort.

The phases involved in the manufacturing process of a garment can be listed chronologically. The diagram of activities in Fig. 1 shows the process involved in making a garment: the process involves everything from designing the garment and cutting the fabric in the factory, through the assembly process and up to the return of the new garments for their corresponding packaging.

In this context, the phase of this entire process that takes on the greatest importance, and from which this research work emerged, was the assembly stage, which is described by the following actions:

- When the packages of cut fabric arrive at the maquiladora, the person who delivers them is responsible for teaching and giving instructions about the garment called "sample" or "pattern" to the assemblers who, in turn, observe it and keep in their memory as many details as possible to later proceed with the assembly process.
- The elaboration of the parts that make up a garment is carried out in series, that is, there is a team of people who perform a specific process such as the assembly of necks, sleeves, backs, and front. After these tasks, the garments are transferred to the people who proceed to finish the garment; this process considers from the placement of buttons, eyelets, closures and/or clasps to the final stage, which is when the remaining threads are removed. The finished garments are transferred to the ironing area, followed by packaging.
- The quality control is carried out in the packaging, because it is precisely at this stage where the garments are arranged to be packaged. That is, at this stage people review the finished garments in detail and, if necessary, they notice some defect in the assembly. This stage is of the utmost importance in quality control, since the presence of the slightest error, usually, will be presented in several garments. The latter is due to the fact that all garments are produced in series and, therefore, the slightest error will also be reproduced in chain.

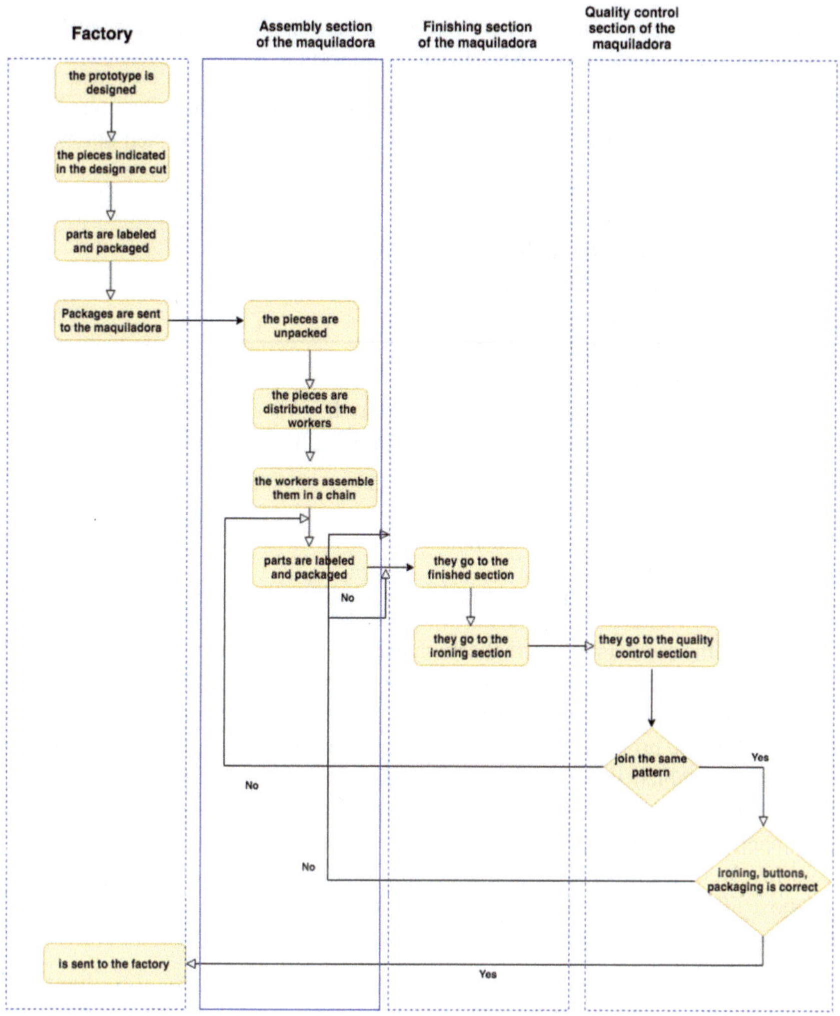

Fig. 1 Process of making a garment. *Source* Author's own elaboration

- A drawback, and perhaps the most important, is that the quality of the garment is reviewed, generally, at the stage of packaging or, what is worse, has happened, that when the garments are delivered to the factory, the personnel of the area who receives the clothes realizes the mistakes. At this stage, the error in the garments is corrected, the correction of the assembly being essential, before the garments are accepted by the warehouse or the parent factory.
- In the assembly industry, there are processes where people have to face real challenges related to the different parts of the product; sometimes there are complex components that have been shaped in the patterns of the garment and rarely consulted by the workers. The workers responsible for processing these complex

patterns have the responsibility to store them in their short-term memory, but on many occasions they may forget extremely important details.

The most frequent difficulties in assembling a garment can be of various kinds, depending on the design of the garment. Some examples are mentioned below:

- The side seams are not properly aligned.
- The stitches are the wrong size.
- Some flaps are not made in the right way and size.
- Buttons are placed on the wrong side.
- The color combination is wrong.
- The bags are not in the right place.
- The number of buttons is wrong.
- The hems do not correspond to what is required.
- The collars can be simple and made with a foot collar or vice versa.
- The bags are not aligned.
- There is confusion in the colors.
- The embroideries are in the wrong place.
- The brooches are inadequate.
- The width of the fold is different from the base sample.

Of course, the errors involved in the manufacture of the garments generally depend on their design. The main problem during the design stage occurs when an error is made in the process of assembling various pieces (sleeves, collars, back, etc.), which, in turn, was made up of several garments. As mentioned above, this is because the pieces are assembled in series that, in consequence, allows an error made in one piece of the garment to trigger others, which can lead to defects in an entire batch of garments consisting of multiple defective garments. Also, the different types of errors involved in the manufacturing process involve significant monetary losses, but, above all, time losses. Some of the errors involved in the manufacturing process of a garment are described below:

- When a batch of defective garments is unsewn to correct the assembly, they must be undone and redone, which multiplies the initial planned time for making the garments. In addition, new yarn is used and, consequently, the yarn used before the error was detected has already been wasted. This disadvantage multiplies the cost of the material resources used to make these garments.
- In some cases, the assembly of certain garments uses fragile and/or delicate fabric, which can lead to a garment being damaged or spoiled in the error correction stage. This will undoubtedly result in new fabric being used to remake that part.
- If the buttons are placed on the opposite side, then they must be removed and placed on the correct side. At this stage, the most complicated part is when the buttonholes have the same error because the procedure becomes considerably more complicated and the loss of resources, both material and time, increases.
- The stage where the finished garments are checked is during quality control. However, if the people responsible for this stage do not notice any errors, then the garments will be packed and then sent to the warehouse of the parent factory. If

an error is subsequently detected in some garments or in a batch of garments, they are returned to the factory for correction, which will undoubtedly generate more costs and inconveniences. This stage is very important, because if the defective garments are not detected before they are presented to a store manager then the loss of that customer can be incurred or, worse, the confidence of the people who buy the garments can be lost due to an error in the quality control of the manufacturing.

These errors, and many others, cause delivery times to be considerably extended, resulting in delayed payments and production losses, and consequently reduced profits. All of the above results in the following question what technological factors influence the improvement of a manufacturing process in a factory that uses AR? To answer this question, the following null hypothesis is posed:

H_0: *The AR is an efficient technological tool capable of helping and improving a manufacturing process that is implemented by a factory that makes garments.*

2.2 Design

For this phase, which impacts the entire manufacturing process, the solution proposed by this research work is the creation or implementation of an AR-based system. This system is intended to serve as a guide and training for the assemblers, but in a complementary manner; the system will be provided to the affiliated maquiladoras, along with the manufacturing materials, so that it is shared among all the members that make up the entire production team. In addition, this system has the advantage that each member of the assembly team can consult it, using only their mobile device, each time they require it. All this is carried out to visualize the virtual model superimposed on the piece of fabric that is want to be built; this virtual model, previously established, includes detailed instructions in audio, image, or video format.

At this stage, the AR is very useful, because it is an emerging technology that is defined as a digital shot or computer-generated information, either with images, audio, video, or tactile sensations and that is generated by overlapping in real time; that is to say, the AR is the mixture between the real world and synthetic (Kipper and Rampolla 2013). In this context, the AR is extremely suitable for the area of maintenance in the industry, being easily implemented in processes that can improve the view of users in different scenarios and that also include visual animations, sounds, written, or static instructions (Novakova 2018).

The production line of this new technology is on the rise. This is due to the fact that at present it has found various applications, among which the following stand out: military use, medicine, design, engineering, robotics, manufacturing applications, maintenance and repair, teaching and learning, entertainment, psychological treatments, among (Azuma 2015). For its part, virtual manufacturing is defined as an integrated synthetic environment that is exercised to improve all levels of decision and control. Furthermore, its classification can be focused on design, production and control (Novak-Marcincin et al. 2013). Based on this type of definition, digital elements are used that combine physical and virtual reality, and where design tools can be used in a wide and varied way, for the construction of the virtual manufacturing system. Under this scheme, the AR presents excellent construction tools, although its design should be simple and easy to understand to avoid the user feeling confused instead of being helped. In this way, and considering these principles for the realization of the proposed AR system, the diagram of activities is presented in Fig. 2.

The garment designer, in addition to making his patterns in 2D images with measurements and assembly instructions, must also draw them in 3D images. The latter will favor the visual display of the details to be shown of the garment through a virtual environment; the 3D images should show quite specific details of the model to be reproduced such as measurements, colors, number of components, instructions issued by audio or text, thus fulfilling a primary objective of the companies, which is to manufacture products that meet the needs of their customers (Sun et al. 2016). It is in this stage of the manufacturing process where the use of videos and text messages that explain and show clearly the procedure for the assembly of a garment is used to its fullest. Undoubtedly, using images, audio, and animation means creating a teaching–learning model that transmits in greater detail the assembly process of each piece of the product, specifically in those where the procedure is more complex.

2.3 *Coding*

The application that managed the digital components was also used to design, build, and implement the manufacturing system, based on AR, combining elements of different formats to achieve a complete digital reality. Currently, there are several AR managers that have similar functions, but differ in their development environments called "IDE", i.e., have different characteristics that translate into advantages and disadvantages for the AR building user. Some of these managers are: Aurasma, Vuforia, Layar, and ArToolKit with its ARTag variants, ATOMIC Authoring Tool, FRARTooKit, NyARToolKit, and ARDesktop.

For the design and implementation of this AR-based manufacturing system, we used Aumentaty software (see http://www.aumentaty.com/indexEN.php), which is a program entirely designed to create innovative content through AR. This software presents the functionality of handling digital elements in the form of video, 2D or 3D images and points of interest with GPS. An add-on to this software is the Scope app, through which AR is displayed on mobile devices. In addition, there is currently a large collaborative community where interested users are allowed to use their projects or technological developments for free (see http://www.aumentaty.com/community/es/). The Aumentaty software has a content manager called Creator, which is quite accessible to build the projects with AR, which can be shared in the cloud and then downloaded to mobile devices to be used.

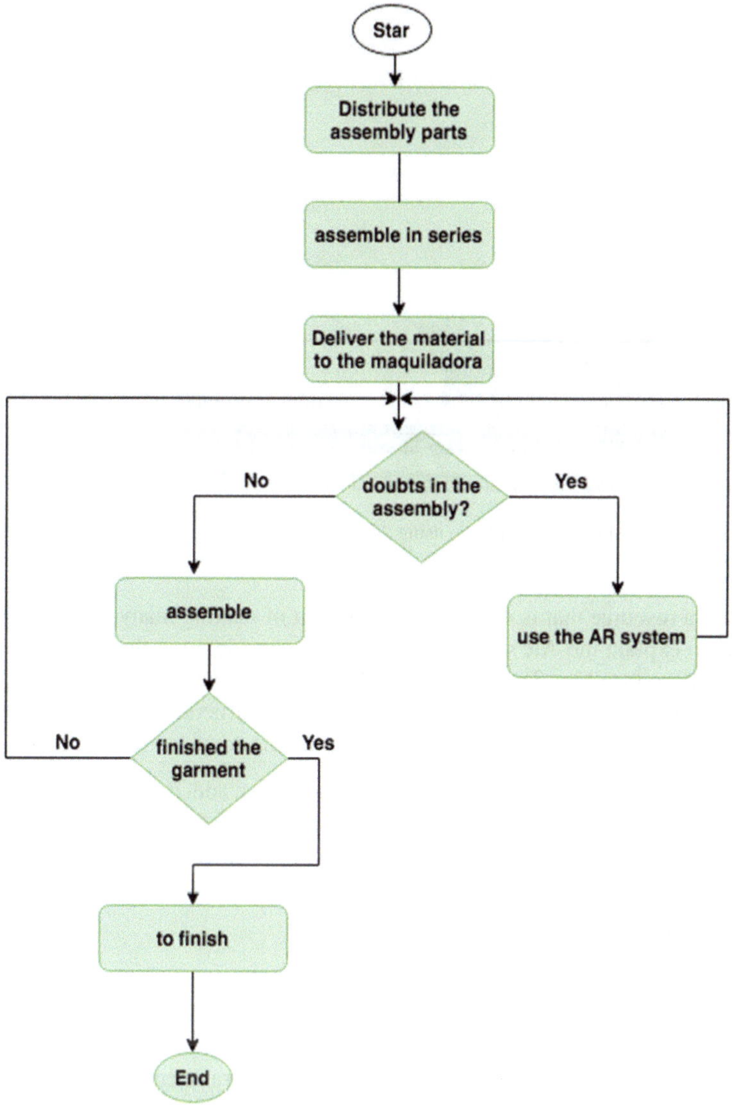

Fig. 2 Manufacturing process of the garments

The AR resources are shown in 3D image and video format, which shows the correct way to do the assembly. These resources allow visualizing step by step the assembly procedure and whose purpose is to provide all the construction instructions to the people responsible for this stage, so that the safety of doing the job in the right way is implicitly adhered to. In this sense, it is recommended that the videos are short, since they are only part of the help that is done through the AR and it is suggested that they have a maximum duration of 2 min. The investment in the time to review

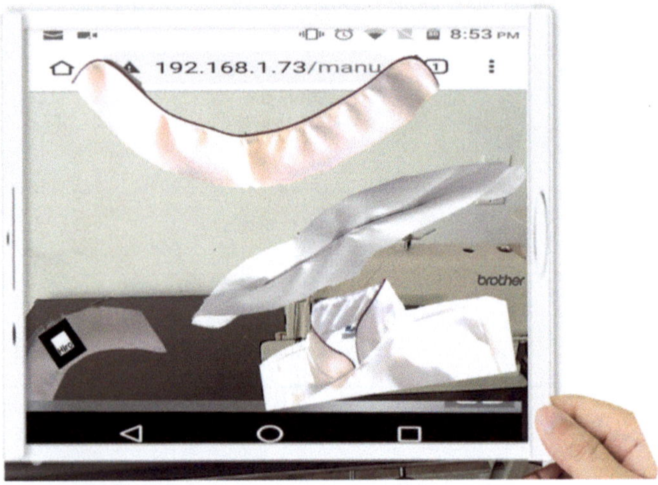

Fig. 3 Sample of a shirt called balloon, using AR

the AR is a resource that is made with the benefit of making many pieces correctly. In order to explain the AR system, we took into consideration the success story of a maquiladora (see Fig. 3), where a specific garment is used, although the system can work with all garments manufactured. For the construction of the manufacturing system, based on AR, the following phases should be considered:

1. The garment is modeled and its general details are added, which will be necessary in the deployment of the AR. These characteristics give the specificity to the garment, and are denoted with the labels of name, color, sizes, and general conditions, as shown below:

   Name: Globe-Shirt.
   Color: White.
   Sizes: 28-42.
   Details: Combination of colors, white and cherry.

2. The construction of a garment is made up of several pieces, which have been designed, cut, and integrated into packages for the construction of several garments. Each package of pieces contains all the necessary elements, for example, a shirt is made up of sleeves, collar, back, and front. It is important that each package matches your size and prototype, as all these elements have different shapes that will be joined together. Figure 4 shows the parts that make up the neck of a shirt called shirt-Balloon.
3. For the development of the system, based on AR, at least the following components must be available:

3D images. The 3D images, which are formed from 2D images and can be photographs or drawings, will serve to model the garments in more detail.

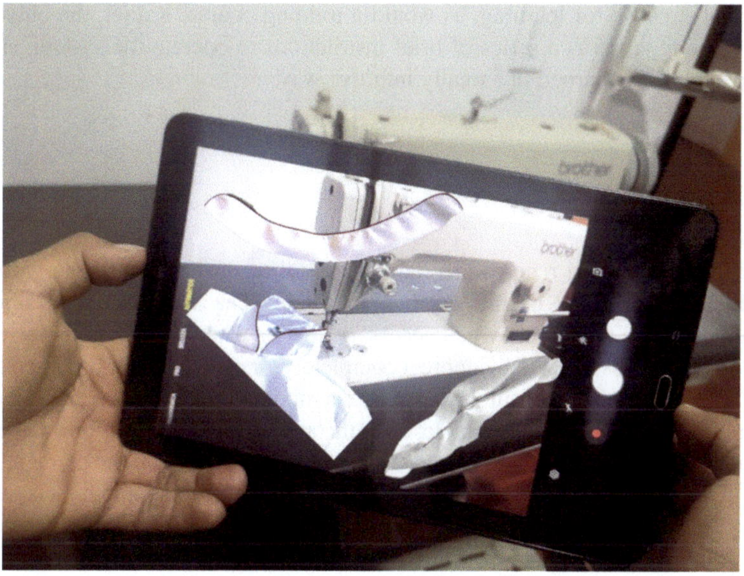

Fig. 4 Assembly parts for the collar of a Balloon Shirt using the AR

Markers. Will function as "triggers" to display the AR. The markers are images that will serve to "trigger" the AR when the camera of the mobile device is focused on them.

Multimedia videos. The videos are films where the most complex processes are interpreted. These films are created separately and trying to be brief. In turn, these short films are intended to resolve a question quickly, which will allow the investment of time to translate into a benefit of improvement for the work.

Audio. The audios are files with instructions in voice format that will be able to guide, in a verbal way, the users during the assembly processes.

Cloud storage. This type of storage serves to make available the digital material used in the system. This can be implicit in existing AR managers and can be freely accessible, since if there are few components up the AR then a plan can be contracted that best suits the needs of the company.

AR management. This application is composed of specific markers, which are only words that make up the name or number of the specific garment to concatenate, in turn, the name with its respective image.

The persons who will use the system will be employees working in the garment manufacturing factories. The manufacturing system, based on AR, is quite didactic, that is, it contains virtual elements that are integrated to the physical reality providing the user an easy and safe interaction. Moreover, it is not necessary for the user to have prior knowledge of this technology, since the movements produced are natural, easy to learn and could even be so simple that it would be like moving a body part or an object (Cordeiro et al. 2015). Therefore, the use of the system will not generate any additional expense for learning, as would a training course. Rather, the only thing that will be required is a series of brief instructions to operate the system, since it was created, to be learned in a totally intuitive way.

2.4 Tests

The maquiladoras will receive the elements for the construction of the garments, including the elements of the AR-based system. The workers responsible for assembly will use the system when necessary, that is, when they do not remember a phase of the process looks like, ignore it completely, or need to be sure of some detail. When the user needs to consult how to make a model, it will be enough to place the mold of a part of a garment on a table, open the application on the mobile device and focus its camera on it mentioned mold. The system will recognize the garment's part will superimpose the associated AR, to instantly solve the doubt about the assembly process. This type of AR was designed to show the basic shapes of a product, such as the combination of colors, measurements, sizes, placement of ornaments that are superimposed, the size of the stitches, and color of the thread. (see Fig. 5).

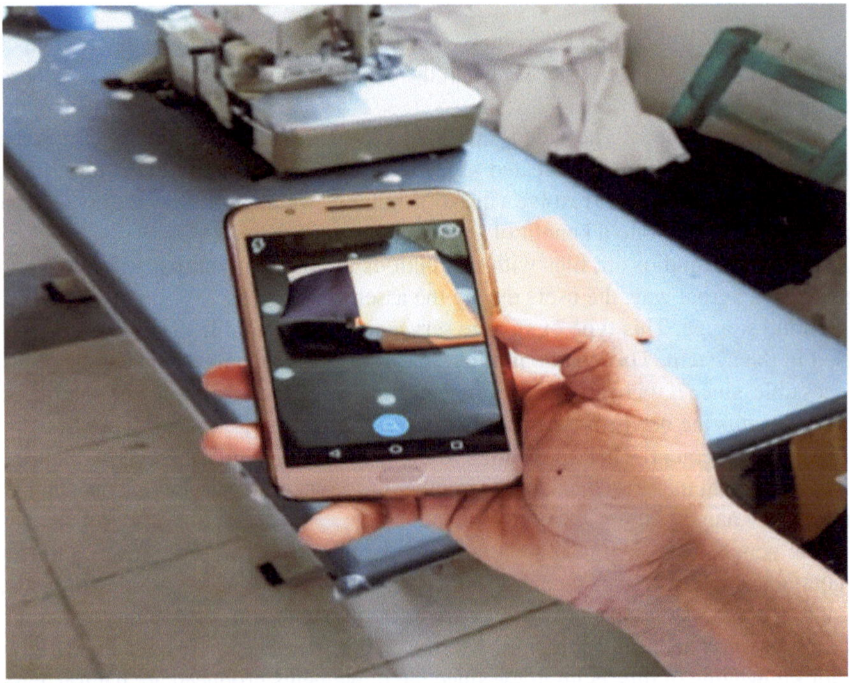

Fig. 5 AR system displaying the sleeve of a garment

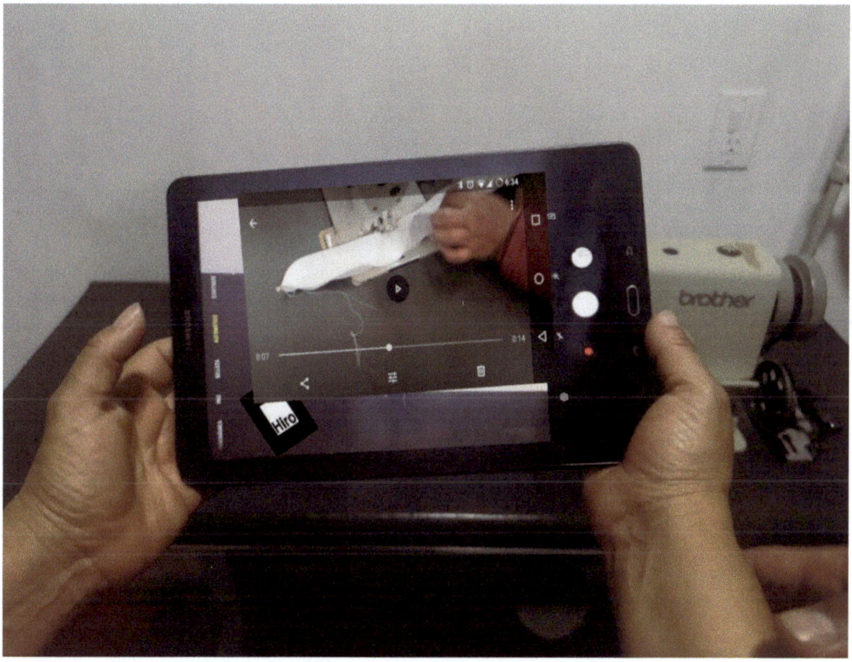

Fig. 6 AR system, showing the assembly process by means of a video

Figure 6 shows the implementation of the system, based on AR, using the digital resource in video format. In this case, the user focuses the camera of the mobile device to the neck of the garment, which serves as a "launcher" of the AR, overlapping the help in the form of a video tutorial that shows, through images and audio, the process of assembling a neck for a shirt. Of course, the complexity of the process will depend on how you want to show the AR.

When making use of a code, which must be the name or number assigned by the factory according to the design of the garment and which will serve as a "trigger" of the AR, what is desired is to show the prototype of a garment, to that users appreciate it and consult whenever they need it (see Fig. 7). In this example, the database has a series of names associated with the garments that are manufactured in the workshop. In this way, when the camera of the mobile device focuses on the name "Shirt-Bona", the AR superimposes the image of the selected shirt showing the complete design.

3 Launch

This system, based on AR, was applied in a microsewing company (located in the State of Tlaxcala, Mexico) where uniforms are made and assembled for various commercial companies. The system was applied for one week and the workers

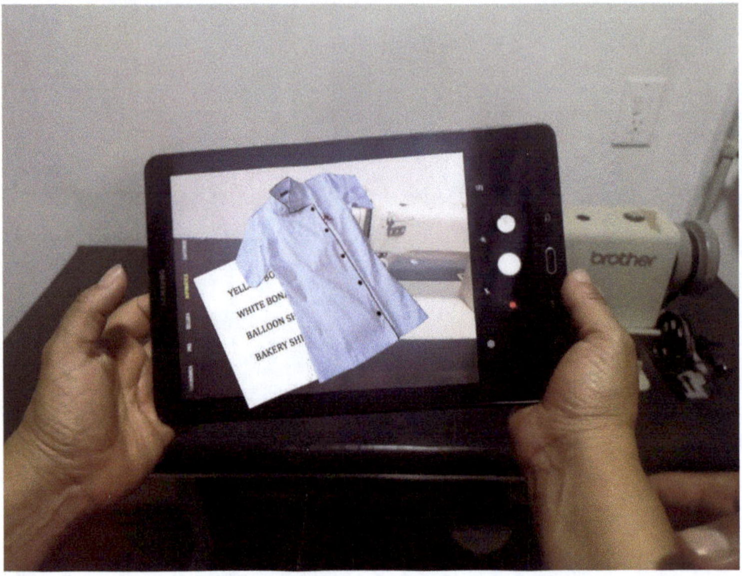

Fig. 7 AR system for a shirt associated with a marker

involved were asked a survey with 15 questions, which were previously designed to obtain the perspective of these users. In other words, the survey was applied to both workers and company management, where the answers to these questions allowed them to know, in a rough way, the activities and/or procedures that correspond to each worker, for example, if they have knowledge about what AR is, if they have a mobile device, what opinion they have regarding the operation of the AR-based system, if they liked the AR-based system, what benefits they believe an AR-based system will provide them, if the manufacturing time is reduced when they applied the AR in the processes they perform The concentration of survey responses is shown in Table 1.

Therefore, the content of Table 1 translates into the following results:

1. On the efficiency of the system, based on AR, users agreed that it is efficient because it allowed them to do their job better. That is to say, when the user has a doubt about some detail then he only has to place the cell phone on the garment to focus the camera and that is when the system shows the 3D image including the specifications and/or instructions for the correct assembly of the garment. In this way, the worker does not waste time asking the people in charge of supervising the garments, which in the worst case are far from the factory or because the employee works in a branch office far from the main workshop or even at home.
2. In terms of accessing AR material, workers commented that the application was stored on their phone and each time they were given new designs then the people responsible for that stage gave them the correct indications, such

Table 1 Concentration of responses through a survey

Procedure	Without AR	With AR	Benefit
Doubts in the assembly	If the employee does the work at home, contact the head of the maquiladora to consult him; In case you do it in the company, wait for the supervisor to answer the question	The employee opens the application, places the camera of the mobile device in the mold or on the sheet where the parts of the garment are shown	Time saving, immediate response, continue process activities
Learn some new process	See the sample and try to memorize the procedure for the worker	Open the application and focus the camera on the name of the piece you want to see and, using the system, appreciate the image or a video through the AR	Train workers step by step and repeat the process as many times as necessary so that they learn it better.
Use of the cell phone	They consulted him just to check their social networks	They consult it to open and consult the AR system, when required	They consult it as a working tool to improve quality in the production process
Use of the cell phone	They consulted him just to check their social networks	They consult it to open and consult the AR system, when required	They consult it as a working tool to improve quality in the production process
Ease of the system	He was not aware of the existence of applications, based on AR, to improve his work	They find it innovative, didactic and easy to apply in their work activities	He was not aware of the existence of these applications for use in his work area

as the name of the new design, to download. Workers found this AR stage quite interesting and innovative because they said that if they had to store all that information in their device memory (photographs and videos) they would easily get confused with the garment models and consequently their search would be longer and more time-consuming. The latter, in the short term, would become complicated and tedious. On the other hand, with the AR system, they would associate the model of a garment in real time.

3. Regarding the functioning of the AR, as a support tool to improve the quality in the correct way of sewing, the results showed in the first instance that, through the AR it is possible to modify in time and form the seams in case their error. So much so, that this system with AR is involved from the precise moment that the assembly of a garment is initiated, that is to say, it serves to consult and to verify the form of assembly, previous to initiating it. In addition, when a damaged piece or garment is detected, the system allows not to damage the others, which results in a correct assembly and to continue with the generation of high-quality garments.

4. With respect to the materials used by the AR, the videos allowed the worker to be shown the correct assembly process for a piece of garment, which included folds, seam adjustments, stitch size, thread color, etc. This phase of the system gave them an excellent understanding of what the garment designer required; workers mentioned that they went to the system whenever they needed to and found the images extremely attractive and easy to appreciate.
5. On the resource of the videos, the workers noticed the advantage of visualizing them through the AR. This consensus was due to the fact that these workers mentioned that if they had to search for a design on their mobile devices, without making use of AR, then they would have to locate them in the gallery of their mobile device where they have their personal videos and photographs, which would generate a search conflict and imply a more extensive search and, consequently, loss of time.
6. Regarding the complexity in the use of the system, the workers agreed that it does not have major difficulties; on the contrary, they found the system quite didactic with a fast and attractive handling.
7. Regarding system compatibility, just to mention that currently almost all workers have an Android smartphone, which meant that none of them had any problem downloading or using the application. It should be noted that the system, based on the AR, also works with iOS.
8. With regard to motivation, it is important to emphasize that the workers were encouraged, happy and excited about the application. They were so enthusiastic that they felt self-sufficient with respect to the assembly of the garments. Although the vast majority of workers interviewed received some training in tailoring, it is always important that their doubts are clarified, so this system makes sense from the very moment of its design because it solved their doubts the moment they were raised and, even more, they were resolved correctly.
9. On the cost, the workers interviewed stressed that they had no conflict with this, as the tool is free and open access. This is undoubtedly an aspect to highlight because it is convenient for both the owners of the maquiladoras and the employees.
10. The time investment to use the application is not expensive, since it only takes a maximum of 3 min if it is a detailed process (video) and one minute if it is a process with AR in image format. Of course, this will depend on the speed at which the data is transmitted.
11. On the learning side, AR is a new, innovative and efficient technology for teaching and training users. The workers interviewed because the vast majority agreed that they learned new learning processes and that supported this result, of course, they did not know about them.

For all the above reasons, there is evidence not to reject the null hypothesis H0: The AR is a technological tool capable of improving the manufacturing process in a garment factory. Likewise, and in order to answer the question ¿What factors affect the improvement of a manufacturing system in a factory using augmented reality?

- Easy and quick consultations: The worker, when using the AR, consults his doubts, related to the manufacturing process, in time and form.
- Time: when the employee makes use of the AR-based system, he does not have to ask the supervisor about the doubt he has, which avoids the loss of time in moving to the maquiladora or the waiting time to be attended.
- Optimization of material resources: When the errors are minimal or nil, then the waste of material is also considerably reduced.
- Motivation: When a worker feels confident to do things or have a support element at hand, he makes the manufacturing process that corresponds to him with more enthusiasm and security, which will result in a greater, but above all, a better productivity.
- Digital resources: The images show how the garment or its parts are displayed, giving the user a comparison of what they are doing versus what they should not do. With respect to the videos, they show all the phases of a process, obtaining abstract details that provide a visual, auditory, and kinesthetic learning to the users.

4 Conclusions

This work shows that AR is an innovative and efficient technology that is applied to various areas of the industry such as manufacturing, the clothing industry, maintenance, quality control, sales and design, just to mention some. In a small family business, a virtual information system, based on AR, was implemented in one of its manufacturing processes and where the main objective was to analyze the results of its implementation. This analysis led to obtaining highly significant deductions such as, the AR based system allowed material losses to be reduced considerably, even these margins became null in a certain production process (such as in fabric cutting used for clothing).

This result was obtained, mainly, because the AR was used to dispel, in real time, any doubts that the workers had regarding their production process. In other words, the AR-based system had a positive impact in order to considerably reduce some errors derived during the entire production process, such as, for example, that no damaged materials were wasted, which were detected by the process involved in the correction of the assembly. For its part, the time spent for a consultation using the AR-based system was relatively low, which positively impacted the entire production process. This reduction of queries, through the AR-based system, resulted in a minimum loss for response time compared to making the query directly with the person in charge of the assembly process. All these results allowed to positively increase the quality of the products derived from a production process and, at the same time, considerably reduce the time of the production line; This premise is supported to recommend the use of this technological tool in other manufacturing companies, since it is expected that a considerable increase in their production will then obtain an increase in their monetary profit.

Likewise, the AR allows the response to a query about a procedure to be almost instantaneous, since by placing a mobile device with the AR-based system on the piece of fabric, then the correct design and the ideal way will be obtained to elaborate. This action is very quick to consult (of course, the speed of the response depends on the transmission of the data), which contrasts if said consultation was made through a wide gallery of photos. This last way of carrying out a consultation, without a doubt, would represent a higher investment of time and, even, a more tedious process for the personnel involved, this without forgetting the confusion that the employee could have, when searching in their gallery photos, the model associated with the article of clothing. Therefore, the AR allows to show the assembly sequence that is essential for the construction of the pieces of clothing in a visual, animated, and auditory way. The implementation of a system, with the characteristics shown in this work, in a manufacturing process is understandable for anyone, even if they have little experience in cutting and making clothing, since AR resources show identically the correct way on how to create each piece of clothing.

AR goes beyond being a tool for solving queries, because today it is also considered as a novel and efficient means for learning and training. In the present work, these results were verified when employees viewed the videos with the intention of learning their processes and repeated the videos, as many times as necessary, in order to train in the construction of a garment. On the other hand, the monetary resources for the implementation of a system of this nature are few, since if it is taken into account that only the videos must be reproduced to show the 3D images and that currently a large part of the workers have a smart phone, then the only cost would fall on the transmission of data, which is already an essential service for a company and even for a person who uses new technologies.

It is important to mention that AR also has areas of opportunity, since the development of systems based on this new technology require specialized personnel in computer processes, who are responsible for designing the images, recording the videos, combining everything in the system, generate the APPs and upload it to the computer cloud and, in addition, carry out the process of distribution of the computer material. The companies that wish to implement, at any stage of their production process, these new technologies, without a doubt, would obtain multiple benefits when hiring this type of specialized personnel, among them the following can be mentioned: the quality of the products would be increased, It would achieve a decrease in production times and, consequently, a substantial increase in profits would be achieved. The challenges of new technologies are many, however, there are other scenarios that are not so easy to detect, such as, for example, there are people involved in a production process who find it difficult to understand and manage mobile devices. In addition, it must be taken into account that some of these people, or at least in the company where the AR system was implemented, are elderly. These people, for obvious reasons, do not have extensive knowledge about the handling of mobile devices and, moreover, they present great barriers to learning new technologies. However, the ease and simplicity of the system, presented in this work, allowed these elderly people to quickly learn to use and manage it.

Another benefit of the AR-based system was reflected in delivery times and, consequently, in payment times and increased production. For its part, this type of system can be implemented in any manufacturing process, since if it is adapted correctly then it will be of great help and even essential in a short time. In this type of technology, mobile devices are in frequent use, so the implementation costs do not require a considerable investment of human resources, much less infrastructure. In addition, mobile devices are of daily use for people, so these tools, in conjunction with new technologies, must be adapted to manufacturing processes, in order to obtain a useful benefit from their use. Thus, and by making use of a system based on AR, it is expected that quality control based on this new technology will achieve better performance and, consequently, processes will be more efficient and dynamic (these results are obtained, during the time it was analyzed, in the company object of this work). This type of systems, based on new technologies, will impact all those companies that decide to use them and whose objective is not only to improve their productivity, since these innovative and efficient processes will allow them to be at the forefront of the technological context, because this digital trends of sorts are already taking place in multiple companies around the world.

An avant-garde and innovative company, without a doubt, will facilitate the creation of other new technologies, since by supporting and promoting this type of new processes, it will motivate many companies to implement and adopt them; which would facilitate the way to the automation of various processes in the manufacturing industry. Another important aspect about the implementation of an AR-based manufacturing system is that this new technology provides workers with several benefits, among which they stand out: the feeling of feeling motivated, encouraged and confident that they are doing their job well, consequently, makes them feel part of an efficient and quality production, the latter, without a doubt, would strengthen the organizational culture of the company. All these aspects of well-being, in turn, lead to production efficiency, which translates into minimizing the loss of staff idle time, optimizing the assembly process, improving the quality of work and, of course, profiting monetary, which is one of the main objectives of all companies. Lastly, AR-based technologies are creating new ways of seeing the world, including their interaction, which is of great benefit to manufacturing companies and in the near future, this interaction may be used to make AR closer to physical reality.

References

Abulrub A-HG, Yin Y, Williams MA (2012) Acceptance and management of innovation in SMEs: immersive 3D visualisation. Procedia Soc Behav Sci 41:304–314

Afteni C, Frumuşanu G (2017) A review on optimization of manufacturing process performance. Int J Model Optim 7(3):139–144. https://doi.org/10.7763/IJMO.2017.V7.573

Alkhamisi AO, Monowar MM (2013) Rise of augmented reality: current and future application areas. Int J Internet Distrib Syst 1(4):25–34. https://doi.org/10.4236/ijids.2013.14005

Awazu Y (2006) Managing technology alliances: the case for knowledge management. Inf Manag 26:484–493

Azuma R (2015) Location-based mixed and augmented reality storytelling. In: Barfield W (ed) Fundamentals of wearable computers and augmented reality, 2nd edn. CRC Press, Boca Raton, pp. 259–276 (chapter 11)

Baratali E, Abd Rahim MHB, Parhizkar B, Gebril ZM (2016) Effective of Augmented Reality (AR) in marketing communication; a case study on brand interactive advertising. Int J Manag Appl Sci 2(4):133–137

Bloching B et al (2015) The digital transformation of industry, how important is it? Who are the winners? What must be done now? A European study commissioned by the Federation of German Industries (BDI) and conducted by Roland Berger Strategy Consultants, pp 1–51

Bottani E, Vignali G (2019) Augmented reality technology in the manufacturing industry: a review of the last decade. IISE Trans 51(3):284–310. https://doi.org/10.1080/24725854.2018.1493244

Bowman D, Kruijff E, LaVoila J, Poupyrev I (2005) 3D user interfaces: theory and practice. Addison-Wesley, Boston, MA

Bravo E, Santana M, Rodon J (2015) Information systems and performance: the role of technology, the task and the individual. J Behav Inf Technol 1(3):247–260. https://doi.org/10.1080/0144929X.2014.934287

Bulearca M, Tamarjan D (2010) Augmented reality: a sustainable marketing tool? Global Bus Manag Res: Int J 2(2/3):237–252

Cirulis A, Ginters E (2013) Augmented reality in logistics. Procedia Comput Sci 26:14–20

Condino S, Turini G, Parchi P, Viglialoro R, Piolanti N, Gesi M, Ferrari M, Ferrari V (2018) How to build a patient-specific hybrid simulator for orthopaedic open surgery: benefits and limits of mixed-reality using the Microsoft HoloLens. J Healthc Eng 2018:Article ID 5435097, 12 p. https://doi.org/10.1155/2018/5435097

Cordeiro D, Correia N, Jesus R (2015) ARZombie: a mobile augmented reality game with multi-modal interaction. In: 2015 7th international conference on intelligent technologies for interactive entertainment (INTETAIN), Turin, 2015, pp 22–31

Cuperschmid ARM, Grachet MG, Fabrício MM (2016) Development of an Augmented Reality environment for the assembly of precast wood-frame wall from the BIM model. Ambiente Construído 16(4):63–78. https://doi.org/10.1590/s1678-86212016000400105

Etonam AK, Di Gravio G, Kuloba PW, Njiri JG (2019) Augmented reality (AR) application in manufacturing encompassing quality control and maintenance. Int J Eng Adv Technol 9(1):197–204. https://doi.org/10.35940/ijeat.A1120.109119

Gren A, Jamarillo B, Kiev V, Shabrov N, Vasiliev D (2018) Development of digital simulation on the basis of technologies of virtual and augmented reality. Emuge-Franken, 91207 Nürnberger Straße 96-100, Germany. Peter the Great St. Petersburg Polytechnic University, 195251 Polytechnicheskaya St. 29, Russian Federation. https://doi.org/10.1051/shsconf/20184400036

Grubor A, Milovanov O (2017) Brand strategies in the era of sustainability. Interdiscipl Descr Complex Syst 15(1):78–88. https://doi.org/10.7906/indecs.15.1.6

Guo B, Wang J, Wei SX (2018) R&D spending, strategic position and firm performance. Front Bus Res China 12:14. https://doi.org/10.1186/s11782-018-0037-7

Henderson SJ, Feiner SK (2007) Augmented reality for maintenance and repair (ARMAR). Final report for June 2005 to August 2007. Columbia University Department of Computer Science, New York, USA, pp 1–70. Recovered from: http://citeseerx.ist.psu.edu/viewdoc/download?doi=10.1.1.149.4991&rep=rep1&type=pdf

Hozdić E (2015) Smart factory for industry 4.0: a review. Int J Mod Manuf Technol 7(1):28–35

Khalek I, Chalhoub J, Ayer S (2019) Augmented reality for identifying maintainability concerns during. Adv Civ Eng 2019:Article ID 8547928, 12 p. https://doi.org/10.1155/2019/8547928

Katrijn G (2012) New products: the antidote to private label growth? J Mark Res 49(3):408–423. https://doi.org/10.1509/jmr.10.0183

Kipper G, Rampolla J (2013) Augmented reality. An emerging technologies guide to AR. Syngree, Elsevier, USA

Koul S (2019) Augmented reality in supply chain management and logistics. Int J Rec Sci Res 10(2A):30732–30734. https://doi.org/10.24327/IJRSR

Lebiedź J, Łubiński J, Mazikowski A (2010) Immersive 3D visualization laboratory concept. In: 2nd international conference on information technology, ICIT 2010 (Gdańsk University of Technology Faculty of ETI Annals, the IT series 18), Gdańsk, Poland, pp 117–120

Martinetti A, Marques HC, Singh S, Van Dongen L (2019) Reflections on the limited pervasiveness of augmented reality in industrial sectors. Appl Sci 9(16):3382. https://doi.org/10.3390/app9163382

Mazikowski A, Lebiedź J (2014) Image projection in immersive 3D visualization laboratory. Procedia Comput Sci 35:842–850

Mekni M, Lemieux A (2014) Augmented reality: applications, challenges and future trends. In: Applied computational science proceedings of the 13th international conference on applied computer and applied computational science (ACACOS), vol 14, pp 23–25

Merlino M, Sproģe I (2017) The augmented supply chain. Procedia Eng 178:308–318

Moghaddam M, Cadavid MN, Kenley CR, Deshmukh AV (2018) Reference architectures for smart manufacturing: a critical review. J Manuf Syst 49:215–225

Nee AYC, Ong SK (2013) Virtual and augmented reality applications in manufacturing. In: 7th IFAC conference on manufacturing modelling, management, and control international federation of automatic control, 19–21 June 2013. Saint Petersburg, Russia

Nie P, Wang Ch, Chen Y, Yang Y (2018) Effects of switching costs on innovative investment. Technol Econ Dev Econ 24(3):933–949. https://doi.org/10.3846/tede.2018.1430

Ning F, Shi Y, Cai M, Xu W, Zhang X (2020) Manufacturing cost estimation based on a deep-learning method. J Manuf Syst 54:186–195

Novak-Marcincin J, Barna J, Janak M, Novakova-Marcincinova L (2013) Augmented reality aided manufacturing. Procedia Comput Sci 25:23–31

Novakova NG (2018) Innovation potential of augmented technologies in industrial context. Int Sci J "Industry 4.0" 4:24–28

Ong SK, Yuan M, Nee AYC (2008) Augmented reality applications in manufacturing: a survey. Int J Prod Res 46(10):2707–2742. https://doi.org/10.1080/00207540601064773

Rabah S, Assila A, Khouri E, Maier F, Ababsa F, Bourny V, Maier P, Mérienne F (2018) Towards improving the future of manufacturing through digital twin and augmented reality technologies. Procedia Manuf 17:460–467

Rao RV (2011) Advanced modeling and optimization of manufacturing processes. International research and development. Springer, London

Rejeb A (2019) The challenges of augmented reality in logistics: a systematic literature review. World Sci News 134(2):281–311

Roschli A, Gaul K, Boulger A, Post B, Chesser P, Love L, Blue F, Borish M (2019) Designing for big area additive manufacturing. Addit Manuf 25:275–285

Sabarinathan K, Kanagasabapathy N, Ambeth Kumar VD, Rishikesh PK, Priyadharshan RV, Abirami A (2018) Machine maintenance using augmented reality. In: 3rd international conference on communication and electronics systems (ICCES), Coimbatore, India, 2018, pp 613–618

Šatanová A, Figuli L, Sedliačiková M (2015) Optimization of production process through selected statistical methods. Procedia Econ Finance 23:959–963

Segovia D, Mendoza M, Mendoza E, González E (2015) Augmented reality as a tool for production and quality monitoring. Procedia Comput Sci 75:291–300

Shao G, Brodsky A, Miller R (2018) Modeling and optimization of manufacturing process performance using modelica graphical representation and process analytics formalism. J Intell Manuf 29(6):1287–1301. https://doi.org/10.1007/s10845-015-1178-6

Siriborvornratanakul T (2018) Enhancing user experiences of mobile-based augmented reality via spatial augmented reality: designs and architectures of projector-camera devices. Adv Multimed 2018:Article ID 8194726, 17 p. https://doi.org/10.1155/2018/8194726

Strozzi F, Colicchia C, Creazza A, Noè C (2017) Literature review on the 'Smart Factory' concept using bibliometric tolos. Int J Prod Res 55(22):6572–6591. https://doi.org/10.1080/00207543.2017.1326643

Sun X, Houssin R, Renaud J, Gardoni M (2016) Integrating user information into design process to solve contradictions in product usage. Procedia CIRP 39:166–172. https://doi.org/10.1016/j.procir.2016.01.183

Turovets Y, Vishnevskiy K, Tokareva M, Kukushkin K (2019) Technology foresight for digital manufacturing: Russian case. IOP: Conf Ser: Mater Sci Eng 497(1):012062. https://doi.org/10.1088/1757-899x/497/1/012062

Uva A, Gattullo M, Manghisi V, Spagnulo D, Cascella G, Fiorentino M (2018) Evaluating the effectiveness of spatial augmented reality in smart manufacturing: a solution for manual working stations. Int J Adv Manuf Technol 94(1):509–521

Zhong RY, Xu X, Klotz E, Newman ST (2019) Intelligent manufacturing in the context of industry 4.0: a review. Engineering 3(5):616–630

Towards an Analysis of the Relationship Between Quality Management and Project Management

Alfonso J. Gil, Mara Mataveli, and Jorge Luis García Alcaraz

Abstract A firm benefits mainly through the success of its projects; therefore, it is important to have the right project management that includes quality management measures. This work aims to examine the relationships between project management and quality management. First, the tools for project management and quality management are studied, and, second, the contents of research on quality management and project management are examined through the database Scopus. Similar features are checked in the quality and project management models. Specific variety is credited in quality research related to project management. Some gaps are found in the literature on quality management related to project management. The most important conclusions are pointed out.

Keywords Quality management · Total quality management · Project management · Scopus

1 Introduction

The development of project management (PM) is essential for organisations to manage various portfolios, programmes and projects (Chofreh et al. 2016). Note (2015, p. 1) defines 'project' as 'a series of unique, multifaceted and related activities with a purpose that must be accomplished at a particular time, within cost constraints and according to specifications'. For its part, PM is the discipline of organising and managing resources, so that a given project is completely completed within the scope,

---

A. J. Gil (✉) · M. Mataveli
Departamento de Economía y Empresa, Universidad de La Rioja, Calle Cigüeña 60, Logroño, La Rioja, Spain
e-mail: alfonso.gil@unirioja.es

J. L. García Alcaraz
Department of Industrial Engineering and Manufacturing, Universidad Autónoma de Ciudad Juárez, Av. Del Charro 450 Norte, Col. Partido Romero, Juárez, Chihuahua, Mexico

Division of Research and postgraduate Studies, Instituto Tecnológico de Ciudad Juárez, Ciudad Juárez, Chihuahua, Mexico

J. L. García Alcaraz et al. (eds.), *Techniques, Tools and Methodologies Applied to Quality Assurance in Manufacturing*,
https://doi.org/10.1007/978-3-030-69314-5_6

time and cost restrictions imposed at its start. It is aligned with the objectives of the organisation and managed according to the practices of the organisation (Picciotto 2019). In this area, quality management (QM) in a project would be understood as a process that ensures that the activities necessary to design, plan and implement a project are carried out efficiently. More specifically, QM focuses on the satisfaction of those interested in the project through the continuous improvement of processes that generate value for all its beneficiaries.

QM is one of the project requirements that should be considered in preliminary project evaluations (Momeni et al. 2019). In reality, the deliverables must meet the quality or standards required by the customer to be accepted as products of the project. Examples exist that show the terrible effects of low quality in projects, such as the bankruptcy of project contractors (Scaringella and Burtschell 2017). Quality management is also considered a beneficial strategy to reduce possible delays and project costs (Momeni et al. 2019).

Despite the scope of quality in PM, managers do not always consider the cost, time and quality triangle (Atkinson 1999); sometimes the quality of the project is summarised in the various 'checkbox' documents. In this regard, managers appreciate the risk of a project due to its uniqueness and complexity, but it seems that they do not always prioritise the existing link between the consequences of risk and the underlying causes supported by the dimensions of quality of the project (Basu 2014).

Therefore, this work aims to analyse the relationship between PM and QM. The work has been divided into four sections. In the first section, PM is studied, and a current PM model is presented. In the second section, QM in PM is studied. In the third section, the literature is analysed, and the relationship between research in PM and research in QM is examined. In the fourth and last section, the most critical conclusions of the relationship between PM and QM are discussed.

2 Project Management

A traditional PM approach, such as PMBOK, APMBOK or PINCE2, emphasises the standardisation of procedures (Wysocki 2014), which guarantees their robustness and applicability in a wide variety of projects, from the smallest and most straightforward to the largest and most complex (Spundak 2014). This traditional standardisation in PM, that is, maintaining the same management patterns for common projects, may be optimal for routine and continuous operations (Note 2015) but would not be the most appropriate for operations that did not have a defined performance pattern (Golini and Landoni 2014).

Quality runs through the three phases of the lifecycle of a project, namely planning, execution and delivery. Basu (2011) defines quality, in general, as what customers expect as a lasting experience; however, the author observes that in the area of project management, the importance of the quality dimension is not so clear, relegated to a position after cost and time. Recognised in the literature as part of the 'iron triangle criteria' (cost, time and quality), or the traditional trifecta (Meredith and Mantel 2003), quality has not received as much attention as the other two dimensions (Turner and Huemann 2002; Williams et al. 2015).

Project success primarily depends on the amount of effort invested in defining the project goals, the functional requirements and the specifications (Dvir 2005). Currently, different aspects of project management have been considered when assessing the achievement of PM goals and, consequently, its success. The evaluation of project success must occur from the perspective of various and different stakeholders (Turner and Zolin 2012; Gemunden 2015), which certainly includes the most important, the viewpoint of the customer.

At present, projects have become progressively more complex, involving more and more interested people, and tasks are increasingly sophisticated and interrelated (Verga Matos et al. 2019). In this context, the general concept of 'agile' and 'adaptive' PM appears (Fernandez and Fernandez 2008). An agile and adaptive PM approach shows greater adequacy in projects with a higher degree of uncertainty, with specific requests, perhaps expected, but more unpredictable, that require more flexible PM (Shenhar 2008). For this, the work teams are organised iteratively, which requires constant collaboration between the beneficiaries, the end-users and other stakeholders in the project (DeCarlo 2004). In this context, communication, the proximity between team members and a set of useful skills of all the people participating in the project are critical success factors in PM (Spundak 2014).

Flexible decision-making allows work teams to balance resources and limitations as contingencies arise (Note 2015). In this regard, Rowlinson and Cheung (2008) emphasised the importance of the participation of stakeholders in the project as a factor that allows for success of the project. As a project identity is formed with shared objectives, the tolerance of interested parties for the possible difficulties encountered in carrying out the project is increased. To prevent a lack of identification of people with the project, it is essential to communicate all aspects of the project, whether good or bad, trying to minimise the negative impacts and maximise the positive ones in favour of all interested parties (Di Maddaloni and Davis 2017). In this sense, the project manager must not stop considering some key actions, such as empathetically communicating the project characteristics and singularities at the beginning of the process (Francisco de Oliveira and Rabechini 2019).

In terms of the success of the project, as Demirkesen and Ozorhon (2017) pointed out, most studies have focused on timely completion, below budget, with quality criteria met, work completed safely and customer satisfaction, although other characteristics intervene in success. The participation of external stakeholders that do not have an official or contractual link with the organisation of the project but can influence and be affected by it, such as regulatory agencies (Winch 2004), is considered an essential factor of success and creation of value, mainly in interorganisational projects (Lehtinen and Aaltonen 2020). However, the involvement of interested external people to generate value for the project is a unique challenge in project management (Eskerod and Huemann 2014). To solve this problem, Lehtinen and Aaltonen (2020) proposed three types of solutions (see Fig. 1).

The solutions that are summarised in Fig. 1 are proposed for interorganisational projects. However, these same solutions can be proposed for all types of projects in which the characteristics that guarantee the success of a project are the commitment of stakeholders, flexibility in decision-making and communication.

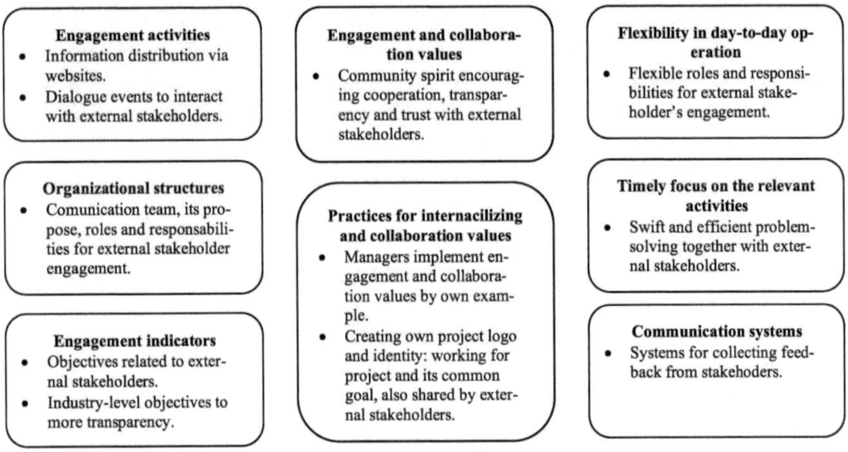

Fig. 1 Code hierarchy and concepts developed from the cross-case analysis. *Source* Lehtinen and Aaltonen (2020, p. 92)

2.1 The Learning Project

As noted, one of the main characteristics that identify current PM is the flexibility and learning capacity that allows projects to adapt to the context of change and to be proactive in anticipating the needs and demands of the project, firm and society. For this, Gil and Mataveli (2018) proposed a project model called 'The Learning Project'. A learning project can be defined as a project that is continuously transformed to adapt to changing contexts and move towards opportunities for improvement. A learning project is identified by four primary dimensions: project leadership, a project structure, the generation of learning opportunities and a project culture (see Fig. 2).

Ethical leadership is proposed; this type of leadership is especially important in project management, which tries to highlight the importance of reducing the behaviours associated with the lack of ethical conduct in companies and professionals. Ethical leaders serve as role models for their followers by setting clear standards and ensuring employees take responsibility for following these criteria.

Fig. 2 Project as a learning project. *Source* Gil and Mataveli (2018, p. 49)

Ethical leadership promotes the behaviour of group learning by providing a basis for the development of norms and productive practices of group learning (Cropanzano and Walumbwa 2010).

The learning culture in a project is defined as the culture oriented to the promotion and facilitation of the learning of those interested in the project, its participation and dissemination, and its contribution to the development and performance of the project as a whole. The values that characterise the culture of a project are not only related to the benefits of learning and innovation but are also distinguished by trust and commitment, especially when considering the uncertainty that sometimes surrounds project management. The risk arises from the factors associated with the project, for example, the uncertainty in the level of performance of the project members, the quality and reliability of the work performed and the alignment of the objectives with each part of the project.

In the design stage, the project structure is configured, especially in significant and complex projects; the main challenge is to coordinate the many specialised technicians involved in the project structure. The coordination of a project requires the exchange of information among the members, which sometimes becomes complex, especially if it involves the coordination of multidisciplinary teams. Communication is an essential factor that contributes to the success of the project; for instance, it is crucial to control time and costs. Inadequate communication between project members is one of the leading causes of complications that arise during the process of a project.

The concept of learning opportunities refers to opportunities related to the training and development of project members. The importance of training is that it enables project stakeholders to acquire new knowledge and skills as well as new ways of working and new tools. The training provides opportunities for those interested in the project to learn throughout their lives and develop professional careers. The importance of training is also due to the increase in competitiveness, productivity and quality in project management. In addition, investment in training is one of the most important aspects when a project is destined to grow, since the training translates into productivity in the generation of human capital, which constitutes an instrument to generate more significant growth and performance of projects, thereby affecting customer satisfaction.

3 Quality Management and Project Management

QM is a useful management tool designed to help companies achieve better performance by continuously optimising production processes and technologies, as well as improving the quality of products and services (Levine and Toffel 2010). QM includes a set of commonly accepted practices and initiatives, such as Total Quality Management (TQM), Lean, and Six Sigma, and a tool that promotes the competitiveness of organisations (Peng et al. 2020). Researchers consider QM to be an essential issue in operations management (Nair 2006).

3.1 Total Quality Management

TQM is a management philosophy and a set of techniques/procedures that imply a total system approach to quality (Zhou and Lee 2000). Deming (1994) argued that TQM requires systems thinking, teamwork and leadership rather than top-down, bureaucratic, indicator-driven management. A convincing synthesis between the two approaches has yet to emerge.

TQM advocates a comprehensive framework for the entire organisation in the form of specific organisational practices, tools, techniques and systems (Alofan et al. 2020). TQM has been recognised for its potential to improve competitive results for organisations through continuous improvement. The justification for the relationship between TQM and competitive advantage is twofold (Yu et al. 2020).

On the one hand, TQM helps companies improve their competitive advantages through sound management practices that improve internal operations. The various components of the TQM process, such as support from senior management, training programmes and the quality management philosophy, help improve competitiveness. Also, on the other hand, companies with TQM are more likely to be market-driven; these companies identify the needs of customers and try to differentiate products and processes to meet them, which helps companies overcome their competitors. Therefore, it has been shown that TQM has a positive impact on company results (Abbas 2019, 2020).

At the same time, it has been pointed out that the achievement of productivity improvement in companies as a consequence of quality improvement is mainly the result of managers' decisions and actions (Singh et al. 2011). It is estimated that 85% of quality improvement depends on the system and the managers, while only 15% depends on worker performance (Djendel et al. 2016). For Nicał and Anysz (2020), the TQM organisational structure consists of (see Table 1):

- Executive Steering Committee (ESC),
- Quality Management Boards (QMB),
- Process Action Teams (PAT).

3.2 International Standard ISO 9001

As Ma et al. (2020) pointed out, a characteristic of QM is that it provides an automatic control mechanism to guarantee the quality of production, with ISO 9001, the most widely accepted QM system. The ISO 9001 Quality Management System is an optional quality management standard for organisations regardless of any consideration such as their size, type of operations, or location (Walaszczyk and Polak-Sopinska 2020). ISO 9001 has established principles and implementation requirements relevant to companies, which include: (1) customer focus; (2) leadership; (3) people's commitment; (4) process approach; (5) improvement; (6) evidence-based

Table 1 Organisational structure of TQM with membership and function description of each level

Level of the structure	Membership	Function
Executive Steering Committee (ESC)	The highest level of management with top managers in the organisation	Identifies strategic goals for organisational quality improvement efforts Prioritizing and listing of organisational goals for quality improvement Determining the effectiveness of changes in meeting the quality needs of customers Providing resources needed to standardise and document these changes
Quality Management Boards (QMB)	Top- and mid-level managers, jointly responsible for a specific product or service	Selecting the organisational areas that might have the most significant impact on the goals Defining indicators of quality improvement and cost reduction (in cooperation with ESC) Organizing PATs Conduction of experiments and identifying the most critical causes of process performance Making changes designed to improve process performance Tracking the performance of the process to determine the impact of the changes on the selected goals
Process Action Teams (PAT)	Staff and/or hourly workers involved in the processes being investigated by the QMBs	Collecting and summarising process data for QMBs Analysing a process and identifying potential areas for improvement by the use of statistical process control (SPC)

Source Nicał and Anysz (2020, p. 579)

decision-making; and (7) relationship management (Fonseca 2015). By these principles and requirements, companies can achieve continuous improvement in quality and economic performance.

The European Foundation for Quality Management (EFQM), Swedish Quality Award (SIQ) and Malcolm Baldrige National Quality Award (MBNQA) are the different quality models that describe the criteria for TQM implementation by considering its core values. The American MBNQA model incorporates soft and hard aspects of TQM and has been proven to be valuable for many organisations in terms of introducing reforms to their management and operational structures (Abbas 2020). This model, which has been examined by different researchers (Sila 2007; Ooi 2014), contains six dimensions, namely leadership, strategic planning, customer focus, process management, human resources management and information and analysis (see Table 2).

Other types of constructs can be joined to the previous description of variables that identify the TQM models, such as continuous improvement and client satisfaction (Ladewski and Al-Bayati 2019). Continuous improvement refers to the propensity of the organisation to vigorously pursue incremental and innovative improvements in the quality and safety of the processes, products and services. Client satisfaction refers to the degree to which an organisation meets the expectations for product or service quality of its external clients (customers) and expectations for job safety while engaged in the production of the product or service by its internal clients (employees).

The literature on PM points to the three criteria of 'time, cost and quality', known as the iron triangle (Meredith and Mantel 2003), for evaluating the success of a project. The first two criteria are relatively easy to evaluate; quality is a little bit more

Table 2 Description of variables and related literature

Constructs	Description
Leadership	Responsible for quality assurance and quality improvement efforts; include top management and focus on quality goals, efforts, and planning to achieve those goals in relation to time and cost
Strategic planning	Vision and mission for quality, policy and strategy development and deployment to achieve organisational goals
Customer focus	Knowledge of customers' demand and market trends; developing and maintaining good relations with customers by ensuring their satisfaction
Process management	Clear division of process, ownership and responsibility; ensuring the perfect product or service design, process control, continuous improvement by self-inspection and automation
Human resource management (HRM)	Effective management of human resource through their active participation in operational issues, contact with top management, empowerment, training, performance recognition and reward; quality responsibility and awareness
Information and analysis	Evaluation and analysis of employees' and managers' performance using information technology and related tools. Giving feedback to them for solving the problems on a timely basis

Source Abbas (2020, p. 3)

complex. Authors such as Turner and Huemann (2002) have identified the quality of a two-dimensional project, the quality of the product and the quality of the process.

3.3 *Quality Management in Project Management*

As Basu (2014) pointed out in reference to quality in PM, quality in a broad context has many meanings, ranging from the concept of excellence to good value for money and even practicality. A generic designation of quality may refer to compliance with the client's requirements in terms of PM. This meaning would refer to (Basu 2014, p. 181):

- 'conformance to requirements,
- fitness for use,
- quality aimed at the needs of the consumer,
- total composite product and service characteristics of the organisation designed to meet the expectations of the customer,
- the totality of characteristics of an entity that bear on its ability to satisfy customer stated and implied need'—ISO 9000: 2000.

Therefore, to the extent that all of the above requirements are met, the quality of the projects would be successfully addressed.

Currently, there are various tools, instruments or QM strategies in projects. For Bamford and Greatbanks (2005), the application of quality management tools has been of great importance for management systems. In the work of Cotrim et al. (2018), a set of tools that can be integrated into projects and programmes to facilitate their implementation are indicated (see Table 3).

Another tool to manage the quality of the projects is the 'Quality Function Deployment' (QFD); it is a systematic method to channel the entry of end-users in the development of products (Hernandez and Aspinwall 2007). Its objective is to develop a quality assurance method that integrates customer satisfaction into the design of a product or service before its manufacture (Bahadorestani et al. 2020). QFD generally includes four phases, and each phase is defined by a matrix in which the relationship between the components of the horizontal and vertical domains (rows and columns) is determined. Domains represent the primary classification of components in a system. The matrices are interconnected and can be analysed using the concept of 'House of Quality' (HoQ). In each matrix, two domains of 'WHAT' and 'HOW' are compared, and their interrelation is weighted. In general, QFD, the 'WHAT', are generally end-user requirements, while the 'HOW' are methods to fulfil them.

Table 3 Description of quality management tools

Quality tool	Purpose
Linear graph	Allows the development evaluation of a data set over time
Responsibility matrix	Used to define the roles and responsibilities of actors during a project
Gantt graph	Facilitates the time distribution of activities
5W2H	Define responsibilities, methods, deadlines, objectives and resources
Flow chart	Represents, by graphic symbols, the steps sequence of a task to facilitate the analysis of processes
Check sheet	Used to quantify the frequency with which certain events occur in a period of time
Stratification	Divide the whole into parts, facilitating the individual observation
Benchmarking	Compares actual or planned project practices to those of comparable projects to identify best practices
Pareto graph	Prioritise problems related to a particular
GUT matrix	Prioritise the elimination of problems, especially if multiple and interrelated
Brainstorming	To launch and detail ideas seeking diversity of opinions from a group creative process
Cause and effect diagram	It helps to search the roots of the problem by raising issues

Source Cotrim et al. (2018, p. 68)

4 Relationship Between Research in Project Management and Quality Management

At this point, the common areas of research on QM and PM were studied. This analysis was intended to examine the relationships between research related to QM and PM, which allows the determination of meeting points and possible research deficits. To develop this work, the Scopus database was used because it has an extensive list of publications, and its search criteria allow a detailed analysis of publications; in particular, it facilitates the comparison of fields or research topics that are the purpose of this study.

Figure 3 shows the publications during the last ten years on QM and PM. Two searches were executed, one with the term 'Quality Management' and the other with the term 'Project Management'. In order to make the search more concrete and not distort it with excessive types of documents, the search was circumscribed in 'document type'—Article, Review and Book chapter—and 'Source type'—Journals, Book series and Books.

As can be seen in Fig. 3, the number of publications with the search term 'Quality Management' is higher than that of 'Project Management', as was expected since the issue of quality management is very extensive and covers full fields of research and application, while project management may have a more specific field of research, although also broad in its subject matter. In relative terms, an increase in quality

Fig. 3 Several publications on 'Quality Management' and on 'Project Management' in the last ten years. *Source* Scopus (2020)

management publications can be observed from 2195 publications in the year 2011 to 4393 in 2019. Constant growth in publications can be seen between 2011 and 2018, the year in which the number of publications suffered a significant decrease from 5404 to 4393. With respect to project management, a constant increase in the number of publications can be observed from 1221 publications in the year 2011 to 1735 in 2019, with a considerable increase from 1571 publications in 2018 to 1735 in 2019.

Figure 4 shows the publications in the last ten years with the combined constructs of 'Quality Management & Project Management'. A single joint search was carried out with the terms 'Quality Management' and 'Project Management'.

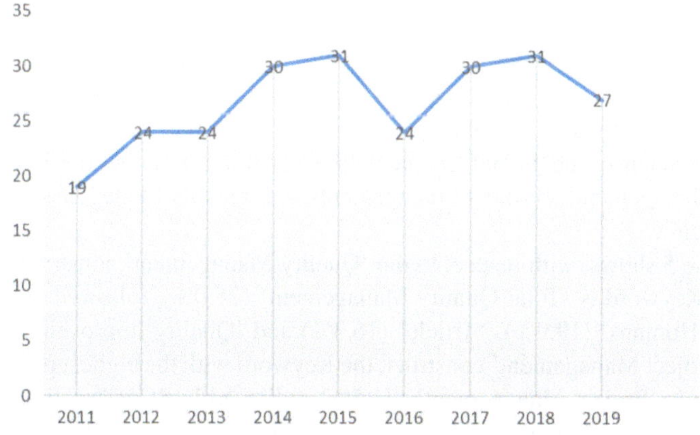

Fig. 4 Publications on 'Quality Management' and 'Project Management' in the last ten years

Table 4 Subject areas of 'Quality Management', 'Project Management' and 'Quality Management & Project Management'

Subject area	Quality Management	Project Management	Quality Management & Project Management
Medicine	25,195 (65.5%)		56 (17.9%)
Nursing	4404 (11.4%)		
Business, Management and Accounting	3525 (9.2%)	6068 (33.1%)	93 (29.8%)
Social Science	2775 (7.2%)		
Engineering	2590 (6.7%)	5843 (31.9%)	101 (32.5%)
Computer Science		2590 (14.1%)	31 (9.9%)
Social Science		2395 (13.1%)	31 (9.9%)
Environmental Science		1411 (7.7%)	

Source Scopus (2020)

As can be seen in Fig. 4, the number of publications that combine the concepts of 'Quality Management' and 'Project Management' has increased in the last ten years, from 19 publications in the year 2011 to 27 works in 2019. In these years, the average number of publications was 26.7, with a standard deviation of 4.18.

For its part, Table 4 presents the knowledge areas of 'Quality Management', 'Project Management' and 'Quality Management & Project Management'.

As can be seen in Table 4, the most significant area of research in 'Quality Management' is 'Medicine' (65.5%), followed by 'Nursing' (11.4%), 'Business, Management and Accounting' (9.2%), 'Social Science' (7.2%) and, finally, 'Engineering' (6.7%).

In 'Project Management', the area of most significant publication is 'Business, Management and Accounting' (33.1%), followed by 'Engineering' (31.9%), 'Computer Science' (14.1%), 'Social Science' (13.1%) and, finally, 'Environment Science' (7.7%).

When the subjects 'Quality Management & Project Management' are studied together, the most researched area is 'Engineering' (32.5%), followed by 'Business, Management and Accounting' (29.8%), 'Medicine' (17.9%) and, finally, 'Computer Science and Social Science' (9.9%). Table 5 shows the keywords of 'Quality Management', 'Project Management' and 'Quality Management & Project Management'.

As Table 5 shows, with respect to the 'Quality Management' construct, the most weighted Keyword is 'Total Quality Management' (25.0%), followed by 'Human' (23.3%), 'Humans' (19.9%), 'Article' (16.3%) and 'Quality Improvement' (15.5). For the 'Project Management' construct, the Keyword with the highest percentage of occurrence is 'Project Management' (71.5%), followed by 'Construction Industry' (10.2%), 'Decision-Making' (6.5%), 'Construction Projects' (6.3%) and 'Quality Management' (5.5%). For 'Quality Management & Project Management', the Keyword with the highest occurrence rate is 'Project Management' (35.0%), followed

Table 5 Keywords regarding 'Quality Management', 'Project Management' and 'Quality Management & Project Management'

Keyword	Quality Management	Project Management	Quality Management & Project Management
Total Quality Management	29,687 (25.0%)		81 (18.1%)
Human	27,657 (23.3%)		116 (25.9%)
Humans	23,559 (19.9%)		41 (9.2%)
Article	19,339 (16.3%)		
Quality Improvement	18,327 (15.5%)		
Project Management		10,438 (71.5%)	157 (35.0%)
Construction Industry		1484 (10.2%)	
Decision-Making		948 (6.5%)	
Construction Projects		918 (6.3%)	53 (11.8%)
Construction Management		799 (5.5%)	

Source Scopus (2020)

by 'Human' (29.5%), 'Total Quality Management' (18.1%), 'Construction Project' (11.8%) and 'Humans' (9.2%).

Table 6 shows the 'Funding Sponsor' of the research in 'Quality Management', 'Project Management' and 'Quality Management & Project Management'.

As Table 6 points out with respect to the 'Funding Sponsor' for 'Quality Management', the institutions that provide the most financing for the projects are in health and natural sciences, notably 'National Institute for Health' (41.6%), followed by 'Agency for Healthcare Research and Quality' (19.7%), 'National Natural Science Foundation of China' (18.9%), 'National Institute for Health Research' (10.2%) and 'Canadian Institutes of Health Research' (9.5%). For 'Project Management', the organisations that finance this type of research are somewhat more diverse, with the 'National Natural Science Foundation of China' (63.1%) the largest funder, followed by 'National Science Foundation' (13.3%), 'Fundamental Research Funds for the Central Universities' (8.8%), 'National Development Council' (7.9%) and 'National Council for Scientific and Technological Development' (6.9%). In the case of 'Quality Management & Project Management', the largest financier is 'National Natural Science Foundation of China' (56.2%), followed by 'National Health and Medical Research Council' (18.8%), 'National Basic Research Program of China (973 Program)' (12.5%) and 'National Basic Research Program of China (973 Program)' (12.5%).

Table 7 shows the Source Title, which allows knowing the journals that publish the most articles with the terms 'Quality Management' and 'Project Management'.

Table 6 Funding sponsor for Quality Management, Project Management and Quality Management & Project Management

Funding sponsor	Quality Management	Project Management	Quality Management & Project Management
National Institutes of Health	840 (41.6%)		
Agency for Healthcare Research and Quality	395 (19.7%)		2 (12.5%)
National Natural Science Foundation of China	381 (18.9%)	494 (63.1%)	9 (56.2%)
National Institute for Health Research	223 (10.2%)		
Canadian Institutes of Health Research	191 (9.5%)		
National Science Foundation		104 (13.3%)	
Fundamental Research Funds for the Central Universities		69 (8.8%)	
European Commission		62 (7.9%)	
Conselho Nacional de Desenvolvimento Científico e Tecnológico		54 (6.9%)	
National Health and medical Research Council			3 (18.8%)
National Basic Research Program of China (973 Program)			2 (12.5%)

Source Scopus (2020)

Table 7 confirms that, in general terms, the journals that publish the most in the field of 'Quality Management' are related to health and health safety, especially the magazine that has published the most articles in recent years. During the last ten years, 'BMC Health Services Research' (26.0%) has published the most, followed by 'Total Quality Management and Business Excellence' (21.2%), a publication whose aim is Total Quality Management. Next are journals in the field of health, 'BMJ Quality and Safety' (19.5%), 'Journal of American College of Radiology' (16.8%) and 'BMJ Open' (16.5%). In the field of 'Project Management', the review that has published the most articles during the last ten years is 'International Journal of Project Management' (40.7%), followed by lower percentages in the 'Journal of Construction Engineering and Management' (16.9%), 'Project Management Journal' (16.3%), 'International Journal of Managing Projects in Business' (13.9%) and 'Construction

Table 7 Source title of the subjects 'Quality Management', 'Project Management' and 'Quality Management & Project Management'

Keyword	Quality Management	Project Management	Quality Management & Project Management
BMC Health Services Research	428 (26.0%)		
Total Quality Management and Business Excellence	350 (21.2%)		
BMJ Quality and Safety	321 (19.5%)		
Journal of American College of Radiology	276 (16.8%)		
BMJ Open	271 (16.5%)		
International Journal of Project Management		961 (40.7%)	5 (20.1%)
Journal of Construction Engineering and Management		399 (16.9%)	5 (20.1%)
Project Management Journal		384 (16.3%)	
International Journal of Managing Projects in Business		329 (13.9%)	
Construction Management and Economics		287 (12.2%)	
TQM Journal			6 (25.0%)
Journal of Civil Engineering and Management			4 (16.7%)
Journal of Management in Engineering			4 (16.7%)

Source Scopus (2020)

Management and Economics' (12.2%). Finally, in the field of 'Quality Management & Project Management', the review that has published the most articles in the last ten years is 'TQM Journal' (25.0%), followed by 'International Journal of Project Management' (20.1%), 'Journal of Construction Engineering and Management' (20.1%), 'Journal of Civil Engineering and Management' (16.7%) and 'Journal of Management in Engineering' (16.7%).

After studying the set of journals that have published the most in quality management and project management, Fig. 5 shows the publication percentages of each of the topics that have been studied.

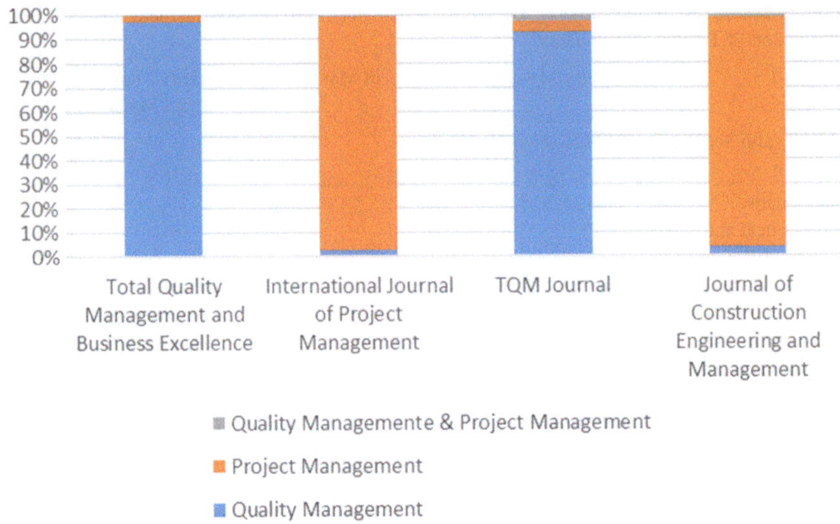

Fig. 5 Four reviews—'Total Quality Management and Business Excellence', 'International Journal of Project Management', 'TQM Journal' and 'Journal of Construction Engineering and Management' and three subjects—'Quality Management', 'Project Management' and 'Quality Management & Project Management'. *Source* Scopus (2020)

As can be seen in Fig. 5, in general and as expected, there is almost complete exclusivity in the topics of analysis depending on the magazine's target. Thus, 'Total Quality Management and Business Excellence' mostly publishes on issues of Quality (97.5%) and, to a lesser extent, Projects (2.2%) and Quality and Projects (0.3%). The 'International Journal of Project Management' magazine mostly publishes on projects (96.8%) and, to a lesser extent, on quality management (2.7%) and on projects and quality management (0.5%). 'TQM Journal' is the magazine that, although mostly publishing on quality management issues (92.5%), dedicates a higher percentage of its publications to project issues (4.8%) and quality and projects (2.7%). 'Journal of Construction Engineering and Management' mostly publishes on Projects (95.0%) and, to a lesser extent, on Quality (4.0%), and Projects and Quality (1.0%).

5 Conclusions

Quality is one of the essential aims of PM (Basu 2014). However, QM in PM is not always dealt with in-depth, resulting in severe problems in implementation and project delivery. For example, it has been proven that the lack of standardisation or useful instructions on the production lines has been generating serious quality problems (Liu 2016) and, consequently, affecting project success.

The objective of this work was to examine the relationships between QM and PM. We wanted to know the possible similarities between the project management models and the models of QM; it was also a question of procuring the constraints in the investigation of the 'Quality Management' and 'Project Management' structures.

To know the possible similarities in the approaches to PM and QM, the PM construct was analysed. The key conclusion that can be drawn from the study of this construct is that, at present, the PM is flexible to make it more adaptable to the current contexts of changes (Verga Matos et al. 2019). In this sense, the literature has proposed some models (Gil and Mataveli 2018) that are based on leadership and learning culture. Diversely, the models or techniques that allow the development of quality, for example, TQM, also suggest that for the implementation of quality principles, leadership and learning are fundamental tools.

In order to verify the differences in the investigation of 'Quality Management' and 'Project Management' structures, an analysis of the Scopus database was carried out. From this analysis, it was found that both types of research in quality management and project management have increased in recent years. However, growth in research on PM is more constant than on QM. In contrast, there are no numerous investigations that address in depth the quality of PM. In this context, the approach to quality in project management is found in the journals more related to PM than those related to QM. This result could indicate a hollow in the investigation, in the sense that from the standpoint of quality management, it would be necessary to investigate in more depth the quality in the management of projects.

In addition, one of the underlying purposes of this work was to propose strategies or tools that allow the development of both PM (e.g. Lehtinen and Aaltonen 2020) and quality management (e.g. Nical and Anysz 2020). At the same time, some proposals (Basu 2014) and tools (Cotrim et al. 2018) that could help facilitate quality in project management have been pointed out.

References

Abbas J (2019) Impact of total quality management on corporate green performance through the mediating role of corporate social responsibility. J Clean Prod 242:

Abbas J (2020) Impact of total quality management on corporate sustainability through the mediating effect of knowledge management. J Clean Prod 244:

Alofan F, Chen S, Tan H (2020) National cultural distance, organizational culture, and adaptation of management innovations in foreign subsidiaries: a fuzzy set analysis of TQM implementation in Saudi Arabia. J Bus Res 109:184–199

Atkinson R (1999) Project management: cost, time and quality, two best guesses and a phenomenon, its time to accept other success criteria. Int J Project Manag 17(6):337–342

Bahadorestani A, Naderpajouh N, Sadiq R (2020) Planning for sustainable stakeholder engagement based on the assessment of conflicting interests in projects. J Clean Prod 242:

Bamford DR, Greatbanks RW (2005) The use of quality management tools and techniques: a study of application in everyday situations. Int J Qual Reliab Manag 22(4):376–392

Basu R (2011) FIT SIGMA: a lean approach to building sustainable quality beyond Six Sigma. Wiley, Chichester, UK

Basu R (2014) Managing quality in projects: an empirical study. Int J Project Manag 32:178–187
Chofreh AG, Goni FA, Ismail S, Shaharoun AM, Klemes JJ, Zeinalnezhad M (2016) A master plan for the implementation of sustainable enterprise resource planning systems (Part I): concept and methodology. J Clean Prod 136(Part B):176–182
Cotrim SL, Filho DAM, Leal GCL, Galdamez EVC (2018) Implementation of cleaner production along with quality management tools. Int J Technol Manag Sustain Dev 17(1):65–85
Cropanzano R, Walumbwa FO (2010) Moral leadership: a short primer of competing perspectives. In: Schminke M (ed) Managerial ethics: managing the psychology of morality. Psychology Press/Routledge/Taylor & Francis, New York, pp 21–52
DeCarlo D (2004) eXtreme project management. Jossey-Bass, San Francisco
Deming WE (1994) Out of the crisis. The MIT Press, Cambridge, MA
Demirkesen S, Ozorhon B (2017) Impact of integration management on construction project management performance. Int J Project Manag 35:1639–1654
Di Maddaloni F, Davis K (2017) The influence of local community stakeholders in megaprojects: rethinking their inclusiveness to improve project performance. Int J Project Manag 35(8):1537–1556
Djendel M, Allaoui O, Bouzid A (2016) Effect of air plasma spraying parameters on the quality of coating. Int J Comput Exp Sci Eng 2(2):1–5
Dvir D (2005) Transferring projects to their final users: the effect of planning and preparations for commissioning on project success. Int J Project Manag 23(4):257–265
Eskerod P, Huemann M (2014) Managing for stakeholders. In: Turner R (ed) Gower handbook of project management. Aldershot, England, Gower, pp 217–232
Fernandez D, Fernandez J (2008) Agile Project Management—Agilism versus traditional approaches. J Comput Inf Syst 49(2):10–17
Fonseca LM (2015) From Quality gurus and TQM to ISO 9001:2015: a review of several quality paths. Int J Qual Res 9:167–180
Francisco de Oliveira G, Rabechini R Jr (2019) Stakeholder management influence on trust in a project: a quantitative study. Int J Project Manag 37:131–144
Gemuenden HG (2015) Success factors of global new product development programs, the definition of project success, knowledge sharing, and special issues of Project Management Journal. Proj Manag J 46(1):1–9.
Gil AJ, Mataveli M (2018) Project management and learning: the learning project. In: Otero-Mateo M (ed) Human capital and competences in project management. Intechopen, London, pp 45–69
Golini R, Landoni P (2014) International development projects by non-governmental organizations: an evaluation of the need for specific project management and appraisaltools. Impact Assess Project Apprais 32(2):121–135
Hernandez DJD, Aspinwall E (2007) Improvement methods in UK and Mexican construction industries: a comparison. Qual Reliab Eng Int 23(1):59–70
Ladewski BJ, Al-Bayati AJ (2019) Quality and safety management practices: the theory of quality management approach. J Saf Res 69:193–200
Lehtinen J, Aaltonen K (2020) Organizing external stakeholder engagement in inter-organizational projects: opening the black box. Int J Project Manag 38:85–98
Levine DI, Toffel MW (2010) Quality management and job quality: how the ISO 9001 standard for quality management systems affects employees and employers. Manag Sci 56:978–996
Liu J (2016) Research on the production quality risk management of the assembly type building products: taking Jinan as an example. Constr Econ 37(11):114–117
Ma Y, Zhang Q, Yin H (2020) Environmental management and labor productivity: the moderating role of quality management. J Environ Manag 255:
Momeni MA, Yaghoub S, Aliha MMR (2019) An optimal control model for analyzing quality investment in the project management. Comput Ind Eng 129:529–544
Meredith JR, Mantel SJ (2003) Project management. Wiley, NJ, USA

Nair A (2006) Meta-analysis of the relationship between quality management practices and firm performance—implications for quality management theory development. J Oper Manag 24(6):948–975

Nicał A, Anysz H (2020) The quality management in precast concrete production and delivery processes supported by association analysis. Int J Environ Sci Technol 17:577–590

Note M (2015) Project management for information professionals. Chandos Publishing, Langford Lane, UK

Ooi KB (2014) TQM: a facilitator to enhance knowledge management? A structural analysis. Expert Syst Appl 41:5167–5179

Peng X, Prybutok V, Xie H (2020) Integration of supply chain management and quality management within a quality focused organizational framework. Int J Prod Res 58(2):448–466

Picciotto R (2019) Towards a 'New Project Management' movement? An international development perspective. Int J Project Manag (in press)

Rowlinson S, Cheung YKF (2008) Stakeholder management through empowerment: modeling project success. Constr Manag Econ 26(6):611–623

Scaringella L, Burtschell F (2017) The challenges of radical innovation in Iran: Knowledge transfer and absorptive capacity highlights—evidence from a joint venture in the construction sector. Technol Forecast Soc Change 122:151–169

Scopus (2020) Document search. https://www.scopus.com/search/form.uri?display=basic

Shenhar A (2008) Unleashing the power of project management. Ind Manag 50(1):1–7

Sila I (2007) Examining the effects of contextual factors on TQM and performance through the lens of organizational theories: an empirical study. J Oper Manag 25:83–109

Singh PJ, Power D, Chuong SC (2011) A resource dependence theory perspective of ISO 9000 in managing organizational environment. J Oper Manag 29(1–2):49–64

Spundak M (2014) Mixed agile/traditional project management methodology—reality or illusion? Procedia Soc Behav Sci 119:939–948

Turner R, Huemann M (2002) Managing quality. In: Turner R, Simister S (eds) Gower handbook of project management. Gower Publishing, Brookfield, VT, pp 315–332

Turner R, Zolin R (2012) Forecasting success on large projects: developing reliable scales to predict multiple perspectives by multiple stakeholders over multiple time frames. Project Manag J 43:87–99

Verga Matos P, Romao M, Miranda Sarmento J, Abaladas A (2019) The adoption of project management methodologies and tools by NGDOs: a mixed methods perspective. J Bus Res 101:651–659

Walaszczyk A, Polak-Sopinska A (2020) The role of leadership in organizations managed in conformity with ISO 9001 Quality Management System Standard. In: Kantola JI, Nazir S (eds) Advances in human factors, business management and leadership. Springer, Switzerland, pp 402–491

Williams P, Ashill NJ, Naumann E, Jackson E (2015) Relationship quality and satisfaction: customer-perceived success factors for on-time projects. Int J Project Manag 33:1836–1850

Winch GM (2004) Managing project stakeholders. In: Morris PWG, Pinto JK (eds) The Wiley guide to managing projects. Wiley, New York, pp 321–339

Wysocki RK (2014) Effective project management: traditional, agile, extreme, 7th edn. Wiley, Indianapolis

Yu GJ, Park M, Hong KH (2020) A strategy perspective on total quality management. Total Qual Manag Business Excell 31(1–2):68–81

Zhou X, Lee CY (2000) Quality management and manufacturing strategies in China. Int J Qual Reliab Manag 17(8):876–899

A Quality Management and Excellence Philosophy from an Islamic Standpoint

Hesham Al-Momani

Abstract Heavenly religious beliefs urged best and truthful authentic practices related to mastering business and economics important dimension of culture, they typically neglected the influence of religion, Islamic primary sources, Qur'an verses, prophetic and holy hadiths (Sunnah) narrations Islamic scholars and Islamic history. To reveal the Islamic total quality management policies and applications in Islamic culture, doctrine, and ideology, this research study and adopts the inductive and descriptive approach to follow the most prominent principles of quality management in Islam's primary sources strengthened with real application cases, and the deductive approach to extract the most prominent principles of quality management derived from them with compared to the modern quality philosophies. Quranic texts and prophetic and holy hadiths declared fourteenth centuries ago the total quality management in Islam culture, doctrine, and ideology and were the first to talk about quality and ordered its application through. The principles of quality management in Islam are in line with the best principles of quality management currently applied, and the ideas of opponents and philosophers of quality are nothing new about Muslims culture and also required by the moral values of the Islam and find a high similarity between the two in the manner, goal, and outcome.

Keywords Total quality management · Islam ideology · Quality gurus · Quality

1 Introduction All Religious

All the heavenly religious urged matters related to mastering business and economics. Nowadays, the quality is considered as an organizational and societal culture followed by a translation that appears on behavior and practices where they are supported by solid religious belief.

H. Al-Momani (✉)
Department of Industrial Engineering, The Hashemite University, Zarqa, Jordan
e-mail: heshamalmomani@hu.edu.jo

J. L. García Alcaraz et al. (eds.), *Techniques, Tools and Methodologies Applied to Quality Assurance in Manufacturing*,
https://doi.org/10.1007/978-3-030-69314-5_7

1.1 Religious Effect on the Economy

Recent empirical research on the determinants of economic growth has typically neglected the influence of religion, and they found that religious beliefs influence individual traits that enhance economic performance (Barro and McCleary 2003). In a study done by Campante and Yanagizawa-Drott (2015), "*Does Religion Affect Economic Growth and Happiness*," they found that some results underscore that religious practices can affect individual behavior and beliefs. Robert and McCleary (2003) found regression suggests that greater belief instills (or goes along with) a stronger work ethic. Many other researchers mentioned that different religious views around the world might translate into a competitive business advantage (Birkner et al. 2018). Barro and McCleary (2003) find that economic growth responds positively to the extent of religious beliefs results in accord with a perspective in which religious beliefs influence individual traits that enhance economic performance. Many scientists claim that enlightenments for economic development should go further to include a nation's culture such as (Huntington 1996; Inglehart and Baker 2000; Landes 1999), while (Weber 1905, 1930) viewed religiosity as an independent variable that could influence economic outcomes to affect the economy by fostering traits such as work ethic, honesty (and hence trust).

McCleary and Barro (2006) study the effects of individual characteristics, such as work ethic, honesty, and thrift, and thereby influences economic performance; however, in Durlauf et al. (2012) study, "Is God in the details? A Reexamination of the Role of Religion in Economic Growth" found little evidence survives that religious variables help to predict cross-country income differences.

2 General Introduction: Islam as a Religion

Islam has many qualities that God deposited with this seal religion, one of the most important of which is inclusion. Islam came as a comprehensive approach to religion and the world, and it includes sound legislation and unique systems that enable it to fulfill the role assigned to guiding people and leading them in every time and place until they inherit God the Earth and those on it. The administrative systems that emerge from the soil of this eternal law come in this context, and these systems have played. Islam as a religion is based mainly on two sources; the first source is the holy Quran—highly esteemed literary source, written in Arabic—which represents the verbatim word of God (Allah) and final revelation to humanity taking into consideration that the Arabic language has a complex and unusual morphology and this makes it explained in many ways. The second source Islam based on is Sunnah (Prophet Muhammad sayings). Sunnah includes interpretations of the holy Quran, teachings, and way of life of the prophet (Al-Kabi et al. 2013) and fulfilled the noble prophetic traditions that alert and draw attention to this matter.

3 Quality Definitions

Quality has many definitions provided according to different researchers and scientists' perspectives. Researchers found that for modern quality definitions, everyone has his insight into the quality, and all of them did not agree on a satisfactory definition. The quality meaning changed from time to time gurus like conformance to requirements. Major quality gurus and others define quality as follows: A predictable degree of dependability and uniformity and product quality defined by the customer and may change in meaning depending on their needs (Shewhart and Deming 1986). Fitness for use and products or service free form deficiencies (Juran 2003) and conformance to requirements (Crosby 1979). Shewhart and Deming (1967), Feigenbaum (1983), Ishikawa (1986), and Taguchi (1986) have made a new differentiation to define "quality." Until then, the quality viewed in a positive approach. Taguchi looks at quality from a negative angle, and a new angle that causes the loss, Taguchi, 1986).

Modern quality is defined in many ways. The American National Institute of Standards and the American Society for Quality Control define quality as "meaning a set of features and characteristics of goods and services capable of meeting specific needs." While the American Federal Quality Institute defines it as "performing the right work correctly from the first time by relying on the recipient's evaluation to know the extent of performance improvement." Cohen and Brand (1993) define it as follows "Management: means developing and maintaining the ability of the organization to improve quality continuously." Gurus general approaches for quality: reduce variability by continuous improvement (Shewhart and Deming 1986), general management approach to quality—especially "human" elements (Juran 2003), prevention, not inspection (Crosby 1979).

Quality defined in Arabic according to Quranic texts and prophetic and holy hadiths called (Sunnah), they were the first to talk about quality and ordered its application through. Therefore, the ideas of opponents and philosophers of quality are nothing new about us Muslims societies. Quality definition in Arabic literature is explained according to different perceptions since the Arabic language is precious, and it is the holy Quran language. Quality in the Arabic language has many meanings, and synonyms like doing things in the best way, i.e., better, and its origin is (goodness), indicating excellent comprehensive performance. It also means mastery, as well as high levels of excellence and creativity. It is the quality of the product, and it is intended to improve the product: making it good (Al-Jawhari 1990, 1/107), and the best of things: against the bad (Al-Zubaidi 1994a, b). Quality among Arabs is an alternative to goodness (beneficence or philanthropy), but it is the end of goodness if its tapes are complete (Al-Ajlouni, Quality Conference 2006, p. 4). Ultimately quality is the result of concern mainly with excellence.

Quality evolution at early twenty-first century, modern quality management and industrial management philosophies and gurus approach mainly aimed toward more significant quality and customer satisfaction and inspection of the product to, and in early stages (the evolution of mass production early twentieth century) to a strategy and essential tool for an organization was found that no clear cut path made to put all the concepts in order, (Kumar et al. 2016), and according to Dahlgaard-Park (2008) evolved from a narrow. Mechanic perspective starting with inspection, moving to statistical process control to a broader and holistic one, known as TQM and business excellence, and where the concept of total quality management (TQM) suggests that everybody in the organization should be involved to improve their skills, knowledge, and perfection to satisfy customers) (Kumar et al. 2016). Modern quality management and industrial management philosophies and gurus' scientists' main themes like Copley and Taylor (1923), who sought to improve industrial efficiency. "Frederick W. Taylor: the father of scientific management application in the industry."

Henry Ford was the first to apply quality practices into operation. Shewhart and Deming (1967)—the father of statistical quality control, founder, and developer of statistical quality control charts, which helps in monitoring a process and to understand what is happening in it. If one can establish a standard for quality, using SQC charts, it becomes easy to estimate the progress of a product or a service, and W. Edwards Deming, 1986, Management Obligations and Pride of Workmanship, stated that a customer is a source to define the quality of a service or a product. Moreover, he clarifies that uniformity in achieving the customers' expectations about quality is the original hypothesis of quality management. Furthermore, he focuses on continuous improvement, which is described by his plan-do-check-act cycle (also called a quality cycle) and who formulated fourteen steps for quality management, and his theory improvement basis continuous to reduce variation. Eliminate goals without methods. Statistical process control (SPC) statistical methods of quality control must be used.

Deming, ideology focuses on "pride in workmanship." Summarily, on the management obligation on (1) constancy in improvement, (2) continuous innovation, (3) education and training, (4) workers' right to take pride in workmanship, and (5) continuity is the process of transformation. Deming's plan-do-check-act (PDCA) cycle initially, instead of Shewhart's plan-do-show-act (PDSA) cycle: (1) Plan: to decide the most effective initiative to work, (2) Do: to follow the plans established in the first stage, (3) Check: to ensure the effects of second stage, and (4) Act: to modify any deviation occurs of attention for managers, and although Deming does not use the term in his book, credited with launching the total quality management movement (Cohen et al. 2011). Create constancy of purpose toward product improvement and service to become competitive and to stay in business.

- Adopt the new philosophy. We are in a new economic age. Western management must cope with the challenge, must learn their responsibilities, and take on leadership for change.
- Cease dependence on inspection to achieve quality. Eliminate the need for massive inspection by building quality into the product in the first place.
- End the practice of awarding business based on a price tag. Instead, minimize total cost. Move toward a single supplier for any one item, on a long-term relationship of loyalty and trust.
- Improve always and forever the system of production and service to improve quality and productivity, and thus consistently decrease costs.

- Institute training on the job.
- Institute leadership (see Point 12 and Chap. 8 of Out of the Crisis). The aim of supervision should be to help people and machines and gadgets do a better job. Supervision of management requires an overhaul as well as supervision of production workers.
- Drive out fear so that everyone may work effectively for the company. (See Ch. 3 of Out of the Crisis)
- Break down barriers between departments. People in research, design, sales, and production must work as a team to foresee problems of production and usage that encountered with the product or service.
- Eliminate slogans, exhortations, and targets for the workforce asking for zero defects and new levels of productivity. Such exhortations only create adversarial relationships, as the bulk of the causes of low quality and low productivity belong to the system and thus lie beyond the power of the workforce.
- Eliminate work standards (quotas) on the factory floor. Substitute with leadership and eliminate management by objective. Eliminate management by numbers and numerical goals. Instead, substitute with leadership.
- Remove barriers to workers of his right to pride of workmanship. The responsibility of supervisors must be changed from sheer numbers to quality.
- Remove barriers that rob people in management and the engineering of their right to pride of workmanship. This means, among other things, the abolishment of the annual or merit rating and management by objectives.
- Institute a vigorous program of education and self-improvement.
- Put everybody in the company to work to accomplish the transformation. The transformation is everybody's job.

Crosby (1979), zero defects and quality are a free concept, associated with the zero defects, which makes the employees make a zero defect program to build an error-free work performance so that a product can be produced with zero defects. He is also famous for his book "Quality Is Free" which is the best seller in the management field, come up with the concept of Quality Is Free by doing it right the first time and conforming to standards. He also raised the need for cost of quality measurement and top management involvement. His quality theory initiated with fourteen steps to quality improvement, improvement basis a "process," not a program. Statistical process control (SPC) rejects statistically acceptable levels of quality that can be improved by looking at the customer's need, and his main consent was in solving a problem and elimination of the causes to those problems, and Crosby known for his important concept "zero defects"' and also a globally known quality leader. e Crosby's approach. The following concepts best understand philosophies toward quality. Moreover, finally, Do It Right First Time-Do It Right Every Time (DIRFT-DIRECT) prevention is the path that leads to achieving the quality Crosby found that it is a prevailing attitude of managers to believe that errors are inevitable, and a standard part of business life and one need to cope up with that.

The zero defects are a high level of error-free work performance. The zero defects program demands the employees and employers to make zero defects a slogan.

The employees take it like a slogan; they digest it and build a no-defect production culture—the four absolutes of quality. The four absolutes are: (1) Quality means conformance to requirements, not goodness. (2) Quality is achieved by prevention, not an appraisal. (3) Quality has a performance standard of zero defects, not acceptable quality levels, and (4) Quality is measured by the price of non-conformance, not indexes.

Where the 6 C's concept that every organization should undergo through a process to maintain a zero defect product or a service that he calls it as 6 C's: (1) Comprehension, which means every employee must understand what quality is. (2) Commitment, which represents each representative in the organization from labor to manager. (3) Competence; education and training to implement the quality improvement. (4) Communication; by this, the understanding of quality by all the people can be achieved. (5) Correction, which focuses on prevention and performance. Finally, (6) Continuance, which emphasized that the process must become a way of life in the organization.

Juran (2003), The Quality Trilogy on three pillars, viz., quality planning, quality control, and quality improvement, and the core ideology is elimination of causes to a problem which minimized the product quality and stated quality as fitness for use; which implies product must be designed to reach the customer's need his theory stands on ten steps to quality improvement basis and project-by-project team approach. Set goals. Statistical process control (SPC) recommends SPLC but warns that it can lead to too driven.

Ishikawa (1986), the one step further, quality circles and cause and effect diagrams are the contributions of which serves in finding out the principle causes and their effect on a problem and solving these problems using a quality circle. Made major contributions for the cause and effect diagram use and preached for the use of quality control at all levels of the organization and the notion of internal customer. Ishikawa's theory focused that quality improvement is a continuous process, and it can always take one step further to stress the importance of total quality control of an organization, rather than just focusing on products and services transferring the Deming's PDCA cycle into a six-step plan. Moreover, introduce seven quality tools, including a control chart, a run chart, a histogram, a scatter diagram, a Pareto chart, and a flowchart. Besides, another great contribution of Ishikawa is the cause and effect diagram, fishbone diagram, and quality circles small teams consisting of employees, who are directed to discuss and solve the quality problems.

Feigenbaum (1983), the hidden plan behind TQC, proposes total quality control as a management tool, which is the most vital element that leads to the growth of the organization. These can be achieved by total attention toward quality and directing this quality as a responsibility, Feigenbaum—Hidden Plant behind TQC theory diagnosing all the customer needs and requirements should be the initial point of the quality chain and ends when the customer remains satisfied after receiving the product or service. Feigenbaum describes quality control as a management tool to achieve quality through the following four steps: (1) Design the quality standards, (2) appraise conformance to the standards, (3) act when standards are exceeded, (4) planning for improvement in the standards. Total quality control is impelling ideology by Feigenbaum; Total: that every individual in the organization maintains the total attention and involvement to ensure quality. Quality: "Quality is, in its essence, a way of managing the organization." Control: control as a responsibility devolution. Feigenbaum defined control as a process Shingo (1990), the Make Mistake-Proof, Not Foolproof (Poka-Yoke). Just-In-Time (JIT), concepts originator, the concept of SMED, Poka-Yoke, and JIT was led by. SMED concept minimizes the changeover of dies and the cost associated with the setup. Poka-Yoke emphasizes 100% inspection to ensure zero defects, and JIT highlights on storing a small quantity of stock which is required for the production, Shigeo Shingo, father of Make Mistake-Proof, Not Foolproof (Poka-Yoke). Just-In-Time (JIT), Shingo is associated with three famous equations.

SMED: SMED led the most effective way to reduce the fixed cost associated with the setup and changeover of dies. Poka-Yoke where it ensures 100% inspection so that no defective products are passed to the stage of the next process. Furthermore, JIT, JIT concept focuses on keeping a small quantity of stock required for the production or a process that can be used with the daily or hourly scheduled delivery fails.

Taguchi (1986): The Practical, Rapid Quality Methods owner, including quality robustness, quality loss function of experiments which is used to optimize the process or product design. He made a different approach in defining the quality, which can be stated simply as "the loss incurred by the society responsible for the loss function concept, the signal-to-noise ratio, and the orthogonal design of experiments methods, in addition to the importance of robust designs and (ASI Mourns 2012) quality function deployment (QFD) (Akao 1994). Taguchi (1986): Practical, Rapid Quality Methods, who revolutionized the manufacturing process in Japan through cost savings and increasing productivity, and he regarded experimental design or design of experiments (DoE) as an important tool for enhancing quality and depend mainly on the following three conceptual features: (1) Quality loss function (QLF) or Taguchi loss function (TLF), (2) quality robust design (QRD) through the design of experiments (DoE), (3) noise and signal-to-noise ratio (SNR), and (4) quality loss function (QLF) or Taguchi loss function (TLF).

Then, holistic initiatives appeared to enhance processes efficacy and product improvement like ISO 9000 series, quality management system (QMS). Toyota Production System (Ohno 1988), Six Sigma (Tennant 2001), design of experiments (Sifri 2014) and failure mode and effects analysis (FMEA) (Marvin and Arnljot 2004), plan-do-check-act cycle (Tague 2005), lean manufacturing (Holweg 2007). total quality management (Ciampa 1992), Top-Down & Bottom-Up Approaches—Leadership approaches to change (Stewart et al. 2015), Kaizen change for the better and make small improvements. Each of these approaches and methods has been met with success but also with failures (Masaaki 1986).

Gurus have a different perspective for the degree of senior employees' effect on quality responsible for 94% of quality problems (Shewhart and Deming 1986), less than 20% of quality problems are due to worker's responsibility (Juran 2003), and responsible for quality (Crosby 1979).

Moreover, for performance standards/motivation, quality has many scales. Use statistics to measure performance in all areas. Critical of zero defects. (Shewhart and

Deming 1986) Less than 20% of quality problems are due to worker's responsibility Avoid campaigns to do perfect work (Juran 2003), Zero defects.

Teamwork has been emphasized where employee participation in decision-making. Deming (1986), team and quality circle approach (Juran 2003). Quality improvement teams. Quality councils (Crosby 1979).

4 Quality in Islam

The administrative approach in Islam is characterized by flexibility and practice. The vision of a quality management system is comprehensive with an Islamic perspective that enhances the value of this system and increases the chances of its application in Islamic societies. Islamic law aimed at preserving the human psyche from all threats and harms. Prevent the danger of the individual and society, and maintain high life quality safety, and preventive and therapeutic methods can achieve that. Islam and new quality science meet where quality concepts in Islam enhance immaculate life. Muslims must employ all their means possible to live an immaculate life and urges to take advantage of modern scientific developments. Islam is called for mastering, improving, and improving the work fourteen centuries ago, and the importance of quality, cleanliness, and fastidiousness is emphasized in the Qur'an; Islamic perspective enhances the value of this system and increases the chances of its application in Islamic societies, and Islam core value that humans are incapable of composing work comparable to the Quran. (Clarke and Cobham 1990): God drew it to harness everything in the universe for a person to rebuild the earth according to his wisdom and will, Almighty said ([It is] the work of Allah, who perfected all things. Indeed, He is Acquainted with that which you do) An-Naml, 88.

Al-Qur'an al-Kareemقال تعالى: "صنع الله الذي أتقن كل شيء" النمل (8)

Also, In the Qur'an, the description of God Almighty said, "Who perfected everything which He created and began the creation of man from clay. As-Sajdah, 7, and this is evidence of knowledge, wisdom, and mastery of making.

قال تعالى: "الذي أحسن كل شيء خلقه" السجدة (7) Al-Qur'an al-Kareem

From the grace of God to the creativity of his creation, God requires a person to be a creative benefactor in his work and profession, whatever it is. In the Almighty saying:

"We have created man in the best of evaluation, or We have indeed created man in the best of stature (Al-tin, 4). A clear indication of the quality of creation and the benevolence and creativity of the Creator

قال تعالى: "لَقَدْ خَلَقْنَا الْإِنسَانَ فِي أَحْسَنِ تَقْوِيمٍ "التين (4) Al-Qur'an al-Kareem

Quality in Islam considered as ideology doctrine and standard lifestyle and exact Muslim model, law and morals—and philosophy for the universe, and its completeness of culture originating from the Qur'an and the Sunnah as seen in art, science, politics, and everywhere else in the Muslim world, and through this we can understand the unlimited dimension in the Almighty saying: This day I have perfected for you your religion and completed My favor upon you and have approved for you Islam as religion (Al-Ma'idah: 3).

قال تعالى: "الْيَوْمَ أَكْمَلْتُ لَكُمْ دِينَكُمْ وَأَتْمَمْتُ عَلَيْكُمْ نِعْمَتِي وَرَضِيتُ لَكُمُ الإِسْلاَمَ دِيناً "المائدة: (3) (Al-Qur'an al-Kareem

Islam is the perfection of quality and creativity, which is the religion of God [Originator of the heavens and the earth. When He decrees a matter, He only says to it, "Be," and it is. (Al-Baqarah, 117).

قال تعالى: "بَدِيعُ السَّمَاوَاتِ وَالأَرْضِ وَإِذَا قَضَى أَمْراً فَإِنَّمَا يَقُولُ لَهُ كُن فَيَكُونُ (١١٧)-Al-Qur'an al-Kareem

5 Lifestyle

Quality in Islam is a lifestyle. Islam was the first and greatest pioneer in laying and building the right and sound foundations that work to build a strong community, and Islam represents the culture and concepts of quality management in daily life affairs through responsibility, accountability, accountability, and the degree of work performance and mastery, and represents a "true Muslim model." That respect science and adhere to apply it in all aspects of life entirely legitimate and considered as lifestyle spread over a full sphere involving all aspects of Muslim communities life, and this lifestyle devoid of quality can form a person's way of thinking to the pleasure they take from life و before quality slogans launched in recent centuries.

Quran establishes the principle of knowledge: Since the Noble Qur'an revealed its verses to Muslims and affirms the necessity of knowledge and its importance in the lives of Muslims, God Almighty says: Allah will raise those who have believed among you and those who were given knowledge, by degrees. "(Al-Mujadila, 11), and the prevalence of the importance of knowledge and its necessity in the lives of individuals is one of the requirements to achieve total quality.

قال تعالى: "يرفع الله الذين آمنوا منكم والذين أوتوا العلم درجات" (المجادلة، أ Al-Qur'an al-Kareem (11)

Moreover, the prevalence of the importance of knowledge and its necessity in the lives of individuals is one of the requirements for achieving total quality. Also established the principle of wisdom: which is knowing the facts of things as they are, and working according to them as in the Almighty saying, “He gives wisdom to whom He wills, and whoever has been given wisdom has certainly been given much good. And none will remember except those of understanding” (Al-Baqarah: 269),

قال تعالى: “يؤتي الحكمة من يشاء ومن يؤت الحكمة فقد أوتي خيرا كثيرا " (البقرة (269) Al-Qur'an al-Kareem

Injury in word and action, and the concept of Islamic wisdom, gives many indications to the concept of comprehensive Islamic quality and enriches it and enriches and strengthens it. Wisdom is a higher degree of knowledge as it includes knowing the truth and working with it, and injury in saying and doing, the concept of Islamic wisdom lends many meanings to the concept of comprehensive Islamic quality, enriches and strengthens.

6 Differences Between Islamic Quality and Modern Quality

Nevertheless, modern quality concepts meet with the Islamic quality instruction, and there are differences between Islamic quality and modern quality, some other concepts mentioned by Islamic sources, first, fundamental difference between the principles is that the goal of quality management in Islam, and the satisfaction of the beneficiary is restricted and linked to the satisfaction of God and its axis for others. Second, quality management in Islam is characterized by the power of the self-censorship of workers. Even if the external sensor is absent, the worker who feels God’s observation will not do anything that harms the institution or causes weak outputs. The third is a cooperation between employees to achieve a common goal considered as prayer and worship act, wherein Hadith (1848) said: “God’s hand is with the congregation (Sahih al Jamie)” Darmakka (و2020).

قال رسول الله النبي محمدﷺ: "يد الله مع الجماعة " صحيح الجامع Sunnah

of individual self-censorship in group work through the achievement of individual responsibility in teamwork, and finally investing in the religious observance of the worker, which requires reducing the working hours of a device cumbersome and costly management and inspection.

The following paragraphs will show more details.

6.1 Quality Considered as Ethical and Religious Requirement

Quality considered as ethical and religious requirements, Prophet Mohammad Peace be upon him, and he said: "Allah will be pleased with those who try to do their work perfectly." God loves if one of you does a job to master it and That God loves us to master our work, (wata 2020) The Final path on the Quality journey is the Business Excellence Models. Like the (JUSE or Deming Prize) established in 1951 in Japan), The EFQM and Malcolm bridge USA all of them assess the performance, learning, and organizational culture, to be the best they can in all aspects by continuous improvement (Dahlgaard 2008). The principle of mastering work perfection and excellence and sincerity in it: Islam urged improving the quality of product and work to best and make this admired by Almighty God said: "God loves if one of you does a job to master it." From Sunnah, the Messenger, may God bless him and grant him peace, urged mastery of the work, and he said: God Almighty loves the worker if he does good. He, may God's prayers and peace be upon him, said: "If one of you does work, God loves to master it" (Al-Bayhaqi), so you have first to be knowing that God is watching you, and every single move you do is monitored,

قال رسول الله النبى محمدﷺ: "إن الله يحب إذا عمل أحدكم عملاً أن يتقنه"(البيهقي).Sunnah

so the conclusion of the Islamic message is the assignment of God to His servants and his demand for them to take the best and optimal position in all their movements and their dwellings, or their overall work and activity.

6.2 Comprehensive Approach and Letter, Quality in Islam (Qur'an and Hadeeth)

Islam is a comprehensive religion for all aspects of life good life, many attributes and conditions required in Islam from God Almighty, and all of them emphasize the necessity of quality performance. Mastery of practice and it is the same methodology adopted in every aspect of Islam to live an immaculate life; it becomes clear to us that the Noble Qur'an is a method of quality and perfection in the generalities of life, its branches, and its details. The Almighty said: (We have not neglected in the Register (Quran) a thing, and or And We did not lose anything in the book) al-an'am, 38

قال تعالى: " (ما فرطنا في الكتاب من شيء الانعام(38) Al-Qur'an al-Kareem

The core concepts of quality in Islam were not developed at the hands of Islamic researchers in the form of an integrated concept in the way that the West emerged and started with the rise of the Islamic community over 1400 years ago. Still, Islamic message is the assignment of God to His servants and his demand for them to take the best and optimal position in all their movements. And their dwellings, or their overall work and activity, calls for ascertaining the quality of the work that a person does, and is free from shortcomings and faults. The importance of quality, method of inclusiveness, quality and mastery in the generalities of life cleanliness, perfectionist and fastidious are emphasized and required by the moral values of the Quran. In Islam, the major quality pillar depends on cultural people factors since they are major change resistance factors, where they are the major factor behind any successes, about such people, Sufian Ath-Thawri narrated that As-Saudi commented, "A man might be thinking of committing injustice or sin. Nevertheless, he abstains when he is told, 'Have Piety (Taqwa) of Allah,' and his heart becomes fearful."

In particular, the term quality is one of the principles mentioned in the Holy Qur'an that has been linked to many vocabularies and concepts related to which are charity, perfection, judgment, reform, and others. We just have to suffice here to talk about one concept, which is the concept of philanthropy or charity or beneficence, where defined as (beneficence defined as an act of charity, mercy, and kindness with a strong connotation of doing good to others including moral obligation. All professionals have the foundational moral imperative of doing right) (Kinsinger 2009). Quranic texts and noble prophetic hadiths that have been transmitted in the individual desire, obedience, and action, through recommendations, and he finds that the purpose of the legislator in charge is not only to submit to and do the work, but only what is desired and required is careful work and charity work Islam is built on quality in every aspect of it, and it calls and incites perfection in all fields,

Charity in language: It is the act of what is good, and the best thing is good for making it, and it is mastery of the work that a Muslim does and exerts an effort to find it to be complete, so if the work is for people, it must be performed in full as if the employer is expert in this work and the worker follows closely: The Islamic message revealed God's assignment to His servants, required by the moral values of the Qur'an asking them to take the best and optimal position in all their movements and dwellings, and Islam considered as inclusiveness and the comprehensive religion. The comprehensiveness of the Islamic curriculum(Qur'an, Hadeeth, Islamic scholars and Islamic history) and its coverage of all aspects of life, all areas of work without allocating or specifying accompanied by invitation to quality and mastery to the extent of the extent and breadth itself, with full sound legislation and unique systems for all aspects of life, as was the fulfillment of the prophets 'messages before and says, peace and blessings be upon him: I was sent to uphold and complement ethical شي " I was sent to fulfill the morals",

- قال رسول الله النبي محمدﷺ: "إنما بعثت لأتمم صالح الأخلاق) رواه أحمد Sunnah

Due to its importance to the individual and family, society, and the nation, the vision of a comprehensive quality management system with an Islamic perspective enhances the value of this system. It increases the chances of its application in Islamic societies. Islam called for the application of quality and rationally ordered its implementation from the executive management from the bottom of the pyramid to the top, and this is known as comprehensive quality management and has clear indications of the application of the quality principle, including the following verses:

From the Quran

In the Qur'an, God Almighty says: God commands justice and charity with شـي Indeed, Allah orders justice and ethical conduct Surah Al-Nahl. 90

قال تعالى: " إِنَّ اللَّهَ يَأْمُرُ بِالْعَدْلِ وَالإِحْسَانِ سورة النحل(90) Al-Qur'an al-Kareem

The words of God Almighty, and do not spoil the earth after its reformation, and call him fear and greed, for God's mercy, is close to the benefactors, Almighty said: And cause not corruption upon the earth after its reformation.

Furthermore, invoke Him in fear and aspiration. Indeed, the mercy of Allah is near to the doers of good. (al-A'raf, 56

قال تعالى: " وَلاَ تُفْسِدُواْ فِي الأَرْضِ بَعْدَ إِصْلاَحِهَا وَادْعُوهُ خَوْفًا وَطَمَعًا إِنَّ رَحْمَتَ اللهِ قَرِيبٌ مِّنَ الْمُحْسِنِينَ)الاعراف(56) Al-Qur'an al-Kareem

The words of God and Hereafter and do not forget your share of the world and the best of the best of God as you do not mischief in the earth that God does not like spoilers, Almighty said "But seek, through that which Allah has given you, the home of the Hereafter; and [yet], do not forget your share of the world. Moreover, do good as Allah has done well to you. And desire, not corruption in the land. Indeed, Allah does not like corrupters." al-Qasas, 77

قال تعالى: " وَابْتَغِ فِيمَا آتَاكَ اللَّهُ الدَّارَ الْآخِرَةَ وَلَا تَنْسَ نَصِيبَكَ مِنَ الدُّنْيَا وَأَحْسِنْ كَمَا أَحْسَنَ اللَّهُ إِلَيْكَ وَلَا تَبْغِ الْفَسَادَ فِي الْأَرْضِ إِنَّ اللَّهَ لَا يُحِبُّ الْمُفْسِدِينَ)القصاص(77) Al-Qur'an al-Kareem

The words of God Almighty in Ali-Imran: So Allah rewarded them the reward of this world and the goodness of the reward of the Hereafter, and God loves the benefactors و Almighty said: So Allah gave good the reward of this world and the good reward of the Hereafter. Moreover, Allah loves the doers of good. Ali-Imran, 148

قال تعالى: " فَآتَاهُمُ اللهُ ثَوَابَ الدُّنْيَا وَحُسْنَ ثَوَابِ الآخِرَةِ وَاللهُ يُحِبُّ الْمُحْسِنِينَ آل عمران(**148**) Al-Qur'an al-Kareem

The words of God Almighty (in An-Nahl, 128): Allah is with those who fear and those who are good. Almighty said: Indeed, Allah is with those who fear Him and those who are doers of good. (An-Nahl, 128)

قال تعالى: " إِنَّ اللَّهَ مَعَ الَّذِينَ اتَّقَوْا وَالَّذِينَ هُمْ مُحْسِنُونَ.سورة النحل(128) Al-Qur'an al-Kareem

The words of God Almighty (in al-‘ankabut) and those who have struggled in us to guide them are our paths, and God is with the good, Almighty said: And those who strive for Us—We will surely guide them to Our ways. And indeed, Allah is with the doers of good al-‘ankabut. 69

قال تعالى: " وَالَّذِينَ جَاهَدُوا فِينَا لَنَهْدِيَنَّهُمْ سُبُلَنَا وَإِنَّ اللَّهَ لَمَعَ الْمُحْسِنِينَ سورة العنكبوت(69 Al-Qur'an al-Kareem (

The words of God Almighty (Ar-Rahman), is the reward for charity other than charity? Almighty said: Is the reward for good [anything] but good? Ar-Rahman, 60

قال تعالى: " هَلْ جَزَاءُ الْإِحْسَانِ إِلَّا الْإِحْسَانُ سورة الرحمن (60) Al-Qur'an al-Kareem

The words of God Almighty to those who do well and increase goodness, and do not overburden their faces with despair, nor is the humiliation of those who belong to Heaven: Almighty said: For them who have done good is the best [reward] and extra. No darkness will cover their faces, nor humiliation. Those are companions of Paradise; they will abide therein eternally, Yunus, 26

قال تعالى: " لِلَّذِينَ أَحْسَنُوا الْحُسْنَى وَزِيَادَةٌ وَلَا يَرْهَقُ وُجُوهَهُمْ قَتَرٌ وَلَا ذِلَّةٌ أُولَئِكَ أَصْحَابُ الْجَنَّةِ هُمْ فِيهَا خَالِدُونَ يونس (26) Al-Qur'an al-Kareem

In the Sunnah, the Prophet Muhammad said in the definition of charity: “To worship God as if you see him, and if you do not see him, he will see you.” “Narrated by Muslim and Ibn Majah.”

6.3 *From the Prophetic Hadiths*

The noble prophetic Sunnah came confirming what was stated in the Noble Qur’an in its affirmation of the principle of total quality through his sayings and honorable deeds in the following hadiths:

- God bless him and grant him peace, “You are all a shepherd, and you are all responsible for his flock” (Al-Albani 1986). (In this verse, God put the right rules of faith perception, and the rules of proper behavior of faith, determines the status of the truthful righteous, value perception and feeling and actions and behavior, establish their impact on the life of the individual and the group), in hadith Ibn’Umar reported that the Prophet, may Allah bless him and grant him peace, “All of you are shepherds, and each of you is responsible for his flock. A man is the shepherd of the people of his house, and he is responsible. A woman is the shepherd of the house of her husband, and she is responsible. Each of you is a shepherd, and each is responsible for his flock.” (Al-Bukhari 1978), so the hadith is a definitive indication of the importance of mastery in work, it is free from deficiency and defects, and the application of instructions and foundations so that it does all its details without negligence or fraud. Or deception, and this calls for sincerity at work.

قال رسول الله النبي محمدﷺ :"كُلُّكُمْ رَاعٍ، وَكُلُّكُمْ مَسْؤولٌ عَنْ رَعِيَّتِهِ، فَالأَمِيرُ رَاعٍ وَهُوَ مَسْؤُولٌ، وَالرَّجُلُ رَاعٍ عَلَى أَهْلِهِ وَهُوَ مَسْؤُولٌ، وَالْمَرْأَةُ رَاعِيَةٌ عَلَى بَيْتِ زَوْجِهَا وَهِيَ مَسْؤُولَةٌ، أَلاَ وَكُلُّكُمْ رَاعٍ، وَكُلُّكُمْ مَسْؤُولٌ عَنْ رَعِيَّتِهِ صحيح (الألباني) Sunnah

The Prophet Muhammad said in defining charity: "To worship God as if you see him, and if you do not see him, he will see you." "Narrated by Muslim and Ibn Majah," inherent continuous self-ethical control that saves a lot from modern control systems. Furthermore, The Prophet (ﷺ) said, "Ihsan (Highest degree of quality) is to worship Allah as if you see Him, and if you do not achieve this state of devotion, then (take it for granted that) Allah sees you."

قال رسول الله النبي محمدﷺ:" أن تعبد الله كأنك تراه فإن لم تكن تراه فإنه يراك "رواه مسلمSunnah

This hadith is an excellent indication that everyone in his responsibility has to master the requirements of his work to succeed in his role and accomplish his mission to achieve quality with what he was assigned to, from the smallest nucleus in society to the largest institution.

He said: ("God wrote good deeds on everything, and Verily Allah has prescribed Ihsan (proficiency, perfection) in all things". Book 12, Hadith 1382 https://sunnah.com/urn/2116440

قال رسول الله النبي محمدﷺ: "إنَّ الله كتب الإحْسَان على كلِّ شيء "رواه مسلم Sunnah

7 Mastering of Work as Worship

In Islam, the mastering of work is considered as worship, and mastery is defined as the source of a master's word, that is to say, the work done with precision and precision. Mastery of work: to do the work to be accomplished and to finish it in the best picture and the fullest, and that the effort, and distance from laxness in work. Again, in Islam, mastery of work considered as pure worship senses the control of God in every step taken in his work, in all its forms, its various permissible areas (i.e., quality seen in art, science, politics, and everywhere else in the Muslim world) and that the work done by his hands in the fullest and best case, it is a legitimate and sincere duty for every employee and worker whatever his job; this worship is rewarded by the Muslim if he does it and performs it with his right, and he is sincere in his work with all the strength and determination that Allah has given him. Many orders from Quran in Islam, Almighty said: "Work, and God will see your work and His Messenger and the believers, and you will be restored to the world of the unseen and the testimony of what is happening شي "Do [as you will], for Allah will see your deeds, and [so, will] His Messenger and the believers. Moreover, you will be returned to the Knower of the unseen and the witnessed, and He will inform you of what you used to do." At-tawbah, 105

قال تعالى: " وَقُلِ اعْمَلُوا فَسَيَرَى اللَّهُ عَمَلَكُمْ وَرَسُولُهُ وَالْمُؤْمِنُونَ وَسَتُرَدُّونَ إِلَى عَالِمِ الْغَيْبِ وَالشَّهَادَةِ فَيُنَبِّئُكُمْ بِمَا كُنْتُمْ تَعْمَلُونَ) التوبه (105) Al-Qur'an al-Kareem

And Almighty said, He has produced you from the earth and settled you in it, (He created you from the land and colonized you in it) Hud, 61

قال تعالى: " هُوَ أَنْشَأَكُمْ مِنَ الْأَرْضِ وَاسْتَعْمَرَكُمْ فِيهَا(هود (61)

God Almighty advancement by relating the work to the pray where the prayer has goals that must be achieved, and that is to elevate the human in the runways. Human perfection (chastity, integrity, ethics, and transparency) is validated by the Almighty saying: And from here was his saying, whose prayers on immorality and evil did not end, he did not increase from God except after Almighty said: Indeed, prayer (include the good work considered as prayer and worship) prohibits immorality and wrongdoing, Al-'Ankabut 45)

قال تعالى: " ان الصلاه تنهى عن الفحشاء والمنكر(العنكبوت(45) Al-Qur'an al-Kareem

Quality accepted and recognized and it must be understood as an essential part of Muslim life worship, the quality and means to achieve quality in product and services are of the essential features and considered as prayer and must of Muslim communities, and quality is, therefore, one of the essential features of Muslim communities.

Enticement, benevolence, and excellent model rewarded by the Muslim, Almighty god saying, "God has written benevolence on everything."

قال تعالى: " إن الله كتب الإحسان على كل شيء حديث Sunnah

Furthermore, mastery here requires one to perform his work to the fullest and to strive to reach him to the stage of human perfection so that he works with all his details without negligence, negligence, fraud or deception, and this calls for complete sincerity in work, and everything here benefits the general, benevolence higher than mastery. (Paschio 2006: 11). (باشيوة، 2006)

Also, The Almighty said while urging a person to be of good quality, charity, and creativity [Al-Anam: 38]

قال تعالى: " (ومن أحسن دينا ممن أسلم وجهه لله وهو محسن الأنعام: (38) Al-Qur'an al-Kareem

Also:

This is also assured in Sunnah that Islam came with full legislation for all aspects of life, as was the fulfillment of the prophets' messages before and says, peace and blessings be upon him: "I was sent to fulfill the morals,

قال رسول الله النبي محمدﷺ: " إنما بعثت لأتمم صالح الأخلاق)رواه أحمد. Sunnah

God urged that there should be a comprehensive understanding of good complete work (quality and excellence) and workers should perform their work as if God is Sergeant them At-Tawba 105

قال تعالى: " وَقُلِ اعْمَلُوا فَسَيَرَى اللَّهُ عَمَلَكُمْ وَرَسُولُهُ وَالْمُؤْمِنُونَ ۖ وَسَتُرَدُّونَ إِلَىٰ عَالِمِ الْغَيْبِ وَالشَّهَادَةِ فَيُنَبِّئُكُم بِمَا كُنتُم تَعْمَلُونَ التوبة (105) Al-Qur'an al-Kareem

8 Consulting (Shura) Principle in Islam

The quantity and quality of employee engagement define the level of effort required to move a business toward excellent performance. The beliefs, perceptions, and values would have engaged employees in reaching organization goal actively, and this is called al Shura in Islam literature.

The principle of Shura: Islam called for adherence to the principle of Shura through consultation with individuals in making decisions and solving problems, the Islam called for adherence to the principle of Shura through consulting with individuals in making decisions and solving problems, and this appears from the Almighty said: "And consult them about the matter" (Al-Imran, 159).

قال تعالى: " وشاورهم في الأمر " أل عمران، الآية:(159) Al-Qur'an al-Kareem

The advice here leads to good and quality work. So, the advice here leads to good and quality work. Islam order command consultancy as principal for organization successes provide a significant amount of value for an organization, The Almighty said and consulted with them on the matter, so if you intend to do so, trust in God ۚ God loves the one who trusts, The Almighty said and consulted with them on the matter, so if you intend to do so, trust in God ۚ God loves the one who trusts in the family of Imran. The Almighty said: "And those who have responded to their lord and established prayer and whose affair is [determined by] consultation among themselves, and from what We have provided them, they spend. Ash-Shuraa, 38

قال تعالى:" وَالَّذِينَ اسْتَجَابُوا لِرَبِّهِمْ وَأَقَامُوا الصَّلَاةَ وَأَمْرُهُمْ شُورَى بَيْنَهُمْ وَمِمَّا رَزَقْنَاهُمْ يُنْفِقُونَ "
الشورى(38) Al-Qur'an al-Kareem

8.1 Recognition and Rewarding Principles

There are many quality rewards to organizations like (EFQM—Excellence Award (Formerly the European Quality Award: European award for Total Quality Management), Deming-Award: Japanese award for Quality management since 1951, and Malcolm Baldrige National Quality Award: US-American Award for performance excellence created in 1987. In Islam, workers will be awarded by both God Almighty for good work and their leaders too. Islam reward and recognized good workers urges us to master the work, calls for improvement, quality, and proficiency in the work, and made for those who improve the work the best reward in the Holy Qur'an and a lot of noble hadiths, and reward the workers according to his law the best reward: "May God reward them with the best of what they were doing" [At-Tawbah: 121],

قال تعالى: "ويجزي العاملين بمقتضى شرعه أحسن الجزاء: " لِيَجْزِيَهُمُ الله أَحْسَنَ مَا كَانُوا يَعْمَلُونَ "
التوبة(121) Al-Qur'an al-Kareem

Furthermore, His Almighty said: indeed, we will not allow to be lost the reward of any who did well in deeds. (Al Kahf, 30) and Those who believe and do good works We do not waste the reward of the best work

قال تعالى: " إِنَّ الَّذِينَ آمَنُوا وَعَمِلُوا الصَّالِحَاتِ إِنَّا لَا نُضِيعُ أَجْرَ مَنْ أَحْسَنَ عَمَلًا الكهف (30) Al-Qur'an al-Kareem

8.2 Continuous Improvement Principles

Islam urges continuous improvement concept, the Qur'an also urged the human being to continuously develop and always ask for the reasons for perfection and gain from it: "Say, Lord, increase my knowledge" (Taha, 114)

قال تعالى: " وَقُلْ رَبِّ زِدْنِي عِلْمًا "طه(114) Al-Qur'an al-Kareem

Perhaps, the most prominent principle of total quality management that the Holy Qur'an preceded is what it benefits The Almighty said: | ("To whom would you, late or late")? Al-muddaththir 37

قال تعالى: " لِمن, شَاء منْكُم أَن يتَقَدم أَو يتَأَخَّر المدثر (37) Al-Qur'an al-Kareem

This verse precedes the adoption of the most prominent principles of total quality management, which is the principle: the necessity of continuous improvement and development, no higher educational institution should stop the continuation of development and improvement on the pretext that it has attained international certificates. Development, because if it stops development, it is necessarily delayed, so the universe does not know to stand, so it must make sure that it is progressing; otherwise, it will be delayed. The total quality management system lies not only in drawing attention to the necessity of implementing the system but also in caring for the continuity of improvement.

Quality improvement: we live in an ever-changing universe. Continue to improve and develop.

Ibn al-Qayyim said: (The slave is still in progress or delay, and there is no standing on the road at all if he is not in progress, then he is in delay and must) ([4]) No institution, company, or supreme body should stop the continuation of development and improvement on the pretext that it has attained the utmost in perfection and quality, and it has attained international certificates in that; because if it stops development, it is necessarily late, so the universe never knows the position, whoever does not progress will inevitably be delayed. Through what we talked about earlier, it becomes clear to us that the Holy Qur'an is a method of quality and proficiency in the generalities of life, its branches, and its details. The Almighty said: (We did not lose anything in the book) [Al-Anam: 38].

قال تعالى: " ما فرطنا في الكتاب من شيء" الأنعام (38) Al-Qur'an al-Kareem

Education that leads to continuous improvement has a significant role in quality management in Islam, and Islam as religion and philosophy of life, considered education that is universal in scope and nature as a basic necessity for the human being, Almighty Allah said in the Holy Quran 'God increases me educationally and request all prophets to seek augmentation of knowledge all fields as fundamental duty and social responsibility for every citizen.

Prophet Mohammed said: To gain knowledge is a duty of every Muslim (2 lbn-e-Maja, chapter: Fadlul Ulama wal hath AI talabil I1m, Hadith, no. 224, vol: 1. p: 81 narrated by AnasRA) and differentiated the person who has the knowledge and the person who is devoid of it, Almighty said: Say, "Are those who know equal to those who do not know?" Only they will remember [who are] people of understanding. Az-Zumar(9)

قال تعالى: " قل هل يستوي الذين يعلمون والذين لا يعلمون إنما يتذكر أولو الألباب" االزمر (9) Al-Qur'an al-Kareem

Urged to "Acquire knowledge and teach other people." Moreover, they said, "Seek knowledge from the cradle to the grave". He said: "Seek knowledge even (you are compelled) to go to China". He said, "People will come to you to acquire knowledge from all directions, teach them good morals."

Perfection and Excellence: People are asked to excel in work and do good deeds in their deeds: Almighty said, "Say to My servants, they say which is the best." and tell My servants to say that which is best. Indeed, Satan induces [dissension] among them. Indeed, Satan is ever, to mankind, a clear enemy. [Al-Isra, 53],

قال تعالى: " قُلْ لِعِبَادِي يَقُولُوا الَّتِي هِيَ أَحْسَنُ " الإسراء(53) Al-Qur'an al-Kareem

Furthermore, to master the work and the need to achieve quality in it and free from defects and strive for a permanent improvement, and the need for the worker to love his work and save in it all sincerity and this appears through the Almighty saying: "We do not waste the reward of the best work" (The Cave: (30)) And the Almighty said: "And you will surely be questioned about what you used to do" (Al-Nahl, 93).

قال تعالى: " ولتسئلن عما كنتم تعملون" النحل (93) Al-Qur'an al-Kareem

Leadership: Leadership is considered an essential factor for successes, and leaders should possess a positive personality that seeks to improve and improve the sayings and deeds of other people: Almighty said You are the best nation produced [as an example] for mankind. You enjoin what is right and forbid what is wrong and believe in Allah, Ali-Imran, 110.

قال تعالى: " كُنْتُمْ خَيْرَ أُمَّةٍ أُخْرِجَتْ لِلنَّاسِ تَأْمُرُونَ بِالْمَعْرُوفِ وَتَنْهَوْنَ عَنِ الْمُنْكَرِ وَتُؤْمِنُونَ بِاللهِ " آل عمران(110) Al-Qur'an al-Kareem

Visionary leadership is urged by Islam, where the vision, mission, goals, and values of the organization are clear and declared, and this is an example. What tho alqarnian had done with a when he declared that the constitution (project fine details plan) commits to Leader shall acquire the Human moral value management and be fair he should not take advantage of administrative powers to the injustice of those without him when tho alqarnian asked if Stazb all people or improved to all people in the country that opened.

Prophet Ḏū al-Qarnayn in Quran "As for one who wrongs, we will punish him. Then he will be returned to his Lord, and He will punish him with a terrible punishment. But as for one who believes and does righteousness, he will have a reward of Paradise, and we will speak to him from our command with ease." Al-Kahaf, 87 88

قال تعالى: " على لسان ذو القرنين: " أَمَّا مَنْ ظَلَمَ فَسَوْفَ نُعَذِّبُهُ ثُمَّ يُرَدُّ إِلَى رَبِّهِ فَيُعَذِّبُهُ عَذَابًا نُكْرًا، وَأَمَّا مَنْ آمَنَ وَعَمِلَ صَالِحًا فَلَهُ جَزَاءً الْحُسْنَى وَسَنَقُولُ لَهُ مِنْ أَمْرِنَا يُسْرًا " الكهف(87 – 88) Al-Qur'an al-Kareem

The leader should always look for the Successes Tools Prophet mouse ask god to support him with his brother because of his keenness on the best methods of calling, he asked God to help him by sending Aaron with a backing: "And my brother Aaron is more eloquent than me for a tongue, so I send him with me a wick that gives me truthfulness, that I fear." This medicine is evidence of one of its advantages, in order to observe it, following the highest standards of quality. It is his keen desire to provide his message with the maximum means of success Moses asks God: And my brother Aaron is more fluent than me in the tongue, so send him with me as support, verifying me. Indeed, I fear that they will deny me. "(Al-Qasas, 34)

قال تعالى: " وَأَخِي هَارُونُ هُوَ أَفْصَحُ مِنِّي لِسَانًا فَأَرْسِلْهُ مَعِي رِدْءًا يُصَدِّقُنِي إِنِّي أَخَافُ أَنْ يُكَذِّبُونِ " القصص(34) Al-Qur'an al-Kareem

Leaders should have the capability of analysis where Analysis, tho alqarnian Then, follow a reason." Almighty said: Indeed, we established him upon the earth, and We gave him to everything away. [Al-Kahf: 84], meaning: Then, he did the reasons Knowledge and reasoning Education, learning and through finding the root causes and (non-stop) Continuous development

قال تعالى: " وَآتَيْنَاهُ مِنْ كُلِّ شَيْءٍ سَبَباً، فَأَتْبَعَ سَبَبَاً " الكهف(84-85) Al-Qur'an al-Kareem

Challenging the obstacles, excellent communication skills, and development of the skill of listening to the beneficiary to know his needs. Tho Qarnain possesses the skill of listening, despite the difficulty of communicating with the people who do not understand a word. However, he listened to them in order to reach the first steps to meet the needs of the beneficiary, which is defining the problem that he suffers from and wants you to solve Almighty said: he found beside them a people who could hardly understand [his] speech. Al-Kahf: 83), where he defined the problem.

قال تعالى: " لا يَكَادُونَ يَفْقَهُونَ قَوْلا الكهف (93) Al-Qur'an al-Kareem

Systematic work and teamwork: Systematic work with some examples appears in order and organization are a bright feature of the soldiers of our master Solomon, despite their abundance and diversity of their varieties, so the Qur'an described them by saying: "Solomon his soldiers are gathered from heaven, mankind, and birds, so they are arranged over," Almighty said: And gathered for Solomon were his soldiers of the jinn and men and birds. They were [marching] in rows. An-naml7

قال تعالى: " وَحُشِرَ لِسُلَيمَانَ جُنُودُهُ مِنَ الجِنِّ وَالإنسِ وَالطَّيرِ فَهُم يُوزَعُونَ "النمل (17) Al-Qur'an al-Kareem

The principle of cooperation: Islam emphasized the necessity of cooperation for the sake of goodness, and this is manifested through the Almighty said: And cooperate in righteousness and piety, but do not cooperate in sin and aggression. “And they cooperated on righteousness and piety, not Cooperate in iniquity and transgression “(Al-Ma’idah, 2), teamwork with team spirit is one of the basic requirements for quality management

قال تعالى: " وتعاونوا على البر والتقوى ولا تعاونوا على الإثم والعدوان " المائدة(2) Al-Qur'an al-Kareem

Teamwork, human resources management, and participation for workers where Quality access is everyone’s responsibility, all have an opinion and best distribution of workers to teams; each group has a goal.

Prophet Ḏū al-Qarnayn in Quran “Bring me sheets of iron”—until, when he had leveled [them] between the two mountain walls, he said, “Blow [with bellows],” until when he had made it [like] fire, he said, “Bring me, that I may pour over it, molten copper.”

قال تعالى: “ آتُونِي زُبَرَ الْحَدِيدِ ۖ حَتَّىٰ إِذَا سَاوَىٰ بَيْنَ الصَّدَفَيْنِ قَالَ انفُخُوا ۖ حَتَّىٰ إِذَا جَعَلَهُ نَارًا قَالَ آتُونِي أُفْرِغْ عَلَيْهِ قِطْرًا "الكهف(96) Al-Qur'an al-Kareem

Rather, even the controversy of the violator should be the better one: almighty said Invite to the way of your Lord with wisdom and good instruction, and argue with them in a way that is best. And “Call for the way of your Lord with wisdom and good advice, and argue with them, which is better” [Al-Nahl, 125].

قال تعالى: " ادْعُ إِلَى سَبِيلِ رَبِّكَ بِالْحِكْمَةِ وَالْمَوْعِظَةِ الْحَسَنَةِ وَجَادِلْهُمْ بِالَّتِي هِيَ أَحْسَنُ " النحل(125) Al-Qur'an al-Kareem

Another example of teamwork, The Messenger of God, practiced it practically: He was on a journey and ordered the repair of a sheep, and a man said: O Messenger of God, I have to slaughter her. Another said: I have to skin it. Another said: I have to cook it. Then, the Messenger of God said: I have to collect firewood. They said: O Messenger of God, we are enough for you. He said: I have known that you are shrouding me, but I hate to be distinguished from you, for God hates His servant to see him distinguished among his companions, and he rose and gathered wood (Al-Safadi, 1931, 1/31). Work, cooperation, and solidarity among members of the community teams ;The Almighty said about Prophet Mousa’s journey “And appoint for me a minister from my family—Aaron, my brother. Increase through him my strength And let him share my task”) (Taha 29-32)

قال تعالى: "واجعل لي وزيرًا من أهلي. هارون أخي. اشدد به أزري. وأشركه في أمري طه(29- 32) Al-Qur'an al-Kareem

The Almighty said (prophet Ḏū al-Qarnayn in Quran): “He said: He said, “That in which my Lord has established me is better [than what you offer], but assist me with strength; I will make between you and them a dam”) ؟Al-Kahaf, 95)

قال تعالى: "قال ما مكني فيه خيرًا فأعينوني بقوة أجعل بينكم وبينهم ردمًا" الكهف(95) Al-Qur'an al-Kareem

He, may God's prayers and peace be upon him, said: "God's hand is with the congregation". Coordination and project management where they apply coaching principles as "a process by which one person helps another person improve performance." The results of individual coaching greatly are multiplied when a whole organization coaches to create willing cooperation among workers as they pursue their personal bests.

قال رسول الله النبي محمدﷺ: "يد الله مع الجماعة " Sunnah

The great leadership ability of the two centuries in coordinating between the various work teams, solving their problems, explaining their powers and areas of work, and selecting the competent for each of them, as will be seen in the following field. He told them that the solution is to build a great backfill.

Prophet Ḏū al-Qarnayn in Quran "I will make between you and them a dam or Make between you and between them a backfill." [Al-Kahaf: 95]

قال تعالى: " أَجْعَلْ بَيْنَكُمْ وَبَيْنَهُمْ رَدْمًا " الكهف(95) Al-Qur'an al-Kareem

Between the ranges of those mountains, and guide them on how to build it. Prophet Ḏū al-Qarnayn in Quran said: "Bring me sheets (core) of iron" [Al-Kahf: 96].

قال تعالى: " آتُونِي زُبَرَ الْحَدِيدِ " الكهف(96) Al-Qur'an al-Kareem

And they piled it between the two dams, and they did so until the iron was a great mound parallel to the mountains, and prophet Ḏū al-Qarnayn in Quran "equated between the two sides or until, when he had leveled [them] between the two mountain walls" [Al-Kahf: 96],

قال تعالى: " " سَاوَى بَيْنَ الصَّدَفَيْنِ " الكهف(96) Al-Qur'an al-Kareem

i.e., he chose the most suitable two coincidental mountains (opposite) in the mountain chain to build the basis of the backfill between them.

Prophet Ḏū al-Qarnayn in Quran said: "he said, "Blow [with bellows]," until when he had made it [like] fire, he said, "Bring me, that I may pour over it, molten copper." [Al-Kahf: 96],

قال تعالى: " انْفُخُوا حَتَّى إِذَا جَعَلَهُ نَارًا قَالَ آتُونِي أُفْرِغْ عَلَيْهِ قِطْرًا " الكهف(96) Al-Qur'an al-Kareem i.e., lit a massive fire such that the pieces of iron are protected and become like a fire, then make the diameter of the molten copper between him and the penetration of copper between the pieces of iron, to form the barrier, so that what was meant was the confinement of the Gog and Magog tribes, and for this prophet, Ḏū al-Qarnayn in Quran said:

"So Gog and Magog were unable to pass over it, nor were they able [to effect] in it any penetration).

قال تعالى: " " فَمَا اسْطَاعُوا أَنْ يَظْهَرُوهُ " أي: يصعدوا ذلك الردم، " وَمَا اسْتَطَاعُوا لَهُ نَقْبًا " الكهف(97) Al-Qur'an al-Kareem

in Sunnah He, may God's prayers and peace be upon him, said: "God's hand is with the congregation." And god hand with teamwork. ami' at-Tirmidhi 2166

قال رسول الله النبي محمدﷺ: "يد الله مع الجماعة حَدِيثٌ." Sunnah

Warns and Punishment for lousy work. On the other side, the Almighty warns people of indulgence and lack of proficiency in their work.

God Almighty said: (He wills. And you will surely be questioned about what you used to do, and you will be asked what you were doing. An-Nahl, 93

قال تعالى: " ولتسألنَّ عما كنتم تعملون) النحل (93) Al-Qur'an al-Kareem

This verse confirms the individual responsibility entrusted to him by God, as he is accountable for every failure he has committed against himself and his nation, everyone will ask about his work inevitably. He asks people to be charitable in their actions: "Indeed, We have made that which is on the earth adornment for it that We may test them [as to] which of them is best indeed." [Al Kahf: 7)

قال تعالى: " إِنَّا جَعَلْنَا مَا عَلَى الأَرْضِ زِينَةً لَهَا لِنَبْلُوَهُمْ أَيُّهُمْ أَحْسَنُ عَمَلاً " الكهف (7) Al-Qur'an al-Kareem

Furthermore, as a general principle rewarding for an excellent job, Punishment for lousy work with Justice and non-arbitrary use of powers has been emphasized in Qur'an.

Prophet Ḏū al-Qarnayn in Quran: Almighty said He said, "As for one who wrongs, we will punish him. Then he will be returned to his Lord, and He will punish him with a terrible punishment. As for one who wrongs, we will punish him. Then he will be returned to his Lord, and He will punish him with a terrible punishment". (Al-Kahaf, 87.88)

قال تعالى: " أَمَّا مَن ظَلَمَ فَسَوْفَ نُعَذِّبُهُ ثُمَّ يُرَدُّ إِلَىٰ رَبِّهِ فَيُعَذِّبُهُ عَذَابًا نُّكْرًا (87) وله تعالى : وَأَمَّا مَنْ آمَنَ وَعَمِلَ صَالِحًا فَلَهُ جَزَاءً الْحُسْنَى وَسَنَقُولُ لَهُ مِنْ أَمْرِنَا يُسْرًا الكهف (88) Al-Qur'an al-Kareem

Customers' Requirements and Wants Mission Clearness. Customers' requirements, wants, mission clearness, and Customer best benefit emphasized in Islam. The clarity of the goals informs these people about the description of the product that seeks to solve their problem, which is: building a fill that is characterized by two main things: it cannot be climbed, and it cannot be punctured. Knowing the real customer's needs, draw the attention of the customer to the possibility of giving him the most excellent and most economical service and products requested.

Prophet Ḏū al-Qarnayn in Quran:] They said, "O Dhul-Qarnayn, indeed Gog and Magog are [great] corrupters in the land. So, may we assign for you an expenditure that you might make between them and us a barrier?" He said, "That in which my Lord has established me is better [than what you offer], but assist me with strength; I will make between you and them a dam". (Al-Kahf 94, 95)

قال تعالى: " فَهَلْ نَجْعَلُ لَكَ خَرْجًا عَلَىٰ أَن تَجْعَلَ بَيْنَنَا وَبَيْنَهُمْ سَدًّا الكهف (94) قَالَ مَا مَكَّنِّي فِيهِ رَبِّي خَيْرٌ
فَأَعِينُونِي بِقُوَّةٍ أَجْعَلْ بَيْنَكُمْ وَبَيْنَهُمْ رَدْمًا الكهف (95) Al-Qur'an al-Kareem

Product fulfillment and suitability of the product with an inspection where 97 So (So Gog and Magog were unable to pass over it, nor were they able [to effect] in it any penetration (Al-Kahaf, 97).

قال تعالى: " فَمَا اسْطَاعُوا أَن يَظْهَرُوهُ وَمَا اسْتَطَاعُوا لَهُ نَقْبًا " الكهف(97) Al-Qur'an al-Kareem

Quality management in Islam is value management, as it urges to give the beneficiary the best outputs, by giving the beneficiary what he requests if you know that there is a better alternative.

For the messenger, Mohamad said: "The best thing is the gain of the worker's hand when he is advised for the best thing, and he should not cheat." And Thus, the justified and peaceful sale of fraud, betrayal, and lying from justified earnings: such as carpentry, blacksmithing, agriculture, and writing, and the like of what a person does with his hand, and acquires in his hand, and constructive works for people that build, or implant, or cultivate, and the like, this is from good earning, If he advised and performed honesty, it is a good deed, a good deed

قال رسول الله النبي محمدﷺ:" خَيْرُ الْكَسْبِ كَسْبُ يَدِ الْعَامِلِ إِذَا نَصَحَ),رواه أحمد Sunnah

Audit, Accounting, and Fair Measurement System. Audit and Accounting through verification of all work activities, and other elements of the system to govern the product conformity with the customer requirements, whatever a small or big issue.

The Concept of Conformity, God Almighty says: {Then We put you, [O Muhammad], on an ordained way concerning the matter [of religion]; so follow it and do not follow the inclinations of those who do not know. And Then He placed you on the law of the matter, so follow it and do not follow the whims). (Al-Jathiyah, 18)

قال تعالى: " ثُمَّ جَعَلْنَاكَ عَلَى شَرِيعَةٍ مِنَ الأَمر فاتبعها ولا تتبع أهواء الَّذِينَ لا يَعْلَمُونَ "الجاثيه(18) Al-Qur'an al-Kareem

Conformity is the balance of the quality of works in its appearance, the standard of performance quality, and the mastery of practice. This represents the principle of conformity with the standards and standards set in a management system.

The Almighty said: "Woe to those who give less [than due], Who, when they take a measure from people, take in full. But if they give by measure or by weight to them, they cause loss. (Al-Mutaffifin1-3)

قال تعالى: " وَيْلٌ لِّلْمُطَفِّفِينَ. الَّذِينَ إِذَا اكْتَالُواْ عَلَى النَّاسِ يَسْتَوْفُونَ. وَإِذَا كَالُوهُمْ أَو وَّزَنُوهُمْ يُخْسِرُونَ " المطففين(1-3) Al-Qur'an al-Kareem

Almighty said: "And the record [of deeds] will be placed [open], and you will see the criminals fearful of that within it, and they will say", "Oh, woe to us! What is this book that leaves nothing small or great except that it has enumerated it?" And they will find what they did present [before them]. And your Lord does injustice to no one. (Al-Kahaf, 49)

قال تعالى: " وَوُضِعَ الْكِتَابُ فَتَرَى الْمُجْرِمِينَ مُشْفِقِينَ مِمَّا فِيهِ وَيَقُولُونَ يَا وَيْلَتَنَا مَالِ هَذَا الْكِتَابِ لا يُغَادِرُ صَغِيرَةً وَلا كَبِيرَةً إِلاَّ أَحْصَاهَا وَوَجَدُوا مَا عَمِلُوا حَاضِرًا وَلا يَظْلِمُ رَبُّكَ أَحَدًا " الكهف (49) Al-Qur'an al-Kareem

Control acts are required to improve work and prove good leadership, like when Prophet Solomon visited the bird community, and his knowledge of the absence of the hoopoe was a guide to the successful leadership that follows the affairs of individuals and their control. The excellent and fair measurement system is also a feature of Muslim life.

Our Master, prophet Shuaib, calls on people to take care of the quality of the commercial standards, Almighty said, "And to [the people of] Madyan [We sent] their brother Shu'ayb. He said, "O my people, worship Allah; you have no deity other than Him. There has come to you clear evidence from your Lord. So fulfill the measure and weight and do not deprive people of their due and cause not corruption upon the earth after its reformation. That is better for you if you should be believers (Al-A'raf, 85)

قال تعالى: " فَأَوْفُوا الْكَيْلَ وَالْمِيزَانَ وَلا تَبْخَسُوا النَّاسَ أَشْيَاءَهُمْ وَلَا تُفْسِدُوا فِي الْأَرْضِ بَعْدَ إِصْلَاحِهَا ذَلِكُمْ خَيْرٌ لَكُمْ إِنْ كُنْتُمْ مُؤْمِنِينَ " الأعراف: (85) Al-Qur'an al-Kareem

The Almighty God control concept is the best control ever where any person must remember that above the control of humankind, there is oversight from God Almighty, which is the highest level of individual and collective questioning.

God Almighty said: "And stop them; indeed, they are to be questioned." (As-Saffat, 24).

قال تعالى: " وقفوهم إنهم مسؤولون" الصافات، (24) Al-Qur'an al-Kareem

The principle of Islamic censorship: whether it is external or self-conducive to ensuring the implementation of the goals and standards set under Islamic Sharia standards, standards and controls, just as the self-censorship of a Muslim is fully responsible toward his actions in this world and the hereafter, and this appears through the Almighty.

Every soul, for what it has earned, will be retained and: "Every soul has gained as a hostage" Al-Muddathther, 38)

قال تعالى: " كل نفس بما كسبت رهينة" المدثر(38). Al-Qur'an al-Kareem

Measuring successes and performance on periodic terms, it is the benefit of him saying: "To whoever wills among you to proceed or stay behind." [Al-Muddathther, 37].

قال تعالى: " لِمَنْ شَاءَ مِنْكُمْ أَنْ يَتَقَدَّمَ أَوْ يَتَأَخَّرَ المدثر(37) Al-Qur'an al-Kareem

The continuation of improvement and development is imperative, and no higher educational institution should stop continuing development and improvement on the pretext that it has attained international certificates in development, because if it stops development, it is necessarily late.

Cheating, In the Qur'an, it is commanded not to cheat people or undermine the value of their possessions.

Almighty said: And to [the people of] Madyan [We sent] their brother Shu'ayb. He said, "O my people, worship Allah; you have no deity other than Him. There has come to you clear evidence from your Lord. So, fulfill the measure and weight and do not deprive people of their due and cause not corruption upon the earth after its reformation. That is better for you if you should be believers. Al-A'raf, 85

قال تعالى: " وَإِلَىٰ مَدْيَنَ أَخَاهُمْ شُعَيْبًا ۗ قَالَ يَا قَوْمِ اعْبُدُوا اللَّهَ مَا لَكُمْ مِنْ إِلَٰهٍ غَيْرُهُ ۖ
قَدْ جَاءَتْكُمْ بَيِّنَةٌ مِنْ رَبِّكُمْ ۖ فَأَوْفُوا الْكَيْلَ وَالْمِيزَانَ وَلَا تَبْخَسُوا النَّاسَ أَشْيَاءَهُمْ وَلَا تُفْسِدُوا فِي الْأَرْضِ بَعْدَ إِصْ
لَاحِهَا ۚ ذَٰلِكُمْ خَيْرٌ لَكُمْ إِنْ كُنْتُمْ مُؤْمِنِينَ سورة الأعراف (85) Al-Qur'an al-Kareem

The Messenger Muhammad has stressed the importance of protecting people against theft and punishing the thief as per Shari'a law. Messenger Mohammed said: O people, those who had gone before you were destroyed because if anyone of high rank committed theft amongst them, they spared him; and if anyone of low rank committed theft, they inflicted the prescribed punishment upon him. By Allah, if Fatima, daughter of Muhammad, were to steal, I would have her hand cut off. In the hadith transmitted on the authority of Ibn Rumh (the words are): "Verily those before you perished."

قال رسول الله النبي محمدﷺ: "قَالَ: إِنَّمَا أَهْلَكَ الَّذِينَ قَبْلَكُمْ أَنَّهُمْ كَانُوا إِذَا سَرَقَ فِيهِمْ الشَّرِيفُ تَرَكُوهُ وَإِذَا سَرَقَ فِيهِمْ الضَّعِيفُ أَقَامُوا عَلَيْهِ الْحَدَّ وَايْمُ اللَّهِ لَوْ أَنَّ فَاطِمَةَ بِنْتَ مُحَمَّدٍ سَرَقَتْ لَقَطَعْتُ يَدَهَا , متفق عليه
Sunnah

In the Islamic history, the famous Islamic (Leader)Umar ibn Al-Khattab God be pleased with him is famous for protecting Muslims' value of their money: when he saw a man who had marred the milk with the water sold, and he shed it on him, and this is proven from Umar bin Al-Khattab, may God be pleased with him.

9 Time Management

One of the essential Islamic principles is to make the best use of times and spend them on what is beneficial in this world and the hereafter.

The Almighty said: {He] who created death and life to test you [as to] which of you is best in deed—and He is the Exalted in Might, the Forgiving—Al-Mulk, 2

قال تعالى: " الَّذِي خَلَقَ الْمَوْتَ وَالْحَيَاةَ يقول: لِيَبْلُوَكُمْ أَيُّكُمْ أَحْسَنُ عَمَلا "الملك (2) Al-Qur'an al-Kareem

Evaluation of work fairness and encourage competition. A clear indication that their warding and penalty relates to how the performance was, whatever it was.

No one of creation knows how God Almighty created the universe with this creativity and competence, and no one knows his wisdom in creating death and life. In the affliction of His servants with them to know at the end of the end, whoever is better and who is offended, he is the world with them from above seven heavens, and in verse is a Quranic call to search for the best, and to emphasize the principle of God observing man bring him to the perfection of work.

And in the Almighty said: that He might test you as to which of you is best in deed (Hud, 7)

قال تعالى: " ليبلوكم أيكم أحسن عملاً " هود(7)Al-Qur'an al-Kareem

Is a clear indication that the penalty relates to how the performance is, regardless of whether this performance?

Almighty said: (Indeed, we have made that which is on the earth adornment for it that We may test them [as to] which of them is best indeed. Al-Kahf, 7

قال تعالى: " إنا جعلنا ما على الأرض زينة لها لنبلوهم أيهم أحسن عملاً) الكهف(7) Al-Qur'an al-Kareem

The caliph Umar bin Al-Khattab (may God be pleased with him) called for accountability of the soul, and evaluated the human person himself better than the evaluation of others for him, "Hold ourselves accountable before you are held accountable, and weigh your deeds before they are weighed against you." So, one of the most critical motives and enhancements of overall quality was the conscience alive and awakening the principle of observation subjectivity, and its meaning applied in a great value of Islam, and it is the principle of benevolence "to worship God as if you see him."

9.1 *Workers' and Employees' Human Rights*

Workers' rights emphasized in Islam culture, the business owners and workers have to be treated humanely, fairly and respect, where always respect a workers dignity: it is worth everything to them and as mentioned always treat your employees exactly as you want them to treat your best, and this is a must in Islam.

Almighty said: "So by mercy from Allah, [O Muhammad], you were lenient with them. And if you had been rude [in speech] and harsh in heart, they would have disbanded from you. So, pardon them and ask forgiveness for them and consult them in the matter. And when you have decided, then rely upon Allah. Indeed, Allah loves those who rely [upon Him] Ali'Imran 159

قال تعالى: " فَبِمَا رَحْمَةٍ مِّنَ اللَّهِ لِنتَ لَهُمْ ۖ وَلَوْ كُنتَ فَظًّا غَلِيظَ الْقَلْبِ لَانفَضُّوا مِنْ حَوْلِكَ ۖ فَاعْفُ عَنْهُمْ وَاسْتَغْفِرْ لَهُمْ وَشَاوِرْهُمْ فِي الْأَمْرِ ۖ فَإِذَا عَزَمْتَ فَتَوَكَّلْ عَلَى اللَّهِ ۚ إِنَّ اللَّهَ يُحِبُّ الْمُتَوَكِّلِينَ " آل عمران (159)
Al-Qur'an al-Kareem

9.2 *Workers Training and Coaching*

Coaching and training considered as important dimensions and a strategy to help workers achieve excellence; training input should be sought from employees to help design quality efforts and develop a sense of ownership training plan of action that should be developed to make it happen over time. It is critical to not do too much at once; workers should be trained enough before doing work, " the necessity of training and qualifying employees is fundamental. The basis for perfection in business in Islam is the availability of knowledge first.

God Almighty's said: "And do not pursue that of which you have no knowledge. Indeed, the hearing, the sight, and the heart - about all those [one] will be questioned" (Al-Isra, 36).

قال تعالى: " ولا تقف ما ليس لك به علم" الإسراء(36) Al-Qur'an al-Kareem

Perfection, progress, and development leading to quality need rehabilitation and training as well as knowledge and experience. Moreover, to reach, the summit needs the mission and the will as well as the confidence and faith.

9.3 Modern Muslims Communities and Quality Management

The crucial question about modern Muslims communities and quality management, since all above mentioned why Muslims communities are not the best why Muslim countries are not the first in quality in the world. However, there are also Muslim communities in some parts of the world that are deficient in that regard, sometimes out of a lack of information and education and sometimes because they learn about Islam from their traditions, rather than from the Qur'an. Those circles seeking to make anti-Muslim propaganda generally portray those images as a "true Muslim model." That is why we often encounter pictures revealing this lack of quality in Muslims in many parts of the world on the television, in newspapers, and videos. These images used for a specific purpose to lead the world into becoming falsely acquainted with Islam.

Therefore, what many people think of when the word "Muslim" mentioned is a loveless and joyless model that rejects science and fails to understand art, has no sense of humanity, has no idea how to dress well, and is far removed from modernity. A way of thinking that takes no pleasure in music that takes a dim view of joyful people that takes no care concerning cleanliness that is unkind and unaffectionate, and that regards women as second or even third-class citizens may be regarded as embodying the basic features of a Muslim in many parts of the world. That perception naturally leads to unease and adverse reactions in the world as a whole and encourages people to

10 Conclusions

From the preceding, we infer through the noble verses and the noble hadiths on the Islamic administration's focus on establishing quality in various businesses and services. That control carried out in the light of defined criteria and standards derived from the Noble Qur'an and the noble Prophet's Sunna, and that Islam was the first and greatest pioneer in laying and building the foundations. The correct and sound work that builds a stable society, just as Islam represented the culture and concepts of quality management in daily life affairs through responsibility, accountability, and the degree of performance and mastery of the work before the slogans of quality launched in the current era and this are evidence that the approach Salami came fully and comprehensively in all fields of work without allocating or specifying.

On personal level: The individual must each of his position to adhere to the quality in his morals, and his dealings with others, whether they are from his skin or others, and his mastery applies it in the practical capacity that the Messenger, may God bless him and grant him peace, set an example for him in his sayings and actions. The Muhammadiyah message came with comprehensive legislation for all aspects of life, as was the fulfillment of the prophets 'messages before, peace and blessings be upon him said:" I was sent to fulfill the morals. "And the positions of the companions, may God be pleased with them, who brought up in the school of (Mustafa) prophet Mohamad, may God bless him and grant him peace. As if you never live and work for the hereafter, Nick is dying tomorrow, "an indication of the importance of perfecting work and sincerity in it. Moreover, the approach of the esteemed companions may God be pleased with them. The righteous followers on the approach of their great Messenger, may God bless him and grant him peace, and take an example in applying the principle of comprehensive quality in all areas of their lives. On the social level: It can be said that the individual's mastery of his behavior following the example of his noble Messenger, may the best of prayers and peace be upon him, has a positive or negative impact on perfection and quality of his dealings with others, whether they are relatives, friends, neighbors, or co-workers. When he takes into account the rights of those around him in dealing, he is a master and master of the art dealing with others, and there are many examples of that, including helping people and standing with them when crises they are exposed to in life and supporting them in their joys and sorrows.

God's prayers and peace be upon him said: "God is in the help of a servant who was not a servant in aid of his brother" (Sahih Muslim). And on Career level: applying the principle of mastering work and sincerity in it: He urged Islam to master the work and the need to achieve quality in it and to be free from defects and seek permanent improvement, and the need for it. The worker loves his work and finds sincerity in it, and this is revealed through the Almighty said: "Indeed, We will not allow being lost the reward of any who did well in deeds. (Al Kahf,30) and God Almighty said: "And you will surely be questioned about what you used to do." (Al-Nahl, 93)

قال تعالى: "إنا لا نضيع أجر من أحسن عملا" الكهف، (30) Al-Qur'an al-Kareem
قال تعالى: "ولتسئلن عما كنتم تعملون" النحل(93) Al-Qur'an al-Kareem

Mastering the work is one of the significant fruits that the individual and the institution earn when adhering to the quality standards, and access to quality must start from the love of the work that the individual does. He must adhere to the instructions directed to him by the person in charge of the work, and from here we see that the quality is the mastery of the success of the work, and it was mentioned in Quranic stories are among the human examples that showed the side of perfection in work. It was mentioned in Surah Yusuf, and peace be upon him when the king lined up. He asked him to leave the treasures of Egypt because he knew and was better able to master his work, and he expressed this as preservation and science as a basis for the success of his work and the reason for his quality and mastery.

Almighty said: Put me on the floor treasures Well knowing "(Yusuf: 55), just as the Almighty mentioned in another verse the importance of being as strong and faithful in the story of the Prophet of God Moses Ali, peace be upon him and the daughter of the Prophet of God, Shoaib, peace be upon him. The Almighty said:" One of them said, Father, hire him. It is better than the one who hires the strong and faithful (Al-Qasas: 26), and these qualities in the two holy verses revolve around the advantages and mastery of work.

References

Akao Y (1994) Development history of quality function deployment. Cust Driven Appr Quality Planning and Deployment 339:90

Al-Ajlouni I (2006) Principles of total quality management in the light of the islamic curriculum. Damascus Univ J Damascus 1:4

Al-Qur'an al-Kareem - القرآن الكريم. (n.d.) Retrieved from https://quran.com/1/1

Sunnah (n.d.) Retrieved from https://sunnah.com/muslim

Al-Albani M (1986) Sahih al-jamia'al-Saghir. On: Beirut, Lebanon: Al-Maktab al-Islami

Al-Bukhari M (1978) Sahih al-Bukhari, Dar Ul-Hadith

Al-Kabi MN, Abu Ata BM, Wahsheh HA, Alsmadi IM (2013) A topical classification of quranic arabic text. In: Taibah University international conference on advances in information technology for the Holy Quran and its sciences, 22–25 Dec 2013, Madinah, Saudi Arabia

Al-Zubaidi M (1994) The crown of the bride by jewels dictionary, investigation: Ali Sherry, (D.T.), Dar Al-Fikr, Beirut

Al-Zubaidi Z-U-D (1994b) Summarized Sahih al-Bukhari. Maktaba Dar-us-Salam, Riyadh

ASI Mourns the Loss of Dr. Genichi Taguchi. Asiusa.com. 2012-03-28. Archived from the Original On 2012-10-19. Retrieved 2012-06-26

Barro RJ, Mccleary R (2003) Religion and economic growth. National Bureau of Economic Research

Barro JR, McCleary RM (2003) Religion and economic growth across countries. Am Sociol Rev 68(5):760–781

Birkner B, Bigler-Münichsdorfer H, Lüttecke H, Gürtler J (2018) Kaufmann/Kauffrau im Gesundheitswesen: Lehrbuch zur berufsspezifischen Ausbildung. Kohlhammer Verlag

Campante F, Yanagizawa-Drott D (2015) Does religion affect economic growth and happiness? Evidence from Ramadan. Q J Econ 130:615–658

Ciampa D (1992) Total quality: a users' guide for implementation. Addison Wesley Publishing Company

Clarke GB, Cobham RT (1990) Descriptions of Lord Cobham's gardens at Stowe (1700–1750). Buckinghamshire Record Society

Cohen MA, Rogelberg SG, Allen JA, Luong A (2011) Meeting design characteristics and attendee perceptions of staff/team meeting quality. Group Dyn Theor Res Pract 15:90

Cohen S, Brand R (1995) Total quality management in government: a practical guide for the real world. Jossey-Bass, San Francisco

Copley FB, Taylor FW (1923) Frederick W. Taylor

Crosby PB (1979) Quality is free: The art of making quality certain. McGraw-hill New York

Dahlgaard-Park SM (2008) Reviewing the European excellence model from a management control view. TQM J

Durlauf SN, Kourtellos A, Tan CM (2012) Is God in the details? A reexamination of the role of religion in economic growth. J Appl Econ 27:1059–1075

Feigenbaum AV (1983) Total quality control

Holweg M (2007) The genealogy of lean production. J Oper Manage 25:420–437

Huntington SP (1996) The West unique, not universal. Foreign Affairs, pp 28–46

Inglehart R, Baker WE (2000) Modernization, cultural change, and the persistence of traditional values. Am Sociol Rev 19–51

Ishikawa K (1986) TQC-total quality control. Estratégia e Administração da Qualidade, 1

Juran JM (2003) Juran on leadership for quality. Simon, and Schuster

Kinsinger FS (2009) Beneficence and the professional's moral imperative. J Chiropractic Human 16:44–46

Kumar N, Kumar H, Khurmi JS (2016) Experimental investigation of process parameters for rapid prototyping technique (selective laser sintering) to enhance the part quality of prototype by Taguchi method. Procedia Technol 23:352–360

Landes DS (1999) The wealth and poverty of nations: why some are so rich and some so poor. Norton, New York

Marvin R, Arnljot H (2004) System reliability theory: models, statistical methods, and applications. John Wiley & Sons Inc, Hoboken, New Jersey, Canada

Masaaki, I. 1986. Kaizen: The key to Japan's competitive success. *New York, ltd: McGraw-Hill*

Mccleary RM, Barro RJ (2006) Religion and economy. J Econ Persp 20:49–72

Ohno T (1988) Toyota production system: beyond large-scale production. CRC Press

Religion And Economic Growth Robert J. Barro Rachel M. McCleary Working Paper 9682 http://www.nber.org/papers/w9682 National Bureau Of Economic Research 1050 Massachusetts Avenue Cambridge, MA 02138 May 2003

Sahih Al Jamie Al Saghir Wa Ziadatuh— صحيح الجامع الصغير وزيادته. (n.d.). Retrieved from https://darmakkah.co.uk/product/Sahih-al-jamie-al-saghir-wa-ziadatuh-الصغير-وزيادته صحيح-الجامع

Shewhart WA, Deming WE (1967) In memoriam: Walter A. Shewhart, 1891–1967. Am Stat 21:39–40

Shewhart WA, Deming WE (1986) The statistical method from the viewpoint of quality control. Courier Corporation

Shingo S (1990) El sistema de producci¢ n de Toyota desde el punto de vista de la ingenier¡ a

Sifri J (2014) How to use design of experiments to create robust designs with high yield. youtube.com. Retrieved 11 Feb 2015

Stewart GL, Manges KA, Ward MM (2015) Empowering sustained patient safety: the benefits of combining top-down and bottom-up approaches. J Nurs Care Qual 30:240 246

Taguchi G (1986) Introduction to quality engineering: designing quality into products and processes

Tague NR (2005.) The quality toolbox. ASQ Press

Tennant G (2001) Six Sigma: SPC and TQM in manufacturing and services. Gower Publishing, Ltd

Weber C (1930) A modification of Sakaguchi's reaction for the quantitative determination of arginine. J Biol Chem 86:217–222

Weber SE (1905) The Charity School movement in Colonial Pennsylvania. Press of GF Lasher

Work performance (n.d.) Retrieved from http://www.wata.cc/forums/forum.php

Optimization in Quality Development

Multi-process Assessment Considering the Error of Measurement Systems Within the Process Capacity Indices

Darwin Young-Vazquez, R. María Blanca Becerra, Elías Carrum-Siller, Cuauhtémoc Sánchez-Ramírez, and Arturo Hernández-Hernández

Abstract The purpose of this chapter is to make available to young researchers and process engineers the importance of integrating measurement systems with multi-process capability analysis systems using Z-values (short and long term) for continuous data considering the error of the measurement systems. Additionally, the modifications of process capacity indicators commonly used in the industrial sectors are shown with the inclusion of the measurement error, determining its significance through the ANOVA analysis. The results obtained show a better contrast with the direct measurements made in the evaluated processes, increasing the effectiveness of the process control.

1 Introduction

In current manufacturing processes, the evaluation of critical variables is decisive for the competitiveness of companies within the industrial sectors. The analysis of process capabilities is generally carried out in isolation for each variable, neglecting

D. Young-Vazquez (✉)
CILSS Consulting, Residencial Los Reales, 25253 Saltillo, Coah, Mexico
e-mail: darwin.young@cilssconsulting.com.mx

R. M. B. Becerra
Tecnológico Nacional de México: Campus Instituto Tecnológico de Querétaro, Av. Tecnológico s/n, Centro, 76000 Santiago de Querétaro, Qro, Mexico

E. Carrum-Siller
COMIMSA-CONACYT, Calle Ciencia y Tecnología 790, Saltillo 400, 25290 Saltillo, Coah, Mexico

C. Sánchez-Ramírez
Tecnológico Nacional de México: Campus Instituto Tecnológico de Orizaba, Avenida Oriente 9 No. 852, Colonia Emiliano Zapata, Apdo. Postal No. 324 Orizaba, 94320 Veracruz, Mexico

A. Hernández-Hernández
Universidad Politécnica de Querétaro, Carretera Estatal 420 S/N, El Rosario, 76240 Santiago de Querétaro, Qro, Mexico

J. L. García Alcaraz et al. (eds.), *Techniques, Tools and Methodologies Applied to Quality Assurance in Manufacturing*,
https://doi.org/10.1007/978-3-030-69314-5_8

in many cases the effects of global interactions or effects on the system as a whole. Research carried out on these aspects is presented by Casaroni (2019) and Ait-Izem (2018), denoting the importance of strengthening processes. Specialists like Triebs (2018) denote the importance of taking production processes from a state of unobserved effects to an observable process. While experts such as Joy and Shunmugesh (2019) make it relevant to develop process capacity indicator systems. The problem addressed in this research work is to identify the effects of not considering the measurement error within the evaluation of process capabilities, considering two working hypotheses:

H1: Including the measurement error within the indicators of process capacity improves the effectiveness to evaluate the quality of the process.

H2: Employ multi-process capacity analysis, evidence the overall performance of the processes to improve decision-making by measuring the impact of measurement errors.

The multi-process system is distinguished by different subgroups of variables to be measured and controlled to maintain the required levels of performance and production, which can be identified as discrete or continuous variables that are related to the integration of different types of data to be analyzed, such as quality, time, cost, resource capacity, productivity, and flexibility in an integrated performance index. The process capability index Z-value determines the ability of a process that is grouped around a target. And you can interact directly with the yield of the process in parts per million non-conforming. When the Z-value is calculated using discrete data is considered, the throughput yield or defects per opportunity for the calculation of probability.

2 Literature Review

The literary review corresponding to this research can be divided into two groups. The first group corresponds to the identification of common indicators of process capacity, which can be improved by considering measurement error. Pearn (2005) evidences that the estimators used when analyzing the data of process samples are contaminated by measurement errors, which severely underestimate the capacity of the process. In investigations carried out by various researchers such as Montgomery (1993), Barrentine (2003), and Fechio (2009), the importance of generating analysis of measurement systems to generate reliable measurements to determine process capacities is presented, without determining the direct impact that generated within the process evaluation when calculating the capacity indicators.

The second group corresponds to investigations within the field of multi-process analysis, which present affectations as the impact of the measurement systems has not been identified. Chen (1994) uses tolerance zones in multivariate process capacity indicators. Boyles (1996) and Wang (2000) employ multivariate processes with principal component analysis or lattice data, considering that the data were adequately taken, but without evidence of the effectiveness of the measurement system. Burlikowska (2005) generates the quality estimation using XR control charts and process capacity indicators, while Chang (2008) uses estimation intervals of the process capacity indicators with asymmetric tolerances, promoting in all cases the importance of a correct measurement to ensure the quality of the analyzes performed.

In general, when calculating the process capability indicators is assumed to have reliable measurement systems and these indicators are calculated without considering the measurement errors. This assumption does not correspond to real situations in which measurement systems can generate dispersion in the accuracy of the data, generating conclusions from unreliable process capabilities. In this paper, we consider the calculation of process capability indicator Z-value considering the measurement errors for real process capability.

In the case study, the mixture of normal distributions is determined for different periods of production being delivered to the client as a mixture of different qualities due to adjustment of equipment or process improvements. Denoting the process capability on real time and capacity delivered to the customer by the mixture of products manufactured at various periods of time.

3 Methodology

The process used to carry out the evaluation of multi-processing capacities allows its replication in a simple way, by developing the following steps:

- Definition of the Z-value process capacity index for short and long term, used within the lean six sigma methodology.
- Modify process capability indicators.
- Determine the analysis of the measurement system for continuous data.
- Define the gage R&R analysis mechanism by ANOVA.

Subsequently, the application of the ANOVA method is shown within the analysis of measurement systems for its correct understanding and application, to complement the study with the application of multi-process evaluation.

3.1 Z-value Process Capability Index

The process capability measured by Z-values in a short and long term has been adopted by several companies that perform processes under the parameters of the six sigma methodology.

$$Z = \frac{x - \mu}{\sigma} \tag{1}$$

Compare the value of the specification, with the process mean and divided by the value of the standard deviation. An advantage of using this indicator is that it can be used with continuous and discrete characteristics. Potential information can be used in the short term, and the real capacity for the application of rational subgroups, considering that at least 80% of the process variation factors.

The Z-value short-term capacity (Zst) is calculated from data taken over a period short enough so that no external influences on the process. While the Z-value long-term capacity (Zlt) is calculated from data taken over a period of time long enough that external factors can influence the process. The PPM defect data and performance data are by nature long-term measures. On average, a typical process, *Z* shift should be around 1.5 which is obtained by considering that the short-term standard deviation is increased by 33.33% over time. *Z* shift is the difference between the short and long term. A process is unstable if the result is negative and cannot do better in a long term than short-term information determines the potential that is the best it can be due to process technology. The *Z* shift indicates the level of control we have over the process. To determine a six sigma process, within bilateral process tolerances, the Z-value in the short term considers the total number of defects on both sides of the specification limits. Figure 1 (left) shows the relationship between defects and the indicator of *Z*, where *P* is the probability of defects in short term.

When considering a long-term process is influenced by external factors of change, which typically causes the standard deviation can reach one third to increase their value. Figure 1 (right) shows the relationship between defects and the indicator Zlt, where *P* is the probability of defects is 3.45 defects per million of opportunities. As noted, the denote 3.4 defects per million is related to the real capacity of the process, not the potential.

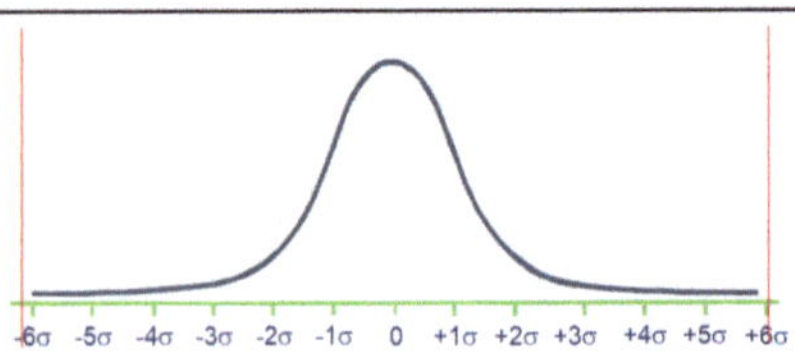

$$Z_{USL} = \left|\frac{6.118\sigma - \mu}{\sigma}\right| = \left|\frac{(6.118\sigma - 0)}{\sigma}\right| = 6.118$$

$$Z_{LSL} = \left|\frac{(-6.118\sigma) - \mu}{\sigma}\right| = \left|\frac{(-6.118\sigma) - 0}{\sigma}\right| = \left|-6.118\right|$$

Probability of defect corresponding to any of the specification limits:

$$P_{LSL} \approx 6.25x10^{-10}$$

Considering both sides

$$P_{Total} \approx 2(6.25x10^{-10}) \approx 1.25x10^{-9}$$

Z-Value (Short Term) is $Zst = 6.0$

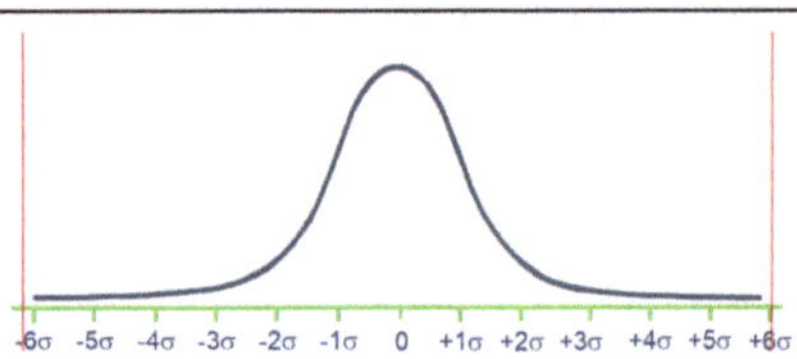

$$Z_{LSE} = \left|\frac{6.118\sigma - \mu}{1.33\sigma}\right| = \left|\frac{(6.118\sigma - 0)}{1.33\sigma}\right| = 4.635$$

$$Z_{LSE} = \left|\frac{(-6.118\sigma) - \mu}{1.33\sigma}\right| = \left|\frac{(-6.118\sigma) - 0}{1.33\sigma}\right| = \left|-4.635\right|$$

Probability of defect corresponding to any of the specification limits:

$$P_{LSL} \approx 1.725x10^{-6}$$

Considering both sides

$$P_{Total} \approx 2(1.725x10^{-6}) \approx 3.45x10^{-6}$$

Z-Value (Long Term) is $Zlt = 4.5$

Fig. 1 Defining a six sigma process: Short term (left), Long term (right)

If the feature is discrete, calculate the defects per opportunity (DPO), which represents the probability that the defect is present. In case of using yields, the value of Zlt is obtained initially with the calculus of throughput yield, these topics will be present in future chapters.

3.2 Modified Process Capacity Indices

The process capability index, C_p, C_{pk}, C_{pm}, and C_{pmk} have been proposed in the manufacturing and service industries providing numerical measures to determine if a process is capable of reproducing items within certain limits of specification. The indices have been defined as follows:

$$C_p = \frac{\mathrm{USL} - \mathrm{LSL}}{6\sigma} \tag{2}$$

$$C_{pk} = \min\left(\frac{\mathrm{USL} - \mu}{3\sigma}, \frac{\mu - \mathrm{LSL}}{3\sigma}\right) \tag{3}$$

$$C_{pm} = \frac{\mathrm{USL} - \mathrm{LSL}}{6\sqrt{\sigma^2 + (\mu - T_0)^2}} \tag{4}$$

$$C_{pmk} = \min\left(\frac{\mathrm{USL} - \mu}{3\sqrt{\sigma^2 + (\mu - T_0)^2}}, \frac{\mu - \mathrm{LSL}}{3\sqrt{\sigma^2 + (\mu - T_0)^2}}\right) \tag{5}$$

where USL is the upper specification limit, LSL is the lower specification limit, μ is the process mean, σ is the process standard deviation, and T_0 is the target value.

The use of Z-values as indicators of process capability can be linked to the C_{pk} index, using sigma level indicator, which denotes the Z-value of the closest limit specification to the mean, as follows:

$$\sigma_{\mathrm{level}} = \min\left(\frac{USL - \mu}{\sigma}, \frac{\mu - LSL}{\sigma}\right) \tag{6}$$

Then,

$$\begin{aligned} C_{pk} &= \mathrm{Min}\left(\frac{\mathrm{USL} - \mu}{3\sigma}, \frac{\mu - \mathrm{LSL}}{3\sigma}\right) \\ &= \mathrm{Min}\left[\frac{1}{3}\left(\frac{\mathrm{USL} - \mu}{\sigma}, \frac{\mu - \mathrm{LSL}}{\sigma}\right)\right] \\ &= \frac{1}{3}\mathrm{Min}\left(\frac{\mathrm{USL} - \mu}{\sigma}, \frac{\mu - \mathrm{LSL}}{\sigma}\right) \end{aligned}$$

$$= \frac{1}{3}\sigma_{\text{level}} \tag{7}$$

As proposed in this chapter, the use of indicator sigma-k, which links similar to the indicator C_{pmk}

$$\sigma_k = \min\left(\frac{\text{USL} - \mu}{\sqrt{\sigma^2 + (\mu - T_0)^2}}, \frac{\mu - \text{LSL}}{\sqrt{\sigma^2 + (\mu - T_0)^2}}\right) \tag{8}$$

The C_p index measures only the spread of distribution, which does not consider the location of the processes mean μ. The indices of C_{pk},C_{pm}, and C_{pmk} take into account the location of the process and the common cause variability, which offset some weakness in the C_p. The major distinction between the levels of C_{pk} and C_{pm} in the relative importance of the specification limits USL/LSL and T_0 goal. An elevated C_{pk} value implies a high processing performance, and a high value *Cpm* is a process of lower expected losses. C_{pmk} index combines the advantages of *Cpk*, and *Cpm*, which is more sensitive to the departure of the process mean from objective value of the other two indexes (Pearn, 2005). The use of Sigma Level and Sigma-k indicators to identify the relationship with measurements made with Z-values with conventional indices and *Cpk* and *Cpmk.*

Usually when determining capacity indicators are not considered errors of measurement within the same indicator, which may occur cannot be differences in the ability to gauge accuracy. The general practice is to make the measurement systems analysis prior to making process capabilities. Montgomery and Runger (1993) denote that the quality of data collection depends largely on the accuracy to show the data, and in case of not having a good measurement system will not have adequate confidence in the capacity indicators and therefore determine the criteria to identify the quality of the measurement system, expressing the gauge capability as

$$\lambda = \frac{6\sigma_{\text{gage}}}{\text{USL} - \text{LSL}} 100\% \tag{9}$$

λ is known as the percentage of variation of tolerance or gage capability, where σ_{gage} is the standard deviation of the measurement system. Table 1 shows the acceptance criteria for this parameter.

Thus by using the variation of measurement system, we can obtain modified expressions for capacity indicators

Table 1 Acceptance criteria for gage R&R

$\lambda < 10\%$	Gage R&R acceptable
$10\% < \lambda < 30\%$	Marginal gage R&R system
$\lambda > 30\%$	Gage R&R unacceptable

$$C_p = \frac{\text{USL} - \text{LSL}}{6\sqrt{\sigma^2 + \sigma_{\text{gage}}^2}} \tag{10}$$

$$C_{pk} = \min\left(\frac{\text{USL} - \mu}{3\sqrt{\sigma^2 + \sigma_{\text{gage}}^2}}, \frac{\mu - \text{LSL}}{3\sqrt{\sigma^2 + \sigma_{\text{gage}}^2}}\right) \tag{11}$$

$$C_{pm} = \frac{\text{USL} - \text{LSL}}{6\sqrt{\sigma^2 + (\mu - T_0)^2 + \sigma_{\text{gage}}^2}} \tag{12}$$

$$C_{pmk} = \min\left(\frac{\text{USL} - \mu}{3\sqrt{\sigma^2 + (\mu - T_0)^2 + \sigma_{\text{gage}}^2}}, \frac{\mu - \text{LSL}}{3\sqrt{\sigma^2 + (\mu - T_0)^2 + \sigma_{\text{gage}}^2}}\right) \tag{13}$$

$$\sigma_{\text{level}} = \min\left(\frac{\text{USL} - \mu}{\sqrt{\sigma^2 + \sigma_{\text{gage}}^2}}, \frac{\mu - \text{LSL}}{\sqrt{\sigma^2 + \sigma_{\text{gage}}^2}}\right) \tag{14}$$

$$\sigma_k = \min\left(\frac{\text{USL} - \mu}{\sqrt{\sigma^2 + (\mu - T_0)^2 + \sigma_{\text{gage}}^2}}, \frac{\mu - \text{LSL}}{\sqrt{\sigma^2 + (\mu - T_0)^2 + \sigma_{\text{gage}}^2}}\right) \tag{15}$$

$$Z = \frac{x - \mu}{\sqrt{\sigma^2 + \sigma_{\text{gage}}^2}} \tag{16}$$

Thus, the ability to process indicators will be sensitive to the measurement system, improving accuracy.

3.3 Measurement System Analysis for Continuous Data

Measurement system analysis evaluates how appropriate a measurement system is for a given application. When measuring the output of a process, consider two sources of variation:

- Variation from part to part
- Variation of the measurement system.

If the variation of the measurement system is large compared to the variation from part to part, the measurements may not provide useful information.

It is important to use a measurement system analysis before collecting data from your process (e.g., to analyze the control or process capacity), use a measurement system analysis to confirm that the measurement system measures consistently and exact, and that properly discriminates between the parties.

The R&R study of the cross-measurement system estimates how much of the total process variation is caused by the measurement system. The total variation of the process consists of the variation from part to part plus the variation of the measurement system. The variation of the measurement system consists of:

- Repeatability— ariation due to the measuring device, or the variation observed when the same operator measures the same part repeatedly with the same device
- Reproducibility—variation due to the measurement system, or variation observed when different operators measure the same part using the same device.

When repeatability is estimated, each operator measures each part at least twice. When reproducibility is estimated, at least two operators must measure the parts. The operators measure the parts in random order, and the selected parts represent the possible range of measurements.

It is advisable to carry out the R&R study of the measurement system through the analysis of variance (ANOVA) since it allows comparing the variation of the measurement system with the total variation of the process or tolerance. If the variation of the measurement system is large in proportion to the total variation, the measurement system may not distinguish adequately between the parts.

σ_{gage} is calculated from the terms of the repeatability and reproducibility variance using the following relationship.

$$\sigma_{gage}^{2} = \sigma_{repeatability}^{2} + \sigma_{reproducibility}^{2} \tag{17}$$

Precision to Total Ratio (% Variation Study). It is used to compare the variation of the measurement and the variation of the process. It is a measurement of how good the measurement system is for use in process control. The variation of the process is determined by the variation of the samples. Therefore, it is extremely important to choose parts that represent the variation of the actual process.

$$\lambda = \%\text{Var} = \frac{\sigma_{gage}}{\sigma_{total}}(100\%) \tag{18}$$

Precision to Tolerance Ratio (% Tolerance). It is used to compare the variation of the measurement with respect to the specification limits. It is a measure of how well the system can determine if a product is outside or within specification.

$$\lambda_{\text{Tol}} = \%\text{Tol} = \frac{6\left(\sigma_{gage}\right)}{\text{Tolerance}}(100\%) \tag{19}$$

The Number of Distinct Categories (Resolution) represents the number of confidence intervals that do not overlap and will cover the range of product variation. It is also considered as the number of groups within their process data that their measurement system can discern. Table 2 shows the acceptance criteria for number of Distinct Categories.

Table 2 Acceptance criteria for number of distinct categories

Number of distinct categories (Resolution)	Decision/value
< 2	Not valid for making process control decisions
2–3	Binary decisions (Go, Not Go)
≥ 4	Enough resolution Acceptable to make decisions about process control (currently AIAG has modified the acceptance criteria to greater than or equal to 5)

$$\text{Resolution} = \frac{\sigma_{\text{product}}}{\sigma_{\text{gage}}}(100\%) \tag{20}$$

3.4 *ANOVA—Gage R&R Method*

The ANOVA-Gage R&R analysis assess the ability to measure the product or process in a precise and consistent way since it allows to determine.

- If the variability of the measurement system is small compared to the process variability.
- How much of the variability in the measurement system is caused by differences between operators.
- If the measurement system is able to discriminate between different elements (parts).

Gage R&R calculates the total variation (VT) from three sources:

- Parts or elements that are measured.
- Appraisers or operators.
- Equipment (gage) or measuring equipment.

Within the gage analysis, we will always consider an operative factor (appraisal) with $\boldsymbol{n_o}$ levels, which represents the number of operators who makes the measurement (or the recording device), an element factor (part) with $\boldsymbol{n_p}$ levels, which represents each of the different elements that are measured, and we will assume a cross design in which each operator measures all the elements, and each of them a number of times, $\boldsymbol{n_{\text{run}}}$.

The total variance can therefore be broken down according to:

$$\sigma^2 = \sigma^2_{\text{error}} + \sigma^2_{\text{part}} + \sigma^2_{\text{operator}} + \sigma^2_{\text{part:operator}} \tag{21}$$

where the variance of the error will be representing the error inherent in the measurement, not explainable by variations between the operators/devices they measure and therefore identifies the consistency in the measurement of the same element by the same operator, that is, the error of repeatability by definition,

$$\sigma^2_{\text{repeatability}} = \sigma^2_{\text{error}} \tag{22}$$

The first step to develop the ANOVA Analysis is to determine the Sums of Squares for the parameters: parts, operators, and runs, determined by the following equations:

$$SS_{\text{part}} = \frac{P_1^2 + \cdots + P_n^2}{n_o \cdot n_{\text{run}}} - \frac{T^2}{n_p \cdot n_o \cdot n_{\text{run}}} \tag{23}$$

$$SS_{\text{oper}} = \frac{At^2 + Bt^2 + Ct^2}{n_p \cdot n_{\text{run}}} - \frac{T^2}{n_p \cdot n_o \cdot n_{\text{run}}} \tag{24}$$

The sum of squares for the interaction of the operator with the parts is determined by,

$$SS_{\text{part:oper}} = \frac{A_1^2 + \cdots + C_n^2}{n_{\text{run}}} - \frac{T^2}{n_p \cdot n_o \cdot n_{\text{run}}} - SS_{\text{part}} - SS_{\text{oper}} \tag{25}$$

The total sum of squares for data set and the sum of squares for repeatability are determined as

$$SS_{\text{Total}} = \sum n_i^2 - \frac{T^2}{n_p \cdot n_o \cdot n_{\text{run}}} \tag{26}$$

$$SS_{\text{repeatibility}} = SS_{\text{error}} = SS_{\text{Total}} - SS_{part} - SS_{\text{oper}} - SS_{\text{part:oper}} \tag{27}$$

The second step is to determine the degrees of freedom (dF) for each concept of sum of squares, using the following equations:

$$\text{dF}_{\text{part}} = n_p - 1 \tag{28}$$

$$\text{dF}_{\text{oper}} = n_o - 1 \tag{29}$$

$$\text{dF}_{\text{part:oper}} = (\text{dF}_{\text{part}})(\text{dF}_{\text{oper}}) \tag{30}$$

$$\text{dF}_{\text{total}} = n_p \cdot n_o \cdot n_{run} - 1 \tag{31}$$

$$\text{dF}_{\text{repeatability}} = \text{dF}_{\text{total}} - \text{dF}_{\text{part}} - \text{dF}_{\text{oper}} - \text{dF}_{\text{part:oper}} \tag{32}$$

The third step of the process is to determine the mean squares that represent an estimate of the population variance. They are calculated by dividing the corresponding sum of the squares by the degrees of freedom.

$$MS = \frac{SS}{dF} \tag{33}$$

In ANOVA, the mean squares are used to determine if the factors (treatments) are significant. As mentioned above, the average square of the treatment is determined by dividing the sum of the squares of the treatment by the degrees of freedom. The mean square of the treatment represents the variation between the means of the samples. The mean square of the error (MSE) is obtained by dividing the sum of the squares of the residual error by the degrees of freedom. The MSE represents the variation within the samples.

The fourth step of the process is to determine the F values, which are obtained by dividing the MS (term) by the MSE, which follows the F distribution with degrees of freedom for the term and degrees of freedom for error.

$$F_{\text{part}} = \frac{MS_{\text{part}}}{MS_{\text{part:oper}}} \tag{34}$$

$$F_{\text{oper}} = \frac{MS_{\text{oper}}}{MS_{\text{part:oper}}} \tag{35}$$

$$F_{\text{part:oper}} = \frac{MS_{\text{part:oper}}}{MSE} \tag{36}$$

The statistical values are determined by tables of the F distribution, considering the level of confidence and the degrees of freedom that determine the term:

$$F\text{value}_{part} = F\left(\alpha, dF_{\text{part}}, dF_{\text{part:oper}}\right) \tag{37}$$

$$F\text{value}_{\text{oper}} = F\left(\alpha, dF_{\text{oper}}, dF_{\text{part:oper}}\right) \tag{38}$$

$$F\text{value}_{\text{part:oper}} = F\left(\alpha, dF_{\text{part:oper}}, dF_{\text{repeatability}}\right) \tag{39}$$

The fifth step of the ANOVA analysis is to determine the variances of the sources of variation defined by each term, using the following equations:

$$\sigma^2_{\text{part}} = \frac{MS_{\text{part}} - MS_{\text{part:oper}}}{n_o \cdot n_{\text{run}}} \tag{40}$$

$$\sigma^2_{\text{oper}} = \frac{MS_{\text{oper}} - MS_{part:oper}}{n_{part} \cdot n_{\text{run}}} \tag{41}$$

$$\sigma^2_{\text{part:oper}} = \frac{MS_{\text{part:oper}} - MSE}{n_{\text{run}}} \tag{42}$$

$$\sigma^2_{\text{repeatibility}} = MSE \tag{43}$$

The sixth step of the ANOVA analysis is to determine the standard deviations of the sources of variation defined by each term:

$$\text{StdDev} = \sigma_i = \sqrt{\sigma_i^2} \tag{44}$$

The study variation is calculated as the standard deviation for each source of variation multiplied by 6 within the lean six sigma standards. When the data follows a normal distribution, approximately 99.73% of the data is within 6 standard deviations from the average.

Another value commonly used as a multiplier of the standard deviation is 5.15 which corresponds to 99% of the data when the standard distribution is used, and which is used by some statistical software and calculation sheets of measurement system analysis.

$$\%\text{Var}_i = \frac{\sigma_i}{\sigma_{\text{total}}} \tag{45}$$

The tolerance percentage is determined by the following equation:

$$\%\text{Tol}_i = \frac{5.15\sigma_i}{\text{USL} - \text{LSL}} \tag{46}$$

The calculation of the number of distinct categories is determined by

$$\text{DistinctCategories} = \frac{\%\text{Var}_{\text{part}}}{\%\text{Var}_{\text{gage}}} \tag{47}$$

To ensure that the results are valid, consider the following guidelines when collecting data, performing the analysis and interpreting the results.

- Operators must measure the parts in random order. To ensure that the order of data collection does not influence the results, each operator must measure all parts randomly within a replica. After all operators measure all parts once, you repeat the process for all runs (Barrentine 2003).
- Operators must measure at least 10 parts for proper study. The variation of the process can be estimated using a large sample of historical data or using the parts included in the study. If you have a historical estimate of the process variation, the usual requirement of 10 parts is acceptable.
- Select the parts that represent the actual or expected range of the process variation. Select parts of the entire process range to increase the probability of having an adequate estimate of the variation between the parts.

- The operator and part factors must be random. A factor is random when the factor has many possible levels, but only a random sample of the levels is included in the data.
- Operators must measure each part at least twice. The variation of the measurements is divided into two components: reproducibility and repeatability. Reproducibility is the variation that occurs when different people measure the same part. Repeatability is the variation that occurs when the same person measures the same part repeatedly. If you use at least 10 parts and at least 3 operators, having each operator measure each part at least 2 times, in random order, allows you to obtain an adequate estimate of repeatability (Barrentine 2003).
- You must have at least 3 operators for a proper study. For best results, include 3–5 operators in the study. You should not have less than 3 operators in the study, unless the number of operators using the measurement system is actually less than 3. If you suspect there are large differences between the operators, you should consider using more than 3–5 operators. If you identify differences between operators, such as an operator whose measurements are lower than those of others, consistency can often be improved with training (Fechio 2009).

4 ANOVA-Gage R&R Example

The present example developed the analysis process of the measurement system using the ANOVA method, considering a quality characteristic with an upper specification limit of 1.1 mm and a lower specification limit of 0.5 mm, which determines a tolerance equal to 0.6 mm.

Ten parts representing the variation of the process are selected, which are measured twice randomly by three operators, complying with the recommendations made in the previous section. The data obtained are shown in Table 3.

Determine the parameters for parts, operators, and runs:

$$\begin{aligned} \text{Parts} \quad n_p &= 10 \\ \text{Operators} \quad n_o &= 3 \\ \text{Runs} \quad n_{\text{run}} &= 2 \end{aligned}$$

First step is to obtain the sums of squares for the terms of parts and operator for gage R&R using Eqs. 23 and 24,

$$SS_{\text{part}} = \frac{4.275^2 + \cdots + 3.725^2}{(3)(2)} - \frac{49.725^2}{(10)(3)(2)} = 1.0244$$

$$SS_{\text{oper}} = \frac{16.650^2 + 16.400^2 + 16.675^2}{(10)(2)} - \frac{49.725^2}{(10)(3)(2)} = 0.0023$$

Table 3 Data set for quality characteristic

Operator	Run	P1	P2	P3	P4	P5	P6	P7	P8	P9	P10	Sum
A	1	0.725	1.000	0.850	0.850	0.650	1.000	0.950	0.800	0.900	0.600	At
	2	0.700	1.000	0.825	0.875	0.650	1.000	0.950	0.825	0.875	0.625	16.650
	Sum (A)	1.425	2.000	1.675	1.725	1.300	2.000	1.900	1.625	1.775	1.225	
B	1	0.675	1.050	0.800	0.825	0.625	1.000	0.950	0.775	0.925	0.625	Bt
	2	0.700	0.975	0.775	0.850	0.650	0.975	0.925	0.800	0.900	0.600	16.400
	Sum (A)	1.375	2.025	1.575	1.675	1.275	1.975	1.875	1.575	1.825	1.225	
C	1	0.725	1.050	0.800	0.825	0.650	1.000	0.950	0.800	0.900	0.625	Ct
	2	0.750	1.025	0.800	0.825	0.675	0.975	0.950	0.800	0.900	0.650	16.675
	Sum (A)	1.475	2.075	1.600	1.650	1.325	1.975	1.900	1.600	1.800	1.275	
		P1	P2	P3	P4	P5	P6	P7	P8	P9	P10	T
Total sum		4.275	6.100	4.850	5.050	3.900	5.950	5.675	4.800	5.400	3.725	49.725

The sum of squares for the interaction of the operator with the parts is determined by Eq. 25,

$$SS_{\text{part:oper}} = \frac{1.425^2 + \cdots + 1.275^2}{2} - \frac{49.725^2}{(10)(3)(2)} - 1.0244 - 0.002 = 0.0089$$

The total sum of squares for data set and the sum of squares for repeatability are determined by Eqs. 26 and 27

$$SS_{\text{Total}} = \left[0.725^2 + \cdots + 0.650^2\right] - \frac{49.725^2}{(10)(3)(2)} = 1.04478$$

$$SS_{\text{repeatibility}} = SS_{\text{error}} = 1.04478 - 1.0244 - 0.0023 - 0.0089 = 0.0091$$

The second step is to determine the degrees of freedom (dF) for each concept of sum of squares, using Eqs. 28–32:

$$dF_{\text{part}} = 10 - 1 = 9$$

$$dF_{\text{oper}} = 3 - 1 = 2$$

$$dF_{\text{part:oper}} = (9)(2) = 18$$

$$dF_{\text{total}} = (10)(3)(2) - 1 = 59$$

$$dF_{\text{repeatability}} = 59 - 9 - 2 - 18 = 30$$

The third step of the process is to determine the mean squares that represent an estimate of the population variance. They are calculated by dividing the corresponding sum of the squares by the degrees of freedom, using Eq. 33,

$$MS_{\text{part}} = \frac{SS_{\text{part}}}{dF_{\text{part}}} = \frac{1.02447}{9} = 0.11383$$

$$MS_{\text{oper}} = \frac{SS_{\text{oper}}}{dF_{\text{oper}}} = \frac{0.00231}{2} = 0.001156$$

$$MS_{\text{part:oper}} = \frac{SS_{\text{part:oper}}}{dF_{\text{part:oper}}} = \frac{0.00894}{18} = 0.000497$$

$$MS_{\text{repeatability}} = MS_{\text{error}} = \frac{SS_{\text{error}}}{dF_{\text{error}}} = \frac{0.00906}{30} = 0.000302$$

The fourth step of the process is to determine the F values, which are obtained by dividing the MS (term) by the MSE, which follows the F distribution with degrees of freedom for the term and degrees of freedom for error, using Eqs. 34–36:

$$F_{\text{part}} = \frac{MS_{\text{part}}}{MS_{\text{part:oper}}} = \frac{0.11383}{0.000497} = 229.25175$$

$$F_{\text{oper}} = \frac{MS_{\text{oper}}}{MS_{\text{part:oper}}} = \frac{0.001156}{0.00497} = 2.32867$$

$$F_{\text{part:oper}} = \frac{MS_{\text{part:oper}}}{MSE} = \frac{0.000497}{0.000302} = 1.64368$$

The statistical values are determined by tables of the F distribution, considering the level of confidence of 95%, using MS-Excel can be obtained with the function DISTR.F.INV:

$$Fvalue_{\text{part}} = F\left(\alpha, dF_{\text{part}}, dF_{\text{part:oper}}\right) = F(0.05, 9, 18) = 2.456281$$

$$Fvalue_{\text{oper}} = F\left(\alpha, dF_{oper}, dF_{\text{part:oper}}\right) = F(0.05, 2, 18) = 3.554557$$

$$Fvalue_{\text{part:oper}} = F\left(\alpha, dF_{part:oper}, dF_{\text{repeatability}}\right) = F(0.05, 18, 30) = 1.96011$$

The significance of each of the terms can be determined by comparing the F values, with their respective table values. As an alternative, the P values for the F distribution can be calculated as is done by most software for analysis of measurement systems. With MS-Excel, the function is used: DISTR.F

$$Pvalue_{\text{part}} = Dist.F\left(F, dF_{\text{part}}, dF_{\text{part:oper}}\right) = F(229.2517, 9, 18) = 0.00000$$

$$Pvalue_{\text{oper}} = Dist.F\left(F, dF_{\text{oper}}, dF_{\text{part:oper}}\right) = F(2.32867, 2, 18) = 0.12606$$

$$Pvalue_{\text{part:oper}} = Dist.F\left(\alpha, dF_{\text{part:oper}}, dF_{\text{repeat}}\right) = F(1.6437, 18, 30) = 0.11124$$

The fifth step of the ANOVA analysis is to determine the variances of the sources of variation defined by each term, using the following equations:

$$\sigma^2_{\text{part}} = \frac{MS_{\text{part}} - MS_{\text{part:oper}}}{n_o \cdot n_{\text{run}}} = \frac{0.11383 - 0.000497}{(3)(2)} = 0.018889$$

$$\sigma^2_{\text{oper}} = \frac{MS_{\text{oper}} - MS_{\text{part:oper}}}{n_{\text{part}} \cdot n_{\text{run}}} = \frac{0.001156 - 0.000497}{(10)(2)} = 0.000033$$

$$\sigma^2_{\text{part:oper}} = \frac{MS_{\text{part:oper}} - MSE}{n_{\text{run}}} = \frac{0.000497 - 0.000302}{2} = 0.000097$$

$$\sigma^2_{\text{repeatibility}} = MSE = 0.000302$$

The variance for reproducibility is equal to the sum of variances of the terms: operator and the parts-operator interaction:

$$\sigma^2_{\text{reproducibility}} = \sigma^2_{\text{oper}} + \sigma^2_{\text{part:oper}}$$

$$\sigma^2_{\text{reproducibility}} = 0.000033 + 0.000097 = 0.00013$$

The variance for gage R&R is determined by the sum of repeatability and reproducibility variances, using Eq. 17:

$$\sigma^2_{\text{gage}} = \sigma^2_{\text{repeatability}} + \sigma^2_{\text{reproducibility}}$$

$$\sigma^2_{\text{gage}} = 0.000302 + 0.00013 = 0.000432$$

The total variance is determined by adding variance of the gage and the variance of the parts, as follows

$$\sigma^2_{\text{total}} = \sigma^2_{\text{gage}} + \sigma^2_{\text{part}}$$

$$\sigma^2_{\text{total}} = 0.000432 + 0.018889 = 0.19321$$

The sixth step of the ANOVA analysis is to determine the standard deviations of the sources of variation defined by each term.

$$\sigma_{\text{gage}} = \sqrt{\sigma^2_{\text{gage}}} = \sqrt{0.000432} = 0.020792$$

In the same way, the standard deviations are obtained for the rest of the terms

$$\sigma_{\text{repeatability}} = \sqrt{0.000302} = 0.017381$$

$$\sigma_{\text{reproducibility}} = \sqrt{0.00013} = 0.011411$$

$$\sigma_{\text{oper}} = \sqrt{0.000033} = 0.005743$$

$$\sigma_{\text{part:oper}} = \sqrt{0.000097} = 0.00986$$

$$\sigma_{\text{part}} = \sqrt{0.018889} = 0.137437$$

$$\sigma^2_{\text{total}} = \sqrt{0.19321} = 0.139001$$

The contribution percentage of each of the sources of variation is obtained by dividing the variance of the term by the total variance.

$$\%\text{Contr}_{\text{gage}} = \frac{\sigma^2_{\text{gage}}}{\sigma^2_{\text{total}}}(100\%) = \frac{0.000432}{0.019321}(100\%) = 2.24\%$$

In the same way, the percentages of contributions of the other terms are obtained

$$\%\text{Contr}_{\text{repeatability}} = \frac{0.000302}{0.019321}(100\%) = 1.56\%$$

$$\%\text{Contr}_{\text{reproducibility}} = \frac{0.00013}{0.019321}(100\%) = 0.67\%$$

$$\%\text{Contr}_{\text{oper}} = \frac{0.000033}{0.019321}(100\%) = 0.17\%$$

$$\%\text{Contr}_{\text{part:oper}} = \frac{0.000097}{0.019321}(100\%) = 0.50\%$$

$$\%\text{Contr}_{\text{part}} = \frac{0.018889}{0.019321}(100\%) = 97.76\%$$

The percentages of variation (% Study Var) and the percentages of tolerance (%Tol) can be obtained for the 6 s parameter, used in six sigma processes, or using 5.15 s as discussed in the previous section.

In the case of the% Study Var, there will be no difference in the results obtained with any of the 6 s or 5.15 s parameters. And only in the case of % tolerance differences are presented.

$$\%\text{Var}_{\text{gage}} = \frac{\sigma_{\text{gage}}}{\sigma_{\text{total}}}(100\%) = \frac{0.020792}{0.139001}(100\%) = 14.96\%$$

$$\%\text{Tol}_{\text{gage}} = \frac{5.15\sigma_{gage}}{\text{USL} - \text{LSL}}(100\%) = \frac{0.107077}{1.1 - 0.5}(100\%) = 17.85\%$$

If the 6 s parameter is used, the % tolerance shows the following result:

$$\%\text{Tol}_{\text{gage}} = \frac{5.15\sigma_{\text{gage}}}{\text{USL} - LSL}(100\%) = \frac{0.124750}{1.1 - 0.5}(100\%) = 20.79\%$$

Tables 4 and 5 show the results of the analysis of the measurement system using both parameters (6S and 5.15S).

For our case study, the dispersion value obtained as a standard deviation of the Gage R&R is used in the indices of process capacities modified by the measurement system error. As this result is observed, it is the same value of standard deviation

Table 4 Results ANOVA-Gage R&R (using parameter 5.15 s)

Source of variation	SS	GL	MS	F	F(Table)	P-Value
Parts	1.02447	9	0.113830	229.25175	2.45628	0.00000
Operators	0.00231	2	0.001156	2.32867	3.55456	0.12606
Part:Operator	0.00894	18	0.000497	1.64368	1.96012	0.11124
Repeatability (Error)	0.00906	30	0.000302			
total	1.04478	59				
Source of variation	Variance	Std. Dev	5.15 s	% Contribution	% Study Var	% Tolerance
Total gage R&R	0.000432	0.020792	0.107077	2.24	14.96	17.85
Repeatability	0.000302	0.017381	0.089510	1.56	12.50	14.92
Reproducibility	0.000130	0.011411	0.058766	0.67	8.21	9.79
Operator	0.000033	0.005743	0.029578	0.17	4.13	4.93
Part:Operator	0.000097	0.009860	0.050780	0.50	7.09	8.46
Parts to part	0.018889	0.137437	0.707800	97.76	98.87	117.97
Total variation	0.019321	0.139001	0.715853	100.00	100.00	119.31
Precision to total ratio (% Variation)					14.96	
Precision to tolerance ratio (% Tolerance)					17.85	
Resolution (number of distinct categories)					9.52	

Table 5 Results ANOVA-gage R&R (using parameter 6 s)

Source of variation	SS	GL	MS	F	F(Table)	P-Value
Parts	1.02447	9	0.113830	229.25175	2.456281	0.00000
Operators	0.00231	2	0.001156	2.32867	3.554557	0.12606
Part:Operator	0.00894	18	0.000497	1.64368	1.960116	0.11124
Repeatability (Error)	0.00906	30	0.000302			
total	1.04478	59				
Source of Variation	Variance	Std. Dev	6 s	% Contribution	% Study Var	% Tolerance
Total Gage R&R	0.000432	0.020792	0.124750	2.24	14.96	20.79
Repeatability	0.000302	0.017381	0.104283	1.56	12.50	17.38
Reproducibility	0.000130	0.011411	0.068465	0.67	8.21	11.41
Operator	0.000033	0.005743	0.034460	0.17	4.13	5.74
Part:Operator	0.000097	0.009860	0.059161	0.50	7.09	9.86
Parts to part	0.018889	0.137437	0.824621	97.76	98.87	137.44
Total variation	0.019321	0.139001	0.834004	100.00	100.00	139.00
Precision to total ratio (% Variation)						14.96
Precision to tolerance ratio (% Tolerance)						20.79
Resolution (number of distinct categories)						9.52

of the measurement system regardless of the parameter used (6 s or 5.15 s) in the statistical software.

The only difference is denoted in the percentage of tolerance within the system, when using the difference of parameters.

5 Multi-process Evaluation

The evaluation of multi-processing allows the flexibility to generate control actions to maintain and improve the regularity of the process, the need for uniformity. The consistency of the process also requires consistency in other aspects of process management, for example, how the team is maintained, information is collected and analyzed the process and how changes are made in the process. The lack of consistency in the process can cause poor performance flexibility.

The process analysis involves the ability to characterize or evaluate products and processes based on engineering specifications with more than one characteristic or quality variables. When these variables are related to the characteristics, the analysis should be based on a multivariate statistical technique. Researchers such as Chen (2008) and Kunlum (2008) presented the capacity of multiple indices to assess the capacity variables. Pearn (2007) proposed multivariate equivalent C_p, C_{pk}, C_{pm}, and C_{pmk} based on principal component analysis, which transforms the original number of variables related to measuring a set of linear functions uncorrected.

For practical applications, most multi-process products are comprised of numerous unilateral and bilateral specifications. Additionally, customers are satisfied when all the attributes of a complete product quality meet the predetermined specifications. The proposed analysis is based on determining the total capacity of product to meet specifications and further evaluation of the critical process variables.

Process control is a basic element of quality production, based on continuous improvement and monitoring of critical parameters of processes and products. The statistical methods are essential and must be complemented by organizational activities of the methods. In some cases, batch processes have been used for producing high-quality foundry products, ceramic and within other industries. Monitoring and fault diagnosis on the process conditions are very important to ensure the safety of facilities and high-quality products. The multi-process statistical process control can be applied to this type of analysis.

When the capacity studies of a process are carried out, the individual results for each quality characteristic can be obtained, and additionally the contribution these characteristics to the analyzed system.

It is recommended to use the process capacity indicators modified by the measurement system error, when there are marginal or inefficient measurement processes, since these will be strong contributors in the acceptance or rejection of parts.

When the measurement systems are acceptable, the contribution of the measurement system to the variation will not be significant, and therefore, the standard process capability indicators can be used with an acceptable level of confidence.

6 Results

In the following, scoreboard is presented process capabilities with the implementation of measurement systems and contrast with conventional indicators.

In Tables 6 and 7, it can be seen that using only the standard deviations of each of the quality characteristics of two processes within a manufacturing company.

When performing the multi-process analysis, it is observed that the total system capability has a Zlt (long term) of 3.52, which represents 212 parts per million defects and corresponds to an equivalent C_{pk} of 1.18. When considering the measurement error considered to the MSA of each quality characteristic and using the modified capability indicators, it is obtained that the process capability of the system has a Zlt value of 2.06 with 19,801 parts per million defects that corresponds to an equivalent C_{pk} of 0.69.

When comparing each quality characteristic, the impact produced by the measurement system can be observed, verifying that a characteristic does not present a significant difference when it has an acceptable measurement system, and as the measurement systems are marginal or unacceptable, its impact has become significant within

Table 6 Multi-process results without gage error

	Continuous data						QTY	CTQ Capacity					
Parameter (KCC)	Units	LSL	Target	USL	Mean	Std. Dev		Defect probability	Z_{LT}	Yield	Z_{LT} Goal	Cpk	Cpk Goal
Construction process													
Characteristic 1	mm	5.7	6	6.3	5.993	0.0596	1	5.97E−07	4.86	99.99994%	4.0	2.18	1.33
Characteristic 2	amp	260	280	300	277.83	4.2815	1	1.58E−05	4.16	99.99842%	4.0	1.85	1.33
Characteristic 3	amp	620	630	640	630.29	2.3422	1	2.27E−05	4.08	99.99773%	4.0	1.84	1.33
Characteristic 4	amp	55	60	65	63.8725	0.2646	1	1.03E−05	4.26	99.99697%	4.0	1.89	1.33
Characteristic 5	amp	85	90	95	88.2675	0.7695	1	1.10E−05	4.24	99.99690%	4.0	1.89	1.33
Enamel process							6	2.27E−03	3.37	99.77%			
Characteristic 1	mls/pulg	6	9	12	9.4185	0.7125	1	1.465E−04	3.62	99.98535%	4.0	1.61	1.33
Characteristic 2	mls/pulg	8	11	14	10.5325	0.7641	1	4.622E−04	3.31	99.95378%	4.0	1.47	1.33
Characteristic 3	gr	1.76	1.78	1.8	1.7819	0.0045	1	2.953E−05	4.02	99.99705%	4.0	1.79	1.33
Characteristic 4	gr	30	32	34	31.8025	0.4469	1	2.807E−05	4.03	99.99719%	4.0	1.79	1.33
Characteristic 5	cm/min	220	230	240	227.075	1.7376	1	2.344E−05	4.07	99.99766%	4.0	1.81	1.33

(continued)

Table 6 (continued)

	Continuous data						QTY	CTQ Capacity					
Parameter (KCC)	Units	LSL	Target	USL	Mean	Std. Dev		Defect probability	Z_{LT}	Yield	Z_{LT} Goal	Cpk	Cpk Goal
Characteristic 6	mm	24	25	26	24.95	0.2345	1	1.581E−03	2.95	99.84185%	4.0	1.35	1.33
System level indicators						212	3.52	99.9788%		1.17			

Table 7 Multi-process result with gage measurement error

	Continuous data								QTY	CTQ capability					
Parameter KCC	Units	LSL	Target	USL	Mean	Std. Dev	Gage	Std Dev Syst		Defect probability	Z_{LT}	Yield	Z_{LT} Goal	Cpk	Cpk Goal
Construction process									5	8.64E−05	4.14	99.99%			
Characteristic 1	mm	5.7	6	6.3	5.993	0.0596	0.0053	0.0598352	1	6.59E−07	4.84	99.99993%	4.0	2.18	1.33
Characteristic 2	amp	260	280	300	277.83	4.2815	1.231	4.4549527	1	3.18E−05	4.00	99.99682%	4.0	1.78	1.33
Characteristic 3	amp	620	630	640	630.29	2.3422	0.0837	2.3436951	1	2.30E−05	4.08	99.99770%	4.0	1.84	1.33
Characteristic 4	amp	55	60	65	63.8725	0.2646	0.0728	0.2744321	1	2.00E−05	4.11	99.99800%	4.0	1.83	1.33
Characteristic 5	amp	85	90	95	88.2675	0.7695	0.0034	0.7695075	1	1.10E−05	4.24	99.99890%	4.0	1.89	1.33
Enamel process									6	2.27E−03	3.37	99.77%			
Characteristic 1	mls/pulg	6	9	12	9.4185	0.7125	0.0167	0.7126957	1	1.470E−04	3.62	99.98530%	4.0	1.61	1.33
Characteristic 2	mls/pulg	8	11	14	10.5325	0.7641	0.0799	0.7682661	1	4.930E−04	3.29	99.95070%	4.0	1.47	1.33
Characteristic 3	gr	1.76	1.78	1.8	1.7819	0.0045	0.0013	0.004684	1	5.735E−05	3.86	99.99426%	4.0	1.72	1.33
Characteristic 4	gr	30	32	34	31.8025	0.4469	0.0461	0.4492714	1	3.073E−05	4.01	99.99693%	4.0	1.78	1.33

(continued)

Table 7 (continued)

	Continuous data								QTY	CTQ capability					
Parameter KCC	Units	LSL	Target	USL	Mean	Std. Dev	Gage	Std Dev Syst		Defect probability	Z_{LT}	Yield	Z_{LT} Goal	Cpk	Cpk Goal
Characteristic 5	cm/min	220	230	240	227.075	1.7376	0.823	1.9226499	1	1.168E−04	3.68	99.98832%	4.0	1.64	1.33
Characteristic 6	mm	24	25	26	24.95	0.2345	0.559	0.6061941	1	2.169E−01	0.78	78.31205%	4.0	0.52	1.33
						System level indicators	1.980E+04	2.06	98.02%		0.69				

the evaluation of the process capability indicator. To clarify the results obtained, the following comparisons are made:

The characteristic 1 of the construction process presented a good measurement system with a standard deviation of 0.0053; therefore, it does not present a significant difference between the analyses of process capacity without using the standard deviation, in contrast to the use of the standard deviation of the measurement system, obtaining a Zlt value of 4.86 versus 4.84. In C_{pk} terms, the variation is not noticeable in the indicator, maintaining a value of 2.18.

In contrast to the characteristic 6 of the enamel process, it shows an inadequate measurement system with a standard deviation of 0.2345, which contributed significantly to the process capacity, going from a Zlt value of 2.95 to 0.78. In *Cpk* terms, the variation of the indicator goes from a value of 1.35 to 0.52.

7 Conclusions

Considering the standard deviation of the measurement systems allows the process indicators to be more accurate, so it is recommended, make the measurement systems and include such variation.

The results obtained allow us to confirm our working hypotheses, within manufacturing systems:

1. Include the measurement error within the process indicators, allow a fine analysis of the critical characteristics of the process, when the percentage of variation is marginal or unacceptable within a process, allowing to know the real impact on the effectiveness of the control quality.
2. The use of a multi-process analysis shows the global performance, allowing corrective actions to be taken after detecting variations caused by errors in the measurement systems and effects on the process.

Additionally, we can conclude that the *Z*-values used as an indicator of the capacity of the process allow determining the probability of non-compliance, in multi-process evaluations, and from which an equivalent C_{pk} can be calculated for the global process, which cannot be calculated directly when using C_p or C_{pk} only.

Using measurement error improves process detection, especially when the need to use process capacity indicators is increasingly relevant, since the competitiveness of companies is based on optimizing processes to reduce production costs. Basically, a process capacity can be specified by the range that limits the values obtained on a quality characteristic, considering that the variation effects related to the measurement system are negligible. However, as it could be observed in this research work, some processes present circumstances that limit the measurement system on a quality characteristic, causing the percentage of variation of the system to be considered within the capacity indices (PCI) as part of the natural variation of the process.

Employing the process capability indicators considering the measurement error is a relevant step within the companies that intend to improve their processes under six sigma techniques since there are marginal processes in their measurement systems, the variation generated by the MSA presents differences in the actual capacity of the process, a situation that a large number of companies do not find an explanation for why there is no contrast between the calculated values and the process situation.

In the same way, it is important to carry out the multi-process analysis of the system to identify the total contribution that exists when relating the quality characteristics, and the relationship between the PCI that are Cp, C_{pk}, C_{pm} and C_{pmk}, S_{pk} and sigma numbers. The effectiveness of employing multi-process analysis, with continuous data presenting normal distributions, may be an opportunity that encourages quality professionals in different industrial sectors to begin to stop relying on independent evaluation of univariate capacity within processes of manufacturing with multiple related variables. Allowing evaluation of improvement efforts through a general evaluation of process performance, regardless of fluctuations in the individual capacities of its variables.

In future investigations, the intention is to carry out the multi-process analysis with the inclusion of discrete variables and mixtures of normal distributions generated by long-term sampling that allow determining the effects of equipment deterioration within manufacturing processes.

Acknowledgements We are grateful for the support of the Industrial and Manufacturing Engineering Departments of the Research center COMIMSA-CONACYT, and the Tecnológico Nacional de México in its campus of Querétaro and Orizaba, as well as the Universidad Politécnica de Querétaro.

References

Ait-Izem T, Harkat M-F, Djeghaba M, Kratz F (2018) On the application of interval PCA to process monitoring: a robust strategy for sensor FDI with new efficient control statistics. J Process Control 63:29–46

Aslam M, Chien-Wei Wu, Azam M, Chi-Hyuck J (2013) Variable sampling inspection for resubmitted lots based on process capability index Cpk for normally distributed items. Appl Math Model 37(31):667–675

Barrentine LB (2003) Concepts for R&R studies, 2nd edn. ASQ Quality Press, Milwaukee, WI, USA, p 76

Boyles RA (1996) Multivariate process analysis with lattice data. Technometrics 38(1):37–49

Burlikowska M (2005) Quality estimate on of process with usage control charts typeX-R and quality capability of process Cp, Cpk. J Mater Process Technol 162–163:736–743

Casaroni G, Kerstens K, Van de Woestyre I (2019) Short- and Long-Run plan capacity notions: definitions and comparison. Eur J Oper Res 275(1):387–397

Chang YC (2008) interval estimation of capability index C pmk for manufacturing processes with asymmetric tolerances. Article in Press, Computers & Industrial Engineering

Chen H (1994) A multivariate process capability index over a rectangular solid tolerance zone. Stat Sin 4:749–758

Chen KS, Chen TW (2008) Multi-process capability plot and fuzzy inference evaluation. Int J Prod Econ 111:70–79

Diamoutene A, Noureddine F, Kamsu-Foguem B, Barro D (2018) Quality control in machining using order statistics. Measurement 116:596–601

Fechio V (2009) Measurement systems analysis MSA—non replicable measurement system. Brasmetal Waelzholz S. A. Industry and Sales, Diadema, Brazil

Joy B, Shunmugesh K, Kumur MP, Arun K (2019) Optimization in turning of llsMn30 through process capability index. Materials Today: Proceedings II(Part 3):961–970

Kunlun Hu, Yuan J (2008) Multivariate statistical process control based on multiway locality preserving projections. J Process Control 18:797–807

Liu PH, Chen FL (2006) Process capability analysis of nonnormal process data using the Burr XII distribution. Int J Adv Manuf Technol 27:975–984

Montgomery DC, Runger GC (1993) Gauge capability analysis and designed experiments, Part II: experimental design models and variance component estimation. Quality Eng 6(2):289–305

Parchami A, Mashinchi M (2007) Fuzzy estimation for process capability indices. Inf Sci 177:1452–1462

Pearn WL, Lin P (2004) Testing process performance based on capability index Cpk with critical values. Comput Ind Eng 47:351–369

Pearn WL, Liao M-Y (2005) Measuring process capability based on CPK with gauge measurement errors. Microelectron Reliab 45:739–751

Pearn WL, Wu C-W (2006) Production quality and yield assurance for processes with multiple independent characteristics. Eur J Oper Res 173:637–647

Pearn WL, Ya-Chen H (2007) Optimal tool replacement for processes with low fraction defective. Eur J Oper Res 180:1116–1129

Pearn WL, Wu C-W (2007) An effective decision making method for product acceptance. Omega Int J Manage Sci 35:12–21

Triebs TP, Kumbhakar SC (2018) Management in production: from unobserved to observed. J Prod Anal 49:111–121

Wang FK, Du TCT (2000) Using principal component analysis in process performance for multivariate data. Omega Int J Manage Sci 28:185–194

Experimentation and Multi-Objective Optimization in Manufacturing of Rubber for Shoe Sole

Armando Mares Castro and Jorge Domínguez Domínguez

Abstract In this chapter, two multi-response optimization approaches are proposed, and they are used in a rubber sole vulcanization process for footwear with the purpose of complying with three important quality characteristics in the material. The design of experiments is a statistical methodology that is applied to improve the quality of the product. With the results of the experiment, a mathematical model is constructed to optimize the process parameters. In the case of several responses, the process is studied with multivariate optimization techniques. A comparative analysis of the methodologies is performed to evaluate the best result. The results obtained in the two approaches were similar, as well as their effect in the vulcanization process.

Keywords Design of experiments · Vulcanization · Desirability function · Multi-objective optimization by genetic algorithms

1 Introduction

Currently, manufacturing companies that want to compete must focus on meeting the quality characteristics of their product, also called the voice of the customer (VoC) (Socconini 2015). The methodologies for quality management and problem solving with Six Sigma, Lean Manufacturing, and AMEF contemplate the improvement in the processes. It can be verified with the implementation of strategies and procedures for obtaining the optimal process parameters. The experimental design is widely used for the improvement of the processes through the obtaining of the optimal parameters associated to the quality characteristics in the product (Myers et al. 2016). In some cases, it is not possible to determine an optimal solution for

A. Mares Castro (✉)
Tecnológico Nacional de México/ITS de Purísima del Rincón, Blvd. Del Valle #2301, Guardarrayas, Purísima del Rincón, Guanajuato. C.P. 36413, México
e-mail: armando.mc@purisima.tecnm.mx

J. Domínguez Domínguez
Centro de Investigación en Matemáticas, Unidad Aguascalientes, F. Bartolomé de Las Casas # 314, Barrio de la Estación, Aguascalientes, Aguascalientes. C.P 20259, México

J. L. García Alcaraz et al. (eds.), *Techniques, Tools and Methodologies Applied to Quality Assurance in Manufacturing*,
https://doi.org/10.1007/978-3-030-69314-5_9

a single quality characteristic. An experiment is carried out in order to find the best condition for several process responses. In this direction, the multi-objective optimization is applied. The desirability function considers all responses with their respective restrictions (Derringer and Suich 1980).

For the modeling of the responses, regression models of linear or quadratic type are fitted. The purpose of the models is to study the relationship between k factors, $X = (X_1, \ldots, X_k)$ control variables, and r responses, $y = (y_1, \ldots, y_r)$, for each of the proposed models. For knowledge and experience in the process, an experimental Box–Behnken design (Box and Behnken 1960) was used in this proposal.

The goals for the responses in the problem can be of three kinds according to the quality characteristic: the bigger is better, the smaller is better, and the objective value is the best. Since normally there is no single solution that generates an optimum simultaneously, to meet all the aims of the problem, a compromise solution must be reached. All the plans are met to an acceptable degree to have the best product, or process performance. Techniques are also adequate for the design of rubber compounds (Nicholson and Pullen 1969).

The optimization issue can be analyzed from different approaches, for example, as a restricted optimization problem. All the targets for the responses must be met for the response variables. There may be a set of linear or nonlinear constraints within the experimental area allowed by the process. The improvement strategy is within the constrained nonlinear optimization scheme and can be approached with metaheuristic techniques, such as multi-objective optimization by genetic algorithms (MOGA) (Arora 2012), which is one of the application techniques in our proposal.

The objective of this chapter is the proposal and evaluation of two schemes for multi-objective optimization applicable to experimental design in manufacturing. The case study deals with the fulfillment of quality characteristics in the rubber for shoe soles. The data were obtained from the test procedures indicated in the standards ASTM D412, Standard Test Methods for Vulcanized Rubber and Thermoplastic Elastomers—Tension (ASTM_D412-06a 2013), and the ASTM_D2240 Standard Test Method for Rubber Property—Durometer Hardness (ASTM_D2240-15e1 2015). The optimization of the scheme was performed using the techniques of the desirability function and MOGA.

The shoe sole manufacturing process requires research. Currently, there is not much information available in the state of the art or standards that indicate the reference values for the different types of sole used. According to the National Institute of Statistics and Geography, the data from the latest economic census indicated that the footwear industry in Mexico employed 112,727 people, contributing 2.4% of the total occupation of manufacturing industries; additionally, it is estimated that the industry generates 17,000 extra jobs in the supply chain. Footwear exports represent 8.7% of production, and the gross domestic product of the footwear industry indicates a constant growth in terms of production and sales, the latest records indicate a production of around 2,300,000 pairs and 14,000 million pesos constant, being at the pair of manufacturing gross domestic product (INEGI 2014). Based on the above, it is observed that it is important to reinforce the applications in the area of research in the sole for footwear. In order to identify and define the ideal configuration that generates the optimal parameters in the processes and that allows the generation of competitive products with high export quality, the focus of this research focuses on the particular analysis of the rubber sole, of which no similar studies were found in the literature.

2 Literature Review

2.1 Quality in Manufacturing

The beginnings of the evolution in modern manufacturing were marked from the division of labor and the invention of the steam engine in 1776, through the contributions of Frederick Taylor with the model of scientific administration and standardization of work. On the other hand, Henry Ford and the creation of the assembly line revolutionized the way of working in manufacturing. At the beginning of the twentieth century, the Toyota production system revolutionized the way to generate products based on contributions from Kiichiro and Eiji Toyoda, Taiichi Ohno, and Shigeo Shingo as pioneers of Lean Manufacturing and the elimination of waste in the manufacturing until reaching the production system called Just in Time (Bhamu and Sangwan 2014). Along with the development of Lean Manufacturing and Just in Time, the foundations of quality were given with the development of statistical methods for the improvement and control of processes by Shewhart, Fisher, Taguchi among others (García-Alcaraz et al. 2014). The quality engineering in the product consists in the application of techniques for the optimization of products and manufacturing processes, starting from the design to the manufacture and monitoring of the product. The ideal quality that a customer can expect is that every time the product delivers the objective performance, every time it is used for all operations destined and throughout its intended life without harmful side effects (Hahn and Doganaksoy 2011).

Additionally, the cost element must be taken into account. The problem of delivering a quality product at a low cost involves engineers, economists, statisticians, and the administration. The operation costs, manufacturing costs, as well as the costs of research and product development must be taken into account. Currently, there is more competition, prices are established by the market, and the way in which companies can survive is reducing costs without neglecting quality. Reducing prices without sacrificing benefits is not possible without improving and ensuring quality in products and processes (Socconini 2015).

2.2 Multi-Objective Optimization in Manufacturing by Desirability Function and Genetic Algorithms

An important problem that arises in the analysis and improvement of manufacturing quality is to find a compromise solution. The input variable levels in a process must meet a set of objectives, criteria, or goals. Because of the conflicting nature and incommensurability of such criteria, a concept of compromise solution, rather than optimal solution, is probably more useful for their analysis (Zelany 1974).

The most widely used techniques in multi-objective optimization in the literature are mainly based on the desirability function approach (Harrington 1965) and search metaheuristic techniques such as multi-objective optimization by genetic algorithms (MOGA) (Zelany 1974) and the swarm of particles (Cao et al. 2017).

Desirability functions have been used extensively to simultaneously optimize several responses. Since the original formulation of these functions contains non-differentiable points, only search methods can be used to optimize the overall desirability response (Del Castillo et al. 1996). In the initial version of the method, the responses involved were treated with the same importance. The currently modified methods allow the assignment of different weightings to the responses in order to assign priorities among the responses. The concept of desirability is a means for complexity reduction of multivariate quality optimization (Trautmann and Weihs 2006). Costa et al. (2011) presented a review on the performance of the desirability function approach.

The application of the desirability function in manufacturing is linked to the modeling of quality responses through the use of experimental design. Islam et al. (2012) presented an optimization proposal based on the desirability function and response surface methodology (RSM) techniques; the experimental responses are important physical and mechanical properties in the quality of the wood for the production of particle board. John (2013) presented a proposal on the optimization of performance characteristics of carbonitrided bushes, a better solution was obtained by the desirability function method compared to the individual optimals for the responses separately. Singaravel and Selvaraj (2016) presented an application of the desirability function for the optimization of the cutting parameters in a steel turning process with the use of an L8 orthogonal array from Taguchi. Ortiz et al. (2004) presented a proposal to use genetic algorithm techniques in conjunction with an unrestricted desirability function.

In multi-objective optimization, a solution method can be designed as either a generating method for obtaining a preferred or compromised solution. The major focus of MOGA has been put on how to generate Pareto solutions (Ziegel 2002).

Within manufacturing technology, the use of MOGA techniques has received special attention for its potential applications that require greater complexity than conventional schemes (Davim and Davim 2012). The most important feature of evolutionary algorithms is that can successfully find globally optimal solutions without getting restricted to local optima. Among the evolutionary algorithms, the non-dominated sorting genetic algorithm (NSGA) and non-dominated sorting genetic algorithm-II (NSGA-II) have emerged as the most efficient algorithms for solving multi-objective problems in manufacturing processes (Kanthababu 2012). Li et al. (2015) presented a detailed review of the details and characteristics of evolutionary algorithms for the analysis of various objectives.

The use of MOGA techniques in conjunction with experimental design has been analyzed in the literature; Köksoy and Yalcinoz (2008) have proposed MOGA optimization and Pareto optimization to robust parameter design methodology.

Mohammad et al. (2018) presented an application of MOGA to the optimal design of thermal conductivity and visibility in nanofuids using the NSGA-II algorithm and the RSM. Nandi et al. (2013) presented a proposal for the application of MOGA optimization based on the NSGA-II algorithm for the optimization of a molded silicone rubber compound for a soft tool process.

2.3 Optimization of Quality Characteristics of Rubber for Shoe Sole

In the literature, there are limited applications to the study of the quality of the material of shoe sole. Mares and Domínguez (2015) presented a study on the improvement of quality characteristics in polyurethane sole by applying the robust design of parameters. Regarding the study of quality characteristics in rubber, Salvatori et al. (2018) presented a study on the optimization of a rubber compound with a ternary polymer mixture using RSM techniques. In addition to the above, there are limited applications in the literature regarding the analysis of mechanical properties in rubber. Available works mainly focus on the analysis of variation in the formulations of the components of the mixture (Onyekwere et al. 2017).

The main contributions and differences of the study presented in this chapter in the current state of the art is that the study focuses on rubber for shoe soles. In this kind of process, the formulation is predefined by the company. The main focus is the analysis of the effect of the factors present in the vulcanization process on three quality responses of the material. Experimental data can be analyzed by means of physical laboratory tests.

For the reason that there are no similar studies available in the state of the art. One of the objectives of this research is that it can serve as a reference in future research on the manufacturing process of the rubber sole for footwear, as well as definition of the most appropriate parameters in the vulcanization process in the material. Likewise, the comparison of the multi-objective optimization techniques of the desirability function and MOGA is presented, in order to verify the results of each method and its effectiveness on the process.

3 Methodology

The proposed methodology for the analysis of the proposed case is shown in Fig. 1. Once the results of the experimental runs are obtained, the analysis of variance and obtaining the most appropriate regression models are carried out, then the multi-response optimization techniques are applied through the desirability function and MOGA with the purpose of contrast the results and determine the best option for the analysis case.

The vulcanization phase in the manufacture of rubber products is the most important process, since this is where the most important quality characteristics of the product are developed. The high competitiveness in the footwear market requires that the product meets certain quality characteristics, which can be determined both by customers and/ or regulations.

The laboratory tests allow to evaluate the mechanical properties in the manufactured material and allow to know the level of compliance with the quality characteristics in the product. Two tests of importance which area applied to vulcanized rubber are the tension test performed in universal machine, which is related to the standard ASTM D412-06a and the material hardness test related to ASTM D2240-15e1.

ASTM D412-06a contains the procedure to assess the tensile properties of vulcanized thermoplastic rubber and elastomers. Vulcanized rubber tension tests are performed in a universal test machine (see Fig. 2), in which the residual elongation of a test sample is evaluated after it has been first stretched, and then, it is allowed to relax in a specified process. The elongation of these materials consists of both permanents (plastic) and recoverable (elastic) components, so it is important to take the time and load into account for the tests. Method A was used for halter-shaped samples with the measurements indicated in the standard.

The jaws used in the tests are manual wedge jaws, which have spring action and are self-adjusting to minimize material slippage.

ASTM D2240-15e1 contains the standard test methods for measuring the hardness property in the material. The standard covers 12 types of rubber hardness meters. This measuring device is used for the tests is a type A hardness tester with shore scale, which is used to measure the hardness of various materials such as polymers,

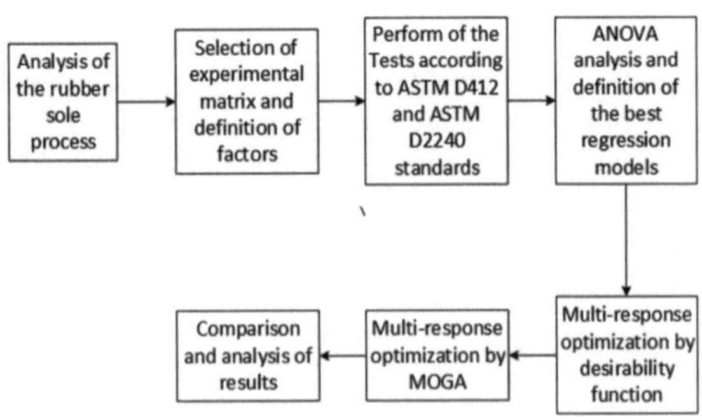

Fig. 1 Research methodology

Fig. 2 Tensile tests performed on the universal machine

elastomers, and rubber. The procedure consists in pressing the tip of the measuring device to measure the depth of indentation in the material created by a standardized force in the device. The higher the value obtained; the material has greater hardness.

The main objective of the study focuses on the optimization of three quality characteristics obtained in the mechanical tests of the specimens: (1) The maximum load, which is a desirable characteristic to maximize in the sole of shoes, (2) The maximum deformation desirable characteristic to maximize, and 3. The hardness, which is desirable to fix at 68 Shore A for a good performance in the product. Experimental tests were performed in the vulcanizing process, varying the levels of temperature, pressure, and time.

3.1 Rubber Sole Manufacturing Process

The rubber sole manufacturing process shown in Fig. 3 is a manufacturing process that involves mixtures of raw materials, as well as a series of physical and chemical processes. The initial part of the process consists of the weighing and mixing of

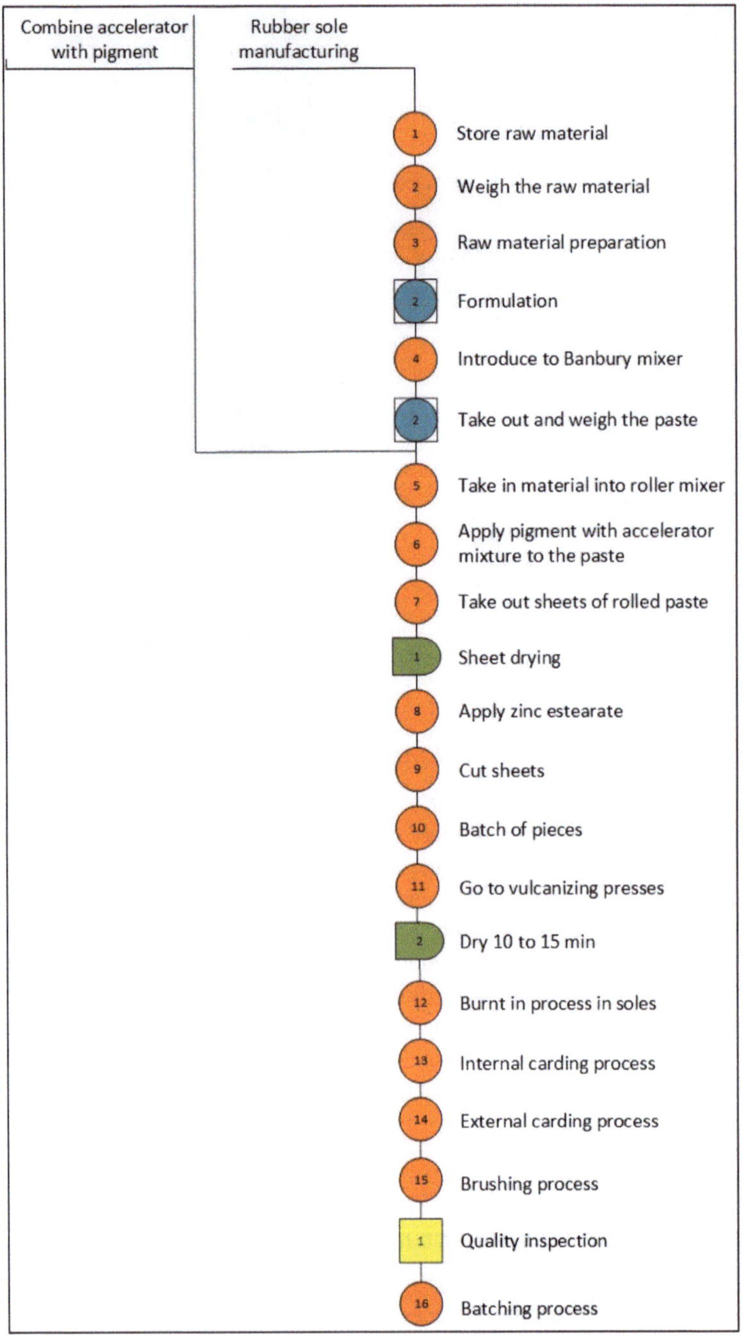

Fig. 3 Rubber sole manufacturing process

materials such as raw elastomeric rubber, carbonate fillers, silicates, oils, waxes, acids, and chemicals (Ciesielski and Limited 1999).

The mixing of the materials is carried out in a mill that works with special blades and temperatures called "banbury mixer." The mixing process involves the use of time and temperature to obtain an adequate mixture of the materials to form a paste.

Once the paste has been weighed, the client's specifications are added by applying pigment or color and accelerators. The mixing is done in machines with large rollers from which the laminated material is formed in the form of strips.

The strips of material pass to a cooling section by means of fans. The strips are cut with a similar shape to the model of soles requested by the customer.

The next stage of the process is the vulcanizing of the sole, which is done in molds that work on the basis of pressure, time, and temperature. Here, material achieves the final shape of the sole, acquiring the characteristic properties of the product.

The last processes of the product are the final finishes in the prefinishing area. The sole is burnt to remove the excess material at the ends of the sole. An internal and external carding is carried out to detail the product. The final process is a brushing that improves the view of the sole to conclude with the batching and inspection of the final product.

3.2 Experimental Design

The design of experiments is a methodology that is used to know, study, and research products and processes in the industry. There are several schemes that are applied according to the objectives that the process engineer wants to achieve. In particular, in this work, the Box–Behnken design is proposed. This design is applied when there are three or more factors; it satisfies statistical properties that make it an efficient design in the analysis, such as orthogonality and rotability. It can be consulted in (Box and Behnken 1960). The points of the design are located in the middle of the edges of the cube centered on the origin. Design does not include the treatments of the vertices that are sometimes extreme and cannot be run. Representation of design is shown in Fig. 4.

With the experimental results of this design, second-order regression models can be fitted. In particular, for the case study in this work, a model is obtained for each response.

So, for r responses, you have r models, and the ith model for the responses Y_i is written as Eq. 1:

$$Y_i = \beta_{i0} + \mathbf{X}'\beta_i + \mathbf{X}'\mathbf{B}_i\mathbf{X} + \varepsilon_i \tag{1}$$

where $\mathbf{X}' = (x_1, \ldots, x_k)$ k factors, β_{i0} the constant, $\beta_i = (\beta_{i1}, \ldots, \beta_{ik})$ one vector of parameters, $\mathbf{B}_i = (\beta_{i11}, \ldots, \beta_{i1k}, \beta_{ik1}, \ldots, \beta_{ikk})$ symmetric matrix of second-order parameters, and $\varepsilon_i \sim N\left(0, \sigma_i^2\right), i = 1, \ldots, r$.

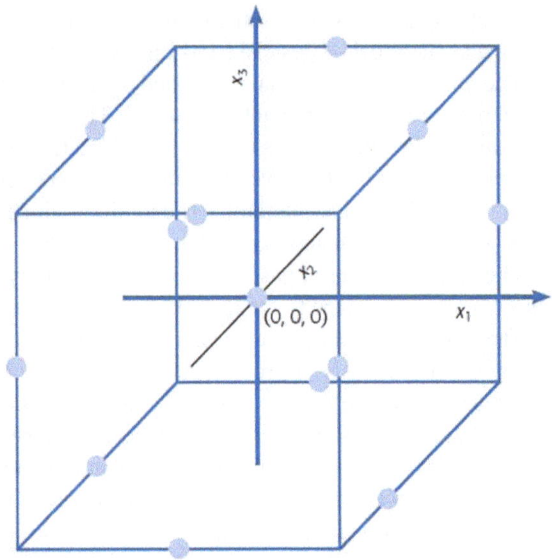

Fig. 4 Box–Behnken design representation for three factors (Gutiérrez Pulido and Salazar 2012)

The statistical analysis of the residuals in each model allows to verify the lack of fit, as well as other relevant characteristics.

The proposed methodology consists in the application of a Box–Behnken design with three factors without blocks, resulting in 15 experimental runs. The factors that were considered in the vulcanizing process are pressure, temperature, and time and are described in Table 1, and the values that were considered for each factor are also shown.

Factor levels were indicated by supervisory staff based on knowledge and experience of the process. The main criterion was the selection of levels in which not much waste is generated. In addition, the variation allowed by the equipment and facilities without risk of damage is considered. Since for example, for the temperature factor, it depends on a boiler that supplies other machines at the same time as the molds.

Table 2 shows the experimental matrix for the Box–Behnken design with the three factors and the three responses; the maximum load and the maximum deformation are obtained from the experimental tests of the stress test of ASTM D412-06a, while the hardness is obtained from the test for the hardness measurement of ASTM D2240-15e1.

Table 1 Factors, levels, and units of measurement for the experiment

Factor	Coded	Pressure	Time	Temperature
Units		psi	min	°C
Low	−1	1200	2.50	145
Mid	0	1250	3.00	150
High	1	1300	3.50	155

Table 2 Experimental matrix* with three factors and three responses

Pressure	Time	Temperature	Max load	Max defor	Hardness
X_1	X_2	X_3	kN	mm	Shore A
1	1	0	0.32	220.5	67.2
−1	0	1	0.4	344	67.5
0	0	0	0.4	304	67.0
0	1	−1	0.38	365	67.3
−1	−1	0	0.42	292	68.0
0	0	0	0.36	288.5	66.2
0	−1	1	0.4	363.5	69.3
0	1	1	0.36	302.5	68.0
1	0	−1	0.36	199	68.7
0	0	0	0.32	249.5	66.5
−1	0	−1	0.44	384	68.2
−1	1	0	0.4	320	65.7
1	0	1	0.34	261.5	68.7
1	−1	0	0.34	312.5	66.0
0	−1	−1	0.36	304	67.2

*The order of runs of the Box–Behnken experiment is randomized

3.3 *The Desirability Function*

Desirability function was introduced by Harrington (1965) and was modified by Derringer and Suich (1980). Individual desirability and compound desirability assess how well a combination of variables satisfies the objectives that were defined for the responses. Individual desirability assesses the way in which the configuration optimizes an individual response, while composite desirability evaluates the way in which the configuration optimizes a set of responses in general. The range of desirability is 0 to 1, the value of 1 indicates the ideal situation while 0 indicates that one or more responses are outside the acceptable limits. The j-th response is maximized when the expected value $\hat{Y}_i(x)$ equals the defined target value M_i, so the value will decrease as it moves away from the target value. Defining V_i and V_s as the extreme values of the process, only solutions within the range $V_i \leq \hat{Y}_i(x) \leq V_s$ will be accepted, and the desirability function d_i for each one of the responses is defined in Eq. 2:

$$d_i\left(\hat{Y}_i(x)\right) = \begin{cases} \left[\frac{\hat{Y}_i(x)-V_i}{M_i-V_i}\right]^s & \text{if } V_i \leq \hat{Y}_i(x) \leq M_i \\ \left[\frac{\hat{Y}_i(x)-V_s}{M_i-V_s}\right]^t & \text{if } M_i < \hat{Y}_i(x) \leq V_s \\ 0 \quad if \ \hat{Y}_i(x)\langle V_i \ or \ \hat{Y}_i(x)\rangle V_s \end{cases} \tag{2}$$

Equation 5 is called the two-sided transformation, which applies when the response variable Y_i has both minimum and maximum constraints, which represent the lower and upper levels, respectively, for the experimental factors. Individual desirabilities are combined using a geometric mean, as shown in Eq. 3:

$$D = (d_1, \times d_2, \times d_3)^{\frac{1}{k}} \tag{3}$$

The value for D represents the assignment of the combined desirability of the response levels, and it is clear that the range of D will be between 0 and 1. One of the properties of D is that if any of the $d_i = 0$, that is, that one of the responses is unacceptable, then $D = 0$, and therefore, the overall product will be unacceptable. Equation 6 is reduced to a geometric mean.

3.4 Multi-Objective Genetic Algorithm

Within metaheuristic search techniques, genetic algorithms are adaptive computational techniques widely used in optimization problems. The main advantage of genetic algorithms over traditional numerical optimization techniques [see Nocedal and Wright (2006)] is that genetic algorithms are more likely to find the global optimum, particularly when highly nonlinear, discontinuous, non-differentiable, or stochastic functions are used. The basis of the genetic algorithms were established by Holland (1984) and have been extensively analyzed by other authors such as Goldberg (1989).

Genetic algorithms can be used to solve problems of restricted or unrestricted optimization. The basis of the technique is the imitation of the natural selection process that happens in biological evolution. The algorithm repeatedly changes a population of individual solutions in an iterative process that involves selection, reproduction, mutation, crossover, and migration techniques. In their mathematical form, populations are matrix arrays, an individual may appear in more than one line of the matrix, and individuals change as new populations called generations occur.

The fitness function is the function that you want to optimize and is equivalent to the objective function of a traditional optimization scheme. The algorithm begins with the creation of a random initial population. In each iteration, sequences of new generations are created that evolve according to the optimization conditions declared in the problem. The creation of a new population is carried out through the evaluation in the fitness function, which is called the fitness score. The fitness evaluations are scaled to make them a range of more appropriate values called expected values. The best individuals with the best fitness value called elite are selected and passed to the new population.

To improve the next generations, the children of the elite are produced by random changes to a single parent (mutation) or by combining entrances of a couple of parents (crossover), the population is replaced with the children to form the next generation. The algorithm will stop when a stop condition determined by the analyst is met.

Multi-objective optimization based on genetic algorithms is related to the minimization of multiple responses or objectives, which may be subject to a set of constraints. There is a vector of objectives (Eq. 4):

$$F(x) = \left[\hat{Y}_1(x), \hat{Y}_2(x), \ldots, \hat{Y}_r(x)\right] \tag{4}$$

which should agree in some way. As the number of objectives increases, the commitments become more complex. Multiobjective optimization based on the use of genetic algorithms is found in the constrained nonlinear optimization scheme. The objective function seeks the minimization of the objective vector within the limits of the experimental region subject to restrictions as shown in Eq. 5:

$$\begin{aligned} &\text{Min } F(x) \\ &\text{s.t} \\ &\hat{Y}_i(x) = M_i \\ &\hat{Y}_i(x) \le M_i \\ &V_i \le x \le V_s \end{aligned} \tag{5}$$

An advantage of this optimization scheme is the possibility of using sets of linear or nonlinear constraints, with which more complex problems can be analyzed. The target vector will contain those functions that you want to maximize or minimize, while the functions that must be matched to a target value will be within the constraints.

The terminology used for the genetic algorithms and MOGA is the same. An additional term that can be mentioned is dominance, which means that a point x dominates a point and for an objective function vector f (see Eq. 6):

$$\begin{aligned} &f_i(x) \le f_i(y) \text{ for all } i \\ &f_i(x) \le f_i(y) \text{ for some } j \end{aligned} \tag{6}$$

Dominance also extends to the rank, which is the location of the solutions. The Pareto front contains the set of solutions that are in the non-dominated set or Rank 1.

The multi-objective optimization scheme for genetic algorithms requires the minimization of the fitness function (7) in which the functions $\hat{Y}_{\text{Max Load}}$ and $\hat{Y}_{\text{Max Defor}}$ are contained, and both responses are required to maximize. Optimization approach allows the inclusion of nonlinear functions as constraints, so the value of the function $\hat{Y}_{\text{Hardness}}$ is fixed as $\hat{Y}_{\text{Hardness}} = 68$ Shore A, thus the optimization scheme is shown in Eq. 7:

$$\begin{aligned}
&\text{Max}\left[\hat{Y}_{\text{Max Load}}, \hat{Y}_{\text{Max Defor}}\right] \\
&\text{s.t.} \\
&\hat{Y}_{\text{Hardness}} = 68 \\
&-1 \leq x_1 \leq 1 \\
&-1 \leq x_2 \leq 1 \\
&-1 \leq x_3 \leq 1
\end{aligned} \tag{7}$$

For the definition of the starting steps for the algorithm, the optimization parameters used in the method can be found in Tables 3 and 4, which show the optimization parameters and the global optimization parameters, respectively.

Table 3 Parameters for the optimization

Parameter	Current value
Maximum number of alternatives	15
Tolerance for generating new alternatives	1.00E−03
Tolerance for constraint feasibility	1.00E−09
Coefficient of the augmentation term	1.00E−03
Difference of ideal and utopian values	1.00E−03

Table 4 Global optimization parameters

Parameter	Current value
Stopping criteria	
Max number of generations	500
Number of last generations examined (n-pop)	100
Difference between the best individual in last n-pop generations	0.01
Other GA parameters	
Population size	101
Tournament size	3
Elitism	1
Crossover rate	0.8
Mutation rate	0.1
Degree of mutation	4
Constraints handling parameters for adaptive penalty	
Initial penalty coefficient	1
Decrease rate of penalty coefficient	3
Increase rate of penalty coefficient	4
Number of best individuals examined	1

4 Results

4.1 Response Analysis and Model Determination

In this section, the regression models obtained for the three responses are presented, as well as the optimization schemes applied to improve the vulcanization process of rubber for shoe sole. Statistical analyzes were performed in Minitab 19 (Minitab Inc 2019).

Analysis for Maximum Load

The ANOVA for maximum load in Table 5 presents a statistically significant element (Pressure) with a *P*-Value of 0.0139, the values of $R^2 = 77.39\%$ and $R^2\text{adj} = 36.69$, respectively.

In order to determine the best model for the analysis, we proceeded to eliminate the less significant terms taking into account the values of R^2 y R^2adj, eliminating linear terms x_2, x_3, the interactions x_1x_2, x_1x_3, and the quadratic terms x_1^2, x_2^2, the regression model with the highest values in $R^2adj = 63.35\%$ y $R^2 = 71.21\%$ is obtained.

In order to maintain a hierarchical model, the terms x_2, x_3 are added since the interaction of these two terms is present in the model, leaving the values of $R^2\text{adj} = 60.78\%$ and $R^2 = 74.79\%$. The Durbin–Watson statistic has a *P*-Value of *P*-Value of 0.1490, which indicates that there is no serial autocorrelation in the residuals with a confidence level of 5%. The model is defined in Eq. 8:

Table 5 ANOVA for maximum load

Source	Sums of squares	Df	Mean square	*F*-ratio	*P*-value
X1: pressure	0.01125	1	0.01125	13.72	*0.0139
X2: time	0.00045	1	0.00045	0.55	0.4921
X3: temperature	0.0002	1	0.0002	0.24	0.6423
$X1^2$	0.000369231	1	0.000369231	0.45	0.5319
X1 * X2	0.0	1	0.0	0.00	1.0000
X1 * X3	0.0001	1	0.0001	0.12	0.7412
$X2^2$	0.0	1	0.0	0.00	1.0000
X2 * X3	0.0009	1	0.0009	1.10	0.3428
$X3^2$	0.000830769	1	0.000830769	1.01	0.3604
Total error	0.0041	5	0.00082		
Total (corr.)	0.0181333	14			

*Significant

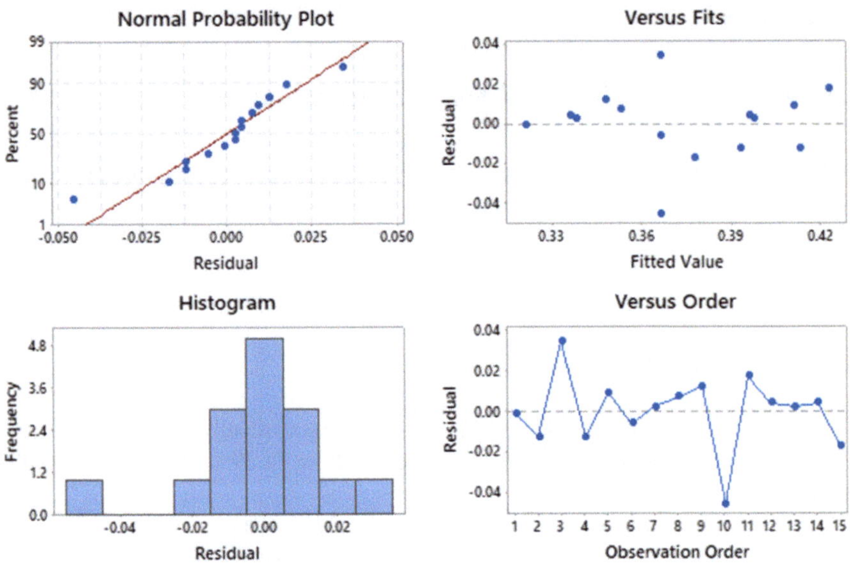

Fig. 5 Residual analysis for the max load model

$$\hat{Y}_{\text{Max Load}} = 0.365714 - 0.0375x_1 - 0.00750x_2 - 0.00500x_3 - 0.015x_2x_3 + 0.0142857x_3^2 \quad (8)$$

The residual analysis for the model (10) is shown in Fig. 5. For the assumption of normality, a regular behavior of the points around the reference line is observed, fulfilling the normality in an appropriate manner, which can also be verified with the Anderson–Darling test. The homoscedasticity analysis also shows a behavior without apparent patterns or funnel shapes. Independence is also adequately fulfilled by not observing apparent patterns throughout the order of the tests. As a complement to the Durbin–Watson test, the proper compliance of the independence assumption can be verified.

Analysis for Maximum Deformation

The ANOVA for maximum deformation in Table 6 presents a statistically significant element (pressure) with a *P*-Value of 0.0199 and the values of $R^2 = 84.86\%$ and $R^2\text{adj} = 52.02$, respectively.

In order to determine the best model for the analysis, we proceeded to eliminate the less significant terms taking into account the values of R^2 and R^2adj, eliminating linear terms x_2, x_3 and the quadratic term x_1^2; a regression model with the highest values is obtained for $R^2\text{adj} = 63.44\%$ and $R^2 = 79.11\%$. In order to maintain a

Table 6 ANOVA for maximum deformation

Source	Sums of squares	Df	Mean square	*F*-Ratio	*P*-Value
X1: pressure	15,007.8	1	15,007.8	11.35	*0.0199
X2: time	512.0	1	512.0	0.39	0.5611
X3: temperature	47.5313	1	47.5313	0.04	0.8571
$X1^2$	889.463	1	889.463	0.67	0.4495
X1 * X2	3600.0	1	3600.0	2.72	0.1599
X1 * X3	2626.56	1	2626.56	1.99	0.2178
$X2^2$	1644.5	1	1644.5	1.24	0.3155
X2 * X3	3721.0	1	3721.0	2.81	0.1543
$X3^3$	3776.0	1	3776.0	2.86	0.1519
Error total	6611.98	5	1322.4		
Total (corr.)	38,586.4	14			

*Significant

hierarchical model, the linear terms $x_2 y x_3$ are maintained since there are interactions that contain them, leaving the values of R^2adj = 54.64% and $R^2 = 80.56\%$.

The Durbin–Watson statistic has a *P*-Value of 0.9300 which indicates that there is no serial autocorrelation in the residues with a 5% confidence level. The model is defined in Eq. 9:

$$\hat{Y}_{\text{Max Defor}} = 271.115 - 43.3125x_1 - 8x_2 + 2.4x_3 - 30x_1x_2 + 25.625x_1x_3 + 22.2981x_2^2 - 30.5x_2x_3 + 33.1731x_3^2 \quad (9)$$

The residual analysis for the model in Eq. 12 is shown in Fig. 6. For the assumption of normality, a regular behavior of the points around the reference line is observed, fulfilling the normality in an appropriate manner, which can also be verified with the Anderson–Darling test. The homoscedasticity analysis also shows a behavior without apparent patterns or funnel shapes. Independence is also adequately fulfilled by not observing apparent patterns throughout the order of the tests. As a complement, the Durbin–Watson test for the proper compliance of the assumption can be verified.

Analysis for Hardness

The ANOVA for hardness in Table 7 presents a statistically significant element (Temperature ^ 2) with a *P*-Value of 0.0102, the value of $R^2 = 84.30\%$ and R^2adj = 56.03. The Durbin–Watson statistic has a *P*-Value of 0.4303 which indicates that there is no serial autocorrelation in the residues with a confidence level of 5%. In order to determine the best model for the analysis, we proceeded to eliminate the less significant terms taking into account the values of R^2 and R^2adj, eliminating the linear term x_1, the interaction x_1x_3 and the quadratic terms x_1^2, x_2^2; you get a

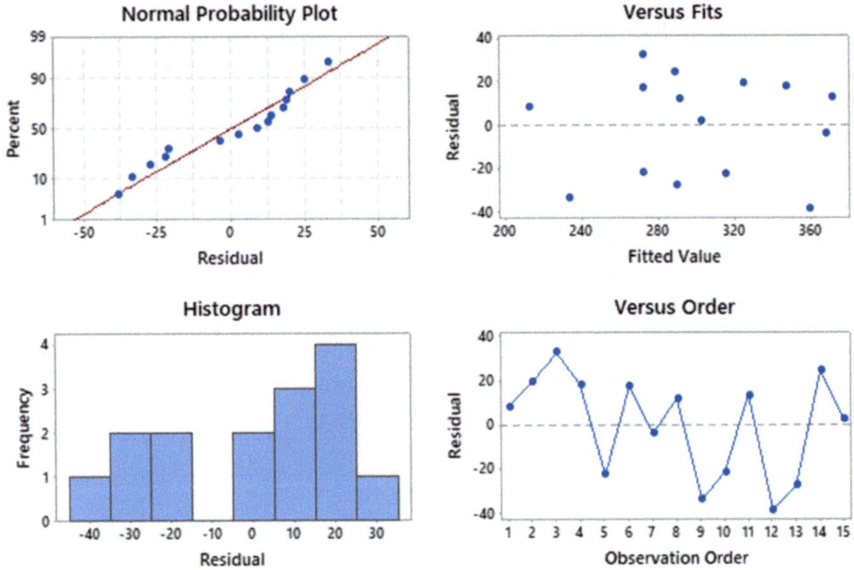

Fig. 6 Residual analysis for the max deformation model

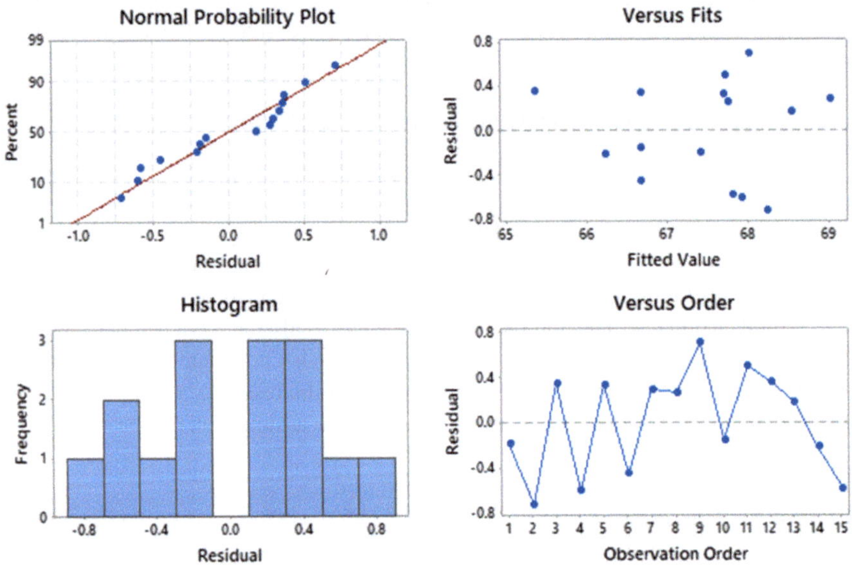

Fig. 7 Residual analysis for the hardness model

Table 7 ANOVA for hardness

Source	Sums of squares	Df	Mean square	*F*-ratio	*P*-value
X1: pressure	0.18	1	0.18	0.37	0.5720
X2: time	0.66125	1	0.66125	1.34	0.2990
X3: temperature	0.55125	1	0.55125	1.12	0.3386
$X1^2$	0.215641	1	0.215641	0.44	0.5376
X1 * X2	3.0625	1	3.0625	6.21	0.0550
X1 * X3	0.1225	1	0.1225	0.25	0.6392
$X2^2$	0.025641	1	0.025641	0.05	0.8286
X2 * X3	0.49	1	0.49	0.99	0.3645
$X3^2$	7.94256	1	7.94256	16.12	*0.0102
Total error	2.46417	5	0.492833		
Total (corr.)	15.6933	14			

*Significant

regression model with R^2adj = 70.06% and R^2 = 80.75%. If another additional element is eliminated, both values begin to decrease.

In order to obtain a hierarchical model, the linear term is maintained x_1 since there are interactions that contain them, leaving the values of R^2adj = 68.32% and R^2 = 81.90%. The Durbin–Watson statistic has a *P*-Value of 0.9300 which indicates that there is no serial autocorrelation in the residues with a 5% confidence level. The model is defined:

$$\hat{Y}_{\text{Hardness}} = 66.6571 + 0.150x_1 - 0.2875x_2 + 0.2625x_3 + 0.875x_1x_2 - 0.35x_2x_3 + 1.45536x_3^2 \quad (10)$$

The residual analysis for Eq. 10 is shown in Fig. 7. For the assumption of normality, a regular behavior of the points around the reference line is observed, fulfilling the normality in an appropriate manner, which can also be verified with the Anderson–Darling test. The homoscedasticity analysis also shows a behavior without apparent patterns or funnel shapes. Independence is also adequately fulfilled by not observing apparent patterns throughout the order of the tests. As a complement, the Durbin–Watson test for the proper compliance of the assumption can be verified.

4.2 *Optimization by Desirability Function*

From Eqs. 5 and 6, the optimization of the responses for the models of $\hat{Y}_{\text{Max Load}}$ (Eq. 8), $\hat{Y}_{\text{Max Defor}}$ (Eq. 9), and $\hat{Y}_{\text{Hardness}}$ (Eq. 10), the maximum load and the maximum deformation are desirable characteristics to maximize in the rubber for shoe sole. Since they are related to the resistance of the material and its performance in the use of footwear, the hardness parameter for the rubber sole for footwear is not defined in reference regulations. Based on the experience of manufacturers of similar products, there is knowledge of values between 65 and 70 Shore A, for the purpose of study, a target reference value of 68 Shore A was set, and thus, the optimization scheme is shown in Eq. 11:

$$\begin{aligned}
&\text{Max} \quad \hat{Y}_{\text{Max Load}} \\
&\text{Max} \quad \hat{Y}_{\text{Max Defor}} \\
&\hat{Y}_{\text{Hardness}} = T \\
&\text{s.t} \\
&-1 \le x_1 \le 1 \\
&-1 \le x_2 \le 1 \\
&-1 \le x_3 \le 1
\end{aligned} \tag{11}$$

Figure 8 shows the results of the optimization. Composite desirability results in

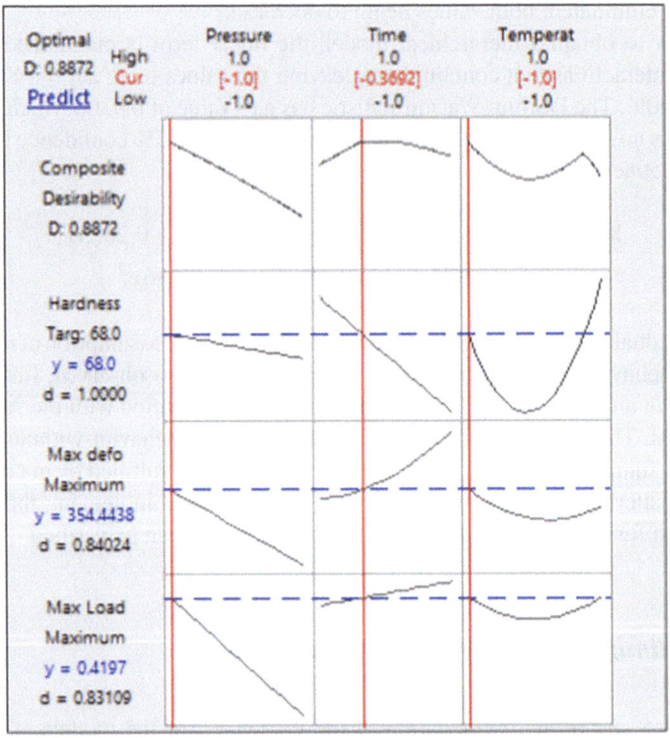

Fig. 8 Optimization results by desirability function

Table 8 Solution for desirability function

Solution	Pressure	Time	Temp	Hardness fit	Max defor fit	Max load fit	Compound desirability
1	−1	−0.3692	−1	68	354.444	0.4197	0.8872
Real values	1200 psi	2.82 min	145 °C				

Table 9 Confidence intervals for responses

Response	Fit	SE fit	95% CI	95% PI
Hardness	68	0.404	(67.068, 68.932)	(66.339, 69.661)
Max defor	354.4	30.5	(279.7, 429.2)	(240.1, 468.8)
Max load	0.4187	0.0147	(0.3864, 0.4530)	(0.3588, 0.4806)

a value of 0.8872 that is considered good convergence. Individual desirabilites are hardness = 1, maximum deformation = 0.84024, and maximum load = 0.83109.

The solution obtained through the desirability function in Table 8 indicates that the pressure and temperature responses must be handled at their low levels. On its actual scale, the pressure should be set at 1200 psi, while the actual value for the temperature is 145 °C. The solution obtained for the time is −0.3692 in coded scale which translated to the actual scale is equivalent to 2.82 min, obtaining a setting of 68 Shore A hardness, 354.4 mm in maximum deformation and 0.4187 kN for the maximum load. Table 9 shows confidence intervals for the three responses.

4.3 MOGA Optimization

The best solution by the MOGA method is shown in Fig. 9. The results obtained are similar to those obtained by the desirability function method and can be consulted in Table 10. In the discussion section, a comparison is made between both methods.

Table 11 and Fig. 10 show the range values for the maximum load and maximum deformation functions. Maximum load values are in a range of 0.4155094 to 0.4197301, while the values for the max deformation function are in a range of 318.6775–354.4792 according to the non-dominated solutions generated on the Pareto front.

4.4 Discussion

The optimization schemes analyzed in this chapter showed good efficiency in obtaining feasible solutions. In the case of the desirability function, good flexibility is noted for the experimental scheme since it allows us to weigh the importance of

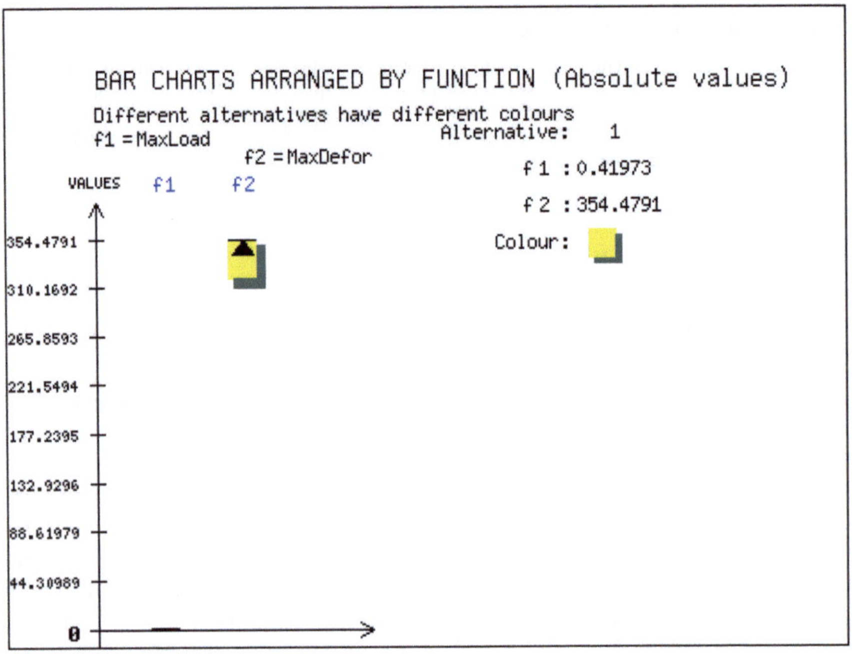

Fig. 9 Absolute values for max load and maximum deformation

Table 10 MOGA solution

Function name	Value	
Max load	0.4197301	
Max defor	354.4792	
Variable name	Value	Real values
x1 (pressure)	−1	1200 psi
x2 (time)	−0.3692788	2.82 min
x3 (temperature)	−1	145 °C

the responses. In the proposed case, the weighting given to each of the responses was 1. A unique solution is achieved that generates good solutions for the optimization approach.

Although some of the results for the process variables are not recommended for its use. When pressure and temperature are used at low levels, there is a risk that they will generate many defects in the sole. The MOGA algorithm does not allow weighting of importance in the variables, but it allows to set the desired variables to the target values of the process. The above can improve the accuracy of the solutions.

In comparison with the desirability function, MOGA allows obtaining a set of non-dominated solutions. The method presents greater flexibility in the results. It

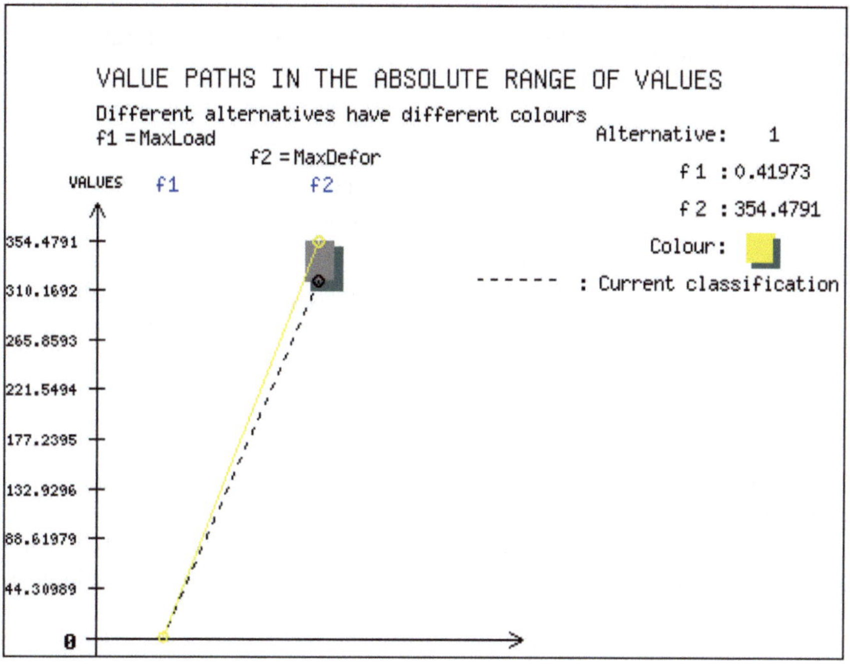

Fig. 10 Value paths in the absolute range of values

Table 11 Function values for MOGA

	Function	Lowest value	Highest value
Max	Max load	0.4155094	0.4197301
Max	Max defor	318.6775	354.4792

allows a more complete analysis with a different set of solutions that can be selected based on the needs of the process.

Table 12 shows a comparison for the optimization methods. It is noted that the same extreme level of process is achieved for the variable by means of the desirability function and MOGA. A value closer to the center is noticed in both methods. The value obtained shows a minimum variation in both methods and is detectable only by precision in the decimals. It is observed that the same extreme level of process is obtained for the variable by means of the desirability function and MOGA.

The resulting data for the pressure variable is the same in both methods. The pressure in the process can be worked at this level, although it is desirable to work a slightly higher pressure to avoid material leaks in the molds. The time shown in the two methods is close to the average experimental value, which is commonly used in the process. This variable must be continuously verified by the operators since they must open the molds at the indicated times to avoid damage to the soles. The level reached for the temperature must be controlled continuously in the process.

Table 12 Comparison of optimum for desirability function and MOGA

Factor	Desirability function	MOGA*
x1	−1	−1
x2	−0.3692	−0.3692788
x3	−1	−1
Pressure	1200 psi	1200 psi
Time	2.82 min	2.82 min
Temperature	145 °C	145 °C
Max load	0.4201 kN	0.4197301 kN
Max defor	384.3 mm	354.4792 mm
Hardness	68 Shore A	68 Shore A

*Best solution in Pareto front

Temperature is a critical factor for vulcanization and can vary throughout the day due to noise factors such as ambient temperature.

The value achieved for the maximum load is higher in the desirability function method. Both values are acceptable from the point of view of the desired quality characteristics. The value for maximum deformation is higher in the desirability function method. MOGA shows lower efficiency for this response. The MOGA algorithm restricts the hardness value at 68 Shore A, which is the desired value. The value obtained by desirability function is also equal to the desired target value.

5 Conclusions

According to our findings, the following conclusions can be proposed:

- Multi-objective optimization in manufacturing is very important since there are many cases in which compromise solutions must be obtained to meet various quality objectives in a product. The desirability function presents the possibility of weighing the importance of each of the answers in the look for a more appropriate solution, and the MOGA technique allows obtaining a set of non-dominated solutions on the Pareto front that can allow the analyst to better selection according to the type of process. MOGA works in a restricted optimization scheme and allows the use of sets of linear or nonlinear constraints that allow the analysis of more complex cases.
- Currently, there are not many applications in the literature on the improvement of physical and mechanical properties in rubber and the effect of vulcanization on quality characteristics. Particularly in rubber for shoe soles, this is an area that requires research to determine quality standards in material properties. In this proposal, results were presented for the sole material of a vans type footwear with export quality, and it would be interesting to compare it with future research.

- Starting from the optimal value found, it will allow process engineers to do sequential experimentation for continuous improvement. In addition, in the search for innovation strategies for future research, the use of robust multi-objective optimization schemes is proposed to reduce the impact of noise factors in the vulcanization process; such as ambient temperature, humidity, among others.

References

Arora JS (2012) Chapter 17—Multi-objective optimum design concepts and methods. In: Arora JS (ed) Introduction to optimum design, 3rd edn. Academic Press, Boston, pp 657–679

ASTM_D412-06a (2013) Standard test methods for vulcanized rubber and thermoplastic elastomers—tension. In: ASTM International, West Conshohocken, PA

ASTM_D2240-15e1 (2015) Standard test method for rubber property—durometer hardness. In: ASTM International, West Conshohocken, PA

Bhamu J, Sangwan KS (2014) Lean manufacturing: literature review and research issues. Int J Oper Prod Manag 34:876–940. https://doi.org/10.1108/IJOPM-08-2012-0315

Box GEP, Behnken DW (1960) Some New three level designs for the study of quantitative variables. Technometrics 2(4):455–475. https://doi.org/10.2307/1266454

Cao B, Zhao J, Lv Z, Liu X, Yang S, Kang X, Kang K (2017) Distributed parallel particle swarm optimization for multi-objective and many-objective large-scale optimization. IEEE Access 1–1. https://doi.org/10.1109/ACCESS.2017.2702561

Cielsielski A (1999) An introduction to rubber technology. Rapra Technology Limited, Shawbury, UK

Costa N, Lourenco J, Pereira Z (2011) Desirability function approach: a review and performance evaluation in adverse conditions. Chemometr Intell Lab Syst 107. https://doi.org/10.1016/j.chemolab.2011.04.004

Davim JP, Davim JP (2012) Computational methods for optimizing manufacturing technology: models and techniques. IGI Global, Hershey PA

Del Castillo E, Montgomery DC, McCarville DR (1996) Modified desirability functions for multiple response optimization. J Qual Technol 28(3):337–345. https://doi.org/10.1080/00224065.1996.11979684

Derringer G, Suich R (1980) Simultaneous optimization of several response variables. J Qual Technol 12(4):214–219. https://doi.org/10.1080/00224065.1980.11980968

García-Alcaraz JL, Maldonado-Macías AA, Cortes-Robles G (2014) Lean manufacturing in the developing world: methodology, case studies and trends from Latin America. Springer International Publishing, Switzerland

Goldberg DE (1989) Genetic Algorithms in search, optimization and machine learning. Addison-Wesley Longman Publishing Co., Inc., Boston, MA

Gutiérrez Pulido H, Salazar R (2012) Análisis y Diseño de Experimentos. McGraw-Hill, México, DF

Hahn GJ, Doganaksoy N (2011) The role of statistics in business and industry. Wiley, New York

Harrington ECJ (1965) The desirability function. 21(10):494–498

Holland JH (1984) Genetic algorithms and adaptation. In: Selfridge OG, Rissland EL, Arbib MA (eds) Adaptive control of ill-defined systems. Springer, US, Boston, MA, pp 317–333

INEGI (2014) Estadísticas a propósito de la Industria del calzado. INEGI

Islam MA, Alam M, Hannan M (2012) Multiresponse optimization based on statistical response surface methodology and desirability function for the production of particleboard. Compos B Eng 43:861–868. https://doi.org/10.1016/j.compositesb.2011.11.033

John B (2013) Application of desirability function for optimizing the performance characteristics of carbonitrided bushes. Int J Ind Eng Comput. https://doi.org/10.5267/j.ijiec.2013.04.003

Kanthababu M (2012) Multi-objective optimization of manufacturing processes using evolutionary algorithms. Computational methods for optimizing manufacturing technology: models and techniques. Hershey, PA, USA, IGI Global, pp 44–66

Köksoy O, Yalcinoz T (2008) Robust Design using Pareto type optimization: a genetic algorithm with arithmetic crossover. Comput Ind Eng. https://doi.org/10.1016/j.cie.2007.11.019

Li B, Li J, Tang K, Yao X (2015) Many-objective evolutionary algorithms: a survey. 48(1 %J ACM Comput. Surv.), Article 13. https://doi.org/10.1145/2792984

Mares A, Dominguez J (2015) Robust design in generalised linear models for improving the quality of polyurethane soles. S Afr J Ind Eng 26:152–166. https://doi.org/10.7166/26-3-1181

Minitab Inc. (2019) Minitab 19 statistical software. State College, PA

Mohammad Hemmat E, Mohammad Hadi H, Somchai W (2018) Pareto optimal design of thermal conductivity and viscosity of $NDCO_3O_4$ Nanofluids by MOPSO and NSGA II Using response surface methodology. Curr Nanosci 14(1):62–70. https://doi.org/10.2174/15734137136661709l4103043

Myers RH, Montgomery DC, Anderson-Cook CM (2016) Response surface methodology: process and product optimization using designed experiments. Wiley, New York

Nandi AK, Deb K, Datta S (2013) Genetic algorithm-based design and development of particle-reinforced silicone rubber for soft tooling process. Mater Manuf Processes. https://doi.org/10.1080/10426914.2013.773022

Nicholson TAJ, Pullen RD (1969) Statistical and optimisation techniques in the design of rubber compounds. Comput Aided Des 1(3):39–47. https://doi.org/10.1016/S0010-4485(69)80084-9

Nocedal J, Wright S (2006) Numerical optimization. Springer, New York

Onyekwere O, Odiakaose C, Uyanga K (2017) Multi response optimization of the functional properties of rubber seed—shear butter based core oil using D-optimal mixture design. Arch Foundry Eng 17. https://doi.org/10.1515/afe-2017-0159

Ortiz F, Simpson JR, Pignatiello JJ, Heredia-Langner A (2004) A genetic algorithm approach to multiple-response optimization. J Qual Technol 36(4):432–450. https://doi.org/10.1080/00224065.2004.11980289

Salvatori PE, Sánchez G, Lombardi A, Nicocia E, Bortolato SA, Boschetti CE (2018) Optimization of properties in a rubber compound containing a ternary polymer blend using response surface methodology. J Appl Polym Sci 135:46548. https://doi.org/10.1002/app.46548

Singaravel B, Selvaraj T (2016) Application of desirability function analysis and utility concept for selection of optimum cutting parameters in turning operation. 15(1):1–11. https://doi.org/10.1142/s0219686716500013

Socconini L (2015) Certificación Lean Six Sigma Green Belt para la excelencia en los negocios. Barcelona

Trautmann H, Weihs C JM (2006) On the distribution of the desirability index using Harrington's desirability function. 63(2):207–213. https://doi.org/10.1007/s00184-005-0012-0

Zelany M (1974) A concept of compromise solutions and the method of the displaced ideal. Comput Oper Res 1(3):479–496. https://doi.org/10.1016/0305-0548(74)90064-1

Ziegel E (2002) Genetic algorithms and engineering optimization. Technometrics 44(1):95–95. https://doi.org/10.1198/tech.2002.s675

A Multi-agent System for the Inventory and Routing Assignment

Conrado Augusto Serna-Urán, Cristian Giovanny Gómez-Marín, Julián Andrés Zapata-Cortes, and Martín Darío Arango-Serna

Abstract In the supply chain management research field, the analysis of collaborative strategies and the joint management of inventory and transport of goods have increased the attention or academics and practitioners due to the everyday bigger amount of freight flows, the complexity of the logistics scenarios and the current changes and tendencies in the goods interchange process. However, the computational complexity and the problems that may arise in the integration processes of different participant actors become the majority of proposals difficult to implement. In this paper, we develop a multi-agent system for solving the joint inventory and routing assignment problem. The proposed multi-agent system facilitates the integration of the distribution processes and the inventory management in a supply network with one depot and n customers. The multi-agent model is based in the autonomy of the actors to manage their capacity and their demand, as well as in the integration of the transport and inventory process using a collaborative strategy. To solve the resulting vehicle routing problem, we design a collaborative behavior that uses as an evaluation tool a local search heuristic with a 2-opt operator. The model for the inventory and routing assignment is implemented on the Java-based software platform JADE. The collaboration-based process in the multi-agent system demonstrates the usefulness of the distributed computing to decrease the total cost in the logistics operation.

Keywords Inventory routing problem · Multi-agent system · Supply chain management · Supply chain collaboration

C. A. Serna-Urán (✉) · C. G. Gómez-Marín
Instituto Tecnológico Metropolitano, Medellín, Colombia
e-mail: conradoserna@itm.edu.co

J. A. Zapata-Cortes
Institución Universitaria CEIPA, Sabaneta, Colombia
e-mail: julian.zapata@ceipa.edu.co

M. D. Arango-Serna
Universidad Nacional de Colombia, Medellín, Colombia
e-mail: mdarango@unal.edu.co

J. L. García Alcaraz et al. (eds.), *Techniques, Tools and Methodologies Applied to Quality Assurance in Manufacturing*,
https://doi.org/10.1007/978-3-030-69314-5_10

1 Introduction

The integration of the distribution and inventory process is one of the main strategies to produce cost efficiency and satisfactory service levels among the actors of the supply chain, which generates a big challenge to the process of the logistics system. The objective of this integration is to reduce the global cost of the distribution process, including the holding and the transport costs (Alkaabneh et al. 2020; Archetti et al. 2007; Zapata-Cortes 2016). Collaborative and integration relationship in the supply chain is a very effective strategy to achieve a competitive advantage. In this context, strategies, such as vendor-managed inventory (VMI), have been successfully implemented in many organizations, which facilitate collaboration between suppliers and their customers through the responsibility assumed by the seller to manage the inventory of their customer (Kim and Shin 2019). Under this methodology, the seller decides the quantity and dates to send the product to the customers, ensuring that they will not run out their supplies (Sainathan and Groenevelt 2019). In this strategy of mutual agreement, suppliers efficiently coordinate the management of orders to respond to the customers' demands, producing savings in transport and inventory costs, as well as increasing the service levels (Lee and Cho 2018; Phan et al. 2019).

However, the implementation of logistics strategies based on VMI is not a simple task since the supplier must solve two problems: a freight transport assignment and the inventory management problem (Kim and Shin 2019; Verma and Chatterjee 2017). This condition gives rise to the problem known in the literature as the inventory routing problem (IRP).

IRP integrates two of the most important components in the supply chain: vehicle routing and inventory management. The objective of this type of problems is to minimize the transport costs associated with the routes of vehicles to deliver the products from the supplier to the customers and the reduction of inventory costs without incurring in stock breaks (Arango-Serna et al. 2016; Shaabani and Kamalabadi, 2016). Considering these two elements and the VMI integration framework, the supplier must answer three fundamental questions in the logistics operation: (1) When must the products be delivered to each customer? (2) How much products must be delivered in each order? and (3) How does the supplier integrate customers on an efficient delivery route? Figure 1 depicts a graphical representation of the IRP model.

The implementation of this type of strategy is difficult due to two different problems: the mathematical complexity of the problem, which is NP-hard, producing a computational challenge to find the exact solution in an appropriate time (Huber and Geiger 2015; Vidović et al. 2014) and the fact that the actors involved in the distribution process normally try to conserve specific and individual objectives without sharing information to other actors of the supply chain (Zapata-Cortes 2016).

As a solution to these difficulties, some authors have found in the multi-agent systems a promising alternative because of its versatility and ease implementation. The use of multi-agent systems facilitates the representation of the supply chain actors in artificial agents with their behaviors, capacities and specific goals interacting in a

Fig. 1 Representation of IRP. *Source* The authors

coordinated and collaborative function system (Arango-Serna et al. 2018). The multi-agent systems consist in multiple agents interacting to solve a common problem, compete for the use of shared resources or simply they coordinate each other to avoid conflicts (Arango-Serna and Serna-Urán, 2016; Serna-Urán, 2016).

Lopes and Coelho (2010) highlight multi-agent systems (MAS) as a good modeling tool for representing decentralized systems with multiple and different levels of representation. Thus, one of the main motivations in the use of multi-agent systems is their ability to solve problems with distributed data, experience or control, as well as the ability to facilitate interaction between different artificial agents and to improve computational performance (Zhao and Zhao 2018). Multi-agent systems have been used to solve real-world problems of industrial and commercial processes, including but not limited to manufacturing processes, process control, air traffic control, telecommunications, information management, business process management, electronic commerce, supply chain management, among others (Böhnlein et al. 2011; Cevirici and Moller-Madsen 2007).

The aim of this paper is twofold: first, to model the inventory and goods routing assignment in the freight distribution process as a multi-agent system, in which each agent has the responsibility of one or more activities of the distribution process and interacts with other agents. This model seeks to reduce the total distribution costs involving the decision of the transportation and inventory plans.

The second objective is to present the main communications protocols that facilitate the interaction and negotiation process among the model agents involved in this distribution process.

The rest of this work is organized as follows: In the first part, the chapter presents a short literature background about the integration of transportation and inventory decision, as well as of multi-agent systems. Then, the proposed multi-agent model is explained and later applied in a median size problem. Finally, the analysis and some conclusions are stated.

2 Background

As two of the main functions of logistics, the inventory management and the routing process are important issues to reduce cost and increase service level that the decision makers must deal with in operational and tactical levels (Soysal et al. 2019). Guerrero et al. (2013) argue that inventory management and distribution decisions are related to each other, since inventory depends on the frequency and time to supply, as well as orders and product costs (Arango-Serna et al. 2015).

The decision of simultaneously assigning the inventory and transportation is carried out by following two types of mechanisms: decomposition and aggregation. The decomposition mechanism separates the problem into two phases, in which the first determines the inventory and the second the transportation routes (Kang and Kim, 2010). The aggregation mechanism finds the solution to the problem simultaneously, producing the inventory and transport decisions directly. To solve this problem, the most studied model is the inventory routing problem (IRP), which is based on the vendor-managed inventory and seeks to combine the transport and inventory allocation problem to reduce the global distribution costs (Arango-Serna et al. 2016).

Bertazzi and Esperanza (2013) and Cho et al. (2014) mention the importance of including the inventory variable within the freight distribution processes. These authors have studied the inclusion of inventory costs in the facilities of suppliers and customers, as a fundamental element in the processes of transport networks optimization in goods distribution, applying collaborations among companies through the application of the vendor-managed inventory model (VMI), in which the resupply decisions are taken by the seller (Archetti et al. 2014; Azuma et al. 2011; Coelho and Laporte 2013).

Some studies have considered to tackle the assignment and routing problems through collaborative process searching to reduce cost and achieve horizontal or vertical logistics integration levels that give answers to the suddenly changes on demand or in the logistics scenarios. Soysal et al. (2018) develop a green inventory for perishable goods model with horizontal collaboration obtaining benefits in performance indexes, emissions, driving time, total routing cost (including fuel and salary cost), inventory and waste cost given an uncertain demand.

Under the framework of collaborative models, some authors propose multi-agent systems which consist of individual software agents with specific functions and capabilities integrated to meet common objectives. Multi-agent systems are based on the distributed computing paradigm, in which computational costs are reduced to deal with combinatorial problems or process integration, such as those found in logistics operations. According to Serna-Urán (2016) and Wooldridge (2002), an agent is a software that virtually represents an entity or service provider, and it is capable of acting autonomously. Each agent has an individual objective that seeks to achieve, as part of a framework of collaboration and coordination with other agents. Decisions are made by performing logical deduction processes. A traditional approach to building an artificial intelligence system suggests that intelligent behavior can be generated in a system that makes a symbolic representation of its environment and its desired behavior and systematically manipulates this representation. In logic-based agent architectures, symbolic representations correspond to logical methods, and the syntactic manipulation is a logical deduction (Serna-Urán 2016).

In this type of architecture, knowledge is represented symbolically. Agents start from an initial state and have the ability to generate plans to achieve their objectives. Agents have a planning system that allows defining the steps that must been taken to achieve the objectives. A deliberative agent contains a symbolic model of the world explicitly represented, which uses logical reasoning mechanisms based on pattern matching and symbolic manipulation to make decisions, aiming to achieve the agent's objectives (Serna-Urán 2016). Figure 2 shows an example of a multi-agent system architecture.

The supply chain is a net of subsystems working together in a dynamic and complex environment. This characteristic allows to model the supply chain as an agents network (Serna-Uran 2016). In recent years, the supply chain management has received much interest to be modeled as a multi-agent system (Aminzadegan

Fig. 2 Multi-agent architecture. *Source* Serna-Urán et al. (2018)

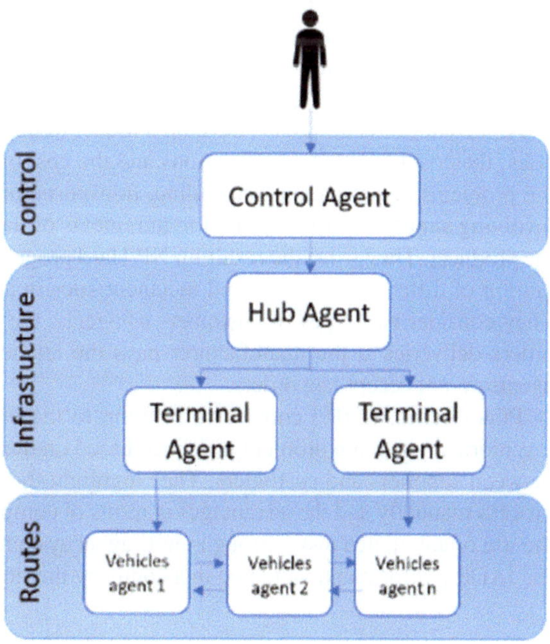

et al. 2019; Avci and Selim, 2016; Ghadimi et al. 2017; Kumari et al. 2015; Pal and Karakostas 2014). In this multi-agent system, each agent acts autonomously, interacts with other agents, reacts to changes in the environment and makes proactive decisions (Böhnlein et al. 2011; Wang et al. 2013).

In the specialized literature, different studies address this topic as presented by Alves et al. (2019), which exhibit a model to evaluate the cost, noise emission and congestion reduction of delivery points in urban freight for an e-commerce process. Arango-Serna et al. (2018) develop a model for urban goods distribution based on an evolutionary multi-agent system that allows the interaction and coordination among multiple suppliers, one hub and multiple customers in the allocations and routing process.

Gómez-Marín et al. (2018) design a multi-agent micro-simulation model for the dynamic urban goods distribution using a multi-layer framework, where the main actors of the distribution are represented and exhibited different behaviors to react to the permanently changing context.

Belykh and Botvin (2018) present a multi-agent system as a possible tool that facilitates in-depth analysis of the behaviors of supply chain members.

Sitek et al. (2014) develop an application of a multi-agent hybrid approach to model and optimize supply chain problems. The authors integrate two environments: mathematical programming (MP) and logical constraint programming (CLP) and two types of agents.

Pal and Karakostas (2014) present a web services multi-agent framework for a collaborative materials procurement system in a supply chain with three types of agents: customer agent, supplier service agent and central service agent.

Avci and Selim (2016) develop a multi-agent model to observe the effects of customer requests parameters in a supply chain with transshipments. They validate the model in a multinational supply chain that presents uncertainty in demand and supply.

Aminzadegan et al. (2019) develop a model to minimize the cost of order deliveries, the cost of resources allocations and the cost of delay penalty by integrating the problems of production scheduling, transportation and resource allocation and involving simultaneously different requirements of the customers and objectives of the producer. They solve the resulting NP-Hard programming problem with the integration of different heuristics and metaheuristics in a multi-agent solution system. They consider two types of customers as agents, the first agents accept a delay in orders deliveries if the manufacturer pays the lateness penalty, while the second agents do not accept late orders.

Plinere et al. (2015) consider a solution to the management of purchases and inventories. The main problem is defined based on the timely coordination of orders between suppliers and customers. They highlight the difficulties of developing this process manually and the advantages in terms of compliance that it would have with the use of automated tasks through multi-agent systems. The model is developed on the JADE platform and validated in a company that manufactures microchips.

Although most of these studies are good approximations to the supply chain, it is still necessary to continue advancing in new models that facilitate the analysis of different functions of the supply chain.

In this sense, the decision process of integrating the minimization of inventory and transportation costs can be modeled as a multi-agent system, with customers, suppliers and transporters interacting to achieve a global objective. Regardless every actor has their interests; they must be balanced to optimize the total costs of the supply chain.

3 Multi-agent System for the Inventory and Routing Decision

Taking advantage of the benefits of distributed computing, the decentralization of tasks and the coordination and integration between supply chain actors, a multi-agent system to solve the combined problem of inventory and routing assignment was developed. This multi-agent system (MAS) is based on the representation of the customers and the transport logistics operator as virtual agents that reproduce the input data from the physical system to the virtual system. The logistic operator develops the information exchange and the coordination processes among all agents in the multi-agent system. Figure 3 depicts the representation of the multi-agent system with their flow of information, and Table 1 presents the description of the different agents.

By using a *collaborative strategy*, the logistic operator (*agent service control*) requests the customers (*agent customer*) for the demand information at the t period, in order to generate the service orders. The *agent customers* may accept or reject the proposal of the *agent service control*. If the proposal is accepted, the *agent service control* request to the *agent route* is to build a route for each service at the t period. Finally, the *agent collaboration* searches for possible improvements by changing the requests on the routes based on the location of the client and the period t of the demand.

The communication and coordination process carried out by the model can be depicted in Fig. 4, and their operation is explained as follows.

3.1 Service Control Agent

This agent sends an offer service to each customer for the period t. Customers with demands on t are accepted. For customers with demands on $t + l$ ($l = 1, 2, 3\ldots, l - T$), a request to integrate the service on period t is sent if

(i) Customers are also in the service on t.
(ii) Customers are near the route in time or distance.

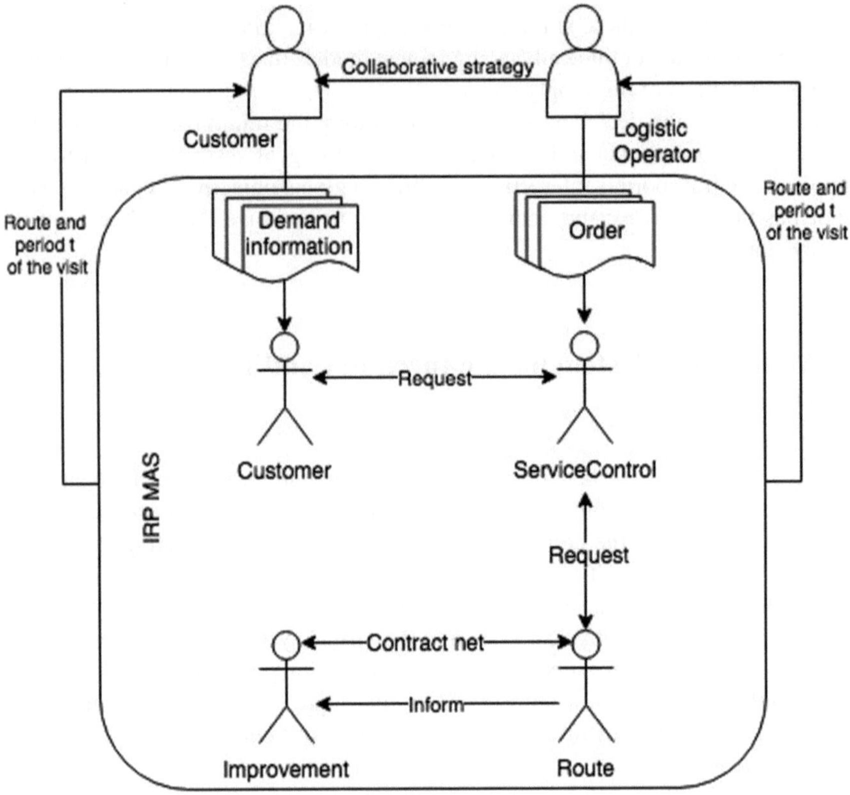

Fig. 3 Multi-agent system for inventory and routing assignment. *Source* The authors

Table 1 Agents description

Agent name	Description
Customer agent	It asks for service order defined by quantity, location and period of time of their demand. The agent customers base their decision process on the inventory cost analysis criterion. They expect to be served in a just in time process
Service control agent	Builds delivery services defining the quantity, period and location of the delivery. Search for reducing the transportation cost and augmenting the use of the vehicles without exceeding their capacity forming the cluster of requests
Route agent	Builds the routes of services using the 2-opt local search heuristic
Collaboration (integrator) agent	This agent explores solutions for transportation cost reduction. The customers with a high ratio between holding and transportation costs are proposed to integrate other service routes. The transportation cost is computing from the cost of inserting the customer to an evaluated route

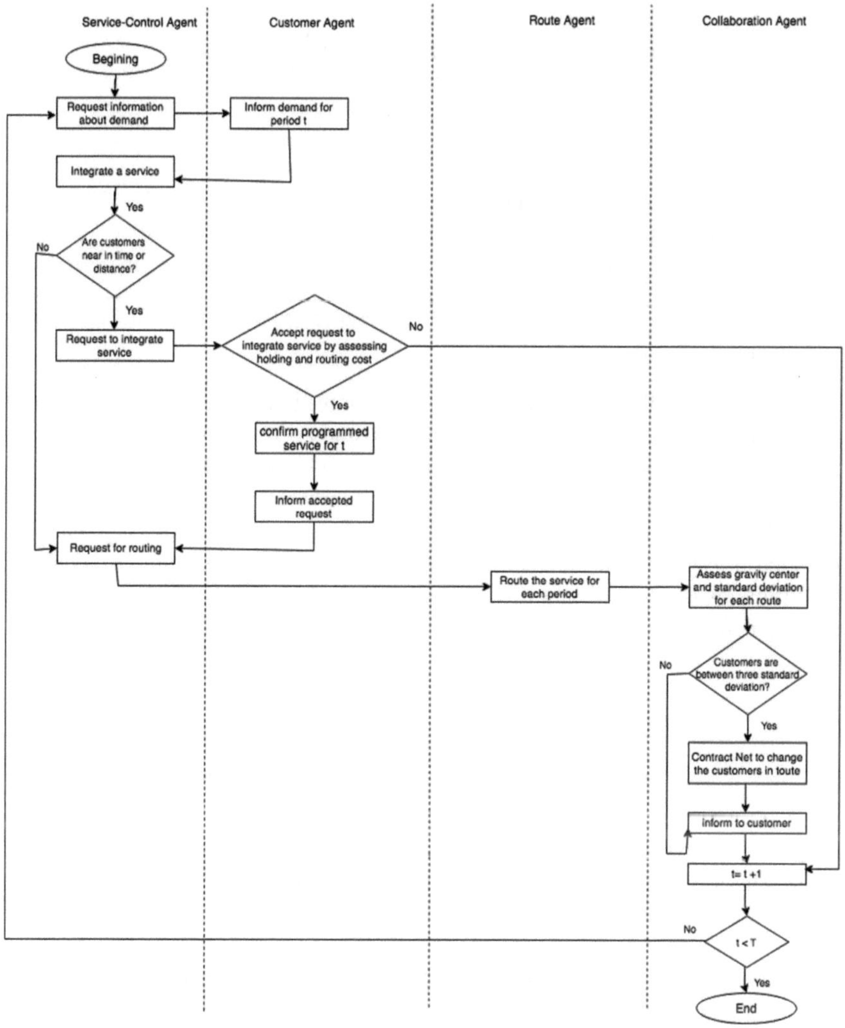

Fig. 4 Communication process at multi-agent system (MAS). *Source* The authors

3.2 *Customer Agent*

Customers with demands on period t confirm services scheduled for the same period. Customers with demands on periods in $t + l$ evaluate the requests to integrate services in t with the following criteria:

(i) Handling cost calculation from the unit period.

The handling cost is calculated using Eq. (1).

$$\text{handling cost(unit} - \text{period)} = h * \sum_{l=1}^{T-t} d_{i(t+l)} * l \tag{1}$$

where:

h	Handling cost of one unit one period
$d_{i(t+l)}$	Demand of customer i on the period $t + l$
l	Number of periods elapsed after the execution of the service in period t.

(ii) Projected route cost from the Euclidian distances between the route's customer location:

The expected route distance is the sum between the total expected travel time and the sum of the service times of each allocated customer. The travel time corresponds to the sum of the times t_{ij} that the vehicle takes to go from a node i to a node j for all i and j in C on a Hamiltonian circuit, in which C represents the number of customers assigned to the distribution center. The expected distance of the tour is calculated as presented in Eq. (2), and the computation of the expected travel time is calculated by multiplying this tour distance by the average travel speed set at 30 km/h for urban settings, as stated in Eq. (3).

$$\text{tour distance} = 2\bar{r} + \left[k\delta^{-1/2}\right]C \tag{2}$$

$$\text{Expected travel time} = \text{tour distance} * 30\,\text{km/h} \tag{3}$$

where $\bar{r}$ is the average distance between the C points allocated to the distribution center, k is a constant that depends on the used metric ($k = 0.57$ for Euclidean metrics), and δ is the customer density function (customers per unit area, $\delta =$ total number of customer in the region/area of the region).

The first term or the tour distance ($2\bar{r}$) corresponds to the distance necessary to reach the center of gravity of the points in the area. The second term $\left(\left[k\delta^{-1/2}\right]C\right)$ corresponds to the distance that must be traveled between one node and another. Note that each stop contributes to the total distance comparable to the separation between neighbor points (Daganzo 2005). Once the tour distance and the expected travel time are determined, the expected tour service time is calculated using Eq. (4), in which t_i is the unloading time at each customer, and the expected tour cost is calculated in Eq. (5) using a cost factor for minute for kilometer.

$$\text{Expected tour service time} = \frac{2\bar{r} + \left[k\delta^{-1/2}\right]C}{30\,\text{km/h}} + \sum_{1}^{C} t_i \tag{4}$$

$$\text{Expected tour cost} = \left(\frac{2\bar{r} + \left[k\delta^{-1/2}\right]C}{30\,\text{km/h}} + \sum_{1}^{C} t_i\right) * \frac{\$1\,\text{min}}{\text{kim}} \tag{5}$$

The acceptance criterion is based on the balance between both costs (handling and expected route cost). It should notice that there is not an integrative cost function of inventory cost plus transport cost, as the multi-agent system has a decentralized process where each agent makes autonomous decisions based on the acceptance criterion, and this is quite different of minimize the sum of the two different cost. The *service control agent* provides information about other customers integrating the service. The *customer agent* analyzes the offer based on the principles of collaboration that governs the relationship customers–distribution center.

3.3 Route Agent

This agent evaluates route cost including customers who accept the offer. The agent implements an algorithm as a behavior to solve a vehicle routing problem with the nodes that make up each of the delivery services. This behavior implements a local search heuristic as used (Gómez-Marín et al. 2019; Guemri et al. 2016; Nguyen et al. 2012) and a 2-opt operator to improve the solutions achieved by the metaheuristic. The results of these operations are the routes necessary to perform the goods delivery minimizing the transportation cost.

3.4 Collaboration (Integrator) Agent Operations

The *collaboration agent* evaluates the total costs of each route and selects those with the greater difference between the unit period handling cost and the transport cost. For the selected routes, it evaluates the possibility of moving a customer request to another service according to criteria of time and distance proximity. This agent calculates the center of gravity for each service. Customers located three standard deviations from the center of gravity of the route that are selected to integrate a route with the nearest center of gravity. The *collaboration agent* also evaluates the possibility of moving customer requests to nearest services temporarily.

This process is done through a contract net protocol between the *collaboration agent* and the *route agent*.

3.5 Protocols Definition

The communication protocols between agents allow to generate flexibility in the multi-agent system, through the structured management of the information that each agent receives and sends. In the designed multi-agent system, the *request* and *contract*

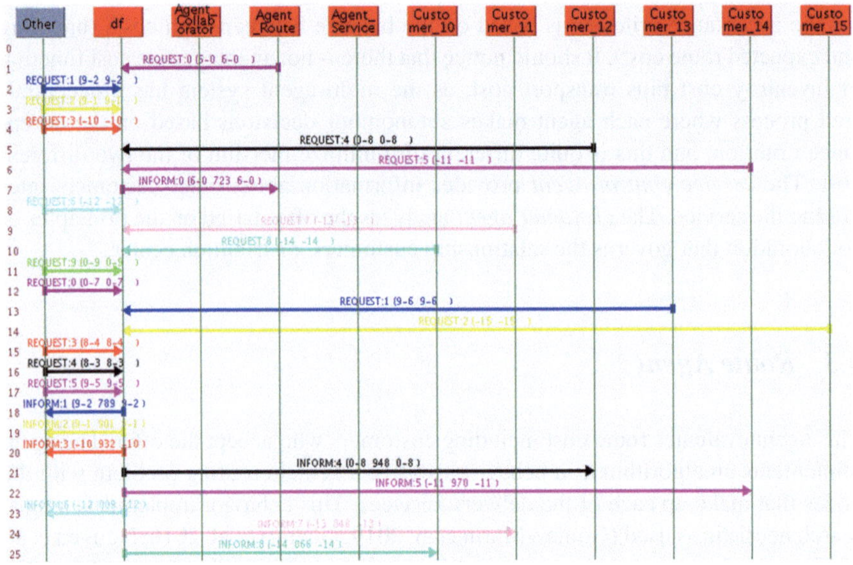

Fig. 5 Communication among agents during simulation. *Source* The authors

net protocols rule the interactions between the agents. These protocols are implemented in JADE®, a Java® agent development framework. Figure 5 represents the protocol communication among agents.

The exchange of information used by the agents in the form of communication protocols follows the Foundation for Intelligent Physical Agents—Agents Communication Language (FIPA-ACL), which are described as follows:

i. *Request protocol (RP)*: The RP is a communication protocol that allows the agents to make information requests and actions to other agents with specific behaviors, capabilities and resources. This protocol allows the multi-agent system to initialize the service offers process by the logistics operator by requesting the demand information. Each customer returns its demand information in period t. With this information, the *control service agent* establishes the service for the period *t*. Figure 6a shows the interaction between *service control agent* and *customer agent using* the request protocol.
ii. *The contract net protocol (CNP)* is a negotiation protocol in which the agent that initiates the process (*collaboration agent*) makes a request to the participating agents (*route agent*) to send the cost proposals of including a customer in a service for the period $t + 1$. The *route agents* evaluate the cost of including the customer in the existing route. With the answers of the *route agents*, the *collaboration agent* selects and assigns the best proposal. Figure 6b shows how the CNP works.

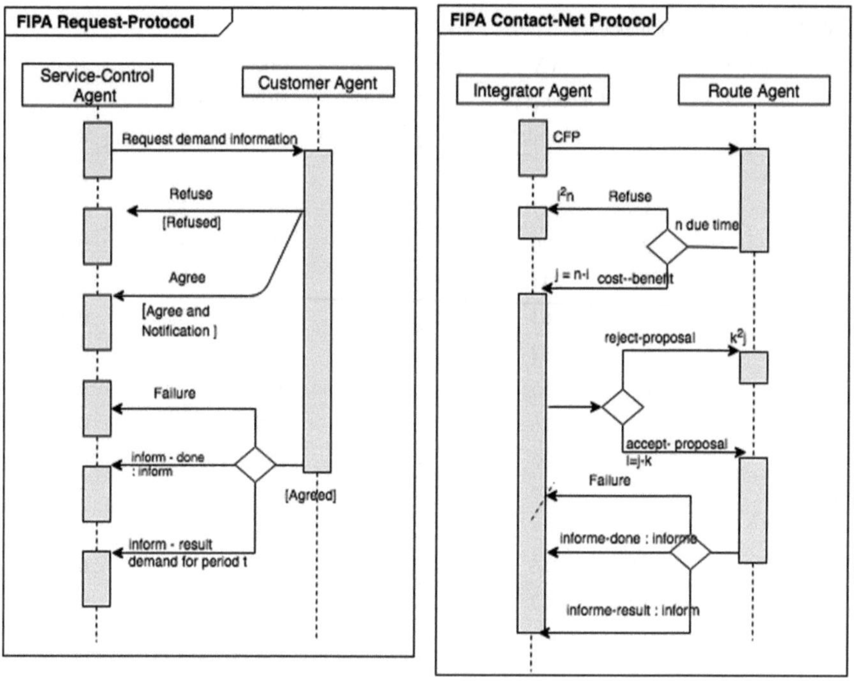

Fig. 6 **a** Request protocol, **b** contract net protocol. *Source* The authors

4 Multi-agent System Application

The MAS was tested using a distribution example which considers 15 customers, one supplier and a three-period time horizon. Table 2 shows the initial data for the tested instance from Archetti et al. (2007). The simulation was carried out using JADE®.

For the first period, all customers can supply their demand from the inventory, in that sense no service has been integrated by the logistic operator. For the second period, there are two different services: The first one is integrated according to the information that the *customers* have sent to the *control service agent*, and the second one is integrated through the interaction between *customer's agents* and *control service agent* to figure out local decision for each customer. Table 3 presents the service integration for the second period with the amount of product to be sent to the different customers.

With these interactions in the second period, the customers that are subscribed for this service are supplied. Then, it is required to compute the needs for the third period. Table 4 shows the service integration for the third period.

The third period is the final one, and there is no more iteration. The customers, in this period, just search to satisfy their needs, reason why their inventory is zero. When the service integration has finished, the agent route starts to build their route using

Table 2 Assessment instance

Logistic operator	Location in X			312	Location Y		363
	Initial inventory			2042	Holding cost		0.03
	Operator available quantity for period						826
Customers	i	x_i	y_1	I_{i0}	U_i	r_i	h_i
	1	237	182	32	64	32	0.02
	2	180	332	72	108	36	0.03
	3	141	388	182	273	91	0.03
	4	163	188	52	104	52	0.02
	5	282	374	152	228	76	0.02
	6	455	296	20	30	10	0.03
	7	326	332	85	170	85	0.,04
	8	235	432	79	158	79	0.04
	9	412	488	22	44	22	0.02
	10	113	46	72	108	36	0.04
	11	266	302	136	204	68	0.02
	12	257	23	46	92	46	0.02
	13	363	22	55	110	55	0.02
	14	158	81	65	130	65	0.03
	15	423	95	146	219	73	0.02

I_{i0} is the initial inventory in customer i; U_i is the maximum inventory that can hold customer i; r_i is the demand for customer i; h_i is the holding costs of each customer

Table 3 Services integration for period $t = 2$

	Customer							
	1	4	7	8	9	12	13	14
First integrated service demand	32	52	85	79	22	46	55	65
Second integrated service	64	104	170	158	22	92	55	130
Inventory cost								U$12.65
Approximated transportation cost								U$10.77

the *2-opt* local search heuristic to compute the cost of the routes for each service in each period. The results of the behavior of this agent are shown in Table 5.

With these results, the *collaboration agent* begins to make their analysis and trigger the contract net protocol with the agent's routes to find a better inventory and transportation cost combination for the system. First, it should compute the gravity

Table 4 Services integration for period $t = 3$

	Customer								
	2	3	5	6	9	10	11	13	15
First integrated service demand	36	91	76	10	22	36	68	55	73
Inventory cost									U$0
Approximated transportation cost									U$ 9.94

Table 5 Routes for each service period

Period	Routes	Cost (U$)	Inventory cost
$t = 2$	0, 1, 4, 14, 12, 13, 7, 9, 8, 0	13,77	12.65
$t = 3$	0, 5, 11, 2, 3, 10, 13, 15, 6, 9, 0	15,14	0
Total cost (Inventory + Transport)		41.56	

center, the average distance and the standard deviation for the bigger and more expensive route. The customers with a bigger distance that three standard deviations to the center of gravity are selected to evaluate a change in the route. The contract net protocol asks for the routes to compute the cost of each route with and without the selected customers, and the *collaboration agent* computes the inventory cost of making the changes in the routes. The result of this behavior is presented in Table 6.

The agent should compare the differences between the reduction in the cost of changing the customer from route $t = 3$ to route $t = 2$ and the cost increase of accepting the customer in the route at $t = 2$. In that sense, the customer 3 is rejected because the savings of changing this customer from service $t = 3$ to $t = 2$ are 0.99, and the cost increase of accepting at $t = 2$ is 4.03.

The final integrated services for periods $t = 2$ and $t = 3$ and their routes are presented in Table 7.

With the *agent collaboration*, the total cost decrease U$3.61 is equivalent to 8.77% less than the initial solution. Figure 7 presents the initial solution, and Fig. 8 is the improved solution obtained with the multi-agent system for the distribution problem.

Table 6 Customers changes costs

	t = 2		t= 3	
Selected customers	Distance change (U$)	Inventory change (U$)	Distance change (U$)	Change accepted?
9	0	+0.44	−1.61	Yes
10	+0.54	+1.44	−2.67	Yes
13	0	+1.1	−1.31	Yes
3	+1.30	+2.73	−0.99	No

Table 7 Final routes after the agent's collaboration

Period	Routes	Transport cost	Inventory cost
$t = 2$	0, 7, 13, 12, 10, 14, 4, 1, 8, 9, 0	14.32	14.09
$t = 3$	0, 5, 3, 2, 11, 15, 6, 0	9.54	0
Total cost			37.95

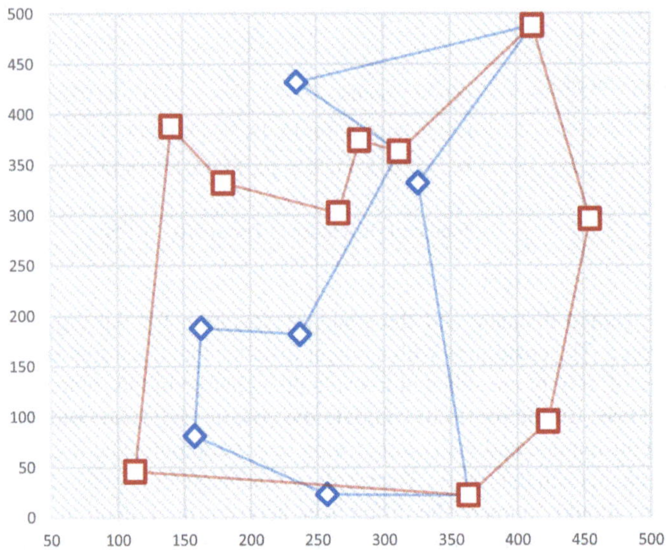

Fig. 7 Routes of initial solution. *Source* The authors

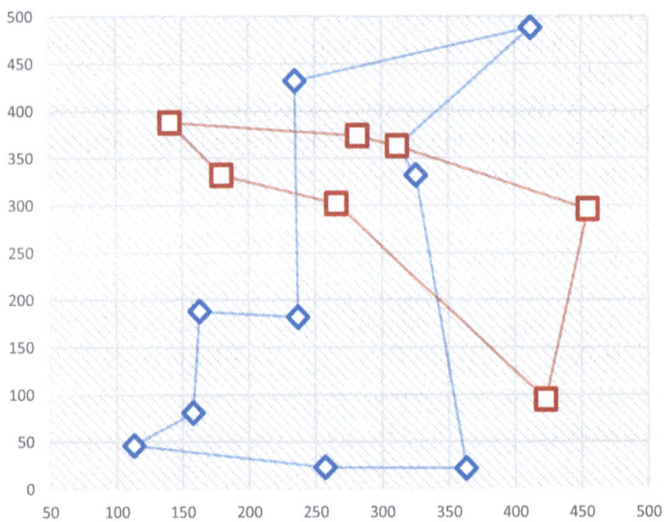

Fig. 8 Improved solution in the MAS. *Source* The authors

5 Conclusions

In this article, the combined inventory and transportation decision process in the supply chain was modeled as a multi-agent system that facilitates the analysis of the relationships between customers and suppliers in a framework of autonomy and collaboration. The model proposes an easy-to-implement and functional architecture in companies in which the inventory administration is not completely centralized. For this, different agents were designed to support the distribution process considering the transport cost and the inventory cost, which integrate functions and interaction protocols to make decisions based on their own resources and interests. However, the agents also evaluate their decisions based on principles of collaboration and global efficiency. The collaboration among all the agents allows the model to support the decision making based on a global objective and not just an individual goal, which generate a commitment of all the actors to reach a more efficient system.

The multi-agent system seeks to improve the solutions in a collaborative framework through the negotiation processes (request and contract net protocols), the agents' behaviors using heuristic procedures and the analysis of the inventory and route cost. The results from the simulation process allow to understand the level of integration that can be achieved with the use of models based on multi-agent systems. The dynamic evaluation of logistic conditions such as travel times, distances and delivery periods, among others is one of the greatest benefits of using this type of models that can be used for short time horizon problems, as in the presented instance, as well as for problems with a greater number of actors (agents) and more planning periods.

As future research work, it can analyze more complex instances of the distribution process and its results should be compared with the optimal solutions produced by optimization methods such as the IRP model, with the aim of deriving how close the results of the multi-agent model are with respect to these optimal distribution plans. In addition, as future lines of research, it is suggested the integration of new agents in the model, which allow the development of simulation process closer to real situations, including suppliers, local administrators and other transport systems such as passengers and private transportation. It could be also interesting to include different metaheuristics as agents' behaviors such as evolutionary algorithms to solve the route and inventory problem assessing the multi-agent performance with these algorithms.

References

Alkaabneh F, Diabat A, Gao HO (2020) Benders decomposition for the inventory vehicle routing problem with perishable products and environmental costs. Comput Oper Res 113: https://doi.org/10.1016/j.cor.2019.07.009

Alves R, da Silva Lima R, Custódio de Sena D, Ferreira de Pinho A, Holguín-Veras J (2019) Agent-based simulation model for evaluating urban freight policy to E-commerce. Sustainability 11(15):4020. https://doi.org/10.3390/su11154020

Aminzadegan S, Tamannaei M, Rasti-Barzoki M (2019) Multi-agent supply chain scheduling problem by considering resource allocation and transportation. Comput Ind Eng 137(August): https://doi.org/10.1016/j.cie.2019.106003

Arango-Serna MD, Gómez-Marín CG, Serna-Urán CA, Zapata-Cortes JA (2018) Multi-agent model for urban goods distribution. Res Comput Sci 147(3):35–44. Retrieved from http://www.rcs.cic.ipn.mx/rcs/2018_147_3/Multi-agentModelforUrbanGoodsDistribution.html

Arango-Serna MD, Romano CA, Zapata Cortés JA (2016) Distribución colaborativa de mercancías utilizando el modelo IRP. DYNA 83(196):204–212. https://doi.org/10.15446/dyna.v83n196.52492

Arango-Serna MD, Serna-Urán C (2016) New contract net negotiation protocol based on fuzzy inference applied to the supply chain. Universidad, Ciencia y Tecnología 20(81):176–187

Arango-Serna MD, Zapata-Cortés JA, Gutíerrez D (2015) Modeling the inventory routing problem (irp) with multiple depots with genetic algorithms. IEEE Latin Am Trans 13(12):3959–3965

Archetti C, Bertazzi L, Laporte G, Speranza MG (2007) A branch and-cut algorithm for a vendor-managed inventory-routing problem. Transp Sci 1(3):382–391

Archetti C, Bianchessi N, Irnich S, Speranza MG (2014) Formulations for an inventory routing problem. Int Trans Oper Res 21(3):353–374

Avci MG, Selim H (2016) A multi-agent system model for supply chains with lateral preventive transshipments: application in a multi-national automotive supply chain. Comput Ind 82:28–39. https://doi.org/10.1016/j.compind.2016.05.005

Azuma RM, Coelho GP, Von Zuben FJ (2011) Evolutionary multi-objective optimization for the vendor-managed inventory routing problem. In: IEEE congress on evolutionary computation (CEC), pp 1457–1464

Belykh DL, Botvin GA (2018) Multi-agent framework for supply chain dynamics modelling with information sharing and demand forecast. In: Alexandrov D, Boukhanovsky A, Chugunov A, Kabanov Y, Koltsova O (eds) Digital transformation and global society. DTGS 2018. Communications in computer and information science, vol 858. Springer, Cham, pp 366–374. https://doi.org/10.1007/978-3-030-02843-5_29

Bertazzi L, Esperanza MG (2013) Inventory routing problems with multiple customers. Eur J Transp Logistic 2:255–275

Böhnlein D, Schweiger K, Tuma A (2011) Multi-agent-based transport planning in the newspaper industry. Int J Prod Econ 131(1):146–157. https://doi.org/10.1016/j.ijpe.2010.04.006

Cevirici A, Moller-Madsen H (2007) Solving logistic problem with multi-agent system. The Maersk Mc-Kinney Moeller Institute—MIP—University of Southern Denmark—SDU

Coelho LC, Laporte G (2013) The exact solution of several classes of inventory-routing problems. Comput Oper Res 40:558–565

Daganzo C (2005) Logistics systems analysis, 4th edn. Springer, Berlin Heidelberg

Ghadimi P, Ghassemi F, Heavey C (2017) A multi-agent systems approach for sustainable supplier selection and order allocation in a partnership supply chain. Eur J Oper Res 269(1):286–301. https://doi.org/10.1016/j.ejor.2017.07.014

Cho DW, Lee YH, Lee TY et al (2014) An adaptive genetic algorithm for the time dependent inventory routing problem. J Intell Manuf 25:1025–1042. https://doi.org/10.1007/s10845-012-0727-5

Gómez-Marín CG, Arango-Serna MD, Serna-Urán CA (2018) Agent-based microsimulation conceptual model for urban freight distribution. Transp Res Procedia 33:155–162. https://doi.org/10.1016/j.trpro.2018.10.088

Gómez-Marín CG, Zapata-Cortés JA, Arango-Serna MD, Serna-Urán CA (2019) An urban supply chain distribution model. Res Comput Sci 148(4):9–18

Guemri O, Bekrar A, Beldjilali B, Trentesaux D (2016) GRASP-based heuristic algorithm for the multi-product multi-vehicle inventory routing problem. 4OR-A Quart J Oper Res 14(4):377–404

Guerrero WJ, Prodhon C, Velasco N, Amaya CA (2013) Hybrid heuristic for the inventory location-routing problem with deterministic demand. Int J Prod Econ 146:359–370

Huber S, Geiger MJ (2015) Dealing with scarce optimization time in complex logistics optimization: a study in the biobjective swap-body inventory routing problem. In: Proceedings of 8th international conference on evolutionary multi-criterion optimization, EMO 2015, March 29–April 1

Kang JH, Kim YD (2010) Coordination of inventory y transportation managements in a two-level supply chain. Int J Prod Econ 123:137–145

Kim SC, Shin KS (2019) Negotiation model for optimal replenishment planning considering defects under the VMI and JIT environment. Asian J Shipping Logistics 35(3):147–153

Kumari S, Singh A, Mishra N, Garza-Reyes JA (2015) A multi-agent architecture for outsourcing SMEs manufacturing supply chain. Robot Comput Integr Manuf 36:36–44. https://doi.org/10.1016/j.rcim.2014.12.009

Lee J-Y, Cho RK (2018) Optimal (z, Z)-type contracts for vendor-managed inventory. Int J Prod Econ 202:32–44. https://doi.org/10.1016/j.ijpe.2018.05.011

Lopes F, Coelho H (2010) Bilateral negotiation in a multi-agent supply chain system. Lect Notes Bus Inform Process (LNBIP) 61:195–206. https://doi.org/10.1007/978-3-642-15208-5_18

Nguyen VP, Prins C, Prodhon C (2012) Solving the two-echelon location routing problem by a GRASP reinforced by a learning process and path relinking. Eur J Oper Res 216(1):113–126. https://doi.org/10.1016/j.ejor.2011.07.030

Pal K, Karakostas B (2014) A multi agent-based service framework for supply chain management. Procedia Comput Sci 32:53–60. https://doi.org/10.1016/j.procs.2014.05.397

Phan DA, Vo TLH, Lai AN, Nguyen TLA (2019) Coordinating contracts for VMI systems under manufacturer-CSR and retailer-marketing efforts. Int J Prod Econ 211:98–118. https://doi.org/10.1016/j.ijpe.2019.01.022

Plinere DS, Borisov AN, Aleksejeva LY (2015) Interaction of software agents in the problem of coordinating orders. Autom Control Comput Sci 49(5):268–276. https://doi.org/10.3103/S0146411615050089

Sainathan A, Groenevelt H (2019) Vendor managed inventory contracts – coordinating the supply chain while looking from the vendor's perspective. Eur J Oper Res 22(1):249–260

Serna-Urán CA (2016) Modelo multi-agente para problemas de recogida y entrega de mercancías con ventanas de tiempo usando un algoritmo memético con relajaciones difusas. Universidad Nacional de Colombia

Serna-Urán CA, Arango-Serna MD, Zapata-Cortés JA, Gómez-Marín CG (2018) An agent-based memetic algorithm for solving three-level freight distribution problems. In: Sánchez-Cervantes, JL, Alor-Hernández, G., Salas-Zárate MP, García-Alcaraz JL, Rodríguez-Mazahua L (eds) Exploring intelligent decision support systems. Springer, Cham, pp 111–131. https://doi.org/10.1007/978-3-319-74002-7_6

Shaabani H, Kamalabadi IN (2016) An efficient population-based simulated annealing algorithm for the multi-product multi-retailer perishable inventory routing problem. Comput Ind Eng 99:189–201. https://doi.org/10.1016/j.cie.2016.07.022

Sitek P, Nielsen IE, Wikarek J (2014) A hybrid multi-agent approach to the solving supply chain problems. Procedia Comput Sci 35(C):1557–1566. https://doi.org/10.1016/j.procs.2014.08.239

Soysal M, Bloemhof-Ruwaard JM, Haijema R, van der Vorst JGAJ (2018) Modeling a green inventory routing problem for perishable products with horizontal collaboration. Comput Oper Res 89:168–182. https://doi.org/10.1016/j.cor.2016.02.003

Soysal M, Çimen M, Belbağ S, Toğrul E (2019) A review on sustainable inventory routing. Comput Ind Eng 132(April):395–411. https://doi.org/10.1016/j.cie.2019.04.026

Verma NK, Chatterjee AK (2017) A multiple-retailer replenishment model under VMI: accounting for the retailer heterogeneity. Comput Ind Eng 104:175–187. https://doi.org/10.1016/j.cie.2016.12.001

Vidović M, Popovi D, Ratković B (2014) Mixed integer and heuristics model for the inventory routing problem in fuel delivery. Int J Prod Econ 147:593–604

Wang G, Wong TN, Wang X (2013) An ontology based approach to organize multi-agent assisted supply chain negotiations. Comput Ind Eng 65(1):2–15. https://doi.org/10.1016/j.cie.2012.06.018
Wooldridge M (2002) An introduction to multiagent systems. UK, Willey. https://doi.org/10.1145/1753171.1753181
Zapata-Cortes JA (2016) Optimización de la distribución de mercancías utilizando un modelo genético multiobjetivo de inventario colaborativo de m proveedores con n clientes. Universidad Nacional de Colombia
Zhao J, Zhao H (2018) Design of prototype system for multi-agent supply chain information sharing benefit distribution management. IseB (0123456789). https://doi.org/10.1007/s10257-018-0386-y

Multi-objective Product Allocation Model in Warehouses

Julián Andrés Zapata-Cortes, Martín Darío Arango-Serna, Conrado Augusto Serna-Urán, and Luisa Fernanda Ortíz-Vasquez

Abstract Warehousing is one of the most important logistics activities in companies, since it generates a great portion of logistics costs related to resource consumption such as energy, machinery, facilities, labor, and inventory costs. However, bad warehouse management can also cause delays in sending orders to customers, which directly affects the service level offered by the warehouse. The adequate allocation of goods inside warehouses is a key factor to reduce handling cost and operation time in such facilities, since a better positioning of goods can reduce the required total distance to pick and store the items, which diminishes the use of forklifts and the distance traveled by the people inside the warehouses, reducing handling costs and time. For that reason, it is important to establish optimal storage positions, seeking better conditions of profitability, and service for the business. This article presents a multi-objective mathematical model to determine the allocation of goods in the different spaces available in warehouses, which simultaneously evaluates operating costs and times required to carry out storage activities. In order to solved the proposed model, a multi-objective genetic algorithm was used, which produces a set of possible optimal goods allocations inside the warehouse (Pareto Frontier), allowing the manager to establish the allocation plan that best fits their preferences, making sure that an optimal relationship between costs and the time required for product storage is being considered.

J. A. Zapata-Cortes (✉)
Institución Universitaria CEIPA, Sabaneta, Colombia
e-mail: julian.zapata@ceipa.edu.co

M. D. Arango-Serna
Universidad Nacional de Colombia, Medellín, Colombia
e-mail: mdarango@unal.edu.co

C. A. Serna-Urán
Instituto Tecnologico Metropolitano, Medellín, Colombia
e-mail: conradoserna@itm.edu.co

L. F. Ortíz-Vasquez
Universidad de San Buenaventura, Medellín, Colombia
e-mail: asistente.investiga@usbmed.edu.co

J. L. García Alcaraz et al. (eds.), *Techniques, Tools and Methodologies Applied to Quality Assurance in Manufacturing*,
https://doi.org/10.1007/978-3-030-69314-5_11

Keywords Warehousing · Product allocation · Multi-objective model · Genetic algorithm · Optimization

1 Introduction

Storage and handling material are logistics activities that account for around 20% of total distribution costs in companies, which force them to establish control measures in order to ensure efficient storage processes without affecting service level (Ballou 2003). The most characteristic activities in warehouses are reception, sorting, material handling, product allocation, storage, deliveries, and control (Chopra and Meindl 2016).

Product allocation is one of the key activities in warehouse optimization, since the appropriate selection of the places where the merchandise must be stored is a key activity to reduce operating costs and time. This is possible due to the reduction of distances and the efforts required to move the goods inside these facilities, which allows diminishing the costs of material handling and labor hours, as well as the time required for internal operations and thus the time for order fulfillment.

In the specialized scientific literature, there are several mathematical models and information communication systems that allow to optimally assign products on warehouses' shelves (Tompkins et al. 2010). However, these models normally only consider the individual optimization of one of the several objectives involved in real operation. This frequently leads to the optimization of the selected objective while negatively affecting other objective functions to be optimized, which is clearly not in accordance with the reality of companies, in which all objectives and variables are important. This is the case of service level, which can be affected when some logistics activities such as warehousing and transportation are optimized. For instance, optimization of operating costs in a warehouse may lead to a decrease in the number of people or material-handling equipment, as well as to a lower utilization of some spaces in the shelves. This can reduce the costs of keeping products but may increase the time required to reach the items due to a lower resource to do the required operations and higher distances to fulfill orders.

This chapter presents a multi-objective optimization model with the aim of simultaneously minimizing product handling costs and the required time to fulfill orders in an industrial warehouse. The model helps the decision maker to define the positions in which the products must be located in the warehouse' shelves, considering both critical objectives for warehouse performance, seeking to find the best cost-time combination for storage activity. Due to the complexity of the optimization process, a multi-objective genetic algorithm of the non-dominated sorting genetic algorithm-II (NSGAII) type was developed to solve the allocation mathematical model.

The model was tested using information from an industrial warehouse that stores three types of different products. The results produced by this genetic algorithm behave according to the model expectations, yielding not a single solution, but a set of possible solutions that simultaneously optimize both objective functions. The decision maker can select any of those solutions based on their preferences, being sure that each solution generates an optimal product allocation plan in the warehouse that considers both functions at the same time.

2 Product Allocation in Warehouses

A warehouse is defined as a structural unit with all the resources and organizational arrangements necessary for the execution of processes related to inventory and storage management (Kappauf et al. 2012). These facilities are essential in the supply chain since they are in charge of buffering the material flow along the supply chain, to adapt companies to the variability caused by factors such as seasonality of product demand and market, batch production, transportation delays, and difficulties in suppliers (Gu et al. 2007). Another important activity that can be carried out in warehouses is consolidation of products from different sources which must be later delivered to customers. Other activities in warehouses include pricing, labeling, and product customization. Research related to logistics performance in warehouses has increased in recent years due to the pursuit of competitiveness by companies, which require to reduce and make their inventory flexible, generate a faster response time from the warehouse (Agarwal et al. 2006) and integrate themselves with a greater number of 3PL providers and multiple customers with diverse needs (Tian et al. 2010).

The main activities in most warehouses are storage, preservation, and movement of products (Zapata-Cortes et al. 2020; Ballesteros-Riveros et al. 2019), which are responsible for causing lots of cost (Departamento Nacional de Planeación 2018) due to the amount of economic, financial, personal, and infrastructure resources required in such facilities (Chopra and Meindl 2016). On the other hand, storage is also responsible for the service level offered to customers, since it affects lead time, orders fulfillment, product quality, and conditions (Frazelle 2016). Both cost and service levels are key factors for the company's success (Chopra and Meindl 2016). Costs optimization and the adequate service level are two of the main objectives pursued by warehouses' managers, which can be achieved through multiple initiatives such as reducing unnecessary distances and movements, improving space use, better equipment and labor utilization, accessibility to all items, among others (Tompkins et al. 2010).

These initiatives can be carried out through the use of information and communication technologies and information systems (Zapata-Cortes et al. 2010), appropriate facilities design (Arango-Serna et al. 2010), and process optimization (De Koster et al. 2007). The use of technologies such as warehouse management system (WMS), barcode, radio frequency identification (RFID), among others, can increase warehouse performance, reduce costs, and improve service levels of storage operations (Zapata-Cortes et al. 2010). An adequate design and the use of the right equipment and technology in handling and storage processes positively impact warehouse performance as well (Tompkins et al. 2010). Optimization refers to the adequate programming and allocation of people, equipment and materials, which can be done through mathematical procedures to determine the necessary resources, the optimal travel distances inside the warehouse for both the movement of people and equipment, the correct product location, the adequate quantities to be stored, among other activities (De Koster et al. 2007; Tomkins et al. 2010).

One alternative to improve costs and response time in warehouses is to reduce the total travel distance required to carry out the operations of positioning products in the shelves and collecting them to serve the customers and the company orders. This distance reduction in the warehouse operation means decreasing the total travel distance in the warehouse (De Koster et al. 2007; Arango-Serna et al. 2010). This can be done by properly selecting the places where the goods should be located, which reduces the time and cost required to reach their storage positions (Zapata-Cortes et al. 2020).

As mentioned, allocation of products in warehouses and distribution centers has a significant impact on costs, and also on storage capacity, inventory tracking and order preparation, and affects the picking and retrieving of products (De Koster et al. 2007). Various factors influence the location of products in the warehouse such as the order preparation method, size and design of the storage system, the material-handling system, product characteristics, demand behavior, turnover rates, and space (Fumi et al. 2013; Gagliardi et al. 2010). Chan and Chan (2011) mention that the problem of storage allocation involves deciding where and how to store a set of elements in order to ensure optimal operation of the logistics system.

According to many authors such as Hausman et al. (1976), Graves et al. (1977), Petersen and Gerald (2004), Tompkins et al. (2010), Pan et al. (2015), and Zapata-Cortes et al. (2020), the most common methods to assign products in warehouses are the dedicated (fixed), the random, the rotation-based, and the product-class methods.

Dedicated storage assigns each product to a fixed location, causing the products to be always stored in the same place. The main advantage of this kind of storage is the easiness to remember the product location, but it requires a high amount of space and results in low utilization in the warehouse. In these cases, occupation of warehouses and distribution centers' total storage spaces is around 50% (Bartholdi and Hackman 2005) due to the place's reservation for a particular product, even for stock-out items.

Random storage assigns products in any available location, but it requires an information system that allows to know the storage position and easy identification of the products. This type of storage is based on the available spaces in the warehouse and means that all the storage places have the same probability of being used; it can reduce space requirements compared to the fixed method. However, it has some disadvantages, such as the fact that workers hardly remember the place where products are located, which increases operating times and requires a robust information system that lets them know the location and quantity of the products randomly stored in the warehouse (Bowersox et al. 2012).

The inventory turnover allocation determines the storage locations of products based on their demand. Frequently requested products obtain the easiest and fastest access locations, usually near the reception/shipment point, while those products with the lowest turnover are located further away. This method requires to know rotation frequencies in advance and seeks to reduce operation costs, since those products with higher movement requirements are the ones that are located at the shortest distance from the reception/shipment points in the warehouse. However, this type of allocation is not as efficient when there are different reception/shipment points, in which the allocation of products is not that easy (Zapata-Cortes et al. 2020).

Class-based storage is an allocation method that divides the warehouse into a number of areas to which a specific product or set of products is assigned, taking into consideration the demand frequency of the products. This method has the same drawback as the rotation-based allocation, in which the assignment is difficult when there are several reception/shipment points, making it necessary to use more advanced techniques to allocate the products in order to reduce operation costs (Zapata-Cortes et al. 2020).

There are different types of costs incurred in product storage in warehouses, among which the most common are the keeping, handling and transportation of products, and administration and facilities cost, which are related among themselves (Orjuela et al. 2016). Handling cost is the one most affected by product location modification, since the distances to be traveled can be reduced or increased. However, handling costs are also affected by other variables such as the diversity and number of references to be stored, facilities configuration, storage infrastructure, material-handling equipment, standardization of loads, seasonality of demand, and product packaging.

But product allocation not only affects costs in warehousing operations but also influences service level, since it reduces the required time to carry out the storage and orders fulfillments operation (Sanei et al. 2011; Bortolini et al. 2015), and can also modify the size of the storage facility (Hou et al. 2010; Fumi et al. 2013; Trab et al. 2015).

Several authors address the product allocation problem in warehouses by proposing optimization methods, such as the ones presented in Quintanilla et al. (2015), Hu et al. (2012), Kovács (2011), Bortolini et al. (2017), Chan and Chan (2011), Ramtin and Pazour (2015), Ross et al. (2017), Ballesteros-Riveros et al. (2019), and Zapata-Cortes et al. (2020).

Many works and models consider the study of the single cost or operating time optimization (Sanei et al. 2011; Kovács 2011; Van Wijk et al. 2013; Zapata-Cortes et al. 2020). Conversely, other studies perform joint optimization processes of several objectives (Ramtin and Pazour 2015; Quintanilla et al. 2015; Hu et al. 2012), which allow to make decisions more in line with reality, where several objectives are important for the adequate performance of the warehouse. However, multiple objective optimization requires to take into consideration several characteristics that differentiate this kind of optimization from the traditional one, as stated in the upcoming section (Zapata-Cortes 2016; Arango-Serna et al. 2018).

3 Multi-objective Optimization

A multi-objective optimization problem (MOP) refers to the simultaneous optimization, either minimization or maximization, of a set of objective functions which can be contradictory among themselves. This contradiction means that when an objective is optimized, another one or the set of the remaining objectives can be deteriorated (Arango-Serna and Zapata-Cortes 2017). In that way, a multi-objective optimization problem can be formulated as stated in Eq. 1, in which the objective functions are subjected to a set of equality and inequality restrictions (Marler and Arora 2004; Zapata-Cortes 2016).

$$\begin{aligned}
&\text{Minimize } F(x) = [F_1(x), F_2(x), \ldots, F_k(x)] \\
&\text{Subject to } g_j(x) \le 0,\ j = 1, 2, \ldots, m \\
&h_l(x) = 0, l = 1, 2, \ldots, e \\
&x \in R^n
\end{aligned} \tag{1}$$

The solution to a multi-objective problem can be carried out with two types of approaches: The one-dimensional approach and the high-dimensional approach (Arango-Serna et al. 2018). The one-dimensional approach is based on the consolidation (aggregation) of the objective functions in a single equation to convert the problem into a mono-objective optimization problem, which is solved by obtaining a unique solution. This aggregation is given by weighting the different objectives and solving the single-objective function that includes all the weighted objectives.

The high-dimensional approach is based on two phases: In the first one, a solution method is used to find a set of optimal solutions (called the non-dominated solutions), which is called the Pareto frontier of the problem. In the second phase, the decision maker selects one solution from the Pareto frontier, according to their preferences regarding the specific problem. In this case, the decision maker can be sure that an optimal solution is being selected which corresponds to one of the best combinations of the objective functions' values, and can also be confident that this solution is not better or worse than any other solution in the frontier (Zapata-Cortes 2016).

Multi-objective optimization differs from conventional single-objective optimization in that it does not generate a single solution but a set of possible optimal solutions, from which the decision maker can select any according to their preferences. This set of optimal solutions implies that the improvement of one objective may result in the decrease of another objective or more objectives (Shenfield et al. 2007). A representation of the Pareto frontier is depicted in Fig. 1, in which some measures of the quality of Pareto solutions, such as pertinence, diversity and proximity, can be observed (Shenfield et al. 2007). Proximity refers to how close a solution is to the Pareto frontier; diversity is the measure of how extensive and uniform the distribution of the set of solutions is; and relevance indicates the importance of the solutions to the decision maker.

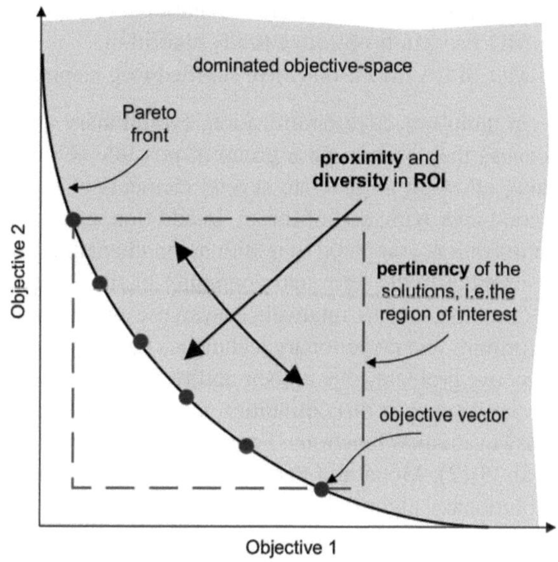

Fig. 1 Pareto frontier and quality measures of Pareto solutions. *Source* Shenfield et al. (2007)

The Pareto frontier can be obtained through conventional (classical) or heuristic techniques (López et al. 2011). Classical methods present several disadvantages in the multi-objective optimization, as the fact that they require a high number of iterations to find the Pareto frontier and a prior knowledge of the problem domain; besides, some of these methods are sensitive to the form or continuity of the Pareto frontier (López et al. 2011). Furthermore, finding a satisfactory solution using classical techniques becomes increasingly complex as the number of objectives increases (Fonseca and Fleming 1995). Heuristic methods allow to face the above-mentioned problems, finding good solutions (close to the optimal) in a reasonable processing time.

Authors such as González (2013) and López et al. (2011) have analyzed the most used metaheuristic techniques to solve multi-objective optimization problems, such as simulated annealing, genetic algorithms, evolution strategies, evolutionary programming, artificial immune system algorithm, and particle swarm optimization (Villalobos 2005). Among these techniques, several of the most common used algorithms to solve this kind of problems are the following (López et al. 2011; González 2013):

- MOGA: Multi-objective genetic algorithm.
- NSGA and NSGA-II: Non-dominated sorting genetic algorithm.
- SPEA and SPEA2: Strength Pareto evolutionary algorithm.
- PAES: Pareto archived evolution strategy.
- PESA: Pareto envelope-based selection algorithm.
- MO-VNS: Multi-objective variable neighborhood search.
- DEPT: Differential evolution with Pareto tournaments.
- MO-TLBO: Multi-objective teaching–learning-based optimization.
- MOABC: Multi-objective artificial bee colony.

- MO-GSA: Multi-objective gravitational search algorithm.
- MO-FA: Multi-objective firefly algorithm.
- MO-SFLA: Multi-objective shuffled frog leaping algorithm.

In multi-objective optimization, evolutionary algorithms are especially suitable because they operate on a group of possible solutions (the population of individuals), allowing to generate several elements of the optimal Pareto set (or at least a good approximation of these). In addition, genetic algorithms for multi-objective optimization, compared to traditional mathematical programming methods, are less susceptible to the form and continuity of the Pareto frontier, require little domain information, and are relatively easy to use and implement (López et al. 2011). These algorithms and evolutionary techniques have proven to be good tools to solve multi-objective problems due to their ability to handle complex problems, involving characteristics such as discontinuities, multimodality, discontinuous feasible spaces, and noisy evaluation functions (Fonseca and Fleming 1995; Tiwari et al. 2002; Vergidis et al. 2012). Moon and Seo (2005) emphasize that the most attractive attribute of evolutionary algorithms in solving multi-objective problems is the flexibility of working with multiple functions using few mathematical requirements.

4 Non-dominated Sorting Genetic Algorithm-II for Solving Multi-objective Problems

The non-dominated sorting genetic algorithm-II (NSGAII) is one of the most relevant evolutionary multi-objective optimization algorithms found in the scientific literature to solve this kind of problems. This algorithm was proposed by Deb et al. (2002) and is based on the non-dominance genetic algorithm (NSGA) developed by Srinivas and Deb (1994). The NSGAII algorithm improves the inconveniences found in the NSGA, which requires a high number of repetitions of classification. The dominance of one solution over another indicates that the first one is not worse than the second one in all the objective functions but it must be better than the second one in at least one objective. The non-dominance Pareto frontier concept means the creation of a set of individuals that are not dominated by other individuals, which means a set of optimal solutions (Zhang et al. 2018).

As part of the genetic algorithm's family, the NSGAII requires all the specific elements of this type of solution techniques, such as chromosome configuration, selection, crossing and mutation operators, fitness functions, elitism decisions, and the termination condition of the algorithm. Figure 2 presents the general operation scheme of the NSGAII algorithm.

In the NSGAII algorithm, the first step is the creation of an initial population P_0 of size N, which is organized according to the non-dominance of its individuals (each individual represents one solution to the multi-objective optimization problem). This process is carried out by an iterative procedure in which different non-dominance levels are determined. These non-dominance levels are obtained by

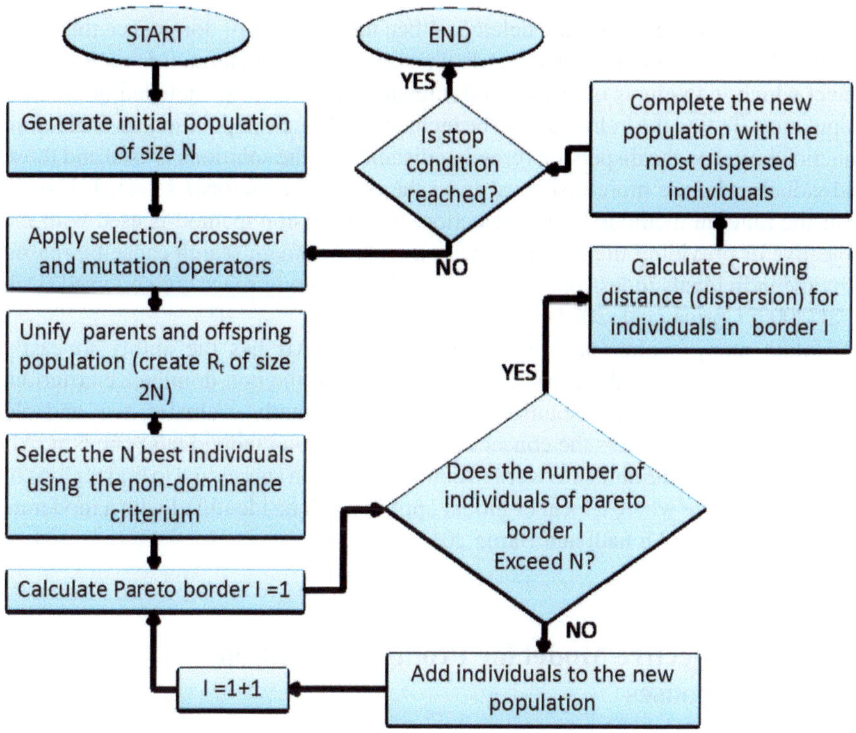

Fig. 2 NSGAII operation scheme. *Source* Zapata-Cortes (2016)

calculating different Pareto borders by separating the non-dominated solutions from the rest of the population. In the first level, the non-dominated individuals of the whole population are extracted and the remaining individuals are also analyzed to determine the second set of non-dominated individuals. Those new non-dominated individuals are extracted from the remaining population and constitute the second level of non-dominance. The procedure continues until there are not more non-dominance levels (Deb et al. 2002; Correa et al. 2008).

From this population, a new offspring population Q_0 with also a size of N is originated through the use of the selection, crossing, and mutation operators. As can be noted, the population of parents is denoted as P_t and the offspring populations as Q_t, with t representing the number of the evolution ($t = 0$ for the initial population). These P_0 and Q_0 populations are combined to create a population R_t of size 2 N, which ensures that all individuals (parents and children) will be considered to create new population, which ensures elitism in the algorithm.

This population R_t is ordered according to the non-dominance levels, from which population P_{t+1} for a new generation is created. The new population P_1 ($t = t + 1$) is constructed by including the individuals belonging to the first levels of non-dominance (first Pareto borders). Since the size of the new population P_{t+1} is N and that of R_t is 2 N, not all individuals can be considered, so the individuals in the worse non-dominated levels are deleted. When the last level of dominance that must be included in new population exceeds the size of the population, it is necessary to select which individuals of such non-dominance level must be included in the new population. To find the individuals to be included in the new population, an evaluation function based on the dispersion (crowing distance) of the solutions is used and those individuals who are more dispersed from the others are the ones selected. Instead of using random methods for the selection, the dispersion method is used with the objective of providing the algorithm with a greater domain search capacity, reason why the individuals in less populated areas of the solution space are preferred (Deb et al. 2002; Correa et al. 2008).

Unlike many evolutionary algorithms, the NSGAII has the ability to easily work with the individuals' classification according to the non-dominance criterion, prevents the loss of good solutions because it constantly evaluates new and old individuals, and considers the concept of diversity among individuals. The NSGAII algorithm also has the ability to work in complex solution spaces and to find regions of the solution space where a local or global optimum can be identified with a moderate level of precision (Ahmadi and Barna 2015).

5 Multi-objective Model for Product Allocation in Warehouses

As previously mentioned, there are different approaches to face product allocation in warehouses, ranging from assigning goods based on fixed locations that can be arbitrarily decided or using more accurate measures such as product demand rotation. Several authors have developed mathematical models to optimize the assignment process; however, many consider only the single objective of cost minimization. In this work, a multi-objective product allocation model is presented, considering the simultaneous optimization of costs and service time.

Product allocation to the different positions in the warehouse can be done by optimizing costs or time through the objective function presented in Eq. 2. This equation represents cost minimization for the allocation problem when there are multiple products and several collecting/delivery points in the warehouse (Tompkins et al. 2010).

$$F_1 = \text{minimize} \sum_{j=1}^{n} \sum_{k=1}^{q} \frac{T_j}{S_j} \sum_{i=1}^{m} p_i c_{ik} x_{jk} \tag{2}$$

x_{jk} is the binary decision variable, which takes the value of 1 if product j is assigned to position k, or zero otherwise. The parameters description in the objective function are:

- n is the number of products.

- q is the number of storage positions.
- m is the number of origin/delivery points in the warehouse.
- T_j is the number of storage trips (input–output) for product j.
- S_j is the number of storage positions required for product j.
- p_i is the percentage of trips to go to i and return.
- c_{ik} is the cost of travel from point i to storage location k.

This model proposed by Tompkins et al. (2010) seeks to minimize the total distance required by the material-handling operation to store and retrieve the products from the shelves. The model is aimed at finding the best product allocation from which a dedicated storage system is established that can be modified every time demands and number of products change (Zapata-Cortes et al. 2020).

Objective function 2 can be modified to optimize the time required to fulfill orders. It can be done by replacing costs (c_{ik}) with the time required to perform this activity (t_{ik}: time required to go from i to the storage position k). This generates the second objective function of the proposed model, which is presented in Eq. 3. The time of every operation (t_{ik}) is calculated considering the distance and speed of the material-handling equipment for both the horizontal and vertical movements required to reach a position k in the shelves and return to the collecting/delivery points j.

$$F_2 = \text{minimize} \sum_{j=1}^{n} \sum_{k=1}^{q} \frac{T_j}{S_j} \sum_{i=1}^{m} p_i t_{ik} x_{jk} \tag{3}$$

The proposed approach considers the simultaneous optimization of costs and time for the expected trips inside the warehouse for any product j, from each origin point to its storage position k, which is represented by the term $p_i d_{ik} x_{jk}$, in which distance is converted to costs and time, as presented in Eqs. 2 and 3.

In addition, each objective function considers movement intensity in terms of storage positions with term $\frac{T_j}{S_j}$ required for each product j. With this, total cost and time for the established operation time T_j are calculated.

Those objective functions are subject to the following constraints:

$$\sum_{j=1}^{n} x_{jk} = 1 \quad k = a, \ldots, q \tag{4}$$

$$\sum_{k=1}^{q} x_{jk} = S_j \quad j = a, \ldots, n \tag{5}$$

As previously mentioned, the probability that each item j travels from point i to each position k is the same for all products. Constraint 4 ensures that product j is assigned to position k only once; that is, only a product j can be assigned to a position k. Constraint 5 indicates that the quantity of products j assigned to positions k must be equal to the storage requirement (demand converted into required spaces) for product j.

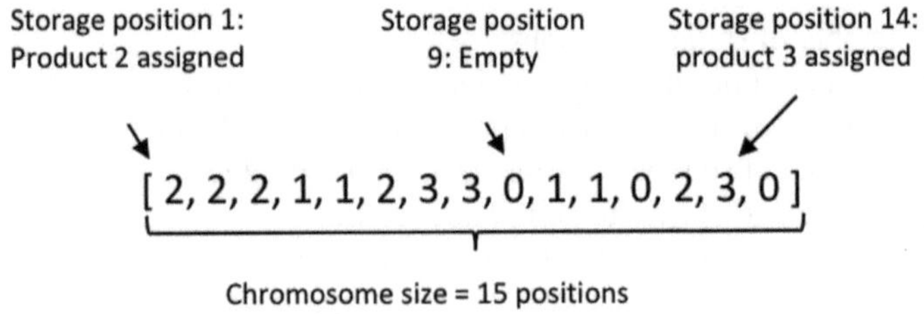

Fig. 3 Chromosome used for the product allocation problem. *Source* The authors

Due to the number of positions that can be found in a warehouse, the amount and variety of products, the several origin/destinations points and the multi-objective treatment, for solving the presented model a NSGAII algorithm is configured as explained in the next section.

6 Multi-objective Genetic Algorithm for Product Allocation Optimization

For the solution of the multi-objective model presented above, a non-dominated sorting genetic algorithm-II (NSGAII) was developed. Every element of the genetic algorithm is specifically designed to the problem as follows.

6.1 Initial Population and Chromosome Representation

Every individual in the NSGAII algorithm is represented as a vector of integer numbers (chromosome), in which each position i represents a storage position, as shown in Fig. 3. To each position i, an integer number from zero to the number of different types of products is assigned, so the sum of positions with the same number is equal to the number of spaces in the warehouse required to store such product. In this representation, a value of 0 means that this space is empty.

6.2 Selection

The selection of the individuals is made by tournament with size m ($m = 5\%$ of the population size), which are randomly selected. These individuals are first compared and selected according to the non-dominance criterion. As it is expected that the new populations are formed by many non-dominated individuals, due to the NSGAII configuration, if two or several individuals are non-dominated, one is selected using

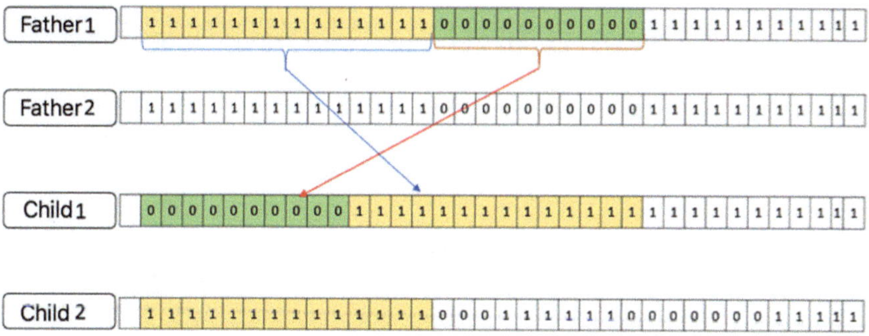

Fig. 4 Crossover operator for product allocation problem. *Source* The authors

the crowing distance criterion. If the individuals also have the same crowing distance value, one is randomly selected.

6.3 Crossover Operator

The crossover operator is performed by determining two randomly selected points, from which the parents' genetic information is exchanged to the children, as presented in Fig. 4 for an individual with 35 positions. In Fig. 4, the crossover points are genes 14 and 24. To ensure the genetic feasibility of each child, a verifying and adjusting process is performed after the crossover, which ensures that total storage quantity is conserved in the children.

6.4 Mutation Operator

The mutation operator is carried out using a randomly selected individual based on the mutation probability set in the algorithm. The mutation consists in two parts: in the first one, two positions are randomly selected and if the products in these positions are different, they are exchanged. If the genes are the same, a new random selection process is carried out until both individuals are different. The second part is similar to the previous one, but ensures that one of the positions to be interchanged corresponding to empty space in the warehouse, which is represented with a zero value. This is done with the aim of giving the algorithm a greater search capability. This mutation operator is presented in Fig. 5

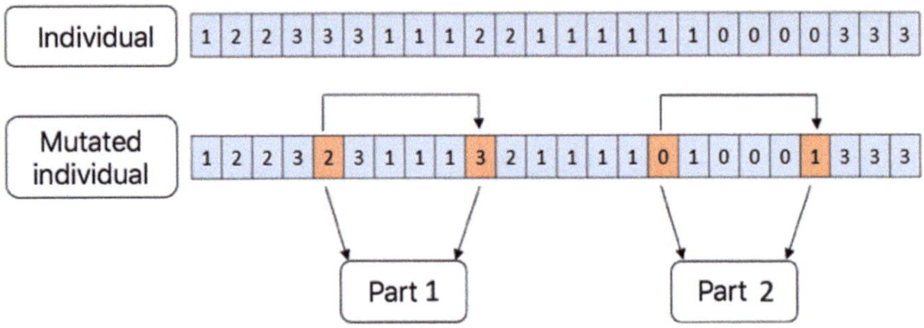

Fig. 5 Mutation representation. *Source* The authors

6.5 Elitism and Fitness Evaluation

Conservation of the best individuals in the upcoming generations is implicit in the NSGAII algorithm, since for the creation of new populations, the union of the parent and children (offspring) populations is required, from which the individuals in the new populations are selected following the non-dominance criterion, which ensures elitism in the algorithm.

The fitness function is evaluated by direct calculation of cost and time to allocate the products in the warehouse position using Eqs. 2 and 3. The non-dominated individuals are those that ensure the dominance criterion in the minimization of the two objective functions.

7 Model Application in an Industrial Warehouse

To analyze the proposed model, we obtained storage information of a food company that has around two thousand different SKUs, which can be grouped in three characteristic types of products: packed cookies, soft drinks in glass bottles, and canned products. These three characteristic products are considered to be analyzed in the validation and testing of the model.

The warehouse has two collecting/delivery docks. In both, the products can enter or exit the warehouse in the company trucks. In those points, the products are sorted and ready to be assigned to their storage position in the warehouse. Those points also receive products retrieved from shelves for further delivery. The company stores products using random assignment according to the visually available spaces in the shelves.

The warehouse has a storage capacity of 2500 positions. In every space, a standard pallet of around one ton can be located. The warehouse layout has 12 double shelves and a single one at the end of the warehouse. Each shelf has 20 storage lines and five height levels, as depicted in Fig. 6

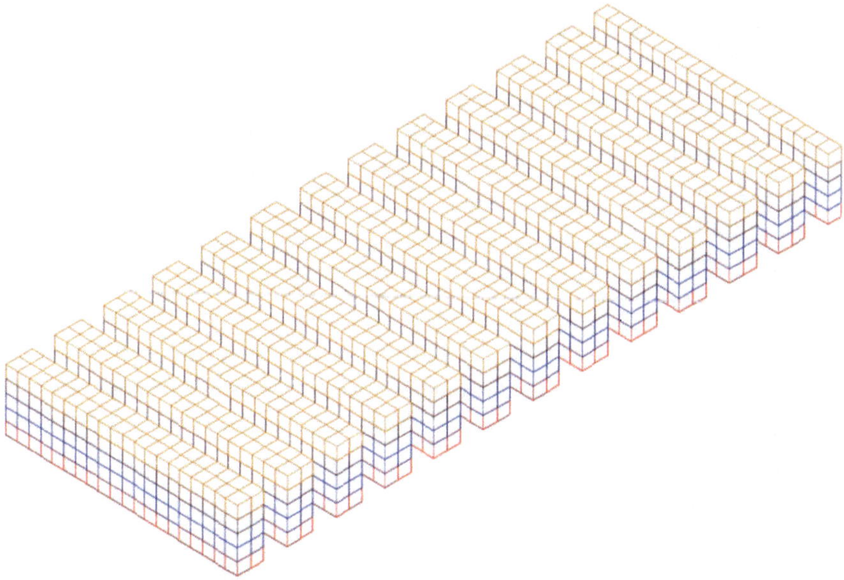

Fig. 6 Proposed distribution center in 3D. *Source* The authors

The NSGAII algorithm was applied to solve the allocation problem in the mentioned company, in which the required quantities to be stored at a specific time in the warehouse are 800 tons of product 1500 tons of product 2 and 300 tons of product 3. The algorithm ran with a population of 100 individuals, 200 generations, and a mutation percentage of 0.2. The number of generations is set at 200 since after that number it was not possible to find better solutions to the problem. The number of optimal individuals in the Pareto frontier produced by the algorithm in the best solution is 12, and their objective function values are presented in Table 1.

Those individuals can be graphically presented in order to observe the Pareto frontier behavior in relation to the two objective functions, as it is depicted in Fig. 7. Each blue point represents one of the 12 non-dominated individuals and on the red line, the extrapolation of the Pareto frontier can be observed.

Table 1 Solution individuals in the Pareto frontier

Individual	Time	Cost	Individual	Time	Cost
1	40,094	22,549	7	39,893	22,600
2	40,001	22,577	8	40,045	22,557
3	39,846	22,729	9	39,913	22,591
4	40,004	22,573	10	39,972	22,578
5	39,920	22,579	11	39,855	22,723
6	40,009	22,563	12	39,888	22,722

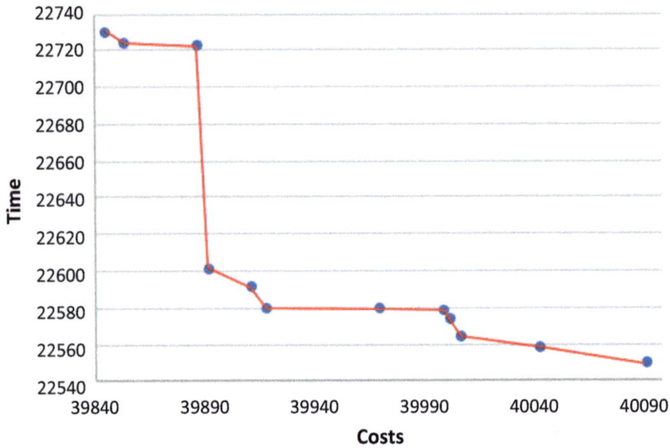

Fig. 7 Pareto frontier. *Source* The authors

Since the non-dominated individuals forming the Pareto frontier are 12 and each is a vector of 2500 positions, as depicted in Fig. 3, they are not presented in this document.

From Fig. 7, it can be observed that both objective functions are contradictory, i.e., the individual that generates the allocation plan with the best costs also produces the worst time and conversely. In this way, all individuals in the Pareto frontier are optimal combination of the two objectives and the decision maker can select the one they consider best suits their preferences and they can also be sure that no solution is worse or better than any other. This behavior is expected when two minimization objective functions are simultaneously optimized.

Each of these individuals generates a product allocation plan in the warehouse, indicating what type of product should be stored in each space, as defined by the chromosome representation in the genetic algorithm. In this way, it is easy for the decision maker to convert the answer of the algorithm (the individuals) into the warehouse allocation plan.

8 Conclusions

This work presented a product allocation model that considers the simultaneous optimization of two fundamental objective functions for the correct operation and performance of warehouses, which are the costs and the time to complete customers' orders. This approach is different from many works found in the scientific literature that consider the single optimization of only one objective function. This is a more realistic way of optimizing the product assignment problem in warehouses, since both time and costs are essential factors for companies' competitiveness. In this way, the proposed model allows managers to find a relationship between these two objective functions ensuring a combination that optimizes both criteria at the same time when product allocations plans are created in the warehouse.

Due to the complexity of the proposed model and the disadvantages of classical methods to solve multi-objective problems, a genetic algorithm NSGAII was developed, which used an easy-to-understand representation of individuals that facilitates the algorithm operation and the conversion of solutions into the product allocation plan in the warehouse. The algorithm ran as expected, generating not an only a single optimal solution but a set of optimal individuals forming the Pareto frontier. From this set of individuals and according to the decision maker's preferences, any solution can be selected, with the assurance of using an optimal allocation plan in relation to the cost and service time in the warehouse, being certain that no one solution is better or worse that the chosen one.

As future research work, it is suggested to analyze other objective functions to optimize, such as the required area for allocation plans in the warehouse and other variables, such as costs and times of order preparation. The model can also be improved by developing a routing algorithm that optimizes costs and time when many products are retrieved from the shelves at the same time. In addition, as future research lines, it is recommended to integrate this model into an information system that not only allows the development of optimal allocation plans but also provides guidance to workers about the location of products.

References

Agarwal A, Shankar R, Tiwari MK (2006) Modeling the metrics of lean, agile and leagile supply chain: an ANP-based approach. Eur J Oper Res 173(1):211–225. https://doi.org/10.1016/j.ejor.2004.12.005

Ahmadi A, Barna L (2015) A process modelling-life cycle assessment-multi-objective optimization tool for the eco-design of conventional treatment processes of potable wáter. J Clean Prod 100:116–125

Arango-Serna MD, Zapata-Cortes JA, Serna-Uran CA (2018) Collaborative multiobjective model for urban goods distribution optimization. In: García-Alcaraz J, Alor-Hernández G, Maldonado-Macías A, Sánchez-Ramírez C (eds) New perspectives on applied industrial tools and techniques. Management and industrial engineering. Springer, Cham

Arango-Serna MD, Zapata-Cortes JA, Pemberthy JI (2010) Reestructuración del layout de la zona de picking en una bodega industrial. Revista Ingeniería Universidad de los Andes 32:54–61

Arango-Serna MD, Zapata-Cortes JA (2017) Multiobjective model for the simultaneous optimization of transportation costs, inventory costs and service level in goods distribution. IEEE Latin Am Trans 15(1):129–136

Ballesteros-Riveros FA, Arango-Serna MD, Adarme-Jaimes W, Zapata-Cortes JA (2019) Storage allocation optimization model in a Colombian company. DYNA 86(209):255–260

Ballou R (2003) Business logistics management, 5th ed. Prentice-Hall International

Bartholdi JJ, Hackman ST (2005) Warehouse and distribution science. Retrieve from http://www2.isye.gatech.edu/~jjb/wh/book/editions/wh-sci-0.96.pdf. Last visit 12 February 2020

Bortolini M, Botti L, Cascini A, Gamberi M, Mora C, Pilati F (2015) Unit-load storage assignment strategy for warehouses in seismic areas. Comput Ind Eng 87:481–490. https://doi.org/10.1016/j.cie.2015.05.023

Bortolini M, Faccio M, Ferrari E, Gamberi M, Pilati F (2017) Time and energy optimal unit-load assignment for automatic S/R warehouses. Int J Prod Econ 190:133–145. https://doi.org/10.1016/j.ijpe.2016.07.024

Bowersox D, Closs D, Cooper B (2012) Supply chain logistics management, 4th edn. McGraw-Hill Companies Inc, New York City

Chan FTS, Chan HK (2011) Improving the productivity of order picking of a manual-pick and multi-level rack distribution warehouse through the implementation of class-based storage. Exp Syst Appl 38(3):2686–2700. https://doi.org/10.1016/j.eswa.2010.08.058

Chopra S, Meindl P (2016) Supply chain management: strategy, planning, and operation, 6th edn. Pearson Education, New York

Correa CA, Bolaños RA, Molina A (2008) Algoritmo multiobjetivo NSGA-II aplicado al problema de la mochila. Scientia et Technica XIV 39:206–211

De Koster R, Le-Duc T, Roodbergen KJ (2007) Design and control of warehouse order picking: a literature review. Eur J Oper Res 182:481–501

Deb K, Pratap A, Agarwal S, Meyarivan T (2002) A fast and elitist multiobjective genetic algorithm: NSGA-II. IEEE Trans Evol Comput 6(2):182–197

Departamento Nacional de Planeación (2018) Encuesta Nacional Logística 2018. Retrieve from https://onl.dnp.gov.co/es/Publicaciones/Paginas/Encuesta-Nacional-Log%C3%ADstica-2018.aspx. Last visit 2 September of 2019

Fonseca CM, Fleming PJ (1995) An overview of evolutionary algorithms in multiobjective optimization. Evol Comput 3(1):1–16

Frazelle E (2016) World-class warehousing and material handling, 2nd edn. McGraw-Hill Education, New York

Fumi A, Scarabotti L, Schiraldi MM (2013) Minimizing warehouse space with a dedicated storage policy. Int J Eng Bus Manag 5(21):1–10. https://doi.org/10.5772/56756

Gagliardi JP, Renaud J, Ruiz A (2010) On storage assignment policies for unit load automated storage and retrieval systems. Retrieve from https://www.cirrelt.ca/DocumentsTravail/CIRRELT-2010-25.pdf Last visit February 14 of 2020

González DL (2013) Metaheurísticas, Optimización Multiobjetivo y Paralelismo para Descubrir Motifs en Secuencias de AND. PhD Thesis. España: Universidad de Extremadura

Graves SC, Hausman WH, Schwarz LB (1977) Storage-retrieval interleaving in automatic warehousing systems. Manag Sci 23(9):935–945

Gu J, Goetschalckx M, McGinnis LF (2007) Research on warehouse operation: a comprehensive review. Eur J Oper Res 177(1):1–21. https://doi.org/10.1016/j.ejor.2006.02.025

Hausman WH, Schwarz LB, Graves SC (1976) Optimal storage assignment in automatic warehousing systems. Manag Sci 22(6):629–638. https://doi.org/10.1287/mnsc.22.6.629

Hou JL, Wu YJ, Yang YJ (2010) A model for storage arrangement and re-allocation for storage management operations. Int J Comput Integr Manuf 23(4):369–390. https://doi.org/10.1080/09511921003642154

Hu W, Wang Y, Zheng J (2012) Research on warehouse allocation problem based on the artificial bee colony inspired particle swarm optimization (ABC-PSO) algorithm. In: 2012 Fifth international symposium on computational intelligence and design, vol 1, pp 173–176. https://doi.org/10.1109/ISCID.2012.51

Kappauf J, Lauterbach B, Koch M (2012) Warehouse logistics and inventory management. In: Logistic core operations with SAP: 99-213. Springer, Berlin, Heidelberg. https://doi.org/10.1007/978-3-642-18202-0_3

Kovács A (2011) Optimizing the storage assignment in a warehouse served by milkrun logistics. Int J Prod Econ 133(1):312–318. https://doi.org/10.1016/j.ijpe.2009.10.028

López A, Zapotecas S, Coelho LC (2011) An introduction to multiobjective optimization techniques. In: Gaspar-Cunha A, Covas JA (eds) Optimization in polymer processing. Nova Science Publishers, New York, pp 29–57

Marler RT, Arora JS (2004) Survey of multi-objective optimization methods for engineering. Struct Multi Optim 26:369–395. https://doi.org/10.1007/s00158-003-0368-6

Moon C, Seo Y (2005) Evolutionary algorithm for advanced process planning and scheduling in a multi-plant. Comput Ind Eng 48(2):311–325

Orjuela JA, Suárez N, Chinchilla YI (2016) Logistic costs and methodologies for supply chain costing: a literature review. Cuadernos de Contabilidad 17(44):377–420. https://doi.org/10.11144/Javeriana.cc17-44.clmc

Pan C-H, Shih P-H, Wu M-H, Lin J-H (2015) A storage assignment heuristic method based on genetic algorithm for a pick-and-pass assignment system. Comput Ind Eng 81:1–13

Petersen CG, Gerald AA (2004) Comparison of picking, storage, and routing policies in manual order picking. Int J Prod Econ 92:11–19

Quintanilla S, Pérez Á, Ballestín F, Lino P (2015) Heuristic algorithms for a storage location assignment problem in a chaotic warehouse. Eng Optim 47(10):1405–1422

Ramtin F, Pazour JA (2015) Product allocation problem for an AS/RS with multiple in-the-aisle pick positions. IIE Trans 47(12):1379–1396

Ross A, Khajehnezhad M, Otieno W, Aydas O (2017) Integrated location-inventory modelling under forward and reverse product flows in the used merchandise retail sector: a multi-echelon formulation. Eur J Oper Res 259(2):664–676. https://doi.org/10.1016/j.ejor.2016.10.036

Sanei O, Nasiri V, Marjani MR Moattar HSM (2011) A heuristic algorithm for the warehouse space assignment problem considering operational constraints: with application in a case study. In: Proceedings of the 2011 international conference on industrial engineering and operations management, Kuala Lumpur, Malaysia, pp 258–264

Shenfield A, Fleming PJ, Alkarouri M (2007) Computational steering of a multi-objective evolutionary algorithm for engineering design. Eng Appl Artif Intell 20:1047–1057

Srinivas N, Deb K (1994) Multiobjective optimization using nondominated sorting in genetic algorithms. Evol Comput 2:221–248

Tian Y, Ellinger AE, Chen H (2010) Third-party logistics provider customer orientation and customer firm logistics improvement in China. Int J Phys Distrib Logistics Manag 40(5):356–376. https://doi.org/10.1108/09600031011052822

Tiwari A, Roy R, Jared G, Munaux O (2002) Evolutionary-based techniques for real-life optimisation: development and testing. Appl Soft Comput 1:301–329

Tompkins JA, White JA, Bozer YA, Tanchoco JMA (2010) Facilities planning 4th ed. Wiley, USA. ISBN 0470444045

Trab S, Bajic E, Zouinkhi A, Abdelkrim MN, Chekir H, Ltaief RH (2015) Product allocation planning with safety compatibility constraints in IoT-based warehouse. Procedia Comput Sci 73:290–297. https://doi.org/10.1016/j.procs.2015.12.033

Van Wijk ACC, Adan IJBF, van Houtum GJ (2013) Optimal allocation policy for a multi-location inventory system with a quick response warehouse. Oper Res Lett 41(3):305–310

Vergidis K, Saxena D, Tiwari A (2012) An evolutionary multi-objective framework for business process optimization. Appl Soft Comput 12:2638–2653

Villalobos MA (2005) Análisis de Heurísticas de Optimización para Problemas Multiobjetivo. PhD thesis. Departamento de Matemáticas. Centro de investigación y de estudios avanzados del instituto politécnico nacional. Mexico

Zapata-Cortes JA, Arango-Serna MD, Serna-Urán CA, Adarme-Jaimes W (2020) Mathematical model for product allocation in warehouses. In: García-Alcaraz J, Sánchez-Ramírez C, Avelar-Sosa L, Alor-Hernández G (eds) Techniques, tools and methodologies applied to global supply chain ecosystems. intelligent systems reference library, vol 166. Springer, Cham, pp 47–70

Zapata-Cortes JA (2016) Optimización de la distribución de mercancías utilizando un modelo genético multiobjetivo de inventario colaborativo de m proveedores con n clientes. Ph.D. Thesis, Universidad Nacional de Colombia, Medellín, Colombia

Zapata-Cortes JA, Arango-Serna MD, Adarme-Jaimes W (2010) Herramientas tecnológicas al servicio de la gestión empresarial. Avances en Sistemas de Información 7(3):87–102

Zhang H, Yue D, Xie X, Hu S, Weng S (2018) Pareto-dominance based adaptive multi-objective optimization for hydrothermal coordinated scheduling with environmental emission. Appl Soft Comput 69:270–287. https://doi.org/10.1016/j.asoc.2018.04.058

Model Design of Material Requirement Planning (MRP) Applied to a Surgical Sutures Company

Fiorella Cerna, Catherine Gómez, Nadia Sánchez, Ana Luna, and Mario Chong

Abstract This case study is restricted to a company in the health area that supplies local and international public and private organizations. Its master product is surgical sutures for the synthesis of tissues and ligation of blood vessels. The company is consolidated as finally for health professionals. Its value-added is to offer its products through demonstrations in the surgery room. For this purpose, sales representatives are present on-site, in the operating room, an average of three times per week, where they demonstrate the use of all their products, including sutures. In addition to this practice, the company has within its facilities a "hands-on room," allowing health professionals to attend and use its facilities to carry out tests on all their products and verify the quality of the manufactured products. In this work, we implement the material requirements planning (MRP) to improve productivity in a company dedicated to the manufacturing of supplies for the health area. The use of material requirements planning improved the estimation of the number of raw materials required and optimized the scheduled deliveries within the company.

Keywords Health suppliers · Surgical sutures · Material requirement planning

1 Introduction

The health sector comprises various industries, sub-industries, and a wide variety of companies, having a strong influence on the global economy. To promote and maintain the health of individuals, this sector develops production and distribution activities and also offers services and treatments. With the advancement of technology, the sector also showed great growth including innovative products aimed at fighting a wide range of diseases and ailments in patients (Ledesma et al. 2014).

According to Deloitte (2018), global healthcare spending is projected to increase at an annual rate of 4.1% in 2017–2021. Aging and increasing populations, developing

F. Cerna · C. Gómez · N. Sánchez · A. Luna · M. Chong (✉)
Universidad del Pacífico, Jr. Sánchez Cerro N° 2141, Jesús María. 15072, Lima, Peru
e-mail: m.chong@up.edu.pe

J. L. García Alcaraz et al. (eds.), *Techniques, Tools and Methodologies Applied to Quality Assurance in Manufacturing*,
https://doi.org/10.1007/978-3-030-69314-5_12

market expansion, advances in medical treatments, and rising labor costs will drive growth in spending. Furthermore, it is estimated that life expectancy will increase by more than a full year between 2016 and 2021, from 73 to 74.1 years. In recent years, the goal of some countries has been to seek to improve living conditions and have broader access to health care and vaccines. On the other hand, rapid urbanization, sedentary lifestyles, changes in diets, and rising levels of obesity are fueling an increase in chronic diseases such as cancer, heart disease, and diabetes that require medical assistance.

Peru is the smallest of the five major healthcare markets in Latin America, after Brazil, Mexico, Colombia, and Argentina. From 2014 to 2016, Peru has demonstrated consistent growth. The Peruvian market exhibited 13% growth from 2014 to 2016, going from 333 million units to 428 million units (Global Health Intelligence 2019). Regarding the pharmaceutical industry, between January and November 2018, Peru exports of pharmaceutical products were recorded for a value of more than 49 million dollars, representing a growth of 4.8% compared to the same period of the previous year (Global Health Intelligence 2019).

Given the importance of the health sector, high-quality care services must be provided, supplying medicines, having optimal and advanced equipment and facilities, and adequate distribution of the required products in precise quantity and time. This is made feasible mainly with efficient inventory management (Ledesma et al. 2014). Within a supply chain, inventory is made up of raw materials, work in process, and finished products. The efficiency and responsiveness of the chain can be significantly modified if any of the links are affected (Chopra and Meindl 2013). Inventory maintenance is expensive, and the costs of storage, handling, and transportation of goods must be assumed. For these reasons, it is of great importance to have efficient inventory management through the adequate planning and control of stocks (Flores and Parra 2007).

The main objective of this research is to improve inventory management in the health sector through the proposal of material requirement planning (MRP). First, we will identify the indicators to identify the adequate quantity of the finished product and raw material within a company in the health sector to avoid large losses. Then, we seek to align sales planning with the supply chain. We tested three scenarios, the first encompasses the implementation of the materials planning framework with the current finished product stock policy (in three months) and the current raw material policy (in six months). The second scenario includes the implementation of the MRP with the current finished product stock policy (in three months) and the new raw material policy (considering the variation in demand and lead time). The third of the scenarios is the implementation of the MRP with a new finished product and raw material stock policy (considering the variation in demand and lead time).

2 Description of the Real-World Instance

It is known that the stock of material and basic supplies within the medical sector is highly demanding, so optimal management of the supply of the most required products is essential, as well as permanent monitoring of the raw material necessary for the production.

After a detailed study and analysis of the company, we identified that during the last three years, its growth increasing in an undesirable disorder. The commercial area expanded the client portfolio, making greater its sales. However, the lack of planning and the inefficient process carried out damaged the area of operations. The impact and consequences were the shortage of raw materials and finished products. In this work, we design an MRP that uses sales outlining as input and in simultaneous allows the company to maintain the desired stock level of raw material and finished product. Based on the inventory and lead time variables of the raw material and the finished product, three MRP scenarios were evaluated. The optimal selection is the one that generates the greatest benefit to the company.

The information is a supply chain driver that the organizations use to become more efficient and responsive (Chopra and Meindl 2013). The MRP is a planning and information system that can improve the decisions in the push supply chain strategy (Chopra and Meindl 2013) focus on the flow of raw materials and components (Mabert 2007) in discrete parts (Graves and Kostreva 1986). Using this approach, the organizations can reduce lead time, work in process inventory, floor space requirements, size of transfer vehicles (Graves and Kostreva 1986), and other inefficiencies. The MRP bases were the shop management proposed by Frederick Taylor during the 20 s, the accounting machines in the 30 s (Wilson 2016). The first MRP systems of inventory management evolved in the 1940s and 1950s, using mainframe computers to extrapolate information from a bill of materials for a specific finished product into a production and purchasing plan. All of these pushed the production planning and control systems to use mainframes in the 60 s (Wilson 2016; Westkämper et al. 2018). After that, MRP systems expanded to include information feedback loops so that production managers could change and update the system inputs as needed. Later, the next generation of MRP, manufacturing resources planning (MRP II), incorporated marketing, finance, accounting, engineering, and human resources aspects into the planning process. The enterprise resource planning (ERP) expanded on MRP with the use of computer technology to link the various functional and transversals areas across an entire business enterprise (Kortabarria et al. 2018). As data analysis and technology became more sophisticated, more comprehensive systems were developed to integrate MRP with other aspects of the manufacturing process. The developments and applications were documented in different industries: semiconductors (Milne et al. 2012), fruit juices (Olaore and Olayanju 2013; Milic et al. 2016), polyurethane coatings (Hualpa and Suarez 2018), papers manufacturing (Lestari and Nurdiansah 2018), such as with other approaches: fuzzy lead times (Diaz-Madroflero et al. 2015), particle swarm optimization (Padron-Cano et al. 2017), demand-driven (Pekarcíková et al. 2019), activity-based costing (Więcek et al. 2020), among others.

The MRP's basic inputs are the master production schedule (MPS), bill of materials (BOM) for every stock-keeping unit (SKU), inventory records, purchase orders, and lead time (Romero et al. 2019). The MPS is the master plan in supplies or material requirements (Olaore and Olayanju 2013). The BOM is the full product description: components, parts, sub-assemblies, and materials (Olaore and Olayanju 2013; Jacobs and Chase 2020). The lead time in the MRP implementation is compound by the transportation time, queueing time, setup time, and processing time, and it is used to determine the intervals, to monitor the progress, and prioritize items (Graves and Kostreva 1986).

The company has more than fifty years of operations in the Peruvian market. Its flagship product is the manufacture and marketing of surgical sutures, among others such as hemostatic collagen sponges and surgical meshes, in addition to medical supplies. The business unit we focus on within the company is that of the laboratory, which represents approximately 84% of the company's income and is in charge of manufacturing and marketing white-label surgical sutures. Also, the company has several certifications such as ISO 9001, ISO 13485, BPM, BPA, CE, and ANMAT, and it is regulated by the Dirección General de Medicamentos, Insumos y Drogas (DIGEMD—General Directorate of Medicines, Supplies, and Drugs), whose main objective is that the population has access to quality medicines.

The scope of this study covers the development of a material requirement planning (MRP) to improve the inventory management applied to the health sector, which is classified into six main industries that include pharmaceuticals, biotechnology, equipment, distribution, facilities, and managed healthcare (Ledesma et al. 2014). In particular, the existence of an inventory in a supply chain is due to a mismatch between supply and demand. Its main function is to meet the demand that can be satisfied by having the product ready and available when the customer wants it. Also, the inventory considers the reduction in costs and the scale reductions that can occur during production and distribution. Because inventory alters assets held, costs incurred, and responsiveness in a supply chain, managing them efficiently is imperative (Chopra and Meindl 2013).

Addressing our case study, Fig. 1 shows that the laboratory unit supply chain had an approximate duration interval of 160 days, of which the supply process represented

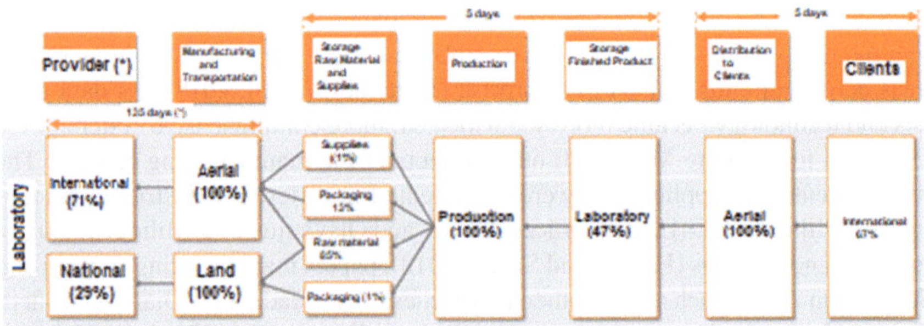

Fig. 1 Supply chain

84% of that time. The determination of the supply quantity was calculated from the historical information of the sales considering the last quarter. This process is the longest one because the suppliers of the company were the ones that manufactured the orders.

The company established the following policy for inventory management:

- The finished product inventory and the obligation must be three months (Fig. 2).
- The inventory of raw materials and supplies must be six months (Fig. 3).

The inventory management process seeks to meet the level of service and availability required by the market, reduce working capital investments without altering the degree of market service, and achieve the desired return on equity investments (Tejero 2007). As each company is different, it has different needs, demands, costs, and specific objectives; there is no single inventory management system for all the companies, but each one must adopt various procedures according to their needs (Ferrero Bécares 2015).

Jacobs and Chase (2020) defined inventory management systems as "the set of policies and controls that supervise inventory levels and certain determining factors that must be maintained, when inventory must be replenished and what size orders

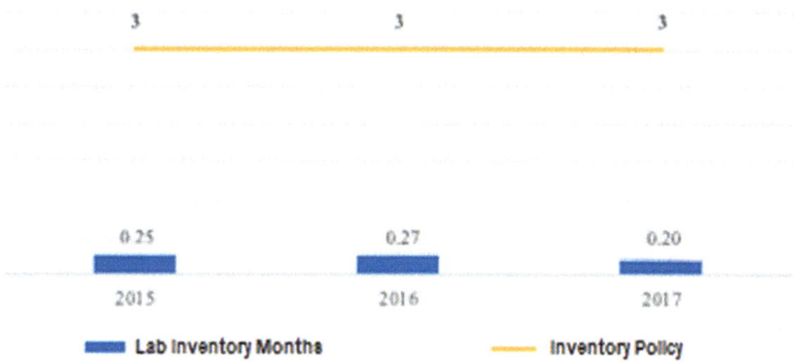

Fig. 2 Inventory's months of the finished product

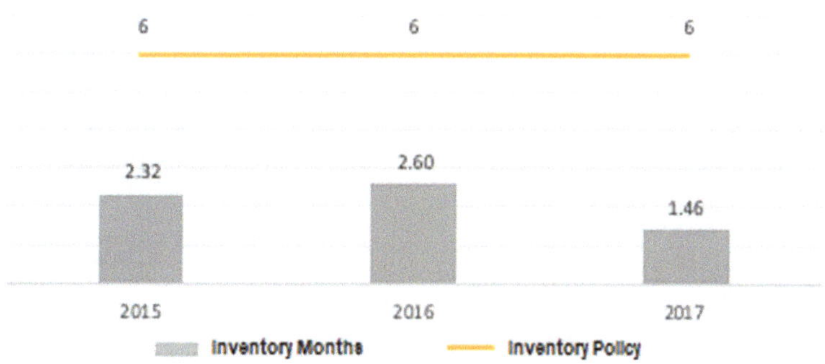

Fig. 3 Inventory's months of the raw material

should be." Within the company under study, the responsible for demand planning is commercial management. In 2018, the required production of 13 million units is estimated. Customer sales orders reached 12 million units, however, the company's sales were 11.6 million units. The main reason for this fact is that the company is not in contact with the finished product and raw material inventory levels to meet the required demand. Furthermore, the supply process is the longest in the supply chain as selected in the previous figures. Due to these shortcomings, it is necessary: the implementation of an MRP to acquire enough raw material to satisfy demand and the correct supply planning based on expected demand.

3 Methodology and Methods

The implementation of the MRP expects improvement in manufacturing systems (Morecroft 1983; White et al. 1982). The production aims align the products to the customer requirements (Elezaj 2019), and its representation of the decision-making process in a typical system could include four units: production control, labor management, material control, and vendor's plan with established policies and procedures (Morecroft 1983; Elezaj 2019). This work is framed within the business unit laboratory since it is the area with the highest income within the company. As an initial phase of project management, the risks that could impact the implementation (Olaore and Olayanju 2013) of the MRP were identified and prioritized (Dumbrava and ladut-Severian 2013; Kassem et al. 2019) (Fig. 4), which combines the impact and probability of occurrence for future escenarios (Chao 2008).

To manage the scope of the proposed project, a test script must be defined where the assessments to be carried out will be defined, as well as those responsible for the MRP. Then, the training will be provided to those responsible for the execution of

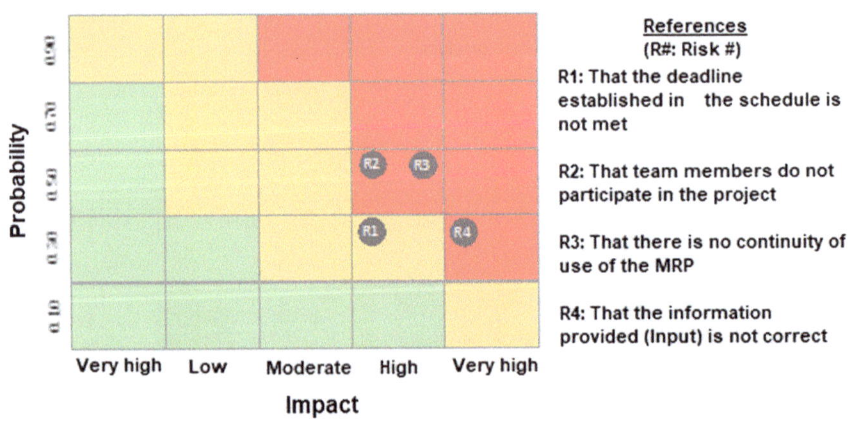

Fig. 4 Project risk assessment

the MRP. The tests will then be run to verify the correct operation of the MRP, and errors will be identified in the settings. In the go live, the communication of the MRP is evaluated, as well as the monitoring of the implementation. Finally, in the project management stage, the actions carried out for the project to proceed as expected will be detailed.

The strategy to attack the identified risks is detailed in Table 1.

The project deliverables will be the work plan, the training material, the project progress report, the MRP process and execution report, the test approval document, and the accompanying minutes. The acceptance criteria for the deliverables will be the conformity of the technical consultancy by the IT area, while for the functional consultancy the conformity of the project sponsor is expected. The scope of the product is to obtain material planning reports following the inventory policy. Product acceptance will be granted by the project manager. To apply a rigorous methodology that allows the development and quantify the benefits of the MRP implementation to a specific company within the health sector, we divide the procedure into various stages:

Step 1: *Identification of the improvement opportunity and definition of the scope of the research*

Table 1 Action plans for risk management

Risk	Responsible	Action plan
R1: That the deadline established in the schedule is not met	Logistics manager	To mitigate The weekly monitoring will be carried out to meet the deadlines and in the event of a deviation of more than 10%, it will be raised to general management
R2: That the team members do not participate in the project	Logistics manager	To mitigate The dedication for the project will be communicated to the heads of the team members, and the project schedule will be validated with said leadership
R3: That there is no continuity of use of the MRP	General management	To mitigate For the issuance of raw material purchase orders, the reports generated by the MRP must be submitted
R4: That the information provided (inputs) is not correct	Head of sales support	To mitigate An indicator will be established in the performance bonus of this department, the degree of accuracy of the planning of the sale. This indicator should be established for all operations and commercial headquarters

In this first phase, an analysis of the current situation of the company is carried out. The cost of sales, demand planning, and inventory management are studied and evaluated. Through a detailed study, we identified 117 finished products with a higher margin, higher volume, and frequency of sales. From this number of products, we selected five and we obtained the production recipe. If the company had at MRP in 2018, its sales would have been approximately $ 83,000. In the last three years, the accumulated sales increased to approximately 50 million dollars, and the return on sales averaged 7.7%. For inventory management, the company's policy was that the inventory of finished products and commodities for the drugstore unit, and the laboratory was three months and for the raw material and supplies six months. However, the inventory policy was not followed. Besides, in recent years, the company did not have enough raw material to satisfy customer needs, resulting in a significant loss of sale. Due to the gap between forecast and annual company sales, the modeling error was evaluated. To calculate the annual cost of orders, we consider annual staff salaries and overhead, including the cost of renting space, utilities, office supplies, software licenses, and more. Likewise, the number of purchase orders issued and the average number of items per purchase order were determined (Carreño Solís 2011). Finally, the number of items purchased and the calculation of the ordering cost per item were obtained. For this last variable, the annual ordering cost was divided by the total number of items purchased.

Step 2: *Inventory holding cost*

The calculation of the inventory holding rate was obtained considering the opportunity cost of capital (the average cost of debt plus the cost of capital of the shareholders), the cost of storage (costs related to personnel costs, fixed costs such as physical space, electricity, water, etc.), and the cost of insurance (the insurance that is contracted to protect the inventory from possible risks of theft, fire, etc.), weighing each of these variables. Then, to determine the cost of maintaining inventory, the annual inventory rate was multiplied by the total cost of the input.

Step 3: *Information gathering*

The creation of future scenarios requires proper planning. For this, it is fundamental to determine and construct the necessary data for the MRP model formulation (Table 2). Information related to purchases (minimum quantity, cost of input, purchase unit, and supplier delivery time), warehouses (finished product inventory, raw material inventory, policies of finished product stock, and raw material), and production (list of materials, percentage of waste, lot size) of the selected business

Table 2 Information required for the construction of an MRP model (Cuatrecasas Arbós 2011)

Supply	Warehouses	Production
The minimum amount of purchase Cost of the input purchasing Supplier lead time Purchase unit	Initial inventory Entries registration of warehouse receipts Exits registration of entrances to the warehouse	Percentage of waste Amount of inputs required according to planning Amount of inputs available for production

Table 3 Scenarios to evaluate the implementation of the MRP

Scenarios	Input for planning	Finished product inventory policy	Raw material inventory policy	Lot size
1	Sales forecast	Current inventory policy (3 months)	Current inventory policy (6 months)	Provided by the company
2			Proposed policy (demand and safety stock according to lead time)	
3		Proposed policy (demand and safety stock according to lead time)		

unit were collected. Likewise, as part of the model's construction, we need to define the cost of storing and the cost of buying (Carreño Solís 2011).

Step 4: *Definition of simulation scenarios*

In order to have the quantity of the finished product and raw material that allows responding to the company's planning, the inventory policy for each scenario shown in Table 3 was defined and evaluated. The main objective of scenario 1 is to apply the proposed framework considering the current finished product and raw material inventory policies. Scenario 2 looks forward to implementing the structure by modifying the raw material inventory policy. Finally, scenario 3 seeks the implementation of the framework by modifying the raw material and finished product inventory policies.

4 Results

The company studied is within the health sector and caused a considerable error between the forecast of demand and annual sales, causing a lack of attention to the orders generated by the customer. These are direct consequences of non-compliance with the finished product and raw material and input inventory policy. Undoubtedly, inventory management plays a fundamental role in the supply chain as it provides responsiveness and enables cost reduction, which ultimately translates into better service to the end-user.

The proposed methodology identifies those important products according to certain criteria and determines the prevailing costs of inventory management so that changes in policies can be proposed to improve the management of existing ones. The first of the analyzes carried out was the comparison between the quantity of raw material purchased and the raw material required for production for each of the scenarios. After identifying the required purchase quantities, it was processed to assess the quantity purchased (cost of purchase, ordering, storing, and managing the raw material) for each of the scenarios. Then, we evaluated the expense of buying, the cost of ordering, and storing of raw materials for the five products in each of the three scenarios, for scenario 1, we obtained 100%, and 127% and 83% for scenarios 2 and 3, respectively.

Finally, we conclude that scenario 3 allows meeting the commercial needs at a lower total cost. It is the scenario that manages a policy considering the lead time of the safety stock is the best because it reduces the loss of sales and reduces the costs of inventory management.

5 Concluding Remarks

During the analysis stage, we calculate the modeling error of the sales forecast. We identified that the sales plan was not used as input for supply chain management to forecast the demand. For this reason, the supply area did not accurately determine the quantity of raw material to purchase based on its history. The insufficient stock of the finished product or raw materials in the company meant large sales losses, in addition to a high percentage of the deviation between the production requirement and the average manufacturing order.

For the supply chain processes, the sales forecast was used as input. The implementation of the MRP sought to reduce lost sales. In this process, three scenarios were analyzed and different inventory policies were evaluated for both raw materials and finished products. The best scenario was the third, in which the finished product stock policy as raw material is modified based on demand in lead time and the standard deviation of demand in lead time. Finally, the MRP model allows the company to level inventory levels with less growth of working capital.

We identified that an improvement aspect would be that the commercial area strategy plans of the company are shared with the supply chain area. Currently, the sales plan is carried out by the business line (sutures, veterinary sutures), which does not allow us to assess compliance with this plan or modifications to it. To keep the information on which the MRP is built up to date, it is suggested to define a person responsible for the update, as well as to establish a process to update/review the information.

References

Carreño Solís A (2011) Logística de A a la Z (Primera Edición ed.). Fondo editorial de la Pontificia Universidad Católica del Perú, Lima, Lima, Perú

Chao K (2008) A new look at the cross-impact matrix and its application in futures studies. J Futures Stud 12(4):45–52

Chopra S, Meindl P (2013) Administración de la cadena de suministro. Pearson educación

Cuatrecasas Arbós L (2011) Logística. Gestión de la cadena de suministros: Organización de la producción y dirección de operaciones. Ediciones Díaz de Santos

Deloitte (2018) 2018 Global health care outlook. The evolution of smart health care

Diaz-Madroflero M, Mula J, Jiménez M (2015) Material requirement planning under fuzzy lead times. In: IFAC-Papers online, vol 28, pp 242–247. https://doi.org/10.1016/j.ifacol.2015.06.088

Dumbravă V, ladut-Severian I (2013) Using probability—impact matrix in analysis and risk assessment projects. J Knowl Manag Econ Inf Technol

Elezaj S (2019) Manufacturing process management. Knowl Int J 26(6):1571–1576. https://doi.org/10.35120/kij26061571e

Ferrero Bécares P (2015) La gestión de inventarios, aplicación práctica en una empresa del sector farmaceútico: el caso de Laboratorios Jiménez, SL. Trabajo de fin de grado. Universidad de León. León, España

Flores C, Parra G (2007) El MRP En la gestión de inventarios. Visión Gerencial 1:5–17

Global Health Intelligence (2019) Latin America medical device and equipment market

Graves S, Kostreva M (1986) Overlapping operations in material requirements planning. J Oper Manag 6(3–4):283–294. https://doi.org/10.1016/0272-6963(86)90004-5

Hualpa ZAM, Suarez RC (2018) warehouse sizing based on material requirements planning for a factory. Ingenieria 23(1):48–69. https://doi.org/10.14483/23448393.11825

Jacobs F, Chase R (2020) Operations and supply chain management: the core. McGraw Hill/Irwin, New York

Kassem A, Khoiry M, Hamzah N (2019) Using probability impact matrix (PIM) in analyzing risk factors affecting the success of oil and gas construction projects in Yemen. Int J Energy Sector Manag 14(3):527–546. https://doi.org/10.1108/IJESM-03-2019-0011

Kortabarria A, Apaolaza U, Lizarralde A, Amorrortu I (2018) Material management without forecasting: from MRP to demand driven MRP. J Ind Eng Manag 11(4):632–650. https://doi.org/10.3926/jiem.2654

Ledesma A, McCulloh C, Wieck H and Yang M (2014) Health care sector overview, Wahington State University

Lestari S, Nurdiansah DD (2018) Analisa perencanaan kebutuhan material pada perusahaan manufaktur kertas dengan metode material requirement planning (MRP). Jurnal INTECH Teknik Industri Universitas Serang Raya 4(2):59. https://doi.org/10.30656/intech.v4i2.956

Mabert VA (2007) The early road to material requirements planning. J Oper Manag 25(2):346–356. https://doi.org/10.1016/j.jom.2006.04.002

Milic D, Lukac-Bulatovic M, Kalanovic-Bulatovic B, Milovancevic Z (2016) Raw material requirements planning in fruit juice production. Ekonomika Poljoprivrede 63(4):1395–1402. https://doi.org/10.5937/ekopolj1604395m

Milne RJ, Wang CT, Yen CKA, Fordyce K (2012) Optimized material requirements planning for semiconductor manufacturing. J Oper Res Soc 63(11):1566–1577. https://doi.org/10.1057/jors.2012.1

Morecroft J (1983) Concepts, theory, and techniques: a systems perspective on materials requirement planning. Decis Sci 14(1):1–18

Olaore R, Olayanju M (2013) Purchasing functions and MRP in foodservice firms. Int Inst Sci Technol Educ (IISTE). 5(13):107–119

Padron-Cano J, Coronado-Hernandez JR, Caicedo-Torres W, Mercado-Caruso N, Ospino-Vadiris FV (2017) Planificación de requerimientos de materiales por medio de la aplicación de un algoritmo basado en enjambre de partículas. Espacios, 38:14

Pekarcíková M, Trebuna P, Kliment M, Trojan J (2019) Demand driven material requirements planning. Some methodical and practical comments. Manag Prod Eng Rev 10(2):50–59. https://doi.org/10.24425/mper.2019.129568

Romero J, Delgado E, Herrera J, Garcia O (2019) Demand forecasting and material requirement planning optimization using open source tools. ICAI Workshops, 97–107

Tejero J (2007) Logística integral: la gestión operativa de la empresa. ESIC editorial

Westkämper E, Mussbach-Winter U, Wiendahl H-H (2018) Material requirement planning (MRP). In: Taschenbuch der Logistik, pp. 87–97. Carl Hanser Verlag GmbH & Co. KG. https://doi.org/10.3139/9783446457676.005

White E, Anderson J, Schroeder R, Tupy S (1982) A study of the MRP implementation process. J Oper Manag 2(3):145–153. https://doi.org/10.1016/0272-6963(82)90002-X
Więcek D, Więcek D, Dulina L (2020) Materials requirement planning with the use of activity based costing. Manag Syst Prod Eng 28(1):3–8. https://doi.org/10.2478/mspe-2020-0001
Wilson JM (2016) The origin of material requirements planning in Frederick W. Taylors planning office. Int J Prod Res 54(5):1535–1553. https://doi.org/10.1080/00207543.2015.1092616

An EWMA Chart with Varying Sample Interval to Monitor Calibration Processes

María Guadalupe Russell-Noriega and Enrique Villa-Diharce

Abstract A measurement assurance program considers the periodic calibration of measuring instruments to verify that their accuracy and precision are kept stable and controlled over time. To monitor the calibration process of a measuring instrument, we may use statistical process control charts, which are very useful tools in the industry. We propose an exponentially weighted moving average (EWMA) control chart with a varying sample interval based on Croarkin and Varner's statistical control chart, which was adopted by NIST. The original Croarkin and Varner chart has lower performance than some charts proposed in recent years; nevertheless, the performance of the modified chart that we propose is comparable to that of better charts, given its detection ability in the face of changes to the calibration process. We evaluate and compare the performance of different charts in phase II through Monte Carlo simulations.

Keywords Calibration processes · EWMA charts · Monitoring of linear profiles · Linear regression

1 Introduction

The monitoring of processes of calibration of measuring instruments is done as part of any measurement quality assurance program to ensure the quality of the calibration results (see Croarkin et al. 1979; Croarkin and Varner 1982). Control charts initially proposed by Shewhart (1931) for the monitoring of processes are univariate charts used for monitoring a product's quality characteristic which is measured periodically

M. G. Russell-Noriega (✉)
Facultad de Ciencias Físico-Matemáticas, Universidad Autónoma de Sinaloa, Ciudad Universitaria, Culiacán, Sinaloa, Mexico
e-mail: mgrussell@uas.edu.mx

E. Villa-Diharce
Department of Probability and Statistics, Centro de Investigación en Matemáticas, A.C. Callejón Jalisco S/N, Col. Valenciana, Guanajuato, Mexico
e-mail: villadi@cimat.mx

J. L. García Alcaraz et al. (eds.), *Techniques, Tools and Methodologies Applied to Quality Assurance in Manufacturing*,
https://doi.org/10.1007/978-3-030-69314-5_13

and whose variation follows a normal distribution. These charts are graphs that have three horizontal lines: a center line located at a height equal to the mean of the normal distribution and two others located at three standard deviations from the mean. These charts graph the values of the characteristic of interest, which are measured periodically by each of the monitoring units. If the graphic pattern of said values does not correspond to a normal distribution, we interpret this as a signal indicating that the process is out of control. When this occurs, we review the process to find and remove specific sources of variation that have caused the process to get out of control. With this operation, we keep the process under statistical control.

The complexity of the processes and quality characteristics has increased over time. Thus, we have products whose quality is expressed in a multivariate form or by an expression of functional type. In the latter case, the quality characteristic is expressed through a functional relationship between a response variable and one or more explanatory variables associated with the monitoring product.

The process of periodic calibration of a measuring instrument can be considered one in which we periodically show the calibration line of the instrument. When there is a significant change in the calibration line, it can be detected by a chart that monitors linear profiles. Timely detection of a change in the calibration line is important to prevent measurement results that are not in accordance with the measurand.

Croarkin and Varner (1982) proposed a control chart for monitoring calibration processes based on deviations of the measured values and patterns identified in Eq. (2). Subsequently, in the literature of statistical process control (SPC), several control charts for linear profiles which have been applied to the monitoring of linear calibration processes in Phase II were published. These charts are identified as T^2 (Mestek et al. 1994), Kim-Shewhart (Kim et al. 2003), EWMA/R (Kang and Albin 2000), and EWMA3 (Kim et al. 2003). Usually, the performance of the charts in Phase II is in terms of run length, defined as the number of samples taken until the occurrence of a signal showing the process is out of control. On comparing the Croarkin and Varner chart with previous ones, it can be seen that the EWMA/R and EWMA3 charts are clearly superior to the first. This advantage is due to the incorporation of the EWMA procedure, which accelerates the detection of small changes in the monitoring process.

For each of the methods used to make the charts, we represent by x_i, the value of the ith level of the reference pattern and by y_{ij}, the jth reading of the ith level of the reference pattern used in the calibration process. When the calibration process is under statistical control, the base model is

$$y_{ij} = \beta_0 + \beta_1 x_i + \varepsilon_{ij}, \tag{1}$$

where $i = 1, 2, \ldots, n$ and $j = 1, 2, \ldots, m$. We assume that the terms ε_{ij} are random, independent, and identically distributed variables, with normal distribution, zero mean, and variance σ^2. Since we are considering Phase II of the monitoring, we assume that the values of the control parameters β_0, β_1, and σ^2 are known.

In monitoring a linear calibration process, we are monitoring calibration lines. In the charts discussed here, changes in calibration lines are detected in one of the following ways:

- Observation of the variation of residuals: $e_{ij} = y_{ij} - \beta_0 - \beta_1 x_i$.
- Observation of the estimated parameters: $\hat{\beta}_0$, $\hat{\beta}_1$, and $\hat{\sigma}^2$.
- Observation of the deviations between the calibration value and the reference pattern: $c_{ij} = \left(y_{ij} - \beta_0\right)/\beta_1 - x_i$.

The T^2 and EWMA/R charts detect changes in the line starting from the point of changes in the residual e_{ij}, while the Kim-Shewhart and EWMA3 charts consider changes in estimates of the parameters β_0, β_1, and σ^2 to detect respective changes in the line. Lastly, the chart proposed by Croarkin and Varner, identified by NIST, is based on the deviations c_{ij} of the calibration values with respect to the values of the reference patterns used in the calibration.

In this paper, we propose a modification of the NIST chart by applying the EWMA procedure to Croarkin and Varner's control statistic, resulting in a chart that we will denote as EWMA/NIST. Likewise, we incorporate the varying sample interval scheme in the EWMA/NIST chart, thus obtaining the VSI-EWMA/NIST chart. We compare the performances of this chart with the VSI-EWMA3 chart proposed in Li and Wang (2010).

In Sect. 2, we describe the NIST, EWMA/R, and EWMA3 charts. In Sect. 3, we present the results of a simulation study where the performances of the NIST chart and EWMA-type charts are compared, such as the T^2 and Kim-Shewhart charts. In Sect. 4, the proposed modifications to the NIST method are presented, incorporating the EWMA scheme and considering varying sample intervals. In this section, we also compare the performance of the EWMA/NIST and EWMA3 charts with fixed and varying sample intervals. Section 5 presents the conclusions.

2 Methods

2.1 *NIST Method*

This technique for monitoring linear profiles from Croarkin and Varner (1982), which we denote as the NIST method, was published in the *NIST/SEMATECH e-Handbook of Statistical Methods* (NIST/SEMATECH 2011). This control chart is also part of the ISO 5725-6 standard (ISO 1994). According to the authors, this system was initially developed to address problems in the calibration of optical image systems. In the making of this control chart, we suppose that the data follow the linear regression model (1). The method is based on the principle of inverse calibration; that is, it is considered to be the inverse regression model that corrects mean values (variable y), immediately deducting the reference value (variable x). In this way, we obtain the statistics we graphed on the control chart at the time of the jth sample:

$$Z_{ij} = \frac{y_{ij} - \beta_0}{\beta_1} - x_i, i = 1, 2, \ldots, n. \tag{2}$$

Croarkin and Varner (1982) recommend taking three measurements ($n = 3$) in the calibration process, two at the extremes and one in the center. The lower and upper limits for the control chart are:

$$\text{LCL} = -s_c z_{\alpha^*}, \text{UCL} = s_c z_{\alpha^*}, \tag{3}$$

where $s_c = \sigma/\beta_1$, σ is the known standard deviation, and β_1 is the known value controlling the slope. The value z_{α^*} corresponds to the upper quantile of the normal standard distribution. We obtain the probability α^* with Bonferroni's correction, $\alpha^* = \{1 - \exp[\log(1 - \alpha)/n]\}/2$, where n is the number of evaluations in each time period, which Croarkin and Varmer propose should be equal to three, and α is chosen so that we have an average run length (ARL) under control using the relationship $ARL_0 = 1/\alpha$. Control limits in (3) are built using the quantile of the normal standard distribution instead of the student's t distribution, as in Croarkin and Varner (1982), since the values of the parameters under control are known.

2.2 *EWMA/R Method*

The proposal of Kang and Albin (2000) is based on the residuals corresponding to the regression model (1) which follow the calibration data. Specifically, an EWMA chart is used to monitor the average value of residuals jointly with an R chart. We obtain regression residuals from the jth sample in the usual way,

$$e_{ij} = y_{ij} - \beta_0 - \beta_1 x_i, i = 1, 2, \ldots, n, j = 1, 2, \ldots, m.$$

The average of residuals for the jth sample is denoted by $\bar{e}_j$, with $\bar{e}_j = \sum_{i=1,\ldots,n} e_{ij}/n$.

The statistic for the EWMA control chart, denoted as $W(j), j = 1, \ldots, m$, is given by

$$W(j) = \theta\bar{e}_j + (1 - \theta)W(j - 1), \tag{4}$$

where $\theta(0 \leq \theta \leq 1)$ is a smoothing constant and $W(0) = 0$. A signal that the process is out of control is detected as soon as W_j is smaller than the lower control limit (LCL) or greater than the upper control limit (UCL), where

$$\text{LCL} = -L\sigma_W \text{ and UCL} = L\sigma_W, \tag{5}$$

and $\sigma_W = \sigma\sqrt{\theta/(2 - \theta)^n}$, with $L > 0$, a constant selected to have a specific ARL under control. Two fundamental reasons led Kang and Albin (2000) to propose an

R chart for use in conjunction with this EWMA chart. The first reason is to detect changes in the variance of the process, since the EWMA chart is not sensitive to changes in the variance of the process. The second reason is that the EWMA chart based on the residual average is not sensitive to some changes in β_0 and β_1, for which the magnitudes of residuals tend to be large but the average tends to be very small. This may occur, for example, when the slope line changes but the mean value of y does not.

For the *R* chart, the sample ranges $R_j = \max_i(e_{ij}) - \min_i(e_{ij}) j = 1, 2, \ldots, m$, are calculated and graphed. The lower and upper control limits for the R chart are

$$\mathrm{LCL} = \sigma(d_2 - Ld_3) \text{ and } \mathrm{UCL} = \sigma(d_2 + Ld_3), \tag{6}$$

respectively, where $L > 0$ is a constant chosen to obtain an ARL in control. The values d_2 and d_3 are constants that are commonly used, depending on the sample size. These tabulated values can be found in (Montgomery et al. 2001). A disadvantage of this proposal is that if $n < 7$, then there is no lower control limit and decreases in the variation of the regression line cannot be detected.

2.3 *EWMA3 Method*

The EWMA3 method (Kim et al. 2003) first codes the independent variable in such a way that its mean value is zero. Estimators of the regression coefficients of each sample are used; that is to say, the estimators of the intercept and slope are used to build two EWMA charts. Also, an EWMA chart of one side is used to monitor the standard deviation of the process as a replacement for the R chart proposed in the EWMA/R method. After encoding the data, there is an alternate form of the basic linear regression model (1), given by the equation

$$y_{ij} = \alpha_0 + \alpha_1 x_i^* + \varepsilon_{ij}, \tag{7}$$

where $i = 1, 2, \ldots, n$, $j = 1, 2, \ldots, m$, $\alpha_0 = \beta_0 + \beta_1 \bar{x}$, $\alpha_1 = \beta_1$, and $x_i^* = (x_i - \bar{x})$.

For the *j*th sample, the least squares estimator of α_0 is $a_{oj} = \bar{y}_j$, while the least squares estimator a_{1j} of α_1 is the same as for β_1. Both a_{0j} and a_{1j} have normal distributions with means α_0 and α_1 as well as variances σ^2/n and σ^2/S_{xx}, respectively. In addition, the covariance between a_{0j} and a_{1j} is zero. Thus, we can employ separate control charts to monitor the intercept and slope without the problem that would occur if the estimates were highly correlated.

Specifically, in this method, three EWMA charts are used: one to monitor the intercept, one for the slope, and one more for the error variance. When at least one of the charts shows that a parameter has changed, this is a signal that the calibration process is out of control. Moreover, having a control chart for each parameter facilitates the diagnosis when any change has occurred in the process.

Intercept Monitoring

In the EWMA chart to monitor the intercept (α_0), we use the estimates a_{0j} to calculate the EWMA statistic given by

$$W_I(j) = \theta a_{0j} + (1 - \theta)W_I(j - 1), \tag{8}$$

with $j = 1, 2, \ldots, m$, the smoothing constant θ, and $W_I(0) = \alpha_0$. A signal that the intercept is out of control happens as soon as $W_I(j) < \text{LCL}$ or $W_I(j) > \text{UCL}$, where

$$\text{LCL} = \alpha_0 - L_1\sigma_{W_I} \text{ and UCL} = \alpha_0 + L_1\sigma_{W_I} \tag{9}$$

and $\sigma_{W_I} = \sigma\sqrt{\theta/(2-\theta)^n}$ and L_1 is chosen to have a specific ARL in control.

Slope Monitoring

The estimator a_{1j} of the slope α_1 is used in the EWMA3 chart to monitor the slope through the statistic

$$W_S(j) = \theta a_{1j} + (1 - \theta)W_S(j - 1), \tag{10}$$

with $j = 1, 2, \ldots, m$, the smoothing constant θ, and $W_S(0) = \alpha_1$. The control limits of this chart are given by:

$$\text{LCL} = \alpha_1 - L_s\sigma_{W_s} \text{ and UCL} = \alpha_1 + L_s\sigma_{W_s} \tag{11}$$

where $\sigma_{W_S} = \sigma\sqrt{\theta/(2-\theta)S_{xx}}$ and $L_S(> 0)$ are chosen to have a specific ARL in control.

Error Deviation Monitoring

Finally, to monitor for error variance $\left(\sigma^2\right)$ in the EWMA chart, we use the mean-squared error based on the residuals corresponding to the fixed line for the jth sample. In contrast to the two EWMA charts described above, this chart uses a control scheme of only one side to detect only increases in the variability of the process. The EWMA statistic is given by

$$W_E(j) = \theta \log\left(E_j\right) + (1 - \theta)W_E(j - 1), \tag{12}$$

where $j = 1, 2, \ldots, m$ and $W_E(0) = \ln\left(\sigma_0^2\right)$. To calculate the control limit of this chart, the authors propose the use of the following approximation, given in Crowder and Hamilton (1992):

$$\mathrm{Var}\left[\log\left(E_j\right)\right] \cong \frac{2}{n-2} + \frac{2}{(n-2)^2} + \frac{4}{3(n-2)^3} + \frac{16}{15(n-2)^5}.$$

A signal that the process is out of control is obtained when the statistic $W_E(j)$ is larger than the control limit given by $\mathrm{UCL} = L_E\sqrt{\frac{\theta}{2-\theta}\mathrm{Var}\left[\log\left(E_j\right)\right]}$ with $L_E(> 0)$, chosen to have a specific ARL in control.

3 Comparison of Charts' Performances

The charts' performances given in Sect. 2 for Phase II are compared through a simulation study in which 10,000 samples are generated under the linear calibration model $y_{ij} = 3+2x_i+\varepsilon_{ij}$ under control, where the ε_{ij} are random normal independent variables with zero mean and variance of one. This model has been used previously in the literature (Gupta et al. 2006; Kim et al. 2003) to see how quickly the different charts detect changes in the parameters of the model under control. In our case, we consider the different shifts shown in Table 1 of Appendix A for the intercept, slope, and standard deviation in order to compare the ARLs for each of the proposed charts, with a constant sample interval, when the parameters change. Table 2 contains the ARL under simultaneous changes of the intercept and slope, as in cases where each of the parameters remains constant relative to the other. In other words, we consider cases where the intercept parameter does not change and the slope does, just like the case where the slope did not change and the intercept did. Table 3 presents the ARL with changes in the deviation.

Figures 1 and 2 show graphically the changes in the ARL for the different intercept shifts while keeping the slope unchanged and for shifts in the slope while keeping the intercept unchanged, respectively. In Fig. 1, we see that of the charts compared, the NIST chart is the one that takes longer to detect changes in the intercept while EWMA-type control charts show a clear advantage over other charts in detecting the changes in the intercept more quickly. In Fig. 2, we see that the NIST and Kim-Shewhart charts have very similar behaviors, as both take longer to detect shifts in the

Table 1 Changes considered in the comparison of methods

Changes in	Notation	Values
Intercept	From β_0 to $\beta_0 + \lambda\sigma$	$\lambda = 0.2, 0.4, \ldots, 2.0$
Slope	From β_1 to $\beta_1 + \gamma\sigma$	$\gamma = 0.025, 0.050, \ldots, 0.250$
Standard deviation	From σ to $\delta\sigma$	$\delta = 1.2, 1.4, 1.6, \ldots, 3.0$

Table 2 ARL values considering interactions, the intercept, and slope shift simultaneously

λ		γ										
0.00		0.00	0.025	0.05	0.075	0.10	0.125	0.15	0.175	0.20	0.225	0.25
	T^2	200.15	173.49	118.12	73.47	45.03	27.92	17.70	11.50	7.63	5.32	3.95
	EWMA_R	201.20	139.07	60.84	27.89	15.46	10.16	7.43	5.83	4.79	4.10	3.57
	Kim_Shewhart	200.20	182.48	138.15	92.03	60.60	38.91	24.93	16.59	11.03	7.68	5.51
	NIST	202.74	181.33	135.05	95.13	61.86	41.68	27.38	18.74	12.76	8.89	6.41
	EWMA_3	200.78	114.83	45.23	21.80	12.71	8.79	6.61	5.31	4.43	3.84	3.36
λ		γ										
0.20		0.00	0.025	0.05	0.075	0.10	0.125	0.15	0.175	0.20	0.225	0.25
	T^2	150.40	100.89	61.93	38.44	23.72	15.17	9.88	6.80	4.87	3.59	2.80
	EWMA_R	87.68	38.30	19.33	11.99	8.45	6.47	5.19	4.39	3.77	3.36	3.01
	Kim_Shewhart	164.04	118.12	76.61	49.59	31.34	20.52	13.25	9.21	6.46	4.74	3.58
	NIST	162.68	121.89	85.64	57.16	37.70	25.59	16.95	11.78	8.42	6.22	4.60
	EWMA_3	73.54	31.03	16.39	10.50	7.56	5.94	4.81	4.11	3.56	3.19	2.89
λ		γ										
0.40		0.00	0.025	0.05	0.075	0.10	0.125	0.15	0.175	0.20	0.225	0.25
	T^2	79.94	50.43	30.90	19.69	12.54	8.64	6.03	4.28	3.26	2.61	2.07
	EWMA_R	25.52	14.48	9.75	7.15	5.68	4.69	4.03	3.53	3.16	2.87	2.63
	Kim_Shewhart	95.56	62.80	38.90	24.84	16.04	10.97	7.59	5.46	4.07	3.19	2.49
	NIST	103.43	72.36	49.88	34.01	22.72	15.71	11.19	8.01	5.88	4.55	3.49
	EWMA_3	21.64	12.76	8.79	6.53	5.24	4.38	3.78	3.34	3.00	2.75	2.52
λ		γ										

(continued)

Table 2 (continued)

λ		γ										
0.60		0.00	0.025	0.05	0.075	0.10	0.125	0.15	0.175	0.20	0.225	0.25
	T^2	38.71	24.89	16.23	10.65	7.41	5.28	3.87	2.95	2.34	1.95	1.63
	EWMA_R	11.18	8.01	6.23	5.09	4.31	3.75	3.33	3.00	2.73	2.52	2.33
	Kim_Shewhart	47.96	30.28	19.55	12.84	8.87	6.36	4.60	3.48	2.75	2.25	1.86
	NIST	61.12	42.12	29.31	20.57	14.34	10.14	7.45	5.52	4.29	3.30	2.64
	EWMA_3	9.97	7.33	5.75	4.73	4.02	3.53	3.15	2.85	2.60	2.42	2.26
λ		γ										
0.80		0.00	0.025	0.05	0.075	0.10	0.125	0.15	0.175	0.20	0.225	0.25
	T^2	19.58	13.19	9.04	6.31	4.61	3.41	2.68	2.17	1.77	1.55	1.36
	EWMA_R	6.87	5.51	4.60	3.97	3.50	3.14	2.84	2.62	2.42	2.25	2.13
	Kim_Shewhart	23.39	15.44	10.62	7.27	5.28	3.89	3.02	2.44	1.97	1.70	1.49
	NIST	34.72	25.27	18.08	13.27	9.44	6.83	5.16	4.02	3.13	2.57	2.11
	EWMA_3	6.30	5.08	4.28	3.74	3.29	2.97	2.70	2.50	2.31	2.18	2.06
λ		γ										
1.00		0.00	0.025	0.05	0.075	0.10	0.125	0.15	0.175	0.20	0.225	0.25
	T^2	10.46	7.48	5.29	4.03	3.04	2.43	2.01	1.70	1.47	1.32	1.22
	EWMA_R	4.91	4.20	3.67	3.27	2.95	2.70	2.50	2.33	2.19	2.08	1.99
	Kim_Shewhart	12.08	8.46	5.95	4.45	3.34	2.66	2.19	1.83	1.59	1.41	1.29
	NIST	21.13	15.58	11.34	8.49	6.23	4.88	3.75	3.00	2.47	2.04	1.75
	EWMA_3	4.58	3.92	3.45	3.09	2.79	2.56	2.39	2.24	2.12	2.01	1.92
λ		γ										

(continued)

Table 2 (continued)

λ		γ										
1.20		0.00	0.025	0.05	0.075	0.10	0.125	0.15	0.175	0.20	0.225	0.25
	T^2	6.15	4.62	3.51	2.75	2.22	1.84	1.58	1.40	1.28	1.18	1.12
	EWMA_R	3.85	3.44	3.09	2.81	2.58	2.40	2.25	2.12	2.02	1.95	1.86
	Kim_Shewhart	6.88	5.06	3.75	2.95	2.37	1.96	1.67	1.47	1.33	1.23	1.16
	NIST	13.01	9.93	7.52	5.87	4.47	3.51	2.82	2.34	1.97	1.69	1.50
	EWMA_3	3.62	3.23	2.91	2.66	2.45	2.29	2.16	2.04	1.95	1.88	1.79
λ		γ										
1.40		0.00	0.025	0.05	0.075	0.10	0.125	0.15	0.175	0.20	0.225	0.25
	T^2	3.85	3.00	2.42	1.99	1.71	1.50	1.35	1.23	1.15	1.10	1.06
	EWMA_R	3.19	2.91	2.67	2.46	2.31	2.19	2.08	1.99	1.91	1.83	1.75
	Kim_Shewhart	4.15	3.20	2.56	2.10	1.79	1.56	1.40	1.27	1.18	1.12	1.07
	NIST	8.42	6.68	5.07	4.02	3.28	2.69	2.25	1.88	1.65	1.46	1.33
	EWMA_3	3.00	2.75	2.53	2.35	2.21	2.10	2.00	1.91	1.83	1.76	1.66
λ		γ										
1.60		0	0.025	0.05	0.075	0.1	0.125	0.15	0.175	0.2	0.225	0.25
	T^2	2.6414	2.1927	1.8221	1.5861	1.4054	1.2838	1.1865	1.125	1.0769	1.0474	1.0328
	EWMA_R	2.7648	2.5615	2.3767	2.2329	2.1149	2.0202	1.9384	1.8716	1.7948	1.7129	1.6341
	Kim_Shewhart	2.8069	2.295	1.89	1.6365	1.4499	1.317	1.2156	1.1465	1.0916	1.0591	1.0406
	NIST	5.849	4.6027	3.7087	3.0284	2.4862	2.0936	1.8109	1.5939	1.422	1.2868	1.2069
	EWMA_3	2.6096	2.4301	2.2668	2.1349	2.0276	1.9432	1.8563	1.7846	1.6972	1.6065	1.5292
λ		γ										

(continued)

Table 2 (continued)

λ		γ										
1.80		0	0.025	0.05	0.075	0.1	0.125	0.15	0.175	0.2	0.225	0.25
	T^2	1.9742	1.6787	1.4843	1.3399	1.2392	1.1472	1.1009	1.0617	1.039	1.021	1.0107
	EWMA_R	2.4341	2.2749	2.1645	2.0645	1.9761	1.8945	1.8226	1.7498	1.6723	1.5841	1.4998
	Kim_Shewhart	2.0426	1.7229	1.5216	1.3604	1.2556	1.1631	1.1128	1.0735	1.0463	1.0249	1.0156
	NIST	4.0554	3.348	2.7677	2.3569	1.9786	1.7198	1.5342	1.3717	1.273	1.1876	1.1187
	EWMA_3	2.3196	2.1761	2.0704	1.9834	1.8956	1.8092	1.7324	1.6516	1.5635	1.4786	1.3874
λ		γ										
2.00		0	0.025	0.05	0.075	0.1	0.125	0.15	0.175	0.2	0.225	0.25
	T^2	1.5577	1.3986	1.278	1.1875	1.1189	1.0789	1.0516	1.0273	1.0182	1.009	1.006
	EWMA_R	2.201	2.0932	2.0092	1.9397	1.8555	1.7896	1.7041	1.6231	1.544	1.4468	1.3585
	Kim_Shewhart	1.5864	1.4243	1.2954	1.198	1.1266	1.0856	1.0563	1.0315	1.0207	1.0123	1.006
	NIST	2.9917	2.5095	2.1376	1.8797	1.6262	1.466	1.3396	1.2323	1.1624	1.1047	1.0716
	EWMA_3	2.1069	2.0069	1.9264	1.8523	1.7647	1.69	1.6012	1.5111	1.4312	1.3361	1.2539

Table 3 ARL values considering shifts in the deviation

Control chart δ	1	1.2	1.4	1.6	1.8	2	2.2	2.4	2.6	2.8	3
T^2	201.2792	40.1093	15.0198	7.8771	5.2044	3.6981	3.0499	2.5053	2.2124	1.9623	1.8228
EWMA_R	201.5284	39.2209	14.8149	7.6465	5.0182	3.6223	2.9238	2.4161	2.0701	1.871	1.7152
Kim_Shewhart	202.7605	42.8075	15.2457	7.5918	4.8458	3.3782	2.7044	2.2174	1.9322	1.7162	1.5735
NIST	201.4676	38.2156	13.926	7.0087	4.5089	3.1752	2.5822	2.1315	1.8524	1.671	1.5402
EWMA_3	197.0065	45.0036	18.3061	10.4036	7.2085	5.381	4.3989	3.6449	3.1824	2.8106	2.5575

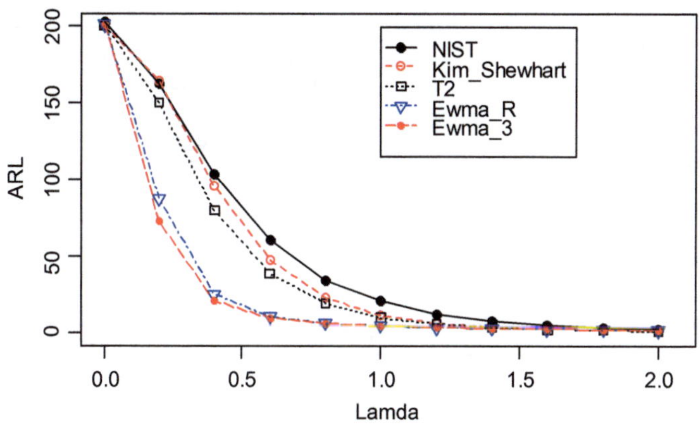

Fig. 1 ARL values for different shifts in the intercept using the NIST, Kim-Shewhart, T^2, EWMA/R, and EWMA3 procedures

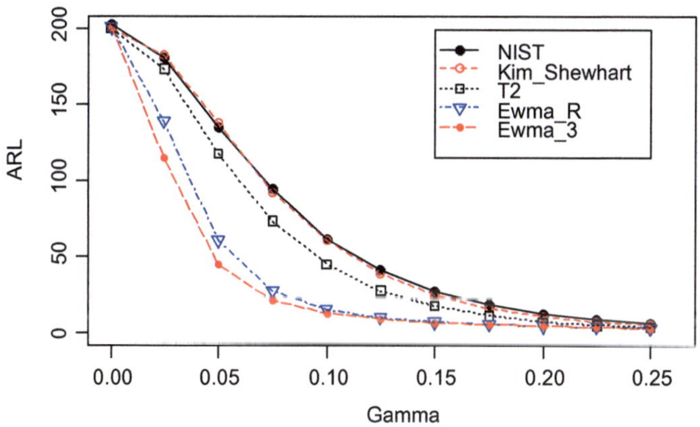

Fig. 2 ARL values for different shifts in the intercept using the NIST, Kim-Shewhart, T^2, EWMA/R, and EWMA3 procedures

slope. The T^2 chart shows a slight advantage over the previous charts, and EWMA-type charts quickly detect changes in the slope. In a global comparison, the EWMA3 chart is the best, as it detects changes in the intercept and slope rapidly. From Table 2, when considering simultaneous shifts of the intercept and slope, we can conclude that for small changes in the intercept, regardless of shifts in the slope, the EWMA-type charts perform best. However, if the displacement is large (greater than or equal to 1.6), the vast majority of charts show similar performance behaviors regardless of the shifts in the slope. In Fig. 3, we present the ARL behavior considering shifts in the standard deviation, shown in Table 3 of the Appendix, where we observe that the four charts behave similarly.

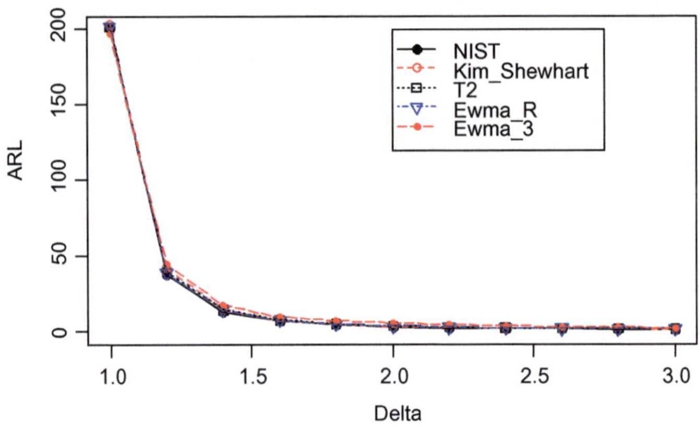

Fig. 3 ARL values for different shifts in the deviation using the NIST, Kim-Shewhart, T^2, EWMA/R, and EWMA3 procedures

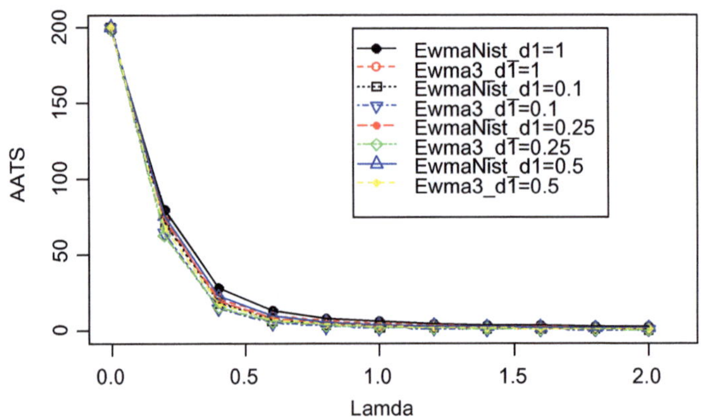

Fig. 4 AATS values for different shifts in the intercept: VSI-EWMA/NIST and VSI-EWMA3 procedures with varying sample interval

4 NIST Method Modifications

We make the first modification to the NIST chart, initially including the EWMA procedure, to increase the ability to detect changes in the parameters of the linear calibration line. This is how we obtain the EWMA/NIST chart. Subsequently, we make a further modification to this chart that consists in considering a varying sample interval and then obtaining the VSI-EWMA/NIST chart, which has an advantage over the previous chart.

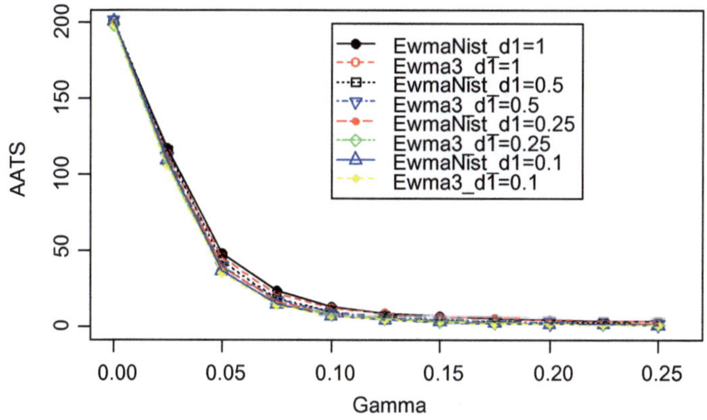

Fig. 5 AATS values for different shifts in the slope: VSI-EWMA/NIST and VSI-EWMA3 procedures, with varying sample interval

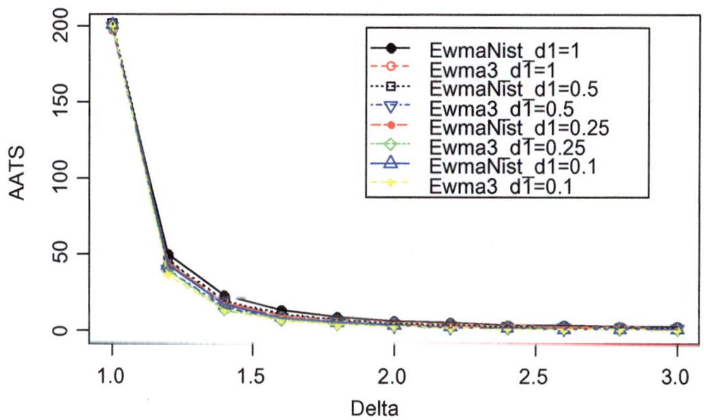

Fig. 6 AATS values for different shifts in the deviation: VSI-EWMA/NIST and VSI-EWMA3 procedures with varying sample interval

4.1 EWMA/NIST Method

In current literature, the NIST method (Croarkin and Varner 1982) is compared with EWMA-type methods, with the latter showing better performance. However, the NIST method was introduced two decades before the EWMA-type methods (EWMA3 and EWMA/R), so it is not surprising that the latter are superior, especially considering that they detect small sharp changes more rapidly than Shewhart-type charts. Our proposal consists of an EWMA from the NIST chart to accelerate the detection of changes in the linear calibration model. In this chart, which we denote EWMA/NIST, we monitor the corrected measured values Z_{ij}, defined in (2), using the EWMA statistic, which is denoted by $W_i(j)$, $i = 1, 2, \ldots, n$, $j = 1, 2, \ldots, m$, and given by

$$W_i(j) = \theta Z_{ij} + (1 - \theta) W_i(j - 1), \tag{13}$$

With $W_i(0) = 0$. As with all charts, θ is a smoothing constant whose value determines the magnitude of the changes it detects. In other words, a signal that the process is out of control is detected as soon as $W_i(j)$ is less than the LCL or greater than UCL where

$$\text{LCL} = -L\frac{\sigma}{\beta_1}\sqrt{\frac{\theta}{(2-\theta)}} \text{ and } \text{UCL} = L\frac{\sigma}{\beta_1}\sqrt{\frac{\theta}{(2-\theta)}} \tag{14}$$

The constant L is chosen such that the ARL in control has a specific value.

4.2 VSI-EWMA/NIST Method

In order to increase the detectability of the EWMA/NIST chart, we make an additional modification which consists in considering a varying sample interval, thus obtaining a NIST chart of EWMA type with a varying sample interval, that is, the VSI-EWMA/NIST chart. In this chart, the sample interval is taken in variable form, with a greater length if the point on the EWMA-type chart falls close to the object (target) value and with a shorter length if the point falls near the control limits, since, in this case, we assume that this means that the process has changed.

Here, l_1 and l_2 are the lengths of the sample intervals, where $0 < l_1 < l_2$. When we take a sample and calculate the EWMA statistic corresponding to $W_i(j)$, the following sample is taken after an interval of time determined by the following function of the sample interval:

$$l = \left\{ \begin{array}{l} l_1 \text{ if } W_i(j) \in R_1 \\ l_2 \text{ if } W_i(j) \in R_2 \end{array} \right\}.$$

The region corresponding to the state control given by (LCL, UCL) is subdivided into two regions, R_1 and R_2, where $R_1 = (-\text{LCL}, -w) \cup (w, \text{UCL})$ and $R_2 = (-w, w)$. When the calibration process starts, it is recommended that the first sample is taken with a small sample interval l_1.

In the case where we compare three magnitudes in the calibration process, which is the most common metrology case, we can base the monitoring on the statistic $\Delta(j)$, which is defined by

$$\Delta(j) = \max\{\text{SE}_1(j), \text{SE}_2(j), \text{SE}_3(j)\},$$

where $\mathrm{SE}_i(j) = \frac{|W_i(j)|}{L\frac{\sigma}{\beta_1}\sqrt{\frac{\theta}{(2-\theta)}}}$.

According to the statistic $\Delta(j)$, the control chart VSI-EWMA/NIST gives a signal that the process is out of control when $\Delta(j) > 1$. The value that this statistic takes also indicates the interval size that should be selected in the next step. The interval function can be defined as follows:

$$l = \left\{\begin{matrix} l_1 \text{ if } w < \Delta(j) < 1 \\ l_2 \text{ if } \Delta(j) < w \end{matrix}\right\}.$$

The conditional probability $p_0 = P(\Delta(j)\langle w|\Delta(j) < 1)$ allows us to determine the value of w. A large value of p_0 indicates that the value of w is close to 1 and thus the number of times that a large sample interval, l_2, is chosen is greater. As the statistical distribution of $\Delta(j)$ is difficult to determine, we obtain the relationship between p_0 and w using Monte Carlo simulation. Li and Wang (2010) recommend using a p_0 value of about 0.8. To make the monitoring schemes with variable and fixed sampling intervals, we use the same sample rate or expected sample time, that is, $(1 - p_0)l_1 + p_0 l_2 = l_0$, where l_0 is the fixed sample length.

In the charts that use a sample interval of constant length, the ARL is used to evaluate the effectiveness of different control schemes. In the case of introducing a sample interval of varying length in a control scheme, the time that elapses until we find a signal that is out of control ceases to be a multiple of the ARL and therefore the ARL is no longer appropriate to assess the efficiency of the control chart with intervals of varying length. In this case, the performance indicators most commonly used are the average time until the signal occurs (ATS), which is defined as the expected value of the time that passes from the beginning of the process until the chart shows a signal that is out of control, and the average adjusted time to the signal (AATS), which is defined as the expected value of time from the occurrence of an assignable cause until the chart indicates a signal that is out of control. The indicator AATS is also known as the steady-state ATS (SSATS).

4.3 Comparisons

Now, we compare the EWMA/NIST chart with the EWMA3 chart, which was the one that had the best performance in the previous comparisons. In Fig. 3, we see that both charts have similar performance, with the EWMA3 chart having a slight advantage.

5 Conclusions

The monitoring of measuring-instrument calibration processes is of great importance because it makes it possible to verify that accuracy and precision remain stable and under statistical control. Often calibration processes are modeled by linear profiles resulting from the functional relationship between measurements of the instrument that is calibrated and the accepted values of a reference standard used in the calibration. We propose and evaluate two modifications to the NIST chart that may be used for the monitoring of calibration processes. The first modification involves incorporating the EWMA scheme for statistical monitoring in the Croarkin and Varner chart. This modification results in the EWMA/NIST chart, whose performance is comparable with that of the EWMA3 chart, which is identified in the literature as having the best performance. The second modification is based on the consideration of varying sample intervals in the EWMA/NIST chart, which results in slight advantages for the performance of this chart under parameter changes. From a practical point of view, it is very useful to consider varying sample lengths as in this way, we can reduce costs in terms of time and money, because when we have information that the measuring instrument is within the limits of accuracy, we can extend the time period between successive calibrations.

Through a simulation study, we evaluate known methods found in the literature (NIST, T^2, Kim-Shewhart, EWMA/R, and EWMA3) plus a method that we propose (EWMA/NIST) for monitoring the calibration process. We then perform the comparison in terms of the average run length (ARL) and find that, while the NIST chart takes longest to detect changes in the intercept and slope, the performance of the modified EWMA/NIST chart is similar to the best-performing charts, EWMA3 and EWMA/R. In the evaluation of the VSI-EWMA/NIST and VSI-EWMA3 charts, which consider a varying sample interval, we find a small advantage over the cases of fixed sample length.

The modifications made to the NIST chart have a very important advantage, since the modified versions detect changes in the calibration process faster than the original version (NIST), as shown in Figs. 7, 8 and 9, furthermore allowing a reduction of calibration costs by using varying sample intervals. In Figs. 7 and 8, we observe that the modified versions of the NIST chart clearly show a better performance than the original NIST chart, whereas in Fig. 9, we notice that these performances are equally good in all versions. This is a general performance that we see in Figs. 4, 5 and 6, and therefore the decision as to the best chart is based on the shifts of the intercept and slope, as well as the deviation.

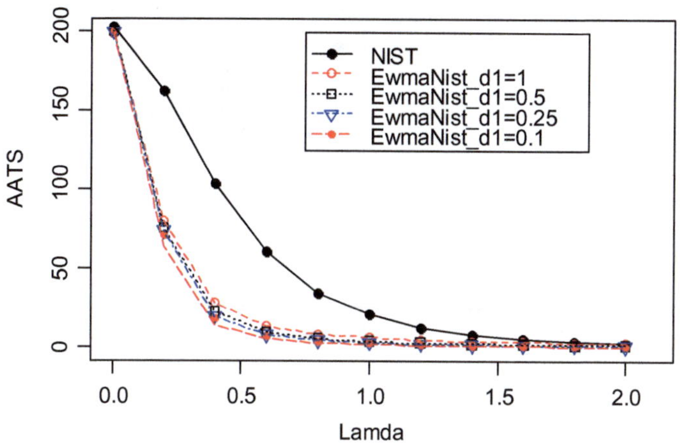

Fig. 7 AATS values for different shifts in the intercept: VSI-EWMA/NIST and VSI-EWMA3 procedures with varying sample interval

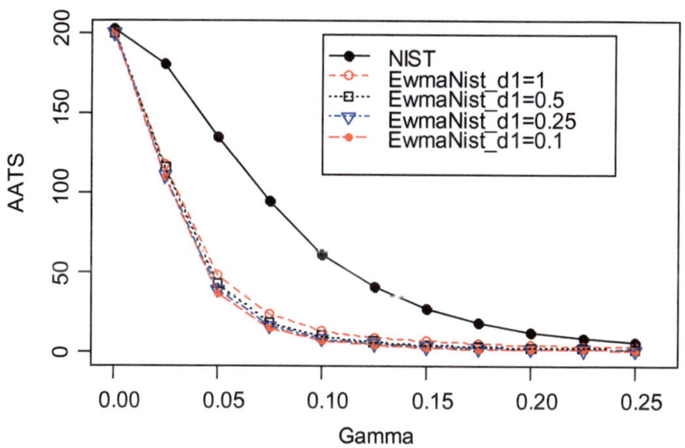

Fig. 8 AATS values for different shifts in the slope: VSI-EWMA/NIST and VSI-EWMA3 procedures with varying sample interval

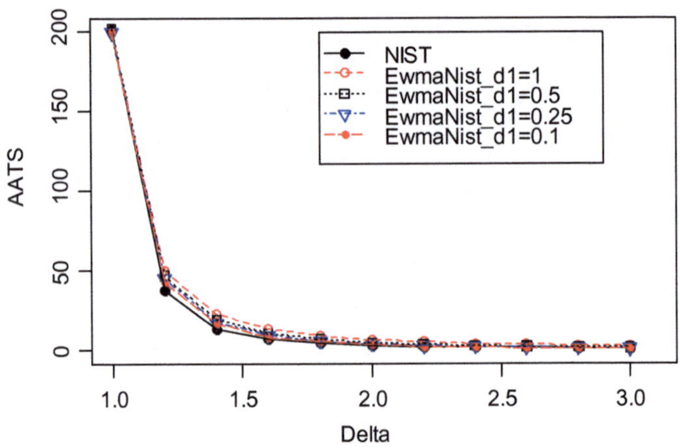

Fig. 9 AATS values for different shifts in the deviation: VSI-EWMA/NIST and VSI-EWMA3 procedures with varying sample interval

Acknowledgements This research was partially supported by the Universidad Autónoma de Sinaloa through the project "Cartas de Control en Muestras Pequeñas" (PROFAPI2015/252).

References

Croarkin C, Beers JS, Tucker C (1979) Measurement assurance for gage blocks. NBS Monograph 163

Croarkin C, Varner R (1982) Measurement assurance for dimensional measurements on integrated-circuit photomasks. NBS Technical Note 1164, US Department of Commerce, Washington DC, USA

Crowder SV, Hamilton MD (1992) An EWMA for monitoring a process standard deviation. J Qual Technol 24:12–21

Gupta S, Montgomery DC, Woodall WH (2006) Performance evaluation of two methods for online monitoring of linear calibration profile. Int J Prod Res 44:1927–1942

ISO 5725-6 (1994) Accurate (trueness and precision) of measurement methods and results—Part 6. International Organization for Standardization: Geneva, Switzerland

Kang L, Albin SL (2000) On-line monitoring when the process yields a linear profile. J Qual Technol 32:418–426

Kim K, Mahmoud MA, Woodall WH (2003) On the monitoring of linear profiles. J Qual Technol 5:317–328

Li Z, Wang Z (2010) An exponentially weighted moving average scheme with variable sampling intervals for monitoring linear profiles. Comput Ind Eng 59:630–637

Mestek O, Pavlik J, Suchánek M (1994) Multivariate control charts: control charts for calibration curves. Fresenius J Anal Chem 350:344–351

Montgomery DC, Peck EA, Vinning GG (2001) Introduction to linear regression analysis. Wiley, New Jersey

NIST/SEMATECH (2011) e-Handbook of Statistical Methods (2011). http://www.itl.nist.gov/div898/handbook/mpc/section3/mpc37.htmDecember
Shewhart WA (1931) Control of quality for manufactured product. Van Nostrand, New York

Application of Constraint Theory (TOC) on Information and Communication Technologies in Quality and Its Impact on the Circular Economy

Manuel Horna and Mario Chong

Abstract The objective of this research is to apply the theory of restrictions (TOC) in the entire business value chain of a bank financial services company, from the highest level (strategy) to the primary activities of the company. To carry out the study, the process of developing products and services (online payments) is chosen in a first stage. Later, the study should be extended to other processes in the chain. For the analysis of payments online, a redesign of the application program interface (API) is carried out, through the application of TOC, to improve the quality of the information and communications technologies (ICT) that supports the payment of telephone services (critical process bottleneck). In this sense, the TOC model is proposed considering an analysis of the problem (design thinking), analysis related to the business processes of the payment line using lean-VSM (virtual stream mapping). Then we apply TOC's Drum Buffer Rope (DBR) method to define time and capacity buffers on the pay line. Likewise, to conclude, we deigned a process to measure the total improved performance of the supply chain. As a future design, the paper recommends developing a new business model by designing a new reusable service based on the service obtained by applying TOC in the entire online payment process.

Keywords TOC · TOC-DBR · ICT quality · Pull supermarkets · Lean-TOC · Pull Kanban

1 Introduction

Software applications based on theory of constraints (TOC) can be considered very important tools for effective production planning, scheduling, and identifying constraints. This theory was mentioned by Ferenčiková (2012). Different methods and software applications have also been developed for the process of planning in the last decade, and most of them use the principles of the theory of constraints (TOC) in the development of algorithms for this process. In addition, Goldratt (2012) maintains

M. Horna · M. Chong (✉)
Universidad del Pacífico, Jr. Sánchez Cerro N° 2141, Jesús María. 15072, Lima, Perú
e-mail: m.chong@up.edu.pe

J. L. García Alcaraz et al. (eds.), *Techniques, Tools and Methodologies Applied to Quality Assurance in Manufacturing*,
https://doi.org/10.1007/978-3-030-69314-5_14

that the systems are normally very complex. These systems must become the sum of several sub-systems with a synchronized flow. The flow must be linear to avoid multitasking processes, what they achieve is to become quality assassins multitasking causes the quality ratios to decrease, the lean time to increase, and the capacity used to increase. On the other hand, 90% of the constraints are not exploited considering the philosophy of the constraint theory. This theory of constraints is a business philosophy that was introduced by Goldratt et al. (Goldratt and Cox 2012). The main idea is that any system, not just a production system, is always limited by at least one restriction, which must be removed to improve the overall process. According to Wang et al. (2014), companies must often deal with multiple restrictions, and the only way to solve them is using sophisticated software applications. In addition, restriction is commonly called the bottleneck (Ferenčíková 2012). Ferenčiková identified the bottleneck or Drum, which comes from the Drum Buffer Rope (DBR) method, because it controls the performance of the entire production or service system. Bottleneck is a resource or process with limited capacity within a system. When optimizing the system, it is important to consider that any loss of time in unrestricted activity does not cause a delay in the total system. Conversely, every minute lost in a restricted activity means a loss in the same amount of time in the total system.

Multiple constraints produce large bottlenecks. This theory was mentioned by Yang et al. (2016) in their paper about the multi-machine maintenance process. On the other hand, TOC has been extended to service scheduling problems, in environments with multiple restrictions. In a follow-up study, we can involve multiple objectives that point out programming problems in the service system. For example, since maximizing customer satisfaction is the highest goal for service companies and may also reflect the essence of scheduling issues, it is beneficial for us to obtain superior scheduling plans with respect to maximizing goals. Achieving the maximum goals impacts the process to maximize the level of customer service, mentioned by Marin (2013). An analysis of the entire supply chain is also required; for that, the theory of constraints is based on the DBM method—Drum Buffer Rope, referring to the synchronization of the entire supply chain. Some other indicators resulting from this synchronization are the reduction of the costs of inventories of raw material, end products, increase in the productivity of the company, and improvement of the fulfillment of the production plans.

To measure the application of the constraint theory, TOC defines a series of indicators that allow its application to be measured throughout the company's supply chain. To this end, some authors (Phruksaphanrat et al. 2011) mention that according to the TOC philosophy, three measures are defined; these should be used to evaluate the performance of any company: performance (TP) = throughput, which is the ratio by which the system generates money through sales. Inventory (IN) is the total amount that the system invests in the purchase of items that the system wants to sell. Operating expenses (OE) are the total amount that the system uses to transform inventory into performance. Yield is calculated by subtracting sales minus the cost of the raw material. The inventory includes products in process, end products, and raw materials. In addition, it includes tools, buildings, capital investment in equipment, and furniture. Operating expenses include direct costs, indirect labor, supplies, external contractors, and interest payments and net income (NP = TP − OE), return on investment (ROI = NP/IN), productivity (PR = NP/OE), and inventory turnover (IT = TP/IN).

Phruksaphanrat et al. (2011) defines the circular economy as a business model that promotes the adoption of closed-loop patterns in order to optimize the use of virgin resources (energy, water…), thus promoting human well-being. The final goal of promoting the circular economy is to decouple environmental pressure from economic growth. In this way, Esposito et al. (2018) mention, in their Introduction to Circular Economy article, the following: "The adoption of the Circular Economy Model requires companies to initiate and develop intensive technologies and business models based on longevity, reuse, repair, upgrade, renovation, capacity sharing, and dematerialization". This means rethinking products and services, as well as creating organizational designs that facilitate the adoption of the circular model through the design of new business models.

In this regard, Weetman (2016) mentions that the circular economy will transform traditional business models and the challenge is how to design and operate each supply network for a circular business model. Focus on green and non-green products, according to Tsan-Ming Choi (2020), permits examining the circular supply chain with an emphasis on optimum product line design, with a special focus on the quality level of the system in the circular economy; information technologies play an important role, considering scientific material and technologies for product design. According to Bruel et al. (2019), the circular economy (CE) is a new model for the production and consumption of goods, which has attracted wide political attention as a strategy toward sustainability. Likewise, a proposal of a collaborative circular economy, by Alexandris et al. (2018), is materialized through the transition of assets between asset operators with a demand-based approach. The demand itself is partially based on the state of the asset, which is described by its circular properties (location condition, availability). This common view of the state of assets among all parties can be enabled by Block Chain and Smart Contracts, which can provide the underlying technology for sharing data with integrity, while offering more efficient interoperability among the participants. A closed-loop supply chain (Daniel and Guide 2009) must be collaborated to integrate all actors of the supply chain, considering the customer requirements about re-manufactured products or services and the internal benefits for companies.

2 Company Strategy Under Study

2.1 Company Description

According to the strategic plan of Banco de la Nacion (2017), this company provides services to state entities, promotes banking and financial inclusion for the benefit of

Table 1 *Source* Planning and development management (value chain)

Secondary activities	Institutional planning	Picture and communication	
	Organizational development	Internal control	
	Budget management	Cash and value management	Documentary management
	Accounting management	Information technology management	Recovery management
	Human resources management	Financial management	Legal support
	Logistic management	Comprehensive security management	
Primary activities	Product and service development	Services	
	Commercial management	Catchments, placements, and investments	Customer service

citizens, complementing the private sector and promoting decentralized growth in the country through efficient and self-sustainable management.

The company's vision is to be recognized as a strategic partnership by the Peruvian State for the provision of quality and innovative financial services, within a management framework based on good corporate governance and human talent management practices. The citizen-oriented services offered by the financial institution are as follows: (1) payment process, (2) check process, (3) transfer process, (4) tele draft/money order process, and other services such as certifications of payment, bank certificates, and collections abroad. The company currently has an entire technological platform that supports the supply chain, which would be the subject of analysis in this work. The bank promotes in its strategic plan 2017–2021, its value chain by identifying its processes, sub-processes, and activities that promote the increase in value to satisfy the needs and the client's requirements. The chain structure explains the logic of the set of primary activities from product development to service deliverys and due to control and monitoring. In Table 1, the value chain is represented.

2.2 *Description of the Sector*

Table 2 shows the evolution of the inclusion indicators of the financial system. Clearly, there is a tendency to increase the number of transactions in the different channels, considering a population of 100,000 adult inhabitants.

Table 2 *Source* Evolution of financial inclusion indicators of the financial system

Access to financial services	2010	2011	2012	2013	2014	2015
Number of offices nationwide	3173	3440	3816	4130	4324	4426
Number of ATMs	5508	6530	8162	8896	12,226	28,976
Number of correspondent ATMs	14,840	17,501	27,906	38,311	54,044	77,857
Number of service points per 100,000 inhabitants	129	148	211	268	362	562
Number of attention channels	15	19	26	34	48	77
Total number of service points	23,821	27,471	39,884	51,347	70,594	109,259

2.3 Strategic Planning

According to strategic plan of Banco de la Nacion (2017), the following processes are included in the macro-process of primary activities: "product and service development" and "services" macro-process. Active products (loans and credit card lines) are oriented to the public sector. Passive products (savings) are oriented to the payment of assets and sectors where the bank is the unique banking offer (UOB). A resume of products and services offered by the company is explained later.

According to Weinberger (2009), the bank follows a differentiation strategy because it makes an in-depth study of its clients' preferences. He constantly coordinates the functions of marketing and research and development of new products or services.

A description of products offered by the bank is the following: (1) multi-red, (2) purchase of credit card debt, studies, (3) bank accounts, (4) saving Accounts, (5) current accounts drawdowns, (6) UOB term deposit accounts, (7) mortgage credits, (8) home purchase, improvement, expansion, and (9) remodeling.

The bank services offerings are the following: (1) payments, tax payment, multi-express payment, card payment, payment of telephone services, light, (2) checks, payment of certified checks, exchange, management checks, and (3) transfers, additional services, and bank draft.

For the initial analysis of the macro-business process, development of products and services, we would focus on the payment process. The online payment process is analyzed using TOC, constraint theory with a global vision of the chain, and then a focus to analyze and eliminate all supply chain restrictions.

2.4 Description of the Supply Chain

This research is related to the main objectives set by the financial banking services company specified in Table 3.

Through the application of TOC, theory of constraints in ICT, information communication technology, and throughout the supply chain, we would be improving the

Table 3 Strategic objectives

Number	Objective 6—improves the use of ICT, information and communication technologies to support the redesigning of internal processes
6.1	Implementing a new model for managing the demand of technological products
6.2	Redesigning the macro-processes
6.3	Formulate and implement a digital transformation plan
Number	Objective 8—strengthening the management of corporate social responsibility
8.1	Support for development management of interest groups
8.2	Financing social responsibility actions
8.3	Synergies in social projects of interest group

Source Strategic plan 2017–2021 own elaboration

quality online services of all the products and services offered by the bank (Objectives 6.1, 6.2). A future digital transformation could be accomplished by implementing a distributed Block Chain system (Objective 6.3). Furthermore, by creating the new circular economy business model, we would be complying with Goal 8.

3 Literature Review

The theories that support the development of the research paper are described, especially the development of the theory of constraints with all the methodology and how its application is. Then, the topic of information and communications systems, the theories on circular economy, and especially the development of new business models are shown.

3.1 Theory of Constraints (TOC)

The methodology for modeling supply chain strategies and restrictions was presented in the book "The Goal," by Goldratt (Goldratt and Cox 2012). They define in their book the following steps to eliminate restrictions or bottlenecks: (a) Identify the (s) system restriction, (b) decide how to exploit the (s) system restriction, (c) subordinate everything else to the decision of the previous step, (d) overcome the restriction of the (s) system, (e) warning! If a restriction has been broken in the previous steps, return to the first step, but do not allow inertia to be the cause of restrictions in the system.

TOC is based on verifying each system or complex environment; it is based on an inherent simplicity and the best way to manage; thus, control and improve the system is to take advantage of its simplicity. That is why the constraints are the leverage points, the support points, and that is why they are so powerful. What should be kept in mind is that this type of approach is a major paradigm shift. People want to do everything before they change their paradigm. From these observations, it can be said that applying this theory, the following requirements must be gotten. First is that there is real pressure for improvement, but by itself this is not enough. The second condition is that it seems obvious to them that there is no solution within the existing paradigm. In other words, they have already tried everything. And the third condition is that there is something to help them to take the first step, for example, the TC method, a simulator or a consultant. Some concepts that are proposed for the application of TOC are these: DBR. Through the theory of constraints (TOC), we know that there are certain processes with restricted capacity, which sets the time of production. The DBR method recognizes this restriction and proposes a production planning system that seeks to reduce control time in the scheduling of operations and avoids the transmission of fluctuations in the process.

Likewise, Marín Marín (2013) defines DBR as buffers of inventories in process and also defines it as an accounting tool for quantifying the impact of the implementation of the TOC Model, on throughput. Goldratt (2012) mentions that DBR consists of three elements: (1) Drum , this element represents the production time dictated by the process restriction. The other processes must respect this rhythm to avoid creating bottlenecks or gaps that harm the development of the established plan. It should be noted that the rhythm dictated by the "drum" corresponds to the time planned for the maximum exploitation of the resource restriction, so it cannot be altered. (2) Buffer, for doing each process, there are certain time frames. Sometimes these can have small variations, also known as fluctuations. Buffers are used to prevent these small lags from affecting the rhythm determined by the process restriction. These are calculated as measures of time. The objective is to avoid that no circumstance must stop the functions of the process restriction. (3) ROPE (Rope), in the DBR method, the rope represents the material or input release program, also called "start of operations." The speed with which the materials are released must be aligned with the rhythm of the processes, determined by the process restriction.

In this sense, Antić (2015) mentions some adjusted concept techniques (lean) to overcome business constraints and describes the similarities between TOC + lean theories. These are the following: (a) observe the company as a comprehensive system including continuous improvement, (b) the business objective is to increase profits and is defined by the customer, (c) the flow of value—FLOW—is a higher concept than that of production (VSM), (d) quality is the key to success, (e) small-scale batch production, (f) the production system based on the order system (make to order), not for the warehouse, (g) minimization of inventories, and (h) participation of all employees in the success of the company.

3.2 Contribution from Lean to TOC

The contribution from lean to TOC is explained by Antić (2015), through the following: (a) It creates an adequate base that clearly indicates the existence of any restriction. By using the value stream map (VSM), administrators will have a clear picture of the performance of the business process and the places where constraints occur.

In this way, they will be able to respond quickly to existing restrictions. They provide a basis for exploiting restrictions using standard procedures and the participation of all employees in the business improvement process. Based on working conditions and the application of best work practices, it guarantees a better use of available resources. (b) They avoid the unnecessary waste of resources by establishing a flow of inputs and the synchronization of labor, machines, and technologies. They provide a basis for increasing business productivity within the limits of existing constraints. Increased productivity implies an increase in the production process and all supporting activities. To apply these concepts successfully, it is necessary to start from the strategic level and promote the lower levels of the organization to accept the basic principles of these·concepts. In fact, the application of these concepts multiplies better results. One of the visual tools to support process improvement is the value stream management (VSM). Finally, the improvement continues an ever better and more Pull flow. It has a sequence integrated by the four basic principles of lean management: VALUE → VALUE FLOW → ACTIVITY FLOW → PULL.

3.3 *Value Flow Map (VSM)*

According to Arbós (2010), the VSM was developed by Toyota, which called it "Material and Information Flow Mapping." With this, it represents, in a very visual way, the current situation and the ideal to be achieved. It includes the sequence of all business processes, the flow of inputs and products, and the flow of information. The entire representation considers the entire flow, from provisioning to the customer. The VSM representation follows some rules and uses specific symbols. Each operation is accompanied by the following data: (1) process orders, which can take the form of cards, Kanban product, production lot (volume), operations of the process to be carried out, (2) transfer batches (which can be Kanban cards), (3) cycle time of each unit of the product, (4) cycle time of each unit of the product, (5) time allocated to manipulations (loading, unloading, transportation), (6) machine preparation time, (7) quality control time per unit of the product, (8) rate or percentage of the product under test, (9) level of quality defects present in the lot, (10) number of workers for the operation, (11) number of machines for the operation, and (12) resulting UPTIME (percentage of the actual work time used to obtain a complete correct product). VSM Kanban follows the techniques for the implied flow of Pull supermarkets and FIFO flow. In Kanban level, it ensures that the flow of activities focuses on what the customer is really demanding. (1) As far as it refers to the flow, it will not always be maintained without interruptions.

There will be points of the flow where it is necessary to maintain stock, which avoids interruptions in it, which may affect the flow downstream (preparation times, quality problems, maintenance, or others). (2) The Pull operation, for only producing what the customer requests (MAKE TO ORDER), to the extent that it is requested, and when it is requested, it may be done from a certain point in the flow, where a total time is assumed until the delivery of the end product, less than the delivery time. (3) The alternative of the Pull flow through supermarkets, which is also included in the TOC philosophy, the necessary stock can be made up of a set of product units arranged in a specific and invariable order, such as FIFO (FIRST INPUT FIRST OUTPUT). The production flow occurs with a sequence of product units already established. The FIFO continues to assume an interleaved stock, the same as the supermarket, allowing the flow to be fully maintained, allowing the effects of flow interruptions to be absorbed. (4) In leveling, the product flow and the Pull operation are integrated continuously.

Regarding the Pull Kanban model, Bozer (Bozer and Ciemnoczolowski 2012) performs a mathematical simulation to determine the number of containers needed in each workstation, according to the valid Pull equilibrium parameters and lean Kanban methodology. Based on the results, a lower limit is also established on the number of Kanban necessary for each type of piece, and the relationship between the consumption rate of containers and the requirement of the workstation is shown. Consumption of a container is also assumed to be constant. For future research, it is important to quantify the relative contributions of the Kanban number and the production capacity of each workstation.

3.4 Aggregate Planning in a Supply Chain

It is important to mention that according to the Guide's explanation (Daniel and Guide 2009) in his paper "Aggregate Production Planning," he considers the variable production capacity and its application in the theory of constraints to detect bottlenecks. The study is carried out for companies in the electricity sector. In addition, the paper aims to determine the function of income, costs, or benefits. For this effect, these variables are selected to be the objective function of PPP problems (aggregate planning). Among these objective functions, the profit function is the most important. Therefore, the net profit (NP) is used as the objective function of the proposed PPP model (linear programming). Net profit (NP) = yield (TP) and operating expense (OE). Two measures of OCD are included in the benefit function. The remaining measure is inventory, which also needs to be considered. In relation to these measures, below are some ratios defined by the author: (a) yield (TP), the rate at which the system generates money through sales, (b) inventory (IN), the total money invested in the purchase of assets, (c) operating expenses (OE), the total money that the system spends to transform the inventory into performance (throughput), (d) net profit = TP − OE, and (e) return on investment (RI) = NP/IN, productivity (PR) = NP/OE, inventory rotation (IT) = TP/IN. The paper proposes for the aggregate production planning process (APP) an APP model with diffuse demand and a variable production capacity of the system.

Regarding the capacity variable, Mendi (Chopra and Meindl 2013) mentions that the planner must trade among capacity costs, inventory, and backlogs. There are three different aggregate planning strategies to balance these costs. These strategies involve bartering between capital investment, size of the workforce, hours of work inventory, and lost/pending sales. The strategies are the following:

Pursuit strategy uses ability as the lever. With this strategy, the production rate is synchronized with the demand rate.

Flexibility strategy manages utilization as a lever. This strategy can be used if there is excess machine capacity, and the workforce shows flexibility in programming.

Level strategy uses inventory as a lever. With this strategy, machine capacity and workforce are kept stable with a constant production rate. The downside is that you can accumulate large inventories and customer orders are delayed. This strategy keeps the capacity and costs of changing capacity relatively low.

Mendi (Chopra and Meindl 2013) mentions the information required to solve aggregate planning problems: (1) forecast of aggregate demand for each period, (2) production costs, (3) labor costs, (4) cost of outsource production, (5) cost of changing capacity, (6) hours of labor/machine required per unit, (7) cost of maintaining inventory, (8) cost of stock out, (9) restrictions, (10) limits on overtime, (11) limits on available capital, (12) limits on shortages, and (13) supplier restrictions for the company. Considering this study of the analysis of ratios for evaluating the effectiveness of the TOC model to be implemented in the supply chain of the bank financial services company, then, we carried out an analysis to study the contribution that the ABC-TOC analysis could give to the study.

3.5 Management Accounting Based on Lean-Just in Time

Activity-based costing (ABC) methodology proposed by Lea and Min (2003) assumes that activities consume resources and products, consume activities contributing to determine the cost of production through an activity report that details the activities and the quantity of each activity consumed in the manufacture of a specific product.

In relation to the theory of constraints (TOC), the author compares the ABC-TOC methodologies as follows: JIT-ABC results with a high profit, better service level, and lower inventories in process. TOC places the inventory of the buffer in front of the bottlenecks and at the intersection of routes, not existing in the bottleneck. ABC can respond better to short-term fluctuations and leads to better short-term performance.

Instead, TOC is not only an alternative technique to determine the cost of production based on marginal cost; according to Moisello (2012), it is a simple and different way of interpreting product cost management. On the other hand, ABC costing is not only a tool to calculate the cost, but also it is a model of reading processes that provides the basis for the control of the determinants of costs and activities. The distinction between a short and a long term, moreover, is not immediate because it is difficult to label a decision in the short or long term when a change implemented in the short term could have implications for long-term profitability. From the point of view of the integration of ABC and TOC, the partial use of the methods characterizes the proposed solutions. When it comes to large companies, with a high production, a company is likely to choose ABC instead of TOC, since the former makes a large part of the costs visible to allow control; the choice is likely to be a TOC-ABC combination when manufacturing of asset-specific products is also limited. In this case, a hybrid TOC-ABC approach offers the highest results in terms of profitability and TOC independently the lowest. When identifying the restriction of the system by means of TOC, the next step is its exploitation, for which it is determined by means of the damper, the time, and the capacity of the process, so that the input data re-supply is synchronized at the entrance and exit of the process.

3.6 TOC-SCRS Supply Chain Replenishment System

In order to balance the capacity measure of each node in the supply chain, effective inventory replenishments methods depend on the flow of sales at a plant; it was mentioned by Jiang (Jiang and Wu 2013). He defines TOC-SCRS as an effective control replenishment in a center or central warehouse with limitations of capacity. The frequency of replenishment is one of the important decision parameters and depends on the flow of sales at this plant. The proposed approach can effectively help managers determine the final optimal setup frequency for each product and facilitate production scheduling. Future work will focus on the implementation of new multi-object search algorithms in system optimization to improve system performance. Additionally, a dynamic TOC-SCRS will be built to make it more efficient and reliable. The regular or moving average method will be used to monitor market changes so that the system of different products is easier to adapt. TOC-SCRS is also defined as the constraint theory (TOC) supply chain replenishment system.

In this regard, in a research, Spencer (2000) mentioned that theory of constraints (TOC) is also applicable to the service sector, and that there is also utility in problem-solving methods used in conjunction with logistics aspects and performance measurements. The TOC literature exposes the balanced line problem and solves it by applying the programming technique called Drum Buffer Rope.

3.7 Circular Economy

Circular economy is defined as a model of production and consumption of goods that promotes the adoption of closed-loop patterns in order to optimize the use of virgin resources; thus, it acts reducing pollution and waste, guaranteeing at the same time the functioning of ecosystems and promoting human well-being. This theory is mentioned by Bruel et al. (2019). The final goal of promoting the circular economy is to decouple environmental pressure from economic growth. On the other hand, Esposito (Esposito et al. 2018) explains that we have been living in a linear economy since the industrial revolution. Our consumer and "single use" lifestyles have made a "taking, doing and disposing world." This refers to a one-way production model: Natural resources provide our factory inputs, which are then the use to create mass-produced goods to be purchased and generally disposed of after single use.

This linear economy model of mass production and mass consumption is testing the physical limits of the world. Unsustainable and a shift toward a circular economy is becoming unavoidable. According to Weetman (2016), one of the challenges in the design of the business models in the circular economy is to have a systematic and clear vision, since these models will be potentially more complex. However, the team must recognize and accept this complexity, in order to work through the system's process map in detail, and it also must discover the full value of materials and energy flows over a product lifetime. In a circular economy, accurate measurement and tracking of lifetime value over life are an essential counterpoint to highlighting the lost value of waste.

In this regard, Stewart and Niero (2018), in the results of their research, concluded that value must be maximized at each stage of the product life. So the French agency for environmental and energy management (ADEME), defines the circular economy as an economic model that values the efficiency of resources at all stages of the value chain, stating that "the circular economy aims to reduce waste of natural resources and further protect the environment (climate change, biodiversity preserving). The transition to this new economy requires the development of new models of production and consumption and the participation of stakeholders at all levels of the Organization."

3.8 *Closed Cycle Supply Chain (LOOP) and Distributed Information Systems*

Re-manufacturing does not make sense if the bottlenecks are not eliminated, according to Guide (Daniel and Guide 2009). Taking a traditional view based on reverse supply chain activity shows key activities with a focus on individual tasks (Fig. 1). Most of the research on reverse supply chains focuses on technical and operational issues. Obviously, re-manufacturing does not prosper if technical/operational bottlenecks cannot be eliminated, so a focus on technical activities would seem a likely place for initial research attention. Even if the event that re-manufacturing is technically feasible, the potential recovery of value owes the costs of the recovery operations. This is a necessary, but not the best condition.

To make re-manufacturing economically attractive, one also needs adequate quantities of used products of the right quality and price, at the right time, as well as, a market for reclaimed products. In other words, one needs to go far beyond technical and operational limits and to take a global perspective of the business process. To move to a process flow perspective, we distinguish three sub-processes (Fig. 1):

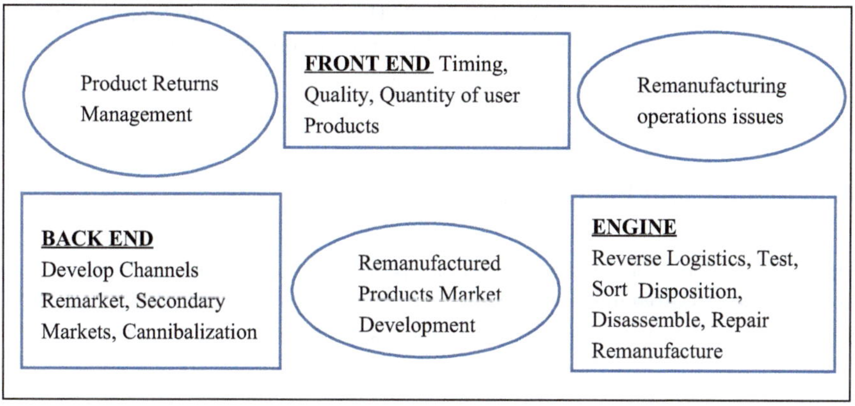

Fig. 1 *Source* Closed cycle supply chain (loop) and distributed information systems. Activities in the reverse supply chain

management of product returns (front end), re-manufacturing of operational problems (engine), and development of the market for re-manufactured products (back end), associating the final product toward environmental sustainability.

As a result of environmental sustainability, the circular economy and the shared economy emerge, concepts that are born from research on production and operations management. In the circular economy, information technologies play an important role, considering scientific material, technologies for new product design. Another important concept refers to shared economy, which connects many ideas through collaborative commerce, the exchange of resources (including the exchange of channels, information) and better and more sustainable operations. The concept of shared economy has some overlaps with the concept of circular economy. The shared economy is generally related to the use of the platform property exchange and information. The activities are explained in Fig. 1.

The development of a circular economy business model, which complements the theme of sustainability by creating value and relying on the use of economic value retained in products after production, is supported by Alexandris et al. (2018). This economic value changes the creation of product value, contributing to sustainability and social and ecological responsibility. This point of view also applies from the point of view of the supply chain. The importance of closed loops (LOOP) in companies as a single unit cannot implement effective closed systems. The correct thing would be to create cooperative closed loops (loops) in process networks. In this sense, this author proposes the development of a system based on Block Chain technology, to form a decentralized network of asset operators, potential operators, and auditors/regulators. According to this systems model, the paper proposes a collaborative circular economy business model, in which the cycle of the circular economy is materialized by the transition of assets between asset operators with a demand-based approach (Pull).

The demand itself is partially based on the state of the asset, which is described by its circular properties (location, condition, availability). This common view of the state of assets between all parties can be enabled by Block chains and Smart Contracts, which can provide the underlying technology for sharing data with integrity, while offering more efficient interoperability among the participants.

4 Methods and Procedures

This paragraph includes the definition of a case where we apply the concept of theory of constraints to solve the online service telephone payment restriction. In addition, we specify the methods to apply theory of constraints in the selected process.

4.1 Case Study

A bank financial service company, since 2017, began to increase the number of physical and virtual channels through which it provides services to its clients. Some of them are correspondent ATMs, Window, WEB, smartphone app, and ATMs. However, the service was provided using the same technological offer and considering old communication interfaces, which required a redesign to improve installed capacity.

In the absence of a global vision of the technological supply chain of services and products, the macro-process of development of products and services, in the case of telephone payments, becomes a serious constraint for the optimal operation of the chain.

In Fig. 2, we show the technological environment and the interface of the application program interface (API), where later, we will apply in model of the theory of constraints for the solution.

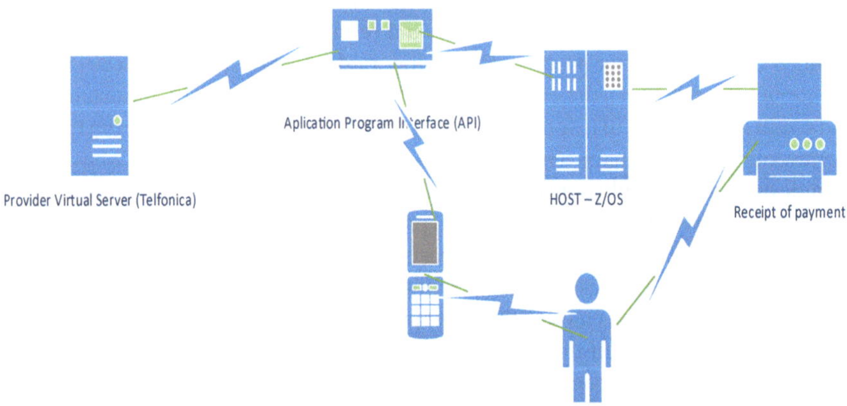

Fig. 2 System architecture for telephone service payment online—own elaboration

API solution establishes the communications between provider database and mainframe HOST with control input control system (CICS) software. This API component has a great restriction which causes a contention in all the payment line service and impact in all multitasking products and technological service of the supply chain.

4.2 Methods

The present study has a multimodal approach, that is, it has a quantitative and qualitative approach for the TOC information systems model, circular economy, with a global approach that would encompass a proposal for the entire supply chain of the company, object of study. For this effect, we start by identifying and defining the SCM strategy. The type of research is correlational because it is intended to show that when applying the TOC model, it significantly influences the quality of information and communications technology systems. It also has an impact of the new technological circular economy business model to be defined in the future. Similarly, the design is experimental because to demonstrate the proposed model, the study of a bank financial services company is carried out. The research work is based on an analysis of the primary and the secondary demand information. For primary demand, a consultancy is carried out in the chosen company, through the development of questionnaires, methods of observing the supply chain, and analyses of the entire supply chain are used. In the secondary analysis of demand, comparative methods are carried out on a sample of companies that have experimented with the proposed or similar model.

Redesign of application program interface (API) will be developed within the proposed TOC model to demonstrate the correlation among the proposed variables and to eliminate the bottlenecks. For the design and development of the model, various methodologies, that will be used to complement the TOC model to be considered, are alternatives for the proposed model as follows:

- Develop a design thinking model to identify constraints at the management level of the supply chain and definition of the SCM strategy, at this stage we define a tree of cause–effect problems to start the study, which will allow us to identify the existing restrictions. In addition, we resort to a LOG (database) where errors are saved at each stage of the chain of the payment services line. This contributes to identify the restrictions in each stage.
- Analysis of processes using lean value stream mapping (VSM) to identify processes, activities, capacities, times, and restrictions of the SCM chain. A process analysis is performed using the value stream mapping (VSM) tool, and the bottleneck processes already identified in the problem tree are indicated. (3) Apply the TOC—DBR method (Drum, Buffer, Rope), to eliminate all the restrictions of the chain and give value to each node of the service production line. Here we apply the concept of time and capacity buffer, exploiting and transferring all resources to the identified major constraint. (4) Design and development of the reverse supply

chain activities model and define the new circular economy business model for the closed cycle supply chain (Daniel and Guide 2009), (Fig. 1). Based on the quality, results are obtained in the application of TOC in the technological line of payments for telephone services. The study is further complemented by calculating the savings in energy and time use and creating a new circular business model based on the final product, which has 80% of processed quality. (5) Apply TOC—theory of constraints, for the calculation of ratios through the concept of throughput (performance). This stage is the complement of the others to establish the performance achieved in the entire supply chain.

The solution proposal in methods (1), (2), (3), and (5) is part of the global TOC solution at the highest level in the entire supply chain proposed in this study. Proposal 4, about the circular economy business model, is derived from the application of TOC in the telephone payment service line, proposed in this study.

5 Results and Discussion

For the analysis of the results of the application of the proposed TOC methodology, the current situation is presented first, and then the proposal of the application of TOC in the telephone payment service line is presented.

5.1 Analysis of the Current Situation

There is an increasing trend in the company in the exponential growth of the demand of products and services, causing bottlenecks in most channels, especially virtual channels (technology); in this way, the demand turns out to be greater than supply. This situation causes a series of operating restrictions, which are solved without a global strategy for the entire supply chain. In this sense, the study will be carried out considering an integral vision of the supply chain, identifying the restrictions of the supply chain in an integrated way. One of the identified constraints is the online payment services process by telephone services. This restriction impacts the entire payment line and in turn produces a containment of services (bottleneck) in the entire SCM network of technological services of the bank. Considering this initial problem, this research study begins. It, for the moment, will have some limitations in terms of line drop statistics and transactions made by channel. In this regard, we proceed to define the main problem of the study. It can be defined as follows: How to apply the theory of restrictions (TOC) to improve the quality of processes and information and communication technologies and its impact on the circular economy?

In Fig. 3, we represent the number of transactions made per year x day, from 2012 until the year 2016, and then we make the forecast until 2019. In Fig. 4, we represent the number of transactions processed per year, per day, and the transactions

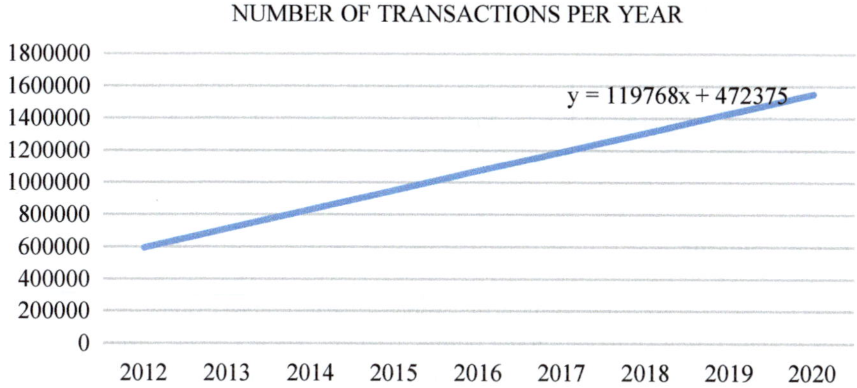

Fig. 3 Taken from strategic plan 2017–2021—own elaboration

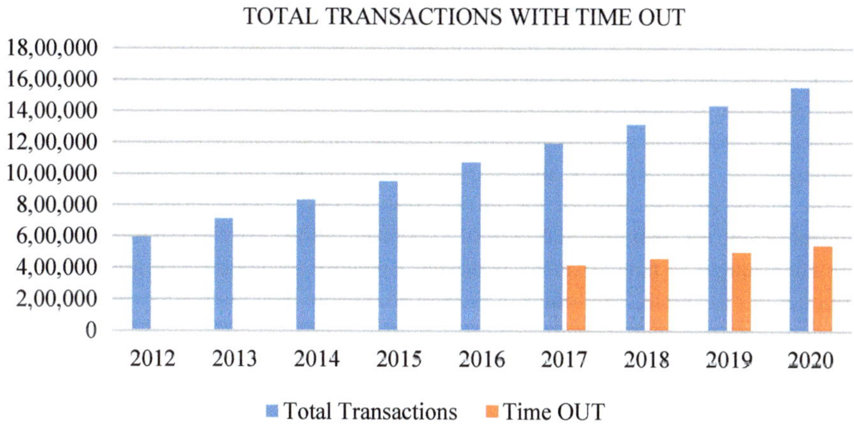

Fig. 4 Taken from strategic plan 2017–2021—transactions with time out—own elaboration

processed for system crashes (time out). The percentage of downtime (time out) was estimated at around 35% per year and has a great impact and contention throughout the bank's technology supply chain.

Average batch of daily transactions, considering the last 3 years, is 3611 transactions. The inventory of products in process due to system failures is 1264 transactions daily.

5.2 *Analysis of the Proposed Situation*

For the analysis, we apply the proposed methodology according to the TOC development cycle expressed in Fig. 5, and then we develop in detail how to apply the redesign of the application program interface (API), applying TOC-DBR.

Design Thinking

In this process, we are going to apply the problem tree technique to identify the restrictions of the current situation on the online payment process of the macro-process, development of products, and services. The exercise is carried out as described in Fig. 6. The first restriction identified in Fig. 6 is number (1). Bill payment demand is greater than available capacity. The second restriction identified number (3), excessive falls of the SNA—system network architecture communication line because time slice permitted in the environment is greater than process time in service line. Constraint 3 is the biggest bottleneck, and the other restrictions (number 4 and number 5) mentioned are solved with the application of the shock absorber in each stage of the chain (line service payment).

The tree performs a cause and effect relationship starting from the lowest to the highest level. This analysis allows us to see the problems that arise at the level of the entire supply chain of the information and communication system. At the highest level, for example, the lack of synchronization processes with the service growth strategy in new channels is arisen. Likewise, problems that arise with the supplier, for example, Telefonica does not provide the bank with adequate advice to

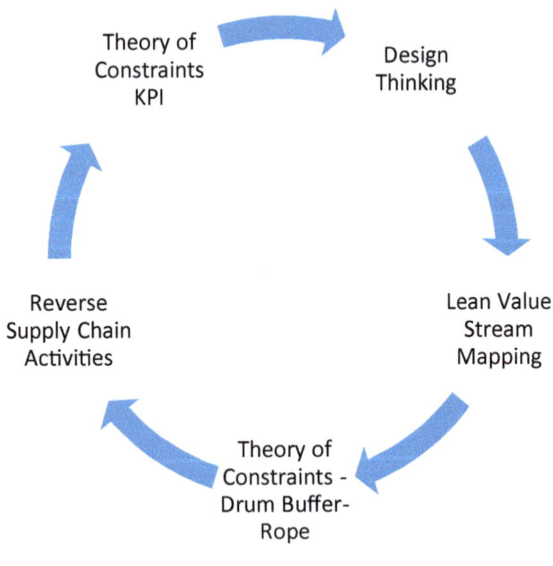

Fig. 5 TOC solution model in a bank financial services company—own elaboration

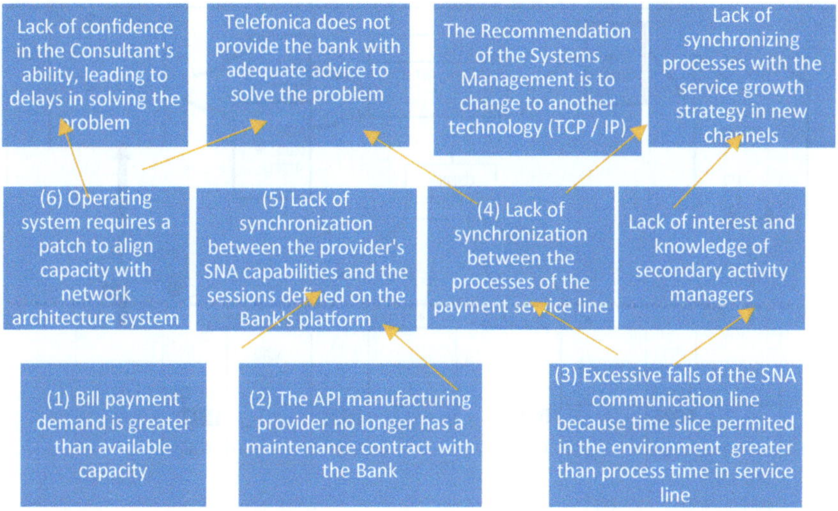

Fig. 6 Problem tree—own elaboration

solve the problem. Using this methodology, we can see that the main bottlenecks are downstream, and as one walks upstream, there will be other restrictions that will be studied later.

In this way, we select different problems, where the DBR, Drum, Buffer, and Rope methodology will be applied. The main problems are (3) excessive falls of the SNA communication line because time slice permitted in the environment is greater than process time in service line, and (1) bill payment demand is greater than available capacity.

Lean Value Stream Mapping (VSM)

In Fig. 7, the value stream mapping (VSM) diagram is shown, to show the virtual stations that are part of the online payment technology service. The current bottlenecks or restrictions are also identified.

The process is carried out as follows: The client gets the bank's window channel and demands payment for a telephone service. The paying receiver enters the system and sends the information to the payment line, the sequence of which is as follows: Allocate, Connect, Send, Send invite, and Receive. The bottleneck is found in the receive statement and in the Allocate statement. SNA means system network architecture, through which the whole conversation takes place. CICS means control input control system and is the environment where resides the API application program interface. Figure 7 represents the current situation and the characteristics of the process. According to the analysis of the problem and the error information in each

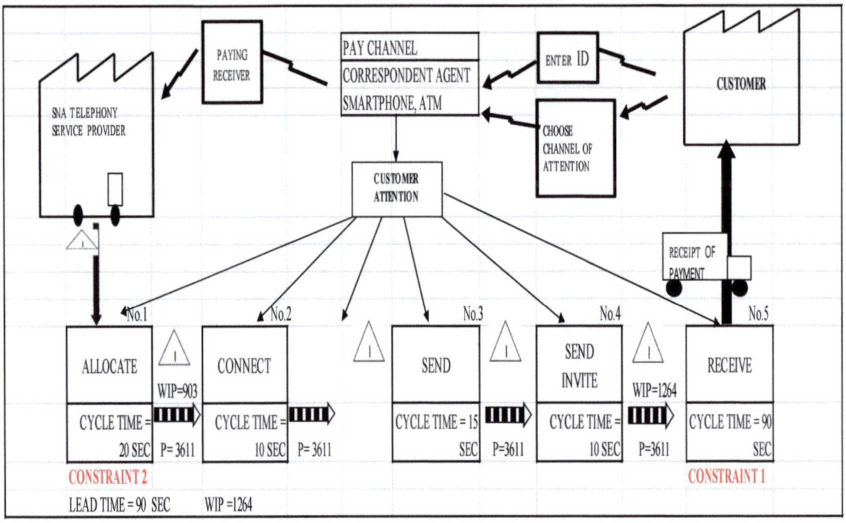

Fig. 7 VSM (current situation value stream mapping)—online payment—own elaboration

link of the payment service line, it is clearly determined that there are two restrictions in the system. At each station, an error occurs every time, and the information is written in a database. This information identifies the constraints and system crashes due to bottlenecks. The first restriction is in the Receive process, and the second (the most important) restriction is in the Allocate process, represented by the longest cycle time, 90 and 20 s.

Process Description

Allocate finds an available space on the net to assign the work session. Synchronization is with the available provider capacity. The installed capacity of the service line is given by the number of reusable fixed sessions established by the service provider. The bank, through its control input control system (CICS) platform, must synchronize the process of its transactions to this capacity. A review found that the installed bank's capacity was 16 sessions and the provider's capacity for online payments was 64. The demand is also greater than the installed bank's capacity. Here one gets the session number to be transferred to each stage of the system line payment.

Connect establishes and synchronizes connection with the bank session and the provider line session.

Send sends the encapsulated data frame in the session to the service provider to advise customer payment. Service provider scans in database the receipt pending payment.

Send invite carries out the same operation, but sends a message to the supplier to indicate that they will receive the payment.

Receive (Major Constraint)—It is the last process, and here it is already understood that the payment has been made. The receipt for the payment of the phone is issued. Currently, neither the capacity nor the processing time adjusted to the number of transactions requested by the payment demand is controlled.

Theory of Constraints (TOC)—Drum/Buffer/Rope Method

In this phase, we apply the DBR methodology considering the analysis of the problem performed and the VSM diagram. The redesign of the API is carried out by applying time and capacity dampers. These buffers are expressed using a COBOL programming language and considering all the stages of the TOC methodology. Once the restriction has been identified, its exploitation is decided, alignment of all the resources considering the rhythm of the bottleneck. Then it is verified if another restriction occurs, and the same procedure is followed. In this case study, what has been done is to include the damper in each digital station in the preceding way, in such a way that it is possible to eliminate all the restrictions of the payment line with a positive impact on the entire supply chain. Below we explain in detail each element of the methodology.

DBR (Drum–Buffer–Rope) in Detail

At this stage, we define the optimized proposed model. A just-in-time synchronized model is understood to eliminate waste of waiting time and over-processing due to lack of capacity in the bottleneck. Below we describe in detail the application of the DBR methodology as follows:

Drum, in this phase, the previously applied methodologies and the system information have allowed identifying the restrictions. For this reason, we will proceed to develop step 2 of TOC, which is to exploit the constraint.

Buffer (damper), in this link, we define a capacity time and state buffer for link Receive (1) and all the previous stages to this. Then, we synchronize the entire line to the installed capacity of the bottleneck link, but also adding a time and condition damper. The state lets you know if the next link is ready to receive payment information. In addition, a time buffer with 30 s was defined as the maximum limit to prevent the line from falling.

Rope aligns demand just in time with supply line service (Pull).

Figure 8 shows the proposal value stream mapping after applying the redesign of the application program interface (API) with TOC DBR.

The COBOL API model is redesigned with the application of TOC – DBR. The redesign includes established dampers of time and capacity in all phases of telephone payment services. In this sense, damper time is represented by WK-VECES (30 s), and damper capacity is represented by WK-STATE (90). Ninety is the maximum state that works on the restriction, and all the stations will be synchronized with this value (Banco de la Nacion 2017). See Table 4, API connects to telephone provider.

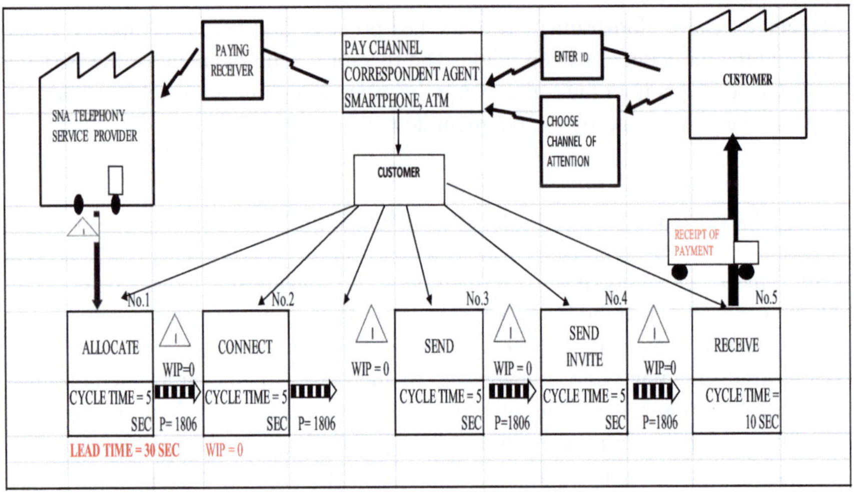

Fig. 8 VSM (proposal value stream mapping)—online payment—own elaboration

6 Results Resume and Discussion

The installed capacity expressed in number of sessions was synchronized with the installed capacity of the number of providers "sessions (Allocate)."

By applying the constraint theory—TOC—DBR method, the main constraint (Receive) and all other preceding constraints were removed from the line of telephone payment services.

It is important to mention that according to the theory of Goldratt (2012), the time buffer should be half of the service delivery time.

The constraint theory was successfully applied by exploiting the main constraint and allocating all resources to the bottleneck. In this sense, the TOC-DBR application is started by placing a time damper and a capacity damper in the Receive constraint where the maximum capacity (state) is 90 s, working at 100% and the time damper is 30 s. Then the same procedure is applied to all the preceding stations of the technological payment line, in order to align all the workstations to the maximum capacity of the restriction (90) and with a time loop to verify the status of the next station before sending information.

To apply TOC DBR, the functions and processes of the API COBOL interface program are redesigned, defining at each stage, through a payment flow programming algorithm, time, and capacity buffers. The automation of TOC DBR, in the API, ensures the quality of service in each station of the information and communication technologies.

By applying the time buffer, the size of the payment service transaction batch flowing on the line is automatically reduced, providing flexibility, quality, and synchronization with customer-requested payment demand (Pull).

Table 4 **API** connects to telephone provider company—own elaboration

API CONNECT	API SEND
2100-APPC-CONNECT. *---------------------------* MOVE 'CON' TO WK-LOG-COMANDO-CICS PERFORM 6000-FECHA-Y-HORA MOVE ZEROS TO WK-VECES MOVE ZEROS TO WK-STATE-C PERFORM 2101-HACER-CONNECT UNTIL WK-VECES GREATER THAN 30 OR WK-STATE-C EQUAL 90 IF WK-VECES GREATER THAN 30 MOVE 1925 TO WK-LOG-LINEA-PROGRAMA MOVE 010 TO RESPONSE PERFORM 8500-TERMINA-SIN-ABEND END-IF 2101-HACER-CONNECT. *---------------------------------* IF WK-STATE-A EQUAL 81 EXEC CICS CONNECT PROCESS CONVID (WK-CONVID) PROCNAME (WK-PROGID) PROCLENGTH (LENGTH OF WK-PROGID) SYNCLEVEL (1) STATE (WK-STATE-C) RESP (RESPONSE) NOHANDLE END-EXEC END-IF ADD 1 TO WK-VECES	2210-APPC-SEND. *---------------------* *===> PARA ENVIAR DATA AL SIX DESTINO <=== MOVE 'SND' TO WK-LOG-COMANDO-CICS PERFORM 6000-FECHA-Y-HORA MOVE WK-TRAMA-IDA (1: LONGITUD) TO BUFFER MOVE LONGITUD TO SEND-LENGTH MOVE ZEROS TO WK-VECES MOVE ZEROS TO WK-STATE-S PERFORM 2211-HACER-SEND UNTIL WK-VECES GREATER THAN 30 OR WK-STATE-S EQUAL 90 * EVALUA SI ESTA EN TIME OUT IF WK-VECES GREATER THAN 30 MOVE 1997 TO WK-LOG-LINEA-PROGRAMA MOVE 010 TO RESPONSE PERFORM 8500-TERMINA-SIN-ABEND END-IF 2211-HACER-SEND. *--------------- IF WK-STATE-C EQUAL 90 EXEC CICS SEND CONVID (WK-CONVID) FROM (BUFFER) LENGTH (SEND-LENGTH) STATE (WK-STATE-S) RESP (RESPONSE) NOHANDLE END-EXEC END-IF ADD 1 TO WK-VECES.
API SEND INVITE	API RECEIVE
2215-APPC-SEND-INVITE.	

(continued)

Table 4 (continued)

```
*---------------------------------
*===> PARA ENVIAR DATA AL SIX
DESTINO <===

   MOVE 'SNI'        TO WK-LOG-
COMANDO-CICS
   MOVE ZEROS        TO WK-VECES
   MOVE ZEROS        TO WK-STATE
   PERFORM 2218-HACER-SEND-INVITE
         UNTIL WK-VECES GREATER THAN
30 OR
            WK-STATE EQUAL 88

* EVALUA SI ESTA EN TIME OUT
   IF WK-VECES GREATER THAN 30
      MOVE 2039      TO WK-LOG-LINEA-
PROGRAMA
      MOVE 010       TO RESPONSE
      PERFORM 8500-TERMINA-SIN-ABEND
   END-IF

 2218-HACER-SEND-INVITE.
*-------------------------------------
   IF WK-STATE-S EQUAL 90
      EXEC CICS SEND CONVID (WK-
CONVID)
              RESP (RESPONSE)
              STATE (WK-STATE)
              INVITE
              WAIT
              NOHANDLE
      END-EXEC
   END-IF
   ADD 1       TO WK-VECES.

2300-APPC-RECEIVE.
*----------------------------
*===> PARA RECIBIR DATA DEL SIX
DESTINO <===
   MOVE ZEROS      TO WK-VECES
   MOVE ZEROS      TO WK-STATE-R

   MOVE 'REC" TO WK-LOG-COMANDO-
CICS
   PERFORM 6000-FECHA-Y-HORA
   MOVE 4096            TO REQUESTED-
LENGTH

   PERFORM 2305-HACER-RECEIVE
         UNTIL WK-VECES GREATER
THAN 30 OR
               WK-STATE-R EQUAL 90
* EVALUA SI ESTA EN TIME OUT
   IF WK-VECES GREATER THAN 30
      MOVE 2139        TO WK-LOG-
LINEA-PROGRAMA
      MOVE 010         TO RESPONSE
      PERFORM 8500-TERMINA-SIN-
ABEND
   END-IF

 2305-HACER-RECEIVE.
*------------------------------*
   IF WK-STATE EQUAL 88
      EXEC CICS RECEIVE CONVID (WK-
CONVID)
              INTO (BUFFER)
              LENGTH (REQUESTED-
LENGTH)
              STATE (WK-STATE-R)
              RESP (RESPONSE)
              NOHANDLE
      END-EXEC
   END-IF
   ADD 1         TO WK-VECES.
```

It is also observed in the results of the API redesign: When verifying if the next station is ready to receive the data (inventory), the concept of lean Kanban supermarkets has been applied automatically.

In order to evaluate the performance of the payment chain, by applying the TOC philosophy, in the conclusions we applied Goldratt's theory (Goldratt 2012) to make a projection of the value achieved in soles.

7 Conclusions and Future Research

7.1 *Conclusions*

Based on the results obtained by the application of the TOC philosophy in the line of telephone payment services, the next step in the research study is to continue with the analysis of the other services and products provided by the bank and identify the other constraints. This implies starting with the business strategy until identifying downstream the other restrictions related to the primary activities.

In order to manage greater flexibility and adaptability of the service chain, it is important that time and capacity buffers be declared as variables in the program interface (API), and they can be adjusted to the chain according to the demand for the service. In this way, we can apply reusability without the need to make further redesigns to the API. This also implies that the batch size is automatically adjusted according to the synchronization of the line and the TAKT time associated with the speed of production to meet just in time with customer demands.

The advantage of implementing TOC in an automated process is the facility to detect and correct the other restrictions of the line automatically after having identified the first one. Also, the flexibility can synchronize each one of the stages of the service line.

By applying TOC in the chain, a series of variables are optimized such as the total execution time of a transaction (lead time). The reduction of the batch of data is transferred through the network, helping to improve the installed line capacity eliminating the risks of falls of the service.

Below we make a comparative table (Table 5) to reflect how the current situation changes to the proposal. According to the results shown in Table 4, the capacity of the service was increased by 64 online sessions. The time buffer according to the data batch was set to 30 s; however, it could drop to 15 s. The lead time was optimized to 30 s. Lot size was optimized in half by improving line timing. Finally, all the services (transactions in process—WIP) were eliminated.

Table 5 Comparative table of the current situation and the proposal—own elaboration

Variable	Current situation	Proposal model
Installed capacity of the service line	16 sessions	64 sessions
Time damper	0 s, time out	Damper time = 30 s, (lead time/2)
Capacity damper	Oversized	90 s
Data batch size	3611 transactions	1806 transactions
Work in process (WIP)	1264 transactions	0 transactions (line synchronous with customer demand)

7.2 Measure the Performance of the Supply Chain According to the Proposal Solution

In Table 5, we make a comparison of the performance of the current situation and the one proposed, applying the indicators and methodology of Goldratt (2012). The information assumed comes from the results and the strategic plan of the financial company considering the years from 2012 to 2019. Below the formulas associated with the comparative performance in Table 6 are shown.

The yield in the current situation is 138,121; and in the proposal, it is 341,235. Net profit is 5016 in the current situation and in proposal 296,995.

The ROI is 0.26 in the current situation and 19.49 in the proposal. Productivity (PR) is 0.04 in the current situation and 6.71 in the proposal. Inventory turnover is 0.07 in the current situation and 0.22 in the proposal.

Table 6 Detail of the formulas applied to calculate the yield of the TOC methodology

Sales	= (D5–D7) * D9 * D3 * D8
Cost of raw materials (energy)	= D15 * D5
Throughput	= D19–D20
Products in process	= D7 * D9
Finished products	= (D5–D7) * D9 * D3
Raw materials (energy)	= D20
Capital investments	= D16 * D27
Equipments	= D17 * D27
Direct cost	= D13 * D5
Indirect labor	= D10 * 8 * D14 * D18
Net profit (NP) = (TP – OE)	= D21–D38
Rate over investment—ROI = (NP/IN)	= (D39/D30)
Productivity PR = (NP/OE)	= D39/D38
Inventory rotation IT = (TP/IN)	= D21/D30

Own elaboration

According to the results, we conclude that by applying the theory of restrictions in the system (API), a series of benefits have been generated in the chain as follows: smaller data batches, elimination of all restrictions, decrease in delivery time energy saving, line speed higher than the current situation, quality in each work station, and better and more sustainable operations (see Table 7).

Table 7 Strategic plan 2017–2021—comparative total improved performance of the supply chain—own elaboration

Assumptions	Current situation	Proposal model
Performance speed	1	2
Transmissions costs	34	34
data batch size (transactions)	3,611	1,264
Lead time (seconds)	60	30
Work in process (WIP) transactions	1,264	0
Average collection free s/	1.5	1.5
Average payment receive amount s/	70	70
Availability of the communication line	35%	100%
Availability of the communication line daily in seconds	10,080	28,800
Total communication channels	7,908	7,908
Daily unit cost of the SNA telephone line s/	35	35
Daily cost of labor s/	200	200
Daily cost of energy s/	30	30
ROE (year 2017)	34.33%	34.33%
ROA (year 2017)	2.90%	2.90%
Average falls out of service	12	0
Sales	246,451	379,155
Cost of raw materials (energy)	108,330	37,920
Throughput (TP)—rendimiento	138,121	341,235
Inventory (IN)		
Product in process	88,470	0
Finished products	164,301	176,960
Raw materials energy	108,330	37,920
Subtotal	361,100	214,880
Capital investments	123,966	73,768

(continued)

Table 7 (continued)

Assumptions	Current situation	Proposal model
Equipments	1,047,190	843,001
Total inventory	1,893,356	1,524,175
Operating expenses		
Direct cost	126,385	44,240
Indirect labor	6,720	0
Supplies (transmissions costs)		
Interest payments		
Total operating expenses	133,105	44,240
Net profit (NP) = TP − OE	5,016	296,995
Rate over investment (ROI) = NP/IN)	0.26%	19.49%
Productivity PR = (NP/OE)	0.04	6.71
Inventory rotation IT = (TP/IN) + B21	0.07	0.22

7.3 Future Research

As a future design, the paper recommends developing a new business model by designing a new reusable service based on the service obtained by applying TOC in the entire online payment process.

References

Alexandris G, Hatzivasilis G, Katos V, Alexaki S (2018) Blockchains as enablers for auditing cooperative circular economy networks. In: 2018 IEEE 23rd international workshop on computer aided modeling and design of communication links and networks (CAMAD)

Antić LČ (2015) Lean concept techniques for overcoming constraints of business. Casopis za Društvene Nauke 39(3): 905–923 (19p)

Arbós LC (2010) Lean Management La gestion competitiva por excelencia. Publidisa, Barcelona

Banco de la Nación (2017) Strategic Planning 2017–2021. Lima, Peru

Bozer Y, Ciemnoczolowski D (2012) Performance evaluation of small-batch container delivery systems used in lean manufacturing. Int J Prod Res 51(2):555–567 (13p) [1 Diagram, 5 Charts, 1 Graph]. https://doi.org/10.1080/00207543.2012.656330

Bruel A, Kronenberg J, Troussier N, Guillaume B (2019) Linking industrial ecology and ecological economics: a theoretical and empirical foundation for the circular economy. J Ind Ecol 23(1):12–21 (10p) [2 Diagrams, 2 Charts, 1 Graph]. https://doi.org/10.1111/jiec.12745

Chopra S, Meindl P (2013) Administración de la cadnea de suministro. D.F. Pearson, México

Guide, VDR, Van WLN (2009) OR FORUM—The evolution of closed-loop supply chain research. Oper Res 57(1):10–18

Esposito M, Tse T, Soufani K (2018) Introducing a circular economy: new thinking with new managerial and policy implications. Calif Manag Rev 60(3):5–19 (15p)

Ferenčíková D (2012) Theory of constraints based information systems in production management. In: Proceedings of the European conference on management, leadership & governance, pp 474–480

Goldratt E (2012) The gestalt of TOC. Change Management/Buy-In, Holistic Implementation, Ongoing Improvement, Decisive Competitive Edge (DCE), Immunizing the Future of the Company, Ever-flourishing, (pág. Full Video: 3hr 10min; Free excerpt: 5 min)

Goldratt E, Cox J (2012) The Goal-La meta un proceso de mejora continua. Lucila Galay, Buenos Aires

Jiang X-Y, Wu H-H (2013) Optimization of setup frequency for TOC supply chain replenishment system with capacity constraints. Neural Comput Appl 23(6):1831–1838 (8p). https://doi.org/10.1007/s00521-013-1376-0

Lea B-R, Min H (2003) Selection of management accounting systems in just-in-time and theory of constraints-based manufacturing. Int J Prod Res 41(13):2879 (32p) [4 Diagrams, 10 Charts]. https://doi.org/10.1080/0020754031000109134

Marín WG (2013) Desarrollo e implementacion de un modelo de Teoria de Restricciones para sincronizar las operaciones en la cadena de Suministro. Revista EIA (19):67–77 (11p)

Moisello AM (2012) Costing for decision making: activity-based costing vs. theory of constraints. Int J Knowl Culture Change Organ Annu Rev 12:1–13 (13p)

Phruksaphanrat BO (2011) Linking industrial ecology and ecological economics: a theoretical and empirical foundation for the circular economy. Int J Ind Eng 18(5):219–231 (13p) [3 Diagrams, 5 Charts]

Phruksaphanrat B, Ohsato A, Yenradee P (2011) Aggregate production planning wih fuzzy demand and variable system capacity based on theory of constraints measures. Int J Ind Eng 18(5):219–231 (13p) [3 Diagrams, 5 Charts]

Spencer MS (2000) Theory of constraints in a service application: the Swine Graphics case. Int J Prod Res 38(5):1101–1108 (8p). https://doi.org/10.1080/002075400189040

Stewart R, Niero M (2018) Circular economy in corporate sustainability strategies: a review of corporate sustainability reports in the fast-moving consumer goods sector. Bus Strategy Environ (Wiley) 27(7):1005–1022 (18p). [1 Diagram, 7 Charts, 7 Graphs]. https://doi.org/10.1002/bse.2048

Tsan-Ming Choi AA (2020) Game theory applications in production research in the sharing and circular economy era. Int J Prod Res 58(1):118–127. https://doi.org/10.1080/00207543.2019.1681137

Wang J-Q, Zhang Z-T, Chen J, Guo Y-Z, Wang S, Sun S-D, Huang G (2014) The TOC-Based algorithm for solving multiple constraint resources: a re-examination. IEEE Trans Eng Manag 61(1):138–146. USA: IEEE Language: English, Base de datos: IEEE Xplore Digital Library

Weetman C (2016) A circular economy handbook for business and supply chains : repair, remake, redesign, rethink, 1st edn. Kogan Page, New York, p 2016

Weinberger K (2009) Estrategia para lograr y mantener la competitividad de la empresa. USAID PERU, Lima

Yang Q, Liu J, Huang Y, Wang Y, Wang T (2016) The dynamic 4S auto maintenance shop scheduling in a multi-constraint machine environment based on the theory of constraints. Int J Adv Manuf Technol 83(9–12):1773–1785 (13p).

Lean Techniques in Quality

Application of Lean Techniques and Queuing Theory in Food Services

Jonnatan Fernando Avilés-González and Sonia Valeria Avilés-Sacoto

Abstract In food services, such as cafeterias, when the servers' capacity is less than the customers' demand, coming with a poorly layout, the generation of large queues with long waiting times, spring out. This scenario is even worsened when the demand's variability appears, creating problems such as the loss of customer, revenue's decrease, and a poorly service quality. Therefore, it is important to look for solutions to revert such problems. Lean principles jointly with queuing theory might be a good approach to address them, by the application of different tools and methodologies that contribute not only to minimize or even eliminate queues but also reduce the waiting times in lines. The study presented herein includes a proposal of a methodology that merges both, Lean and queuing theory, applied to a Mexican university cafeteria, which during rush time it blows up. The methodology includes three phases: the current process analysis, the arrival analysis, and the improvement analysis. By applying different tools and using simulation scenarios in each phase, an improvement in the development of the activities in the cafeteria is reached, and several strategies are designed to enhance the behavior of the system.

Keywords Queues · Lean · Queuing theory · Cafeteria · Waiting time

1 Introduction

Queuing is part of our everyday life, and it can be found in different places where a demand for a good or a service is required. Examples of such situations are queues at the checkout counters at supermarkets, pharmacies or restaurants, banking services, purchase of tickets, payment of government services, purchase at fast-food

J. F. Avilés-González
Departamento de Ingeniería de Producción y Operaciones, Universidad del Azuay, Cuenca, Azuay, Ecuador

S. V. Avilés-Sacoto (✉)
Departamento de Ingeniería Industrial, Colegio de Ciencias e Ingenierías, Universidad San Francisco de Quito, Quito, Pichincha, Ecuador
e-mail: svaviless@usfq.edu.ec

J. L. García Alcaraz et al. (eds.), *Techniques, Tools and Methodologies Applied to Quality Assurance in Manufacturing*,
https://doi.org/10.1007/978-3-030-69314-5_15

restaurants, or cafeterias, among others. Usually, a queue has a negative connotation because it exposes that the demand exceeds the server capacity. The variability in arrival patterns and service times are responsible of long queues. As a result, the customers' satisfaction is reduced and leads them to seek service elsewhere. This setting becomes worst, when a poor layout is also present.

The food services at universities, well known as cafeterias, are not exempt of such situation. The problem of demand variability is noticeable, and it is worst during rush time. The result is an overcrowding situation due to the growth of lines in front of food bars and cashiers, giving rise to disorder and confusion among different customers. The consequences are reduction of customer satisfaction, negative opinions regarding to the service, and loss of income. It is here, where the interaction among customers and servers' gains importance, and a systems performance evaluation is necessary.

A possible approach to find a solution to such problem is Lean manufacturing, a philosophy that seeks the continuous improvement by the elimination of waste activities during a process. Several tools from Lean manufacturing are implemented to enhance human effort, inventory management, time efficiency, and layout distribution. The results might modify the response to demand, while keeping efficiently the processes. On the other hand, queuing theory analyzes the behavior of queues, by the evaluation of arrivals and services time along the system, the queue length, the expected number of customers to be served, and the arrival of customers.

As mentioned before, long waiting lines are also evident at restaurants and cafeterias especially during rush time. This is the case of a cafeteria in a Mexican university, which face the problem of queues and long waiting times, because of the arrivals of a high number of customers, who cannot be served immediately in the system.

It is here, where queuing theory and Lean techniques constitute an effective approach to address this problem, especially because both can identify the activities from a systematic and organized path, improving, and adjusting the processes simultaneously. An analysis of operations based on Lean features was developed, being the focus on the waiting time and the reduction of non-add value activities.

Additionally, a queuing theory analysis was performed at the same time with a frequency arrival rate study, with the desire to analyze the arrival rate behavior and the number of servers required to improve the service.

2 Literature Background

For several decades, Lean manufacturing has been a key initiative for the industry, helping to achieve operational efficiencies, enhancing their competitive profile, and contributing to customers' satisfaction. An extensive literature highlights the benefits of Lean techniques in diverse aspects of the manufacturing environment (Juran et al. 1998; Liker and Morgan 2011; Ohno 1988; Psomas et al. 2018). However, the literature also points out the failure of sustaining the benefits of Lean (Taylor et al. 2013). It has been reported that significant benefits from Lean implementation tend to dissipate over the years (Jayaram et al. 2010). There have been efforts to explain this phenomenon for several years. Examples of these studies are outlined below:

- Forrester (1995) explains that improving performance can produce conflicts if the implementation is not evaluated from a system perspective.
- Narasimhan et al. (2006) conclude that Lean is affected by the dynamic production schedule due to demand fluctuations.
- Seddon (2011) states that people become disappointed because Lean results fail to meet expectations with reality.
- Keyser and Sawhney (2013) explained the need for reliability analysis due to the high variations into several systems.
- Pearce et al. (2018) study how different factors modify the behavior of the enterprises; the authors conclude that the knowledge is a key factor that contributes according to the maturity level and stakeholder engagement.
- Belhadi et al. (2018) show how through Lean practices and tools, applied to SMEs, might achieve the majority of the operational results, but it is not enough to sustain the program in the time.

The importance of several tools and techniques has been widely applied and studied. A detailed example is shown by Zhenyuan et al. (2011), who identify the advantages and disadvantages of the tools under diverse considerations.

Following the same line, Chong et al. (2013) have reported that in models focused to SMEs, the variability is still a problem, affecting the diverse management style even when a degree of maturity is present in the system.

On the other hand, queuing theory studies the waiting lines known as queues, and considering customers' behavior when they are approaching and/or are in a line (Cope and Syrdal 2011). Examples include hotel check outs, waiting for a medical service, getting a car washed, or eating lunch.

The arrival process, however, in many cases, tends to be very complex, especially during rush time. The consequence, when that waiting time is longer than their expected time, is generally a negative effect on customer satisfaction in terms of the service quality (Chadha et al. 2012). Therefore, changes should be applied either in the behavior of the arriving units and/or in the service facilities, in order to sustain growth and competitiveness (Chadha et al. 2012).

Queuing theory has been developed from different perspectives and its degree of complexity changes according to the needs of the systems. As an example, the study of Pourvaziri and Pierreval (2017) explains how queuing network theory is able through metaheuristics to optimize multilevel problems for dynamic facility layouts, without losing the essence of modeling the arrival and waiting times.

With these ideas in mind, some studies demonstrate that Lean practices can help to manage and to reduce the number and length of queues. In manufacturing, for example, it is well known that Lean practices are very helpful in eliminating work in progress "queues." The most common Lean method applied to manufacturing to eliminate queues is batch size reductions, this technique can also be applied to services jointly with other tools (Monden 1993).

From a "Lean Service" perspective, waste reduction involves the decrease or the elimination of the non-add value activities through elimination or reduction techniques (Li et al. 2017). In analyzing processes, where the customer is in contact with the servers, both participants can generate waiting time; for example, during periods of high demand, customers have to wait for providers, while in periods of low demand, occurs the opposite, the servers wait for customers. This discrepancy between the customer arrival pattern and the capacity to meet customer demand, gives rise to queues (Li et al. 2017).

Li et al. (2017) describe different Lean techniques to reduce waiting waste in services; these include cross-training of multifunctional employees and teams, the introduction of high and low contact services, and various time-varying staffing models. In the same line, a study done by Brown et al. (2010) and from Sarkar et al. (2014), both mention the benefits of using queuing models, due to the significant impact on capacity analysis.

An example of how to merge Lean principles and queuing theory is explained in a study by Chadha et al. (2012). In this study, these authors expose, through a health care case analysis, their model Lean-HC. They describe how to apply Lean techniques and some queuing theory principles to the arrival of hospital patients. They also emphasize how their methodology can reduce waiting times and create a favorable response in terms of quality service.

Another example of how Lean and queueing are linked is proposed by Weiss and Tucker (2018). The authors expose a framework based on three principles, eliminate or reduce waiting time, managing expectations through communication with the customer, and enhance the waiting experience. The authors apply simultaneously Lean techniques and its waste reduction philosophy as a base for these principles. These studies might be analyzed initially based on observation and data collection tools, as is described in the work of Olfert et al. (2019).

3 Proposed Methodology and Application

The model Lean-HC done by Chadha et al. (2012) constitutes a good approach to tackle the problem of large queues in cafeterias or fast-food business. Using this research as a baseline, a methodology was proposed considering three phases, in which both Lean techniques and queuing theory analysis are integrated to identify and propose solutions in order to improve the systems efficiency. Figure 1 summarizes the different actions done to reach the objective.

3.1 Phase 1: Current Process

At this phase, the problems are identified, and "wastes" are distinguished in the process. Some tools are used to get tangible and quantifiable terms of those problems,

Phase 1 Current Process Analysis	**Phase 2 Arrival Analysis**	**Phase 3 Improving Analysis**
VSM Service Layout Analysis Route Diagram Flow Chart Analysis	Customer Arrival Rate Customer Demand Rate Service Process Rate Simulation Scenarios	Possible Solutions

Fig. 1 Macro view for the method applied

with a specific description. These metrics will reflect the present situation, as well as those activities that are impacting the proper functioning of the system. Tools such as value stream mapping, layout analysis, route diagram, and operational analysis are useful to determine the problem.

3.2 *Phase 2: Arrival Analysis*

The second phase involves a series of measures that describe the capacity of the systems to attend the demand. At this point, clear and quantifiable metrics are analyzed to define those faults that are occurring in the process and are contributing to its deficiency. The best way to carry out the monitoring of these metrics is through a gathering data, like customer arrival, demand rate, and service process rate. These results will be used to simulate scenarios to better understand the processes behavior.

Phase 1 and Phase 2 describe all the possible causes and errors found in the system, which must be corrected. Furthermore, these phases give an insight on the gap between the current level of performance and the expected level by the recognition of the multiples deficiencies and those elements that are generating the problem.

3.3 *Phase 3: Improving Analysis*

Once the problems and possible causes are found from the previous phases, the next phase is the proposal of innovative improvement solutions. Some of those solutions can be basic other can require more elements to achieve the desired level.

4 Study Case Using the Proposed Methodology

As it was mentioned before, queues can appear in cafeterias, either at the cashiers and/or at food bars. This is the case of a Mexican university cafeteria, which has large queues and long waiting times, due to the variable arrival rate, especially during rush hours, resulting that customers cannot be served immediately. As a result, there is a loss of customers, which not only affects the reputation of the cafeteria, but also impacts the financial revenues.

4.1 Phase 1: Current Process Analysis

At this phase, the objective is to identify how the cafeteria is functioning and what will be the scope of the analysis. By doing so, it is possible to determine those potential wastes that are contributing to the creation of queues in both food and cashier lines. The goal is to reduce queues and eliminate the "waste" in the service processes. Following, the current process will be described.

The cafeteria under study provides food services through three food bars. There is a vegan food bar, a general food bar, and a special food bar. Additionally, there are self service areas, those are the salad bar, a beverages section, and some shelters with snacks and fruits. The layout of the cafeteria is shown in Fig. 2. The service is available 24/7 for their clients that is 24 h a day and 7 days a week. However, this multi-server queueing system blows up during rush time (lunch time, from 12:00 to 14:30), especially during the working days. Usually, at this time, the generation of unnecessary queues and the long waiting times generate confusion and desertion. Therefore, this study will be focus on this period of time and will provide solutions to this problem.

Although the layout shows available spaces, it has not been properly used. By an improvement in this usage, the service time and permanence of customers inside the system can be decreased noticeably. Therefore, more customers can be served without harming service quality.

The first step to reach such change is an analysis of the customer's behavior using a route diagram. Figure 2 details the different "routes" done by customers arriving to the cafeteria. The majority of customers arriving to the system present a similar pattern, which will be following described.

The layout includes three zones, namely the cashier, the special orders and food service, and the vegan zone. For purposes of the study, customers are classified in three colors, red, green, and purple. The red customers visit the cafeteria for a snack or drink, and their permanency in the system is around 20–25 s. Green and purple customers stay longer in the system, around 68–93 s, respectively, because they walk around in the system looking for a specific product, and this circumstance causes unnecessary traffic and queues.

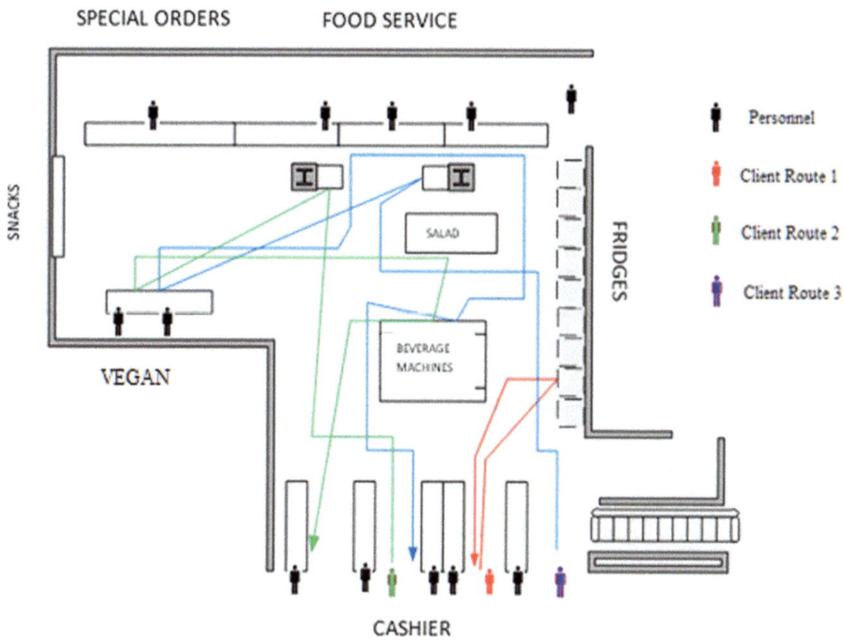

Fig. 2 Layout and route diagram

The presence of waste in this system is evident. Therefore, a Lean waste analysis is required. The following steps describe such analysis.

(1) **Activities Analysis**: For this step, both the customer and the service provider perspectives were analyzed, considering the diverse layout areas, several workers, and customer's routes. A small part of the flowchart process diagram used in this analysis is shown in Fig. 3. The final results show that 34.28% of customers are walking around the systems, generating traffic in a "short space."

(2) **Value Stream Mapping (VSM)**: Since the study applies to a service, it is more appropriate to use a value stream focused on services. For the construction of the VSM, the following information is required:

- Customers routes
- Average time for each process
- Customer arrival rate
- Activities flowcharts.

Once the information is obtained, it is possible to design the VSM (see Fig. 4). The VSM gives a macro-picture of the current system behavior. The data shown is a result from a sample of 507 customers that enter to the cafeteria during rush time. As well, the information of the services time is exposed.

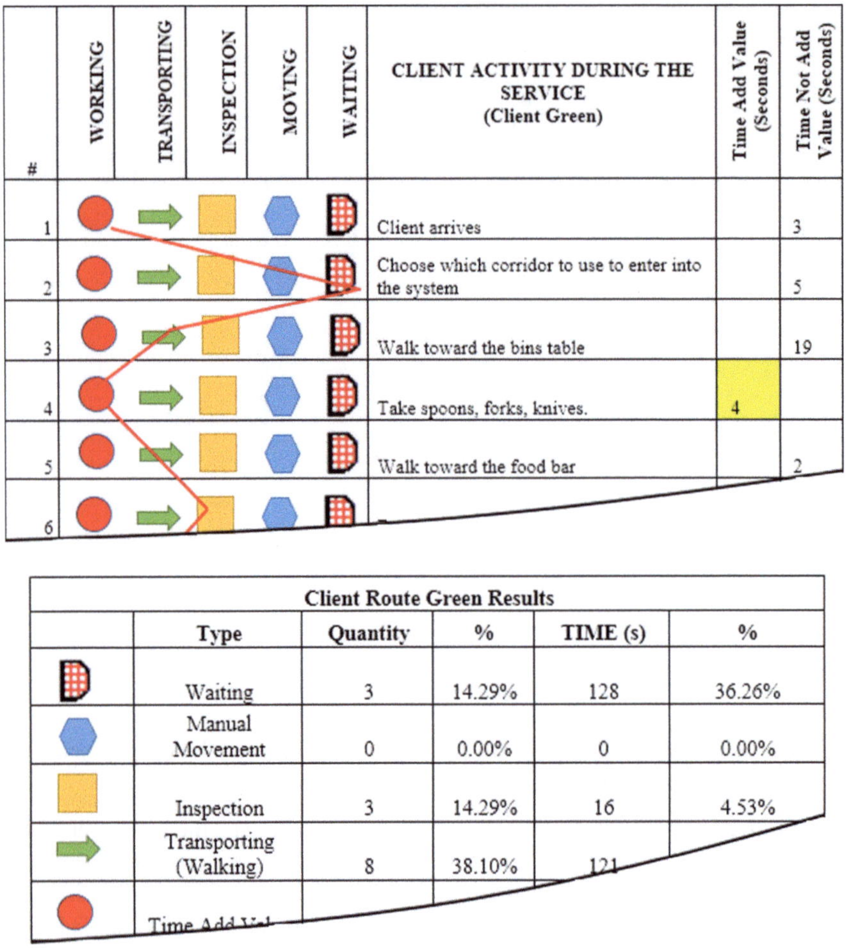

#	WORKING	TRANSPORTING	INSPECTION	MOVING	WAITING	CLIENT ACTIVITY DURING THE SERVICE (Client Green)	Time Add Value (Seconds)	Time Not Add Value (Seconds)
1						Client arrives		3
2						Choose which corridor to use to enter into the system		5
3						Walk toward the bins table		19
4						Take spoons, forks, knives.	4	
5						Walk toward the food bar		2
6								

Client Route Green Results					
	Type	Quantity	%	TIME (s)	%
	Waiting	3	14.29%	128	36.26%
	Manual Movement	0	0.00%	0	0.00%
	Inspection	3	14.29%	16	4.53%
	Transporting (Walking)	8	38.10%	121	
	Time Add Val				

Fig. 3 Sample of a flowchart process. Client/Customer point of view

Having this information, the next step is an analysis of the demand flow rate arriving to the system and its behavior.

4.2 *Phase 2: Arrival Analysis*

It is well known that the demand for a product is a discrete random variable, and it can be assumed to behave as a Poisson process with a known arrival rate. Since the objective of this study is to analyze the system during rush time, customers' arrival and their behavior should be evaluated.

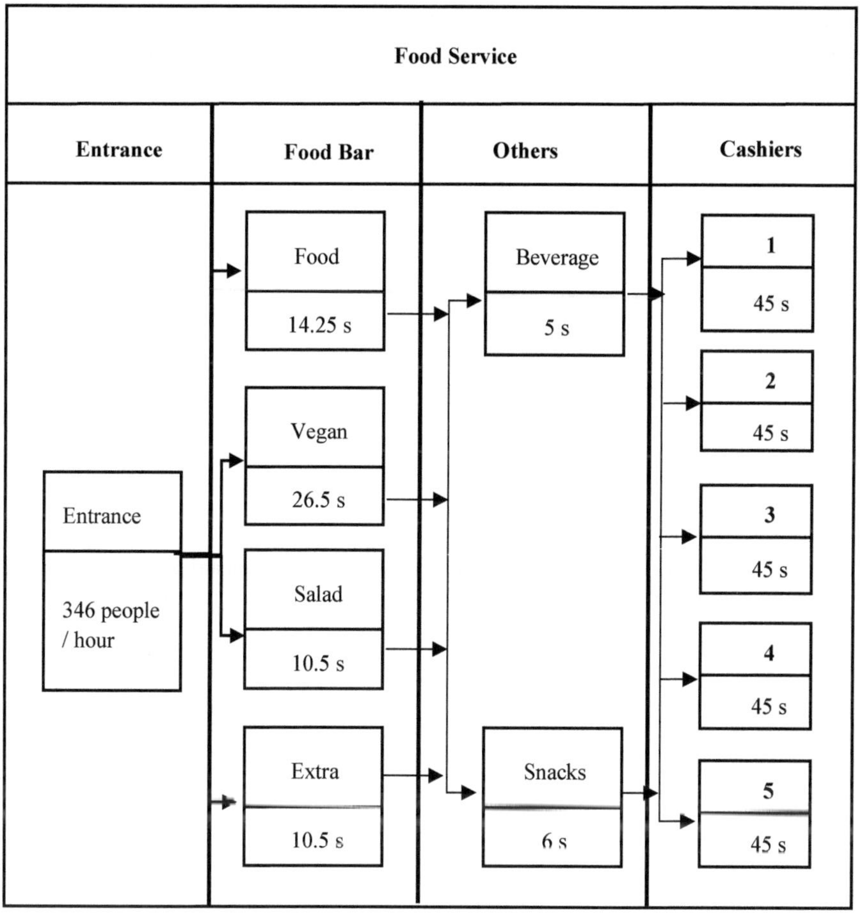

Fig. 4 Value stream mapping for services

Arrival Rates Analysis

In this study, the denomination "system" refers when a customer enters to the cafeteria during rush time. The customer can go to the food bar, the vegan bar, or both. Subsequently, there are cases in which the customers can either go to the salad area or beverages section (see Fig. 2), creating confusion and disorder. The information was collected, and a part of the sample is shown in Table 1. Here, the starting point is the moment when a customer enters to the system within a specific time.

Using the information from Table 1, gathered during an hour, it is possible to conclude that the percentage of customers arriving to the system, who directly go to the food bar and vegan bar is approximately 68.33%. Some details of the different observations are shown in Table 2.

Table 1 Number of customers entering the system

Observation Number	Time	Customers Quantity entry	Accumulated number of customers
1	1:00:00	0	0
2	1:05:00	27	27
3	1:10:00	45	72
4	1:15:00	48	120
5	1:20:00	27	147
6	1:25:00	54	201
7	1:30:00	41	242
8	1:35:00	72	314
9	1:40:00	74	388
...	...	...	...

Table 2 Customers arrival during rush time

Arrivals to the system taken each 5 min during rush time	Food bar arrivals	Vegan bar arrivals	Other options
27	11	7	9
45	18	12	15
48	20	13	15
27	11	7	9
54	22	15	17
41	17	11	13
72	30	20	22
74	30	20	24
43	18	12	13
35	14	10	11
32	13	9	10
...	...	...	...

Output arrival rate–Using the data collected, it was possible to compute the average arrival value. Customers were classified into two groups, those who made a purchase and those who not. In total, the average number of customers in the system was 507 during the rush time, with an average of 346 customers per hour/per day. The arrival customer rate and average time between customer arrivals to the system are represented by λ and t_a, respectively, in Eq. (1).

$$\text{Average Customer Arrival Rate} = \lambda = 346.108\frac{\text{customers}}{\text{hour}} = 0.096\frac{\text{customers}}{\text{second}} \tag{1}$$

Table 3 Customer arrival data to the system

Customer arrival rate to the system = λ	346.108 Customer/h
Average time between arrivals = t_a	0.00289 h
Standard deviation = σ_a	0.00289 h
C_a^2	1

Table 4 Arrival of customer rate to bars

Description	Food bar arrivals	Vegan bar arrivals
Average customer rate	252.66 customer/h	93.45 customer/h
Average time between arrivals*	0.0039 h	0.0107 h
σ*	0.0039 h	0.0107 h
Squared coefficient of variation	1	1

*Exponential distribution

It is known that:

$$\text{Arrival Customer Rate} = \lambda = \frac{1}{t_a}, \tag{2}$$

Therefore, the time between customer arrivals for each of the observations, and its average time was calculated. For purposes of the analysis, it will be assumed that the average time between arrivals follows an exponential distribution (Cinlar 2013). This implies that the average value is equal to the standard deviation. A summary about the arrival rate information obtained is shown in Table 3. The squared coefficient of variation for arrival C_a^2 is represented by Eq. (3).

$$C_a^2 = \frac{\sigma_a^2}{t_a^2} = \frac{(10.404)^2}{(10.404)^2} = 1 \tag{3}$$

Using these data, an analysis of customers' behavior at the different bars were carried out, and the results are detailed in Table 4. It can be concluded that approximately 27% of customers who bought food or snack-food, as well go to the vegan bar, while the rest bought only at the food bar.

Customer Demand Rate and Service Process Rate

The analysis is done under the assumption that there is no limit on the number of customer arrivals, however, the number of servers is limited. Simulation scenarios are used to identify the behavior through the data collected; both, the sales and takt time were computed.

Table 5 Results for server 1 and 2 in the system

Description	Server 1	Server 2
Average of server service time	0.00394 h	0.003958 h
Standard deviation server	0.00198 h	0.00277 h
Square coefficient of variation of service time	0.2519	0.4905
Capacity	253.52 customer/hour	252.66 customer/hour
Utilization	0.997	0.999

The takt time reflects the time between single outputs in a smoothly running system, while production quota is the demand to be met during the period (Hopp and Spearman 2008). In this scenario, takt time can be defined as the average unit production time needed to meet customer demand. The takt time seeks to control the pace of the production line and can be understood as a control technique toward synchronous manufacturing (Iriondo et al. 2016).

The number of customers per day is computed considering the arrival rate to the system and historical data.

$$\begin{aligned}&\#\text{Customers who buy in the food bars per day}\\&= 2700\frac{\text{Customers}}{\text{day}} \times 68.33\%(\text{Customer that buy in the system})\\&= 1844.91\frac{\text{Customers}}{\text{day}}\end{aligned}$$

The takt time is computed using Eq. (4).

$$\text{Takt Time} = \text{Available working time per shift/Demand per shift} \tag{4}$$

$$\text{Takt Time} = 0.104\ \text{Hours/Customers}$$

Simultaneously, the service rate is studied. First the servers' activities are analyzed, then the service process rates are studied, following a similar method as the arrival rate previous mentioned. Some results are shown in Table 5.

Simulation Scenarios

The use of simulation software helps to design, improve, and validate systems in a variety of knowledge's areas (Villarreal 2014). This is useful, especially, when the background in mathematics and probability theory are complex. Thus, by using such approach, it is possible to appreciate queues in both scenarios, current and future. Therefore, for this study, the scenarios were created using the data obtained in the previous section.

Following the recommendations and steps mentioned by Vendramini et al. (2016), a model was developed based on the workflow observations of the current situation and other related variables. It is relevant to clarify that when designing a system, the alternatives generated about its structure could be numerous, depending on the number of variables of interest. For this reason, if simulation is applied, it will be necessary to have the values of those variables (Villarreal 2014).

A scenario is presented in Fig. 5, comparing current behavior with a possible solution. This figure shows the current situation and its corresponding queues in front of the bars and in the servers' locations. Through the simulation, some scenarios were proved, and some solutions were stated (see phase 3).

- E-marketing

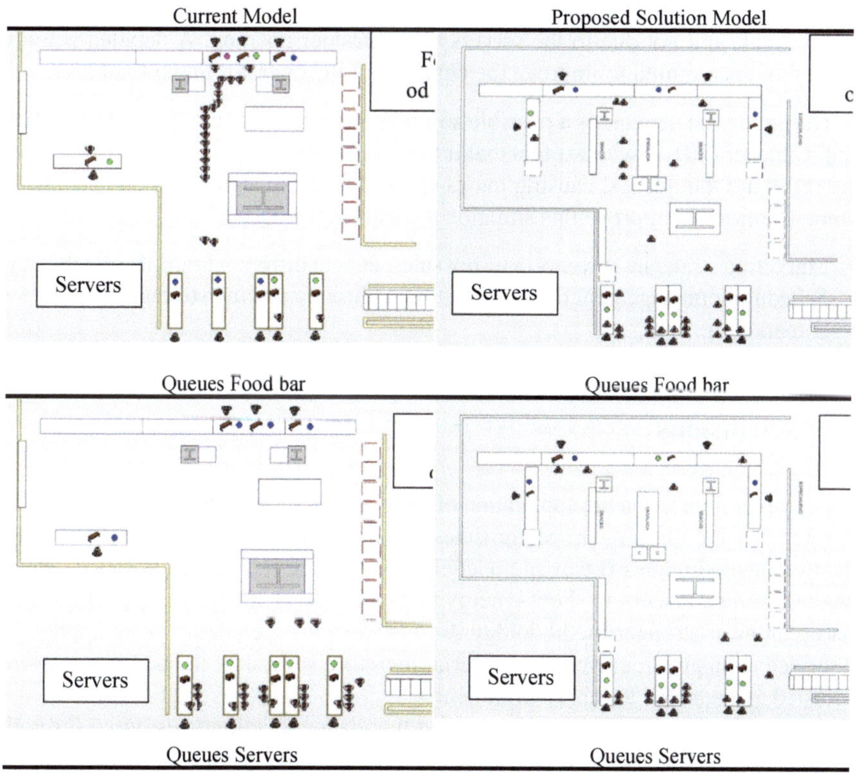

Fig. 5 Simulation scenarios comparing current model against proposed solution

4.3 Phase 3: Improving Analysis

Once the problems were identified, possible strategies are proposed to solve them, based on previous steps:

- Redesign the layout, considering a "U" layout. The result is shown in Fig. 5. Using this layout and considering simulation times, the number of queues, in front of the food bars, was reduced. It was used the same arrivals rates in both scenarios.
- The use of signals is required to identify the entrance and exit in order to avoid chaos.
- The different food bars (vegan, food, special) were unified into a general food bar, saving space and movements. Several snacks were placed in shelves along the customer's route, thereby improving their sales.
- The study revealed that the problem of wasting time is due to the food bars divisions, and not due to the servers in the cashier position. A detailed activity analysis is required to improve the behavior of the operator in the food bars.

Designing the demand is a possible solution, in fact this is mentioned by Klassen and Rohleder (2010), who explain that if the server capacity is less than the demand, the result are the queues, causing losses in sales, and dissatisfaction on customers. Some solutions to improve this situation are as follows:

- Marketing strategies, promotions, hot sales, etc., in different hours during the day.
- Scheduling management during rush times, through pricing strategies.
- E-marketing.

5 Conclusions

This chapter focuses on the application of queuing theory and Lean techniques along a system that is collapsing in specific times. It is important to note that along this study, the two methodologies (Lean and queueing theory) are not treated as independent, on the contrary, it seeks to show synergy among all the elements used, and how they create an even stronger methodology. In this case, the techniques were applied in a service company, a university cafeteria, therefore several of the techniques were adjusted and adapted to this environment.

As a starting point, it can be concluded that data collection is essential for both queuing theory and Lean, because they help to identify problems in the system. The data gathered for this study includes customer arrivals in specific times, waiting times, and services times; with this information, it was possible to identify what Lean tools or techniques were useful for this particular system.

The fact of studying a real case like this one, makes possible to understand the reasons for the system's collapse, concluding the following:

1. Rush time causes that the system blows up, no matter how well designed it is. Therefore, strategies for workers and the system activities to engage this specific event should be designed.
2. The demand uncertainty is one cause of the high variability in the systems, and this situation is considered as high level of randomness. However, it can be controlled through marketing techniques (discounts for hours, promotional events, customer contact, customer relationship management, among others), thus controlling the flow of demand into the system.
3. The disorder and bad habits in the system are another cause for the chaos. The use of standard techniques and clear and specific methods will be the solution for this problem.

It is also possible to conclude that Lean manufacturing principles, along with queuing theory, can contribute positively to the reduction of one of the most undesirable wastes in services, "waiting times."

It is necessary to mention that simulation constitutes a valuable tool in this type of problems, in order to identify results of possible actions under scenarios of rush time, or scenarios of normal behavior. This would allow to manage the strategies based on the dynamic behaviors of the system.

Finally, both, queuing theory and Lean, require that the system has as basic attributes of flexibility and adaptation to change, in order to tolerate the different techniques and strategies.

References

Belhadi A, Touriki FE, El Fezazi S (2018) Benefits of adopting lean production on green performance of SMEs: a case study. Prod Planning Control 29(11):873–894. https://doi.org/10.1080/09537287.2018.1490971

Brown SM, Hanschke T, Meents I, Wheeler BR, Zisgen H, Wheeler BR (2010) Queueing model improves IBM' s semiconductor. Interfaces. https://doi.org/10.1287/inte.1100.0516

Chadha R, Singh A, Kalra J (2012) Lean and queuing integration for the transformation of health care processes a lean health care model. Clin Gov Int J 17(3):191–199. https://doi.org/10.1108/14777271211251309

Chong MY, Chin JF, Loh WP (2013) Lean incipience spiral model for small and medium. Int J Ind Eng 20(2000):487–501

Cinlar E (2013) Introduction to stochastic processes. Republished Courier Corporation, Mineola, New York

Cope R, Syrdal H (2011) Innovative knowledge management at disney: human capital and queuing solutions for services. J Serv Sci 4(1). https://doi.org/10.19030/jss.v4i1.4268

Forrester R (1995) Implications of lean manufacturing for human resource strategy. Work Study 44:20–24. https://doi.org/10.1108/00438029510146944

Hopp WJ, Spearman ML (2008) Factory physics, 3rd edn. Waveland Press, Long Grove

Iriondo R, Lasa S, Vila DC (2016) Takt time as a lever to introduce lean production in mixed engineer-to-order/make-to-order machine tool. Int J Ind Eng 23(2):94–107

Jayaram J, Das A, Nicolae M (2010) Looking beyond the obvious: unraveling the toyota production system. Int J Prod Econ 128(1):280–291. https://doi.org/10.1016/j.ijpe.2010.07.024

Juran JM, Godfrey AB, Hoogstoel RE, Schilling EG (1998) Juran's quality handbook. Mc Graw-Hill

Keyser RS, Sawhney R (2013) Reliability in lean systems. Int J Qual Reliab Manag 30(3):223–238. https://doi.org/10.1108/02656711311299818
Klassen KJ, Rohleder TR (2010) Combining operations and marketing to manage capacity and demand in services. Ser Ind J 21(2):37–41. https://doi.org/10.1080/714005019
Li G, Field JM, Davis MM (2017) Designing lean processes with improved service quality: an application in financial services. Qual Manag J. https://doi.org/10.1080/10686967.2017.11918497
Liker JK, Morgan J (2011) Lean product development as a system: a case study of body and stamping development at ford. Eng Manag J 23(1):16–28. https://doi.org/10.1080/10429247.2011.11431884
Monden Y (1993) Toyota production system. an integrated approach to just it time, 4th edn. Institute of Industrial Engineering, Boca Raton
Narasimhan R, Swink M, Kim SW (2006) Disentangling leanness and agility: an empirical investigation. J Oper Manag 24:440–457. https://doi.org/10.1016/j.jom.2005.11.011
Ohno T (1988) Toyota production system: beyond large-scale production. Productivity press, Tokyo
Olfert MD, Hagedorn RL, Clegg EN, Ackerman S, Brown C (2019) Choice architecture in appalachian high schools: Evaluating and improving cafeteria environments. Nutrients 11(1):1–12. https://doi.org/10.3390/nu11010147
Pearce A, Pons D, Neitzert T (2018) Implementing lean—outcomes from SME case studies. Oper Res Perspect 5:94–104. https://doi.org/10.1016/j.orp.2018.02.002
Pourvaziri H, Pierreval H (2017) Dynamic facility layout problem based on open queuing network theory. Eur J Oper Res 259(2):538–553. https://doi.org/10.1016/j.ejor.2016.11.011
Psomas E, Antony J, Bouranta N (2018) Assessing lean adoption in food SMEs: evidence from Greece. Int J Qual Reliab Manag 1–5
Sarkar A, Mukhopadhyay AR (2014) Productivity improvement by reduction of idle time through application of queuing theory. Opsearch. https://doi.org/10.1007/s12597-014-0177-2
Seddon J (2011) Lean is a waning fad. Manage Ser 34–36. Retrieved from http://web.b.ebscohost.com/ehost/pdfviewer/pdfviewer?vid=14&sid=4c24f179-953f-40ff-a64e-caa7faf882f8@sessionmgr115&hid=112
Taylor P, Taylor A, Taylor M, Mcsweeney A (2013) Towards greater understanding of success and survival of lean systems. Int J Prod Res 51(August 2015):37–41. https://doi.org/10.1080/00207543.2013.825382
Vendramini M, Cagnin F, De MC, Sérgio M, Dos C (2016) Simulation and lean principles: a case study in a public service in Brazil. Int J Performability Eng 12(2):103–113
Villarreal B (2014) A simulation approach to improve assembly line performance. Int J Ind Eng 18(6):283–290
Weiss EN, Tucker C (2018) Queue management: elimination, expectation, and enhancement. Bus Horiz 61(5):671–678. https://doi.org/10.1016/j.bushor.2018.05.002
Zhenyuan J, Xiaohong L, Wei W, Defeng J, Lijun W (2011) Design and implementation of lean facility layout system of a production line. Int J Ind Eng 18(5):260–269

Dynamic Study of Soil Improvement for Sugarcane Cultivation in Colombia

Dayhanna Stephanía Vargas-Mesa, Luz Karime Torres-Lozada, Juan Carlos Osorio-Gómez, and Patricia Torres-Lozada

Abstract Sugarcane cultivation is one of the main crops in Valle del Cauca, Colombia. However, intensive agricultural production leads to soil depletion (which in the long run can seriously affect production) and, therefore, the economic development of the region and the country. System dynamics is a discipline that allows long-term study of the behavior of complex systems from the feedback relationships between variables, by studying both the structure and behavior of systems. It is particularly recommended for complex systems with nonlinear relationships, and in cases which exhibit a time delay between the actions taken and the expected consequences of those actions. In this study, it was proposed to evaluate through the system dynamics methodology, and the possible long-term impacts that this crop could generate in the Valle del Cauca's soils. The simulation's model was applied using Vensim DSS software and it explored soil recovery scenarios using compost, which is produced from sugarcane residues composting (cachaça and bagasse). It was evident that the utilization of this by-product can represent an important contribution in the soil's loss and degradation reduction, plus economic and environmental

D. S. Vargas-Mesa
Faculty of Engineering, Logistic and Production Research Group, Dayhanna Stephania Vargas Mesa, Universidad del Valle, Calle 13 #100-00, Cali, Colombia
e-mail: dayhanna.vargas@correounivalle.edu.co

L. K. Torres-Lozada
Faculty of Natural Sciences and Engineering, Luz Karime Torres Lozada, Fundación Universitaria Unisangil, Calle 7 No. 20-63, Yopal, Colombia
e-mail: ltorres2@unisangil.edu.co

J. C. Osorio-Gómez (✉)
Faculty of Engineering, Logistic and Production Research Group, Juan Carlos Osorio Gomez, Universidad del Valle, Calle 13 #100-00, Cali, Colombia
e-mail: juan.osorio@correounivalle.edu.co

P. Torres-Lozada
Faculty of Engineering, Study and Control of Environmental Pollution –ECCA Research Group, Patricia Torres Lozada, Universidad del Valle, Calle 13 #100-00, Cali, Colombia
e-mail: patricia.torres@correounivalle.edu.co

J. L. García Alcaraz et al. (eds.), *Techniques, Tools and Methodologies Applied to Quality Assurance in Manufacturing*,
https://doi.org/10.1007/978-3-030-69314-5_16

benefits. Although the proposed model has been applied to the specific case of sugarcane, it can be replicated in other types of crop, thus becoming a valuable tool for the decision-making process involved in crop planning.

Keywords Systems dynamics · Land use · Sugarcane · Soil improvement

1 Introduction

Over the past 50 years, agricultural technology advances and increased demand due to population's growth have raised the soil's pressure. In many countries, intensive agricultural production has led to soil depletion, threatening the productivity of soils and the ability to meet the needs of future generations (FAO and ITPS 2015).

Colombian's agriculture is diversified, and traditional crops continue to occupy the largest proportion of the sowed area, highlighting those used as a raw material in the production of Colombian's most consumed foods. This is the case of sugarcane, which has approximately sown 352,786 for 2019, (from this amount, 200,499 belong to the Valle del Cauca's department) (Dane 2014; ASOCAÑA 2020). For this reason, it is necessary to conduct studies that contribute to an adequate management in the producer's side of the supply chain to identify and establish improvements in the production processes, and even more so in developing countries (Monterroso 2000; OECD/FAO 2017).

The excessive use of commercial fertilizers has contributed to reduce the organic matter content (OMC) and different physicochemical soil's properties, resulting in its quality's decrease, acidification, and contamination (Guo et al. 2010; Chuanchuan et al. 2017). Based on the problems caused by commercial fertilizers, organic compost use has led to a growing worldwide interest in the utilization of organic materials (Asses et al. 2018; Mardomingo et al. 2013), as it has been proved that these can increase soil's fertility and improve physical, chemical, and biological properties (Abbasi and Khizar 2012) being a safer and more effective alternative for nutrient recovery (Abbasi and Khizar 2012; Gilly and Eghball 2002) and leading to a commercial fertilizers usage reduction in regards to crop production (Paterson et al. 2011).

An alternative to organic amendment is cultivation and processing by-products that come from sugarcane, such as cachaça and cane bagasse (Meunchang et al. 2005; Bohórquez 2014). Due to its high impact in the country, and even more in the Valle del Cauca's department (where there is a suitable cropping area of approximately 400,618) (SIPRA 2020), this study assessed the applicability of the system's methodology dynamic, which is a modeling approach based on systemic thinking and the usage of feedback and delays information-based perspective. This can be used to understand the dynamics of complex behavior on physical, biological, and social systems (Aracil 1995; Serra 2016).

In this study, the impact of applying compost (which results from cachaça's and cane bagasse's composting) over the cultivation of sugarcane, was evaluated over a period of 40 years using tools from system dynamics.

2 Literature Review

A literature review was conducted in the SCOPUS and Web of Science databases with the search equation "Systems Dynamic" Agriculture OR "Systems Dynamic" Crop, which had 57 articles as results, and a bibliometric analysis was performed using VosViewer software with the keywords and the years in which the publications were made, obtaining the results of Fig. 1 identifying that the system dynamics has been used in models for decision making mainly in the management of water resources in the agricultural sector.

Then, two filters were made; the first in the RefViz software and five clusters were found where two additional keywords were determined the resource "soil" and the resource "water" which are the main decision-makers for ensuring the development and growth of the agricultural sector (Fig. 2).

From these articles are found some objectives for the simulation from the dynamics of systems in crops such as cereals, energy crops and other; the important criteria are shown in Table 1.

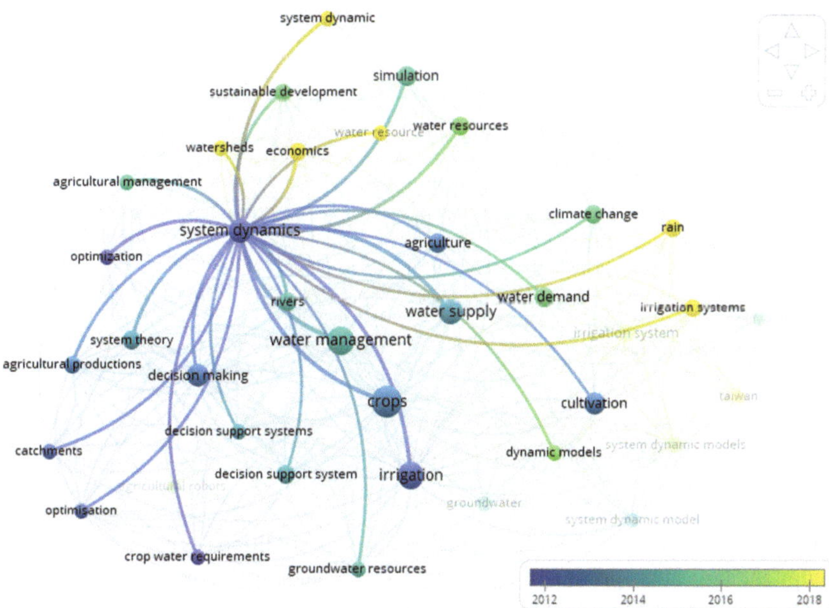

Fig. 1 Results of the bibliometric analysis VosViewer

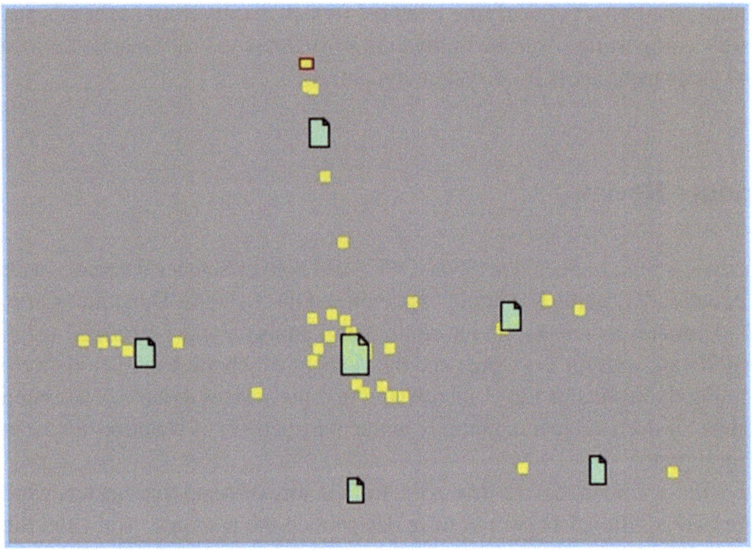

a.

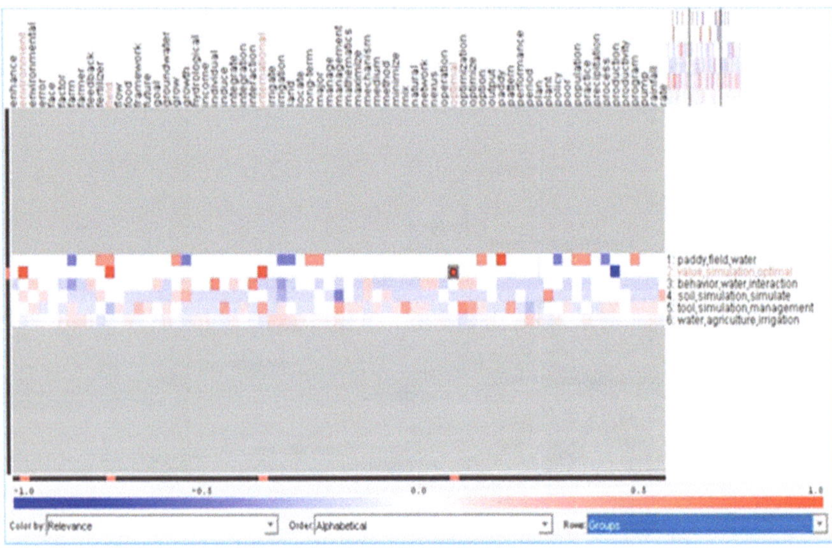

b.

Fig. 2 Results of the bibliometric analysis RefViz

Colombian agriculture is diversified; however, traditional crops continue to occupy the largest proportion of the planted area, characterized by those crops used as raw material for the production of the foods most consumed by the Colombian population, as is the case with sugarcane, with approximately 298,357 cultivated hectares (DANE 2019).

Table 1 Use of system dynamics for agricultural policy evaluation

System simulated	Simulation objective	Criteria	Reference
Complex socio-ecological systems	Environmental sustainability and human wellbeing	Resources: climate water Ecological: salinity social Economical: crops shrimp and mangrove production	El Gafy et al. (2017)
Intercropping of cereals and legumes	Determine the accumulation of dry matter and the effect of the Nitrogen supply on the penetration and depth of crop roots	Aerial biomass Nitrogen accumulation Nitrogen fixation Root depth	Corre-Hellou et al. (2007)
Hydraulic infrastructure policies and agricultural practices in dry seasons	Evaluate policies regarding available water resources over time	Availability of water Soil degradation Nutrient availability Socio-ecological well-being	Gies et al. (2014)
Rice crops and methane emissions	Estimate methane emissions from rice fields in India up to the year 2020	Reduce methane emissions in rice production The use of rice varieties with low methane emissions Water management The fertilizer amendment	Anand et al. (2005), Cui et al. (2009), Xu et al. (2018)

(continued)

In report #1 by The National Planning Department, Green Growth Mission, on the study of the bio-economy—as a source of new industries based on capital (Diagnosis and definition of strategic sectors for Colombia)—the consulting team in the company of strategic actors considered that the sugarcane sector should be prioritized in the agricultural field, to take advantage of the opportunities for biotechnology innovation, in order to find new levers for growth and entry to more sophisticated markets, where it is suggested to include applications such as the development of new varieties, agricultural bio-inputs, generation of germplasm banks, use of varieties for soil bioremediation, extraction of metabolites and active ingredients from agricultural

Table 1 (continued)

System simulated	Simulation objective	Criteria	Reference
Integrated system to support hydrological economic decision making for the use of groundwater considering climate change	Define the best combination of crops and the total land use required to maximize total monetary income of the sub-watershed	Land use by harvest Total net benefit of the sub-basin Total withdrawal of water from the sub-basin Water withdrawal limit according to river discharge The total number of wells needed to irrigate the entire area	Tromboni et al. (2014), Kotir et al. (2016)
Wheat cultivation and the use of water resources	Identify the water footprint of harvest, production and consumption	The population Per capita consumption harvest Harvest business patterns Harvest yield Impact of climate change	Wu et al. (2007, 2009, 2011, 2017, 2018, 2019), Jackson et al. (2007), El-Gafy (2014), Beddek et al. (2005), Elmahdi et al. (2005)
Coffee crop	Determine the growth of the coffee plantation on different substrates	Climate Soil Vegetal material	Díaz-Ambrona et al. (2008)

waste, development of biorefineries and bio-products, management of stillage for alcohol production fuel and production of biofertilizers, among others (DNP 2018). This recommendation is also associated with studies that provide opportunities for improvements in processes such as the proper use of soil in sugarcane cultivation and the effect of this factor upon crop production (Fernández 2013; MINAMBIENTE 2016; FAO 2018).

It is estimated that 95% of our food is produced directly or indirectly in our soils, which means that soils are the foundation of the food system and are the basis of agriculture and the environment in which almost all plants grow. Furthermore, the quality of soils is directly related to the quality and quantity of food (FAO 2015).

Food availability depends on soils: good quality and nutritious food and feed cannot be produced if they are not healthy and living soils. Over the past 50 years, advances in agricultural technology and increased demand caused by population growth have put increasing pressure on soils. In many countries, intensive agricultural production has led to a depletion of soils that has endangered their productive capacity and the possibility of meeting the needs of future generations (FAO 2015).

Four fifths of the food needed for more than nine billion people in 2050 will come from existing land through the intensification of agricultural production, and a large percentage of this food demand is expected to be met by the countries of Latin America and the Caribbean (LAC). Coincidentally, the high vulnerability of many countries in the region, accentuated to a great extent by the adverse effects of climate variability and climate change, pose greater threats to agricultural production, food and nutrition security, and sustainable development (Montiel and Ibrahim 2016).

FAO estimates (2011) indicate that a quarter of the planet's land shows a high tendency to degradation or is heavily degraded land. According to Gardi et al. (2014), more than half of the 576 million hectares of arable land in Latin America, particularly 74% in Mesoamerica and 45% in South America, is affected by degradation processes due to changes in land use, overexploitation, climate change and social inequity. The report also highlights the climatic vulnerability that exists in the region to typically be more critical for small producers.

The Integrated Soil Management (ISM) approach can lessen these negative effects by highlighting the physical properties of soil and organic matter to improve fertility, water availability, vegetation cover, optimization of nutrient cycles, and conservation techniques (Montiel and Ibrahim 2016).

For this reason and the fact that it is one of the most representative crops in Colombia, the cultivation of sugarcane is taken as a case study. Furthermore, there exists the need to evaluate the soil use conditions to mitigate environmental impacts and evaluate the possible effects these can generate on sugarcane production.

3 Methodological Approach

The proposed methodological approach is based on Aracil (1995) and Sterman (2000) and comprises a sequence of steps in which one can regress a step to fine-tune the model as shown in Fig. 3. Here it is important to highlight that the modeling process is an iterative process by means of which the different conceptual and operational elements provided by system dynamics are combined to reach an acceptable model of the process under study as a final result. This is the working scheme when simulation models are made with system dynamics.

At this point then, we begin with the definition of the problem and its conceptualization. This conceptualization is achieved from the causal diagram, in which the main variables associated with the problem and the existing influence relationships between them are represented. Depending on whether the relationships are direct or inverse, then, we identify them as reinforcement or compensation feedback loops.

From these loops, the first approximation to the expected behavior of the system is obtained. Later, this diagram of influences is taken to a Forrester diagram which, once the mathematical equations that relate the variables are incorporated, can be simulated in a specialized language.

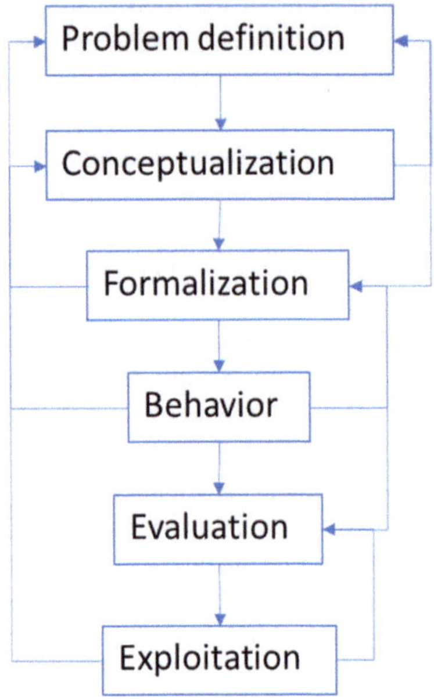

Fig. 3 Methodological approach (Aracil 1995)

These models are validated to see if they effectively represent the real situation, and scenarios are proposed that involve the study of different policies or strategies to be implemented, to determine whether these strategies achieve better system performance based on the indicators of interest.

4 Results

4.1 Causal Loop Diagram

Figure 4 shows the relationship of the feedback loops, whether positive or negative, of the different variables involved in the appropriate land use for crop production (specifically for sugarcane).

Each positive feedback or reinforcement loop (R1, R2 and R3) is described below and is understood as the variation of a propagating variable, reinforcing the initial variation and tending to generate both growth behavior and negative feedback or balancing (B1 and B2). These indicate that the variation of a variable determines what counteracts the initial transition and tends to generate equilibrium behavior (Morlán 2010; Sterman 2000).

Positive feedback or reinforcement loops are:

R1: represents that within greater availability of resources (pH, climate, Topography, microorganism, organic matter and texture), there is greater soil recovery,

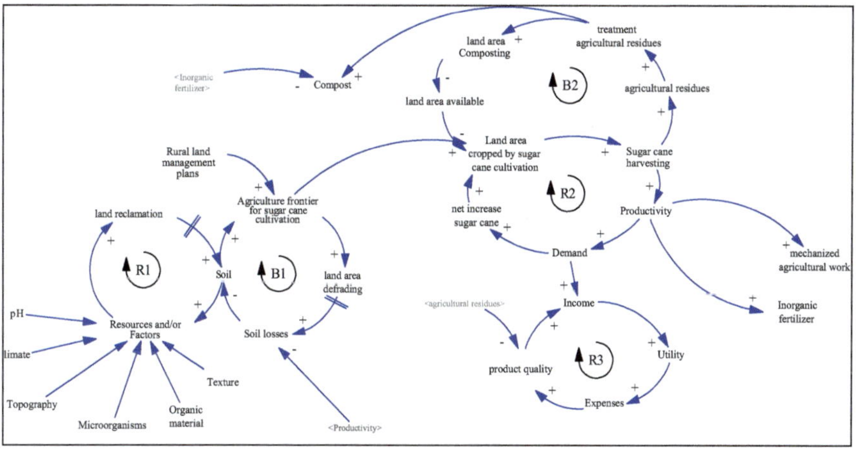

Fig. 4 Sugarcane cultivation influence diagram

leading to more soil availability over time (with the conditions for use in agricultural activities) (FAO and ITPS 2015; Scialabba and Hattam 2002) (Fig. 5).

R2: represents that the greater the sugarcane cultivation's area sown, the greater the yield will be, which leads to an increase in productivity. This will be bound to the product's increased demand, resulting in a rising effect on the cane net's increase (ASOCAÑA 2020; Melgar et al. 2012) (Fig. 6).

R3: demonstrates that higher incomes with the cultivation of sugarcane lead to the generation of greater profits, being one of the most representatives and with the greatest contribution of gross domestic product (GDP) in the agricultural sector. Therefore, the sugar sector has implemented improvement strategies and processes

Fig. 5 R1

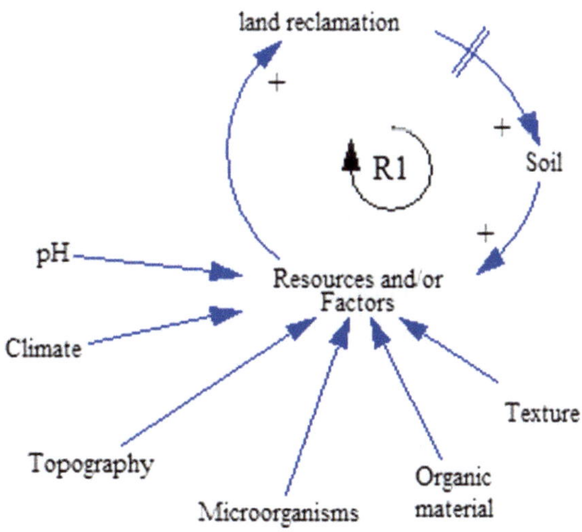

Fig. 6 R2

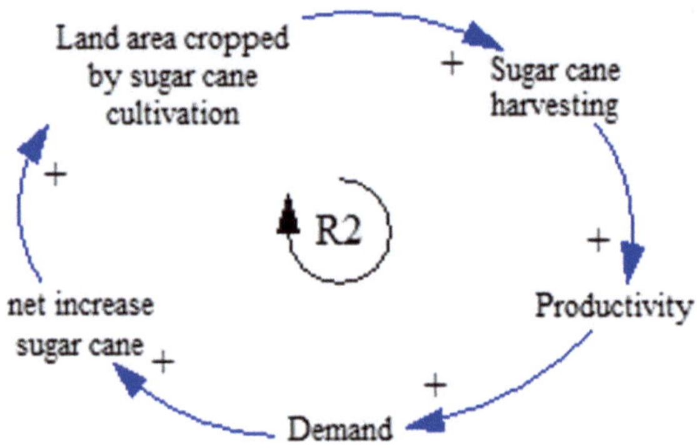

Fig. 7 R3

investments, which in turn have improved the crop's productivity and the product's quality (Bohórquez et al. 2014; Quiroz and Perez 2013; DNP 2018) (Fig. 7).

Negative feedback loops, or balancing ones are:

B1: shows that the higher the land use, the more degradation problems will arise, leading to gradual losses of soil and making the resource less available, with soil's over-use or overexploitation. Land and soil degradation refer to the negative decrease or alteration of one or more ecosystem's offerings and environmental goods, services and/or functions. This is caused by natural or man-made processes which, in critical cases, may cause the loss or destruction of the environmental component (IDEAM 2004) (Fig. 8).

B2: shows the relationship between the use of organic waste and the area available after the soils are recovered. The more waste is used as compost for organic waste, the more can be returned as fertilizer, conditioner, or supplement for fertilization. As a result, the physical, chemical, and biological properties improve, leading to a more environmental alternative, reducing costs at the level of fertilization, and minimizing the impact generated by soil degradation (Daza et al. 2015) (Fig. 9).

Fig. 8 B1

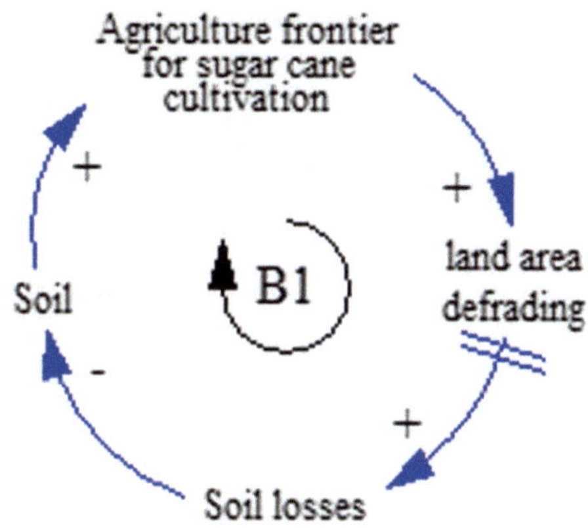

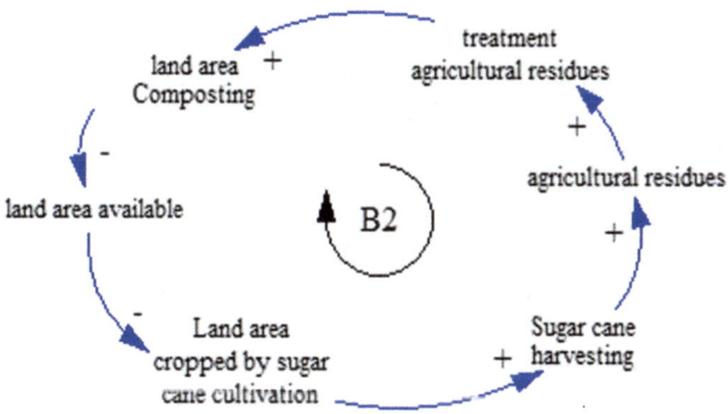

Fig. 9 B2

4.2 *Forrester's Diagram*

The study was applied to the context of sugarcane cultivation in Valle del Cauca's department, Colombia. A causal loop diagram was drawn up and based on this, a Forrester diagram was proposed to design two hypothetical scenarios for a period of 40 years, from 2001 to 2041.

The scenarios have the following characteristics:

Scenario 1—Sugarcane cultivation with compost application: The compost application rate varies according to the type of soil, crop, and season. The World Health Organization (WHO) states that 100–300 Ton/Ha of compost is usually required per year (Mara and Cairncross 1989). This study assumed a requirement of 0.3 Ton/Ha compost from the cachaça's and bagasse's composting by-products, whose processing time is in the order of 90 days (Bohórquez et al. 2014; Quiroz and Pérez Vázquez 2013). In addition, a percentage of soil loss due to compaction was considered. This is the most evident factor in the sugarcane production process due to the

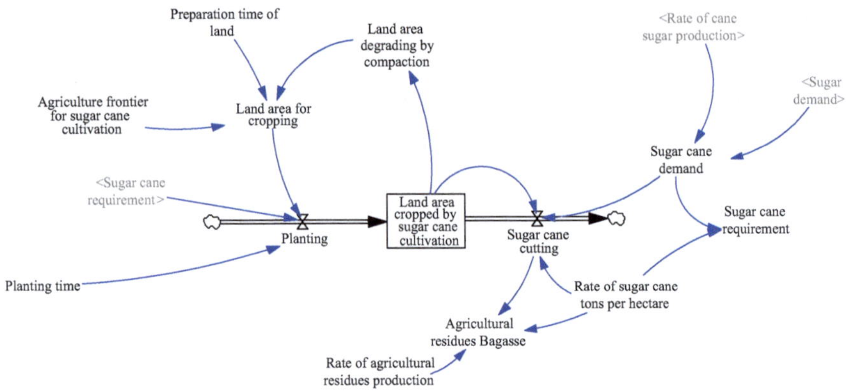

Fig. 10 Forrester diagram on sugarcane planting

agricultural work generated by the agricultural machinery (Torres 2006), which was 1% (Wischmeier and Smith 1978; González del Tánago 1991).

Scenario 2—Sugarcane cultivation without compost application: It is evaluated without the use of compost. In this case, a soil degradation rate of 4% is proposed (Morales and Parada 2005; Castro Mendoza 2013).

In both cases, it is assumed that the productions reported by the different sugarcane entities are based on traditional fertilization, which is made of chemical fertilizers.

Figures 10, 11, 12, and 13 show the basic variables that give the movement to the system and are indicated as "Level", which were taken to perform the simulation associated with the production of sugarcane cultivation and its relationship to the use of land suitable for the development of such crops:

- Sugarcane planting is associated with factors such as cultivation area, preparation time of the soil, and the area that is destined for the agricultural vocation. In

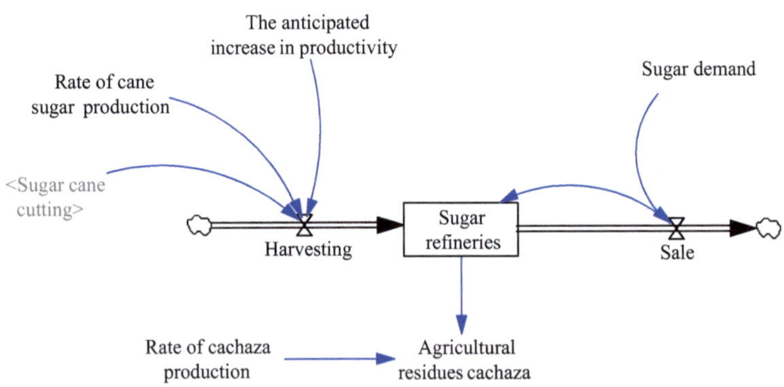

Fig. 11 Forrester diagram on sugarcane crop production

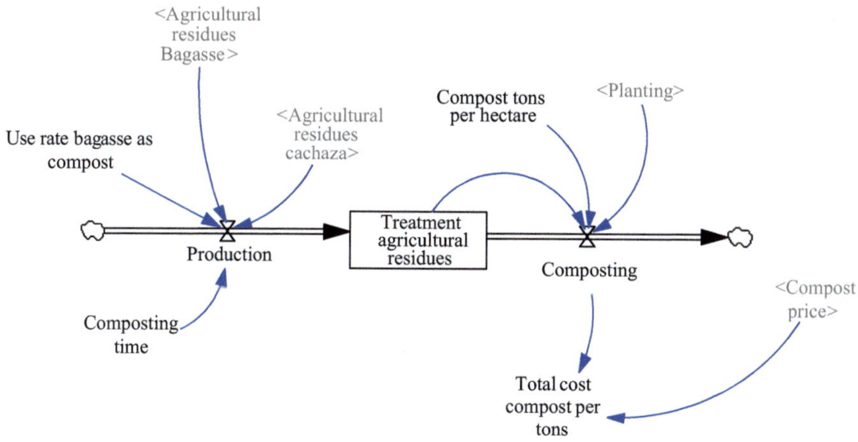

Fig. 12 Forrester diagram on the use of organic waste

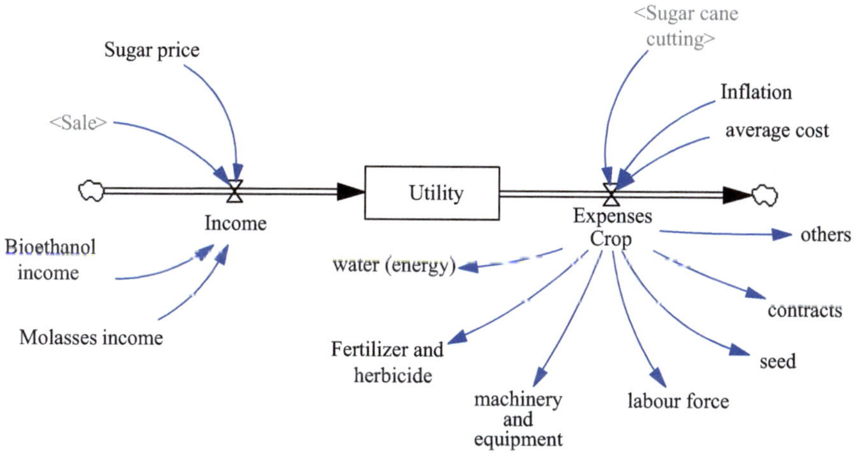

Fig. 13 Forrester diagram on utilities

addition, the area of soil degraded by the compaction factor was considered, which is associated with the demand of the crop and its harvest (see Fig. 10).

- For the sugarcane's production and harvesting process, the response is given to the demand's satisfaction and production's sales (see Fig. 11).
- The use of organic wastes in the sugarcane's production involves the generation of wastes from the process of composting that will be used in the soil for their improvement and up to their processing (see Fig. 12).
- Some variables associated with the income and expenses that are considered in the sugarcane's production process were considered for the profit flow (see Fig. 13).

Table 2 Variables for the simulation of the two scenarios

Variable	Scenario 1 and scenario 2	References
Agriculture frontier for sugarcane cultivation	400,618 ha	SIPRA (2020)
Expenses crop Water (energy) Fertilizer and herbicide Machine and equipment Labor force Seed Contracts Others	Proportion 22.22% 19.66% 10.64% 8.66% 3.09% 3.35% 32.3%	MINAGRICULTURA (2015)
Processing time composting	90 days	
Income		
Sugar ton	$2,878,000 (Ton)	ASOCAÑA (2020)
Bioethanol	$918,855,255,000	
Molasses	$210,460,140,000	

Table 3 Variables considered for scenario 1 and scenario 2 simulation

Variable	Scenario 1	Scenario 2	Reference
Rate land area degrading by compaction	4%	1%	Tauta Muñoz et al. (2018), Torres (2006)
The anticipated increase in productivity (year)	4%	7%	MINAGRICULTURA (2015)
Compost tons per hectare (year)	0.3 Tons	0 Tons	Mara and Cairncross (1989), Garcés (2019)

Now, the result of the two evaluated scenarios will be presented: where A represents the income-to-expenses ratio discharges, B shows the relationship between the costs of inorganic fertilizers and those produced by the resulting composting process, C shows the correspondence between the area to be cultivated and the sown area, and D shows the generation's residues behavior and how the compost is consumed when it is used as organic fertilization (Tables 2 and 3).

Scenario 1—Sugarcane Cultivation Without Compost Application

Figure 14 shows the result stated in scenario 1 for the Forrester diagram in sugarcane cultivation.

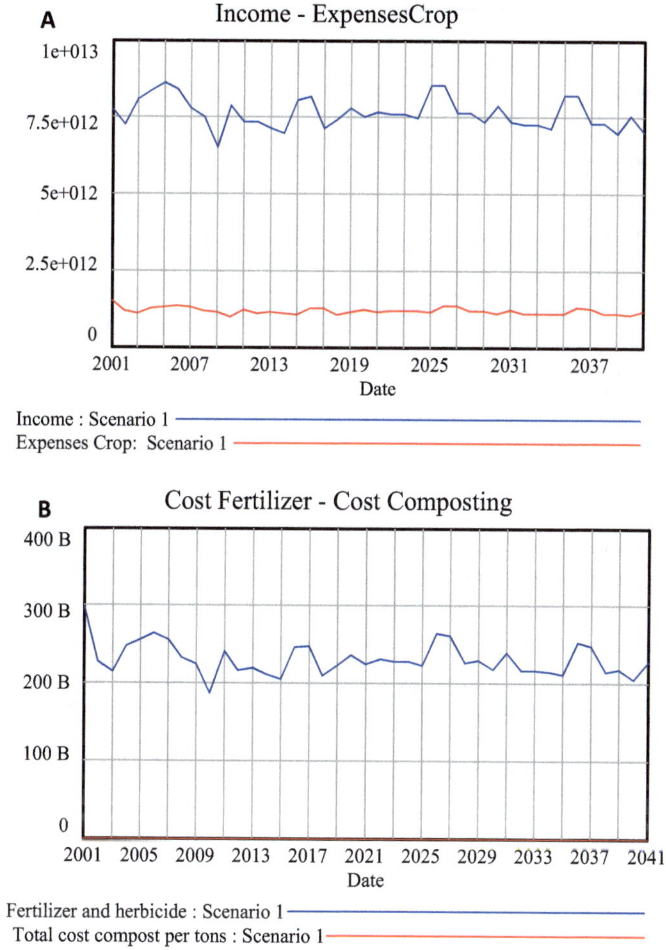

Fig. 14 Scenario 1—forrester diagram in sugarcane crop

Scenario 2—Sugarcane Cultivation with Compost Application

Figure 15 shows the outcome of the proposed approach for scenario's 2 outcome.

When comparing the two scenarios, it can be shown that a positive impact is generated by making use of the by-product resulting from the organic's composting waste generated in the sugarcane's cultivation. By compaction, there is a soil degradation reduction and, as well, it favorably affects the economic aspect since the cost associated with the use of compost is lower than the costs incurred for inorganic fertilization. We also observe the close relationship between the generation and consumption of compost within the process, which means that all the compost generated can be used in soil recovery tasks through organic fertilization.

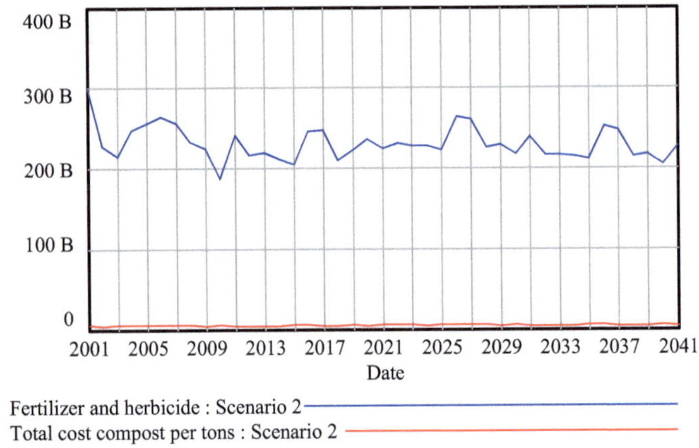

Fig. 15 Scenario 2—Forrester diagram in sugarcane crop

Figure 16 shows the influence of the organic waste's usage on the soil's degradation variable by compaction. For this case, with a crop such as sugarcane, it is evidenced that the use of compost from the cachaça's and bagasse's composting contributes to the mitigation of the gradual's degradation impact in the 40-year observation window over land use.

Figure 17 shows that the cost of composting ranges from 0.3 to 1.5 (percentage of the compost's cost proportion versus the fertilizers and herbicides cost), indicating that its cost is of greater benefit than inorganic fertilizers. Additionally, the area to be sown meets the demand's requirements and leads to costs reduction assumed by inorganic fertilization.

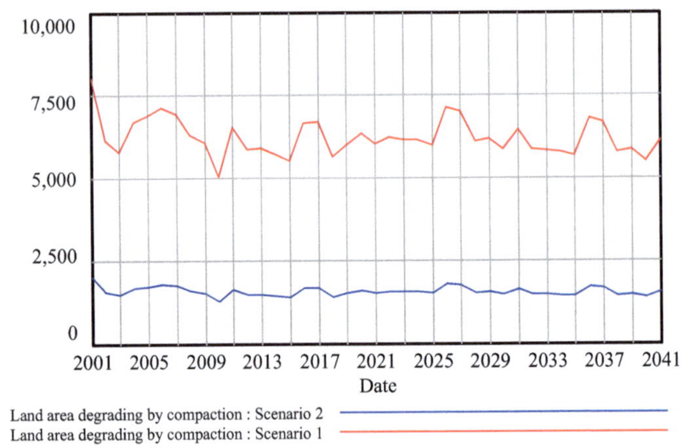

Fig. 16 Approach of soil's degraded area by compaction for both scenarios

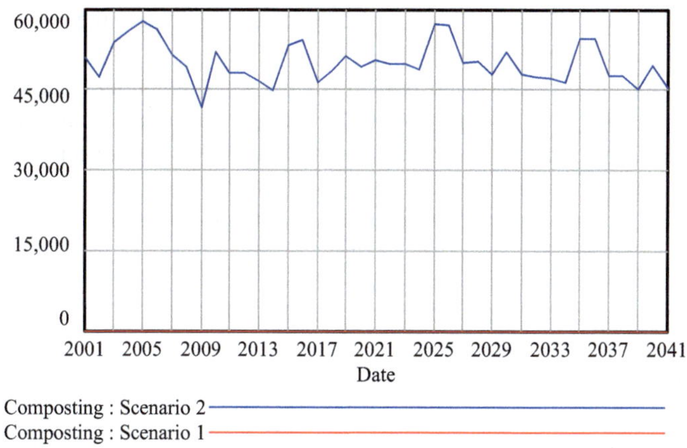

Fig. 17 Composting used in the scenarios

In a review of related topics, it was evident that there are no similar studies focused on the use of soil associated with the organic waste's usage versus production and improvement of costs compared to the production processes of sugarcane cultivation. Nevertheless, certain studies were found such as the application of system dynamics in the evaluation of the Bioethanol Production Potential from Panel-Cane: dynamics between contamination, food safety, and land studies (Castro Mendoza 2013). In this study, the Colombian's panel-cane was characterized for producing agro-fuels in Colombia and it evaluated the influence of oil's price in sugar and panel.

Another case study found is The Systemic Approach to Sustainability in Bioethanol Production, which interconnected the production's process indicators to qualitatively highlight the main attributes of the system (Morlán Santa Catalina 2010).

5 Conclusions

The utilization of the cachaça's and bagasse's composting by-product resulting from the sugarcane production process may contribute greatly to reducing the soil's loss and degradation (currently occurring in the world due to extensive cultivation practices such as that of sugarcane in Colombia).

The incurred costs to carry out the composting process can be recovered gradually over time, since carrying out the soil's restoration process with the same by-products leads to the fertilizer's investments cost reduction. In addition, it was evident that the generated waste was consumed entirely in the process, which means that there is no loss of the obtained composting. According to the requirements established by the crop, the compost generated would be entirely consumed and would not meet the needs of the crop, so the use of compost is more a complement than a replacement for chemical fertilization.

Future work should include the associated cost for commercial fertilizers and the reduction in costs for the use of composting and the nutritional contribution of the same in the cultivation of sugarcane. This is also part of the concept of the green economy, where environmental risks are substantially reduced, and this use with organic products can improve the diet of the population by minimizing the use of chemicals.

References

Abbasi MK, Khizar A (2012) Microbial biomass carbon and nitrogen transformations in a loam soil amended with organic-inorganic N sources and their effect on growth and N-uptake in maize. Ecol Eng 39:123–132

Anand S, Dahiya RP, Talyan V, Vrat P (2005) Investigaciones de las emisiones de metano del cultivo de arroz en el contexto indio. Medio Ambiente Internacional 31(4):469–482

Aracil J (1995) System dynamics. Alianza Editorial, Madrid

ASOCAÑA (2020) Sustainability Report 2019–2020. Cali, Colombia

Asses N, Farhat A, Cherif S, Hamdi M, Bouallagui H (2018) Comparative study of sewage sludge co-composting with olive mill wastes or green residues: process monitoring and agri-culture value of the resulting composts. Process Saf Environ Prot 114:25–35

Beddek R, Elmahdi A, Barnett B, Kennedy T (2005) Integration of groundwater models within an economical decision support system framework. In: MODSIM 2005 International Congress on modelling and simulation, Melbourne, Australia (unpublication)

MINAMBIENTE—Ministerio de Ambiente y Desarrollo Sostenible (2016) Política para la Gestión Sostenible del Suelo. Grupo de Divulgación de Conocimiento y Cultura Ambiental. Bogotá, D.C. Colombia

Bohórquez A, Puentes YJ, Menjivar JC (2014) Quality assessment of compost produced from agro-industrial sugar cane byproducts. Corpoica Cienc Tecnol Agropecu 15(1):73–81

Castro Mendoza I (2013) Estimation of soil loss due to water erosion in Madín Dam micro-basin, Mexico. Hydraul Environ Eng XXXIV (2), 3–16

Chuan-chuan N, Peng-dong G, Bing-qing W, Wei-peng L, Ni-hao J, Kun-zheng C (2017) Impacts of chemical fertilizer reduction and organic amendments supplementation on soil nu-trient, enzyme activity and heavy metal content. J Integr Agric 16(8):1819–1831

DANE (2014) Third National Agricultural Census. Colombia

Corre-Hellou G, Brisson N, Launay M, Fustec J, Crozat Y (2007) Effect of root depth penetration on soil nitrogen competitive interactions and dry matter production in pea–barley intercrops given different soil nitrogen supplies. Field Crops Res 103(1):76–85

Cui J, Huang G, Tang Y (2009) System dynamic simulation of three crops per year in paddy field. World J Model Simul 5(2):144–150

DANE (2019) Boletín Técnico Encuesta Nacional Agropecuaria (ENA) Primer semester. Colombia, p 2019

Daza MC, Oviedo ER, Marmolejo LF, Torres P (2015) Selection of agro-environmental systems with potential use of municipal bio-waste compost. Agronomic Act 64(2):134–145

Gónzalez del Tánago M (1991) The universal equation of past, present and future soil losses. Ecology 5:13–50

Díaz-Ambrona CGH, González de Miguel C, Martínez-Valderrama J (2008) Three layer coffee plantation model. In: IV International symposium on applications of modelling as an innovative technology in the agri-FOOD-CHAIN: Model-IT. vol 802, pp 319–324

DNP (2018) Study on bioeconomy as a source of new industries based on Colombia's natural capital No 1240667, Phase I. National Planning Department. Colombia

El-Gafy IK (2014) System dynamic model for crop production, water footprint, and virtual water nexus. Water Resour Manage 28(13):4467–4490

El Gafy I, Grigg N, Reagan W (2017) Dynamic behaviour of the water–food–energy Nexus: focus on crop production and consumption. Irrig Drainage 66(1):19–33

Elmahdi A, Malano H, Etchells T, Khan S (2005) System dynamics optimisation approach to irrigation demand management. In: MODSIM 2005 International Congress on modelling and simulation, modelling and simulation society of Australia and New Zealand, pp 196–202

FAO (2011) El estado de los recursos de tierras y aguas del mundo para la alimentación y la agricultura. La gestión de los sistemas en situación de riesgo. Mundi-Prensa, Madrid

FAO (2015) Los suelos sanos son la base para la producción de alimentos saludables. Organización de las Naciones Unidas para la Alimentación y la Agricultura. Roma, Italia

FAO (2018) Guía de buenas prácticas para la gestión y uso sostenible de los suelos en áreas rurales. Organización de las Naciones Unidas para la Alimentación y la Agricultura. Bogotá, Colombia

FAO and ITPS (2015) Status of the World's Soil Resources (SWSR)—Main Report. Food and Agriculture Organization of the United Nations and Intergovernmental Technical Panel on Soils, Rome, Italy

Fernández M (2013) Efectos del cambio climático en la producción y rendimiento de cultivos por sectores: evaluación del riesgo agroclimático por sectores. Fondo Financiero de proyectos de desarrollo FONDA e Instituto de Hidrología, Meteorología y estudios ambientales IDEAM. Recuperado de https://www.ideam.gov.co/documents/21021/21138/Efectos+del+Cambio+Climatico+en+la+agricultura.pdf/3b209fae-f078-4823-afa0-1679224a5e85

Garcés E (2019) Guía para labores de cultivo de la caña de azúcar en la empresa Garcés Eder SAS. Disponible en: https://red.uao.edu.co/bitstream/10614/11279/12/T8670F.pdf

Gardi C, Angelini M, Barceló S, Comerma J, Cruz Gaistardo C, Encina Rojas A, Jones A, Krasilnikov P, Mendonça S, Brefin ML, Montanarella L, Muñiz Ugarte O, Schad P, Vara Rodríguez MI, Vargas R (2014) Atlas de suelos de América Latina y el Caribe, Luxembourg, Comisión Europea-Oficina de Publicaciones de la Unión Europea, p 176

Gies L, Agusdinata DB, Merwade V (2014) Desarrollo y evaluación de políticas de adaptación a la sequía en África Oriental utilizando modelos hidrológicos y de dinámica de sistemas. Peligros Naturales 74(2):789–813

Gilly JE, Eghball B (2002) Residual effects of compost and fertilizer applications on nutrients in runoff. Biol Syst Eng 45:1905–1910

Guo JH, Liu XJ, Zhang Y, Shen JL, Han WX, Zhang WF, Christie P, Goulding KWT, Vitousek PM, Zhang FS (2010) Significant acidification in major Chinese croplands. Science 327:1008–1010

IDEAM—Institute of Hydrology, Meteorology and Environmental Studies (2004) Annual report on the state of the environment and renewable natural resources in Colombia. Bogotá

Jackson TM, Khan S, Ahmad A (2007) Exploring energy productivity for a groundwater dependent irrigated farm using a system dynamics approach. In MODSIM 2007 international congress on modelling and simulation. Modelling and Simulation Society of Australia and New Zealand, pp 156–162

Kotir JH, Smith C, Brown G, Marshall N, Johnstone R (2016) A system dynamics simulation model for sustainable water resources management and agricultural development in the Volta River Basin, Ghana. Sci Total Environ 573:444–457

Mardomingo JI, Soler RP, Casermeiro MÁ, De la Cruz MT, Polo A (2013) Seasonal changes in microbial activity in a semiarid soil after application of a high dose of different organic amendments. Geoderma 206:40–48

Mara D, Cairncross S (1989) Guidelines for the safe use of wastewater and excreta in agriculture and aquaculture. Measures for public health protection. World Health Organization, Geneva, Switzerland

MINAGRICULTURA- Ministerio de Agricultura y Desarrollo Rural (2015) Cadena productiva de la caña de azucar. Asocaña–procaña-ministerio de agricultura y desarrollo rural. Colombia

Morlán Santa Catalina, I. (2010a) System Dynamics Model for the Implementation of Information Technologies in University Strategic Management. Editorial Service of the University of the Basque Country /Euskal Herriko Unibertsitatearen Argitalpen Zerbitzua, San Sebastian.
Melgar M, Meneses A, Orozco H, Pérez O, Espinosa R (2012) Sugar cane cultivation in Guatemala. Artemis Edinter, Guatemala
Meunchang S, Panichsakpatana S, Weaver RW (2005) Co-composting of filter cake and bagasse; by-products from a sugar mill. Biores Technol 96(4):437–442
Monterroso E (2000) Logistics process and supply chain management. Bogotá
Montiel K, Ibrahim M (2016) Manejo integrado de suelos para una agricultura resiliente al cambio climáticosistematización del ciclo de foros virtuales: Año Internacional de los Suelos (AIS) 2015 (No. P40). IICA, San José, Costa Rica
Morales C, Parada S (2005) Poverty, desertification and degradation of natural resources. Cepal, Santiago de Chile
OECD/FAO (2017) Guide to responsible supply Chains in the agricultural sector. París
Paterson E, Neilson R, Midwood AJ, Osborne SM, Sim A, Thornton B, Mil-lard P (2011) Altered food web structure and C-flux pathways associated with mineralization of organic amendments to agricultural soil. Appl Soil Ecol 48(2):107–116
Quiroz Guerrero I, Pérez Vázquez A (2013) Vinasse and cachaça compost: effect on the quality of the soil cultivated with sugarcane. Mex J Agric Sci 5:1069–1075
Scialabba N, Hattam C (eds) (2002) Organic agriculture, environment and food security (No. 4). Food & Agriculture Organization, Roma
Serra F (2016) The systemic approach and system dynamics as a methodology of the NTE for the study of complex phenomena. Civilizing Commun Sci 3(3):119–134
SIPRA (2020) Rural agricultural planning information system. https://sipra.upra.gov.co/. Last accessed 8 Sept 2020
Sterman J, Oliva R, Linderman K, Bendoly E (2015) System dynamics perspectives and modeling opportunities for research in operations management. J Oper Manag 39–40:1–5
Tauta Muñoz JL, Camacho-Tamayo JH, Rodríguez Borray GA (2018) Estimación de erosión potencial bajo dos sistemas de corte de caña panelera utilizando la ecuación universal de pérdida de suelos. Revista UDCA Actualidad & Divulgación Científica 21(2):405–413
Torres A (2006) Crop management under green cane conditions. CENICAÑA, Cali, Colombia
Tromboni F, Bortolini L, Morábito JA (2014) Sistema integrado de apoyo a la toma de decisiones hidrológico-económicas para el uso de aguas subterráneas frente a las incertidumbres del cambio climático en la cuenca del río Tunuyán, Argentina. Medio Ambiente, Desarrollo Y Sostenibilidad. 16(6):1317–1336
Wischmeier WH, Smith DD (1978) Predieting rainfall erosion losses. USDA Agr. Res, Servo Handbook, p 537
Wu R, Zheng Y, Liu J (2009) A dynamic simulation study of the impacts of enhanced UV-B radiation on winter wheat photosynthetic production and dry matter accumulation. Agrofor Syst 77(2):123–130
Wu R-S, Li M-H, Chen S-W (2011) Estimation and modeling the surface flow system for irrigation regions. J Taiwan Agric Eng 57(1):76–91
Wu RS, Liu JS, Chang SY, Hussain F (2017) Modeling of mixed crop field water demand and a smart irrigation system. Water 9(11):885
Wu R-S, Liu J-S, Chang S-Y, Su C-S, Chen P-Y (2018) Developing an irrigation model using both surface water and groundwater for a mixed paddy rice and upland crops field. J Taiwan Agric Eng 64(1):60–90
Wu R-S, Hsieh C-L, Ma C-C, Liu J-S (2019) Regulating and analysis of farm pond irrigation system in Taoyuan. J Taiwan Agric Eng 65(4):12–25
Xu P, Lin Y, Zheng J, Yang S, Luan S (2018) Policies scenario analysis based on the farmers' individual behavior for nitrogen and phosphorous nutrient controlling of Pearl River Basin. Hupo Kexue/J Lake Sci 30(1):44–56

Lean Manufacturing Implementation in Management of Residues from Automotive Industry—Case Study

Mercedes Estefanía Pérez-Naranjo, Sonia Valeria Avilés-Sacoto, and Galo Eduardo Mosquera-Recalde

Abstract Lean manufacturing is a system that improves the processes in an organization by using principles and tools to reduce several types of waste. This system has been applied to a wide number of areas and activities in different companies. Having said that, still there are certain companies operating in the recycling industry, which do not consider waste processes as a problem. It is here where a controversy is born: when a recycling company treating waste from other industries has "waste" in its processes. Lean manufacturing techniques tackle this problem by diminishing or eliminating the waste. This is the case presented herein, the Ecuadorian company AV. CORP., which experienced an improvement in its activities to treat waste materials coming from an automotive industry due to the application of lean manufacturing tools. All has been done by following the DMAIC methodology. Many benefits were obtained, such as the reduction in the occupied space in the plant, the reduction in movements of workers and others, the learning of lean culture in different levels of the enterprise and the general picture of the plant. An improvement of seventeen percent was achieved in terms of space, and it was experimented a reduction of a fifty percent in terms of variability in regard to time of disassembly of wooden pallets.

Keywords Waste · Recycling industry · Lean · Efficiency · Automotive · Space optimization

1 Introduction

During the Industrial Revolution, it was thought that consumption had a similar meaning as progress. Under this belief, industrial leaders exploited a lot of natural resources. This exploitation sparked a large amount of waste. In the following decades, the emphasis on waste reduction throughout production systems started to gain importance, and the result is the rise of lean manufacturing (Bradley et al. 2018).

M. E. Pérez-Naranjo · S. V. Avilés-Sacoto (✉) · G. E. Mosquera-Recalde
Departamento de Ingeniería Industrial, Colegio de Ciencias e Ingenierías, Universidad San Francisco de Quito, Quito, Pichincha, Ecuador
e-mail: svaviless@usfq.edu.ec

J. L. García Alcaraz et al. (eds.), *Techniques, Tools and Methodologies Applied to Quality Assurance in Manufacturing*,
https://doi.org/10.1007/978-3-030-69314-5_17

Lean manufacturing is not a new management model; actually, its origins can be traced back to the 1930s, but it has lasted in the world until now (Villareal et al. 2009). Worldwide, many enterprises use lean manufacturing in their processes; several companies in different fields such as automotive, food, construction, foundries, and even in the services sector have had great benefits of using lean manufacturing. Currently, care for the environment had acquired a new relevance, and lean tools can help to get it.

Enterprises can adopt at the same time two key strategies "lean and green"; both of them help to "create an environment stance that is a driver for reduced costs and risks, increased revenues, and improved brand image" (Fercoq et al. 2013). In the twenty-first century, it is commonly used the concept of "triple-bottom-line", a paradigm which combines environment, society, and economy aiming not only to reduce, reuse, and recycle different types of objects like pallets, tanks, metal structures, etc., but also to recover materials, redesigning legacy, and remanufacturing of products (Bradley et al. 2018). Among the different industries that manages this concept is the automotive industry; just to give an idea, the automotive sector is the biggest remanufacturing sector in the USA, with a market share of 33% (Butzer et al. 2016). "Waste management in automotive manufacturing has moved beyond the simple recycling of materials toward a concept known as 'the circular economy'" (McCandless 2017). These means that materials considered as scrap are given a "new and another" life, instead of throwing them away, and those will be used as raw material for new products. AV. CORP., for example, is an enterprise, whose activities focus on give the scrap materials a new use.

AV. CORP is an Ecuadorian enterprise dedicated to the logistic of waste management. It manages a vast range of projects concerning to waste management; one of these projects is related to automotive industry and involves different types of materials like cardboard, wood, plastics, gaskets, etc. Regarding the project of the automotive industry, AV. CORP does not have updated documentation related to the processes, so the knowledge on this subject is only in the worker's heads. Since there is a lack of standardization of processes, the presence of inefficiency among the activities is evident, originating a high percentage of "waste," being noticeable the unnecessary movements in the treatment of the different materials in process, long waiting times of trucks, accumulated inventory in the storage zone because there is not enough demand; therefore, the streams of recovered materials are slow, and overproduction in the terms that large quantities of materials are ready, but they have not been yet ordered by clients. In this context, some questions spring up: How to reduce waste in AV. CORP.'s solid residues storage plant by the use of lean tools to increase their profitability and to reduce their inventory? What are the non-added value activities in the critical processes in AV. CORP.? And what are the key performance indicators (KPIs) of the AV. CORP.'s critical processes? The right path to solve these questions is the correct application of lean manufacturing concepts and tools, which will be described in this study.

2 Literature Review

In the industrial production processes, different types of waste appear. Waste in the processes can be defined as activities that add cost but no value to processes, such as overproduction, waiting, excess transport, excess processing, excess stock on hand, excess movement, making defective product (Villareal et al. 2009). Additionally, "the under-utilization of creativity of employees is considered as eight waste and environmental waste is considered as ninth waste" (Vinodh et al. 2011). It is possible to reduce waste by eliminating non-value adding steps or instead minimizing the time and resources spent on them (Järvenpää and Lanz 2019).

A brief description of each waste will be detailed as follows; *Overproduction* means to produce goods not yet ordered, and in short, it is produce more than what actually is required (Lewis 2005). *Waiting* is referred to the waste of time on hand, for example, an operator who waits for the machine to finish its process or a line stop caused by defective parts entered in the process (Lewis 2005). Waste in *Transportation* is associated with the movement of components and products between different locations (Lewis 2005). In respect of *Over-processing*, it means to put more than it is required to the product, which causes slower processes and higher costs (Lewis 2005). In regard to waste of *Stock on hand*, it takes place when there is more inventory than is needed to satisfy customer demand (Lewis 2005). Waste of *Movement* is associated with the excess of movement that workers make while they are performing their day-to-day work such as squatting, standing up, moving round a table, and lifting loads. (Lewis 2005). Finally, making *Defective product* occurs when someone makes a mistake in a stage in the process or when a machine fails; this in turn causes inaccuracy in inventory counts (Lewis 2005).

However, all these waste problems can be tackled by the use of lean manufacturing. Lean manufacturing has its origins in 1930s, when Henry Ford revolutionized the car manufacturing by the introduction of techniques related to the mass production in Toyota, a Japanese automotive manufacturer (Villareal et al. 2009). During the last decades, the use of terms such as rationalization, optimization, or lean management by many companies gained importance, and it includes the industrial segment of remanufacturing (Butzer et al. 2016). Accordingly, Sutharsan et al. (2020) mention that lean manufacturing is the elimination of waste from all aspects of an organization, which occurs systematically and where waste is any use or loss of resources that are not directly related to the creation of the product or service. It must also be clear, as stated by Mudhafar et al. (2017) that "...one of the major challenges of Lean implementation is guiding the change journey, where guidance is the responsibility of the top management and leadership within an organization" (Mudhafar et al. 2017; Sutharsan et al. 2020).

According to Abdul Rahman et al. (2013), the objectives of lean manufacturing focus to reduce waste in human effort and inventory which allows to reach the market on time and to manage manufacturing stocks while quality products are produced in the most efficient and economical way. To achieve that the lean uses different tools to optimize waste removal. Some of the most famous and widely adopted tools to succeed in lean are 5S's, value stream mapping, SMED, standardized work, poka-yoke, and others (Samant and Prakash 2020). To clarify how these tools work, "5S's is a Japanese management technique for workplace organization and cleanliness" (Beck 2016), and it consists of five steps: sort, set in order, shine, standardize and sustain—from the Japanese Seiri, Seiton, Seiso, Seiketsu, Shitsuke, respectively. When this tool is performed well, it is possible to get a safer, more efficient, and more profitable organization (Beck 2016). Just to explain about the 5S's: *Sort* is the stage where all the unwanted tools materials and clutter—that is not required in the area—are removed (Beck 2016); *Set in order* is about arranging those items that are needed, in such a way that they can be used in the moment they are required, and in this stage, it is important to establish a place for everything (Beck 2016); *Shine* is the phase where it should be cleaned and refurbished the workplace as well as the equipment in the work area (Beck 2016); *Standardize*, as the name suggests, means to set schedules and standards, and to assign tasks to different job roles, and in short, it means cleaning the workplace every single day (Beck 2016); and finally *Sustain* is the stage where it is necessary to maintain the standards mentioned before and keep ongoing reviews (Beck 2016). All of this is done in order to promote continuous improvement.

In this field of lean manufacturing, there are a variety of tools that helps to improve processes, making them more efficient. These tools are described with more details below:

The value stream mapping (VSM) "is a technique used, especially for diagnoses, implementation, and maintenance of a lean approach" (Dal Forno et al. 2014). According to Sutharsan et al. (2020), VSM tool allows to visually display the whole flow of materials and information during all stages of the production process. It is key to emphasize that value stream maps should reflect exactly what is happening in the process rather than what is assumed to happen in that way different improvement opportunities can be determined.

Single minute exchange die (SMED) is another lean tool used to reduce setup time—times required for changeover and tuning—as well as to provide quick equipment changeover and rapid die exchange (Monteiro et al. 2019). Their benefits include the increase in production capacity without the necessity of purchasing new machines, reducing scrap and rework, quality improvement, lowering inventory, and increase the responsiveness and flexibility to customer (Mohammed et al. 2013).

The Kanban is another important lean tool, which tries to maintain the minimal inventory and reduced costs; Kanban system is made up of operational decisions that are taken in the production lines (Abdul Rahman et al. 2013). In the same field, kaizen, which means "continuous improvement" is a Japanese production method that encourages everybody in the factory to give ideas to make improvements and to modify the improperly production activities (Tsao et al. 2015). A kaizen event is a meeting that takes place in different days in which people from different levels in an organization contribute with ideas to solve a defined problem (Vinodh et al. 2011). Having frequent kaizen events allows the team to be connected and aligned with regard to the detected problems, which not only provides control in the improvement project, but also efficiency in the results (Ferreira et al. 2019).

Another important pillar of lean manufacturing is Jidoka; it can be translated into "autonomation" as mentioned by Järvenpää and Lanz (2019). The term "autonomation" indicates that it can stop itself when something wrong happens in the process and defects occur. The two principles of Jidoka are built-in quality and stopping production at the moment that defects are detected. This tool allows processes to have their quality self-assessment; Jidoka was invented by Sakichi Toyoda; he created a gadget with a self-assessment warning of failures or delays (Nelson 2016).

Some successful cases related to lean implementation include the Mexican producer of refrigerated food "*SIGMA Alimentos*" to reduce waste in transportation (Villareal et al. 2009). They saved around 12.3 million pesos in the future budgeted investment because they improved the vehicle capacity utilization and availability of vehicles (Villareal et al. 2009). Another example in the services field is related to the firm "*IC Insurance,*" which provides insurance service for homes, vehicles, pets, and travel (Smith et al. 2018). The principal focus of lean in this enterprise was "solving customers problems during their first contact, therefore avoiding multiple transfers between agents and the associated delay and increased opportunity for error"; all of this is a waste reduction in terms of lean methodology (Smith et al. 2018).

Although there are multiple successful case studies in many areas, in the recycling industry, there is evident the lack of lean techniques to help their processes. This mentioned industry focuses their activities, especially in recovering materials considered as waste in different processes that take place in industry and then transforms that waste into useful materials for other industries. Since nowadays the soaring prices for raw materials get consumer awareness, it is important to focus on remanufacturing, recycling, and collection (Thürer et al. 2016). "Recycling has several advantages. It enables to assign value to the waste generated and its return to the value chain. Activities such as the collection, sorting, and cleaning of material requires labor and little capital equipment, which promotes the inclusion of vulnerable populations" (Pardo Martínez and Piña 2017). In this line, Butzer et al. (2016) indicated that remanufacturing processes can be divided into five steps: disassembly, cleaning, inspection and diagnosis, reconditioning, and reassembly. The order in which the steps are followed depends on the type of product; for example, in the case of electronic products, it is needed first to make an entrance diagnoses before the disassembly step (Butzer et al. 2016).

The commonly used methodology in lean and six sigma projects is DMAIC. According to Cabrita et al. (2016), "the DMAIC—Define, Measure, Analyze, Improve and Control—cycle is an organized and effective methodology for defining and implementing improvement opportunities, creating at the same time a global mentality of continuous improvement." This cycle is recommended for the systematic handling of any project; for that reason, it will be used in this study. It is worth to mention here that "DMAIC helps expose and kill the root causes of defects in the process and thereby helps achieve higher customer satisfaction by means of improved quality" as Anshula et al. (2019) mentioned (Anshula et al. 2019).

3 Methodology

It is well known that the commonly used methodology in lean and six sigma projects is DMAIC. By the different stages "the DMAIC—Define, Measure, Analyze, Improve and Control—cycle, it is possible to implement improvement opportunities and create a culture oriented to the continuous improvement."

First phase: *Definition*, the problem related to inefficiency is identified, and after that, it is established the aspects of the processes that will be improved. These problems include aspects related to the seven MUDA, a Japanese term which means uselessness or misuse: defects, overproduction, transport, waiting, motion, inventory, and excessive processing (Fercoq et al. 2016). Additionally, in this phase, it is important to consider critical characteristics of products that are key for customers. Those characteristics will be monitored later in the following stages with the purpose to improve them (Cabrita et al. 2016). It is in this stage where improvement opportunities have to be clearly defined; they will be exploited in the following stages (Srinivasan et al. 2014). Many meetings are called with team members and stakeholders such as suppliers, with the purpose to get a better understanding of processes, define the project focus, duration, scope, and also expected result (Gupta et al. 2016).

Second phase: *Measure*, this phase includes collection of data and data evaluation (Cabrita et al. 2016). Here, it is necessary to develop a baseline which includes the functioning of the process, system, or issue of interest. Cabrita et al. (2016) also mention that "The goal of Measure phase is to establish a clear understanding of the current state of the process the company wants to improve." To reach that goal and to measure performance of different processes, key performance indicators (KPIs) are used, such as the number of finished products manufactured in a day, amount of work in progress in a day, the number of defective products, the number of deliveries, the number of customer's complaints, and minutes/hours lost as a result of inefficiency. In this stage, a value stream mapping (VSM) is essential because it helps to gather all information related to the execution of tasks to be performed (Cabrita et al. 2016).

Third phase: Analyze, the main causes of the problems and different sources of variation are detected: variance between what was planned and what was achieved, problem's effects, customer needs, etc. (Ferreira et al. 2019). In this phase, the tool cause-and-effect diagram is constructed by doing a brainstorming between all members of the project team (Gupta et al. 2016). This cause-and-effect diagram allows to determine the causes that are causing those problems related to 6 M's: Methods, Machinery, Materials, Manpower, Mother Nature, Measurement. Tools such as Kanban card have also been used in this stage to try to solution some problems found (Cabrita et al. 2016). The Pareto diagram is an important tool as well. It helps to note the relative importance of different factors of a problem; it allows to identify the few vital factors from the many trivial factors, so the vital factors will be first analyzed (Kumaravadivel and Natarajan 2013).

To handle correct measures, a sample size of the group of study is required, and when standard deviation of the population is unknown, the proportions sample size formula is used (Banks et al. 2005). In regard to sample methods, a good alternative

Table 1 Summary of implemented tools in the case study

Stage	Objective	Implemented lean tools
Define	Identify the problems	Kaizen event
Measure	Measure the current situation	Value Stream Map
Analyze	Identify the root causes of problems	Cause-and-effect diagram Pareto diagram
Improve	Make changes in the current situation	Poka-yoke Jidoka 5S's
Control	Control the implemented activities/processes	Out of the scope, under study

to use when researchers have barriers in collecting data is the availability method also known as convenience sampling. It is a non-probability sampling which consists of a collection of data depending on the availability of the investigator. The researcher selects the closest people or subjects of study (Showkat and Parveen 2017).

Fourth phase: *Improve*, it has the objective of enhancing quality in terms of reduction of variability, process improvement and product performance. Here, it is important to consider that 5S's is the first step to improve any process. Additionally, to analyze more than one critical process parameters and their interaction, several methods of design of experiments can be used (Kumaravadivel and Natarajan 2013). In previous studies, a key part in this stage was to determine the route to be taken by the forklift and for workers for handling products (Cabrita et al. 2016).

Fifth phase: *Control*: It is necessary to control if the changes are positive and if the performance is adequate (Cabrita et al. 2016). According to Gupta et al. (2016), "Control measure involves repeated sampling of current process and measuring the process performance to establish the process mean and variance." At the moment that deviations from the limits are detected, they have to be recorded and then analyzed. Tools for this step are kaizen events, meetings between the stakeholders, control charts, among others.

The following table summarizes the lean tools used in each phase of DMAIC in this study; more details will be described in the case study section (Table 1).

4 Case Study

AV. CORP. is an Ecuadorian enterprise dedicated to the logistic of waste management that involves different types of materials. The company was founded in 1996, and it was intended to become a real alternative in respect of waste management and recovery. All activities that take place in the enterprise are aligned with the commitment to perform operations with responsibility. AV. CORP. is formed by systems for treatment and revalorization of waste, so it can be transformed into available by-products to be reused in other industries. AV. CORP. handles projects with different companies which include the transportation of hazardous wastes, fuel production, wood treatment, and so on. In the study herein, it will be covered the project related to the management of solid residues from an automotive industry, by using the DMAIC methodology. The main types of waste generated in the enterprise are waiting, transportation, movement, and stock on hand.

4.1 Define

In the case of AV. CORP., the biggest observed problem was excess of worker's movements along the plant. Critical characteristics for AV. CORP's customers are to get the product for which they paid; it means for example functional pallets, wooden planks without nails or staples, complete cardboard boxes, metals without galvanized coated, and so on. Additionally, a study about the workers was conducted, and it was found that the majority of them are between twenty to forty years and that the 33% of them attended only to primary school, which means that for them it is necessary a very clear and simple explanation when implementing the different tools. Figure 1 gives an explanation related to the workers in AV. CORP.

First of all, a "White Belt" course was given to the Chief Executive, Project Leader, Field Leader, and Administrative Assistant with the purpose of make them part of this lean implementing project in the enterprise. Then, a kaizen event with AV. CORP.'s administrative and operative personnel took part with the purpose of determining the main problems that affect the processes. It was identified that the main problems are the lack of space for storage of materials, the classification of materials, and the under motivated employees. Table 2 has six columns: "aspect" referring to the big problem, "author" is the person or people who described the idea to solve the mentioned problem, "criteria" is a brief description of the problem, "ideas" exhibits the possible solutions for that problem, "classification" means how important is to implement that idea, and finally "responsible" is the person who has the responsibility to execute that idea.

In general, while touring the plant, it was noticed that there is a lack of signalization in the whole plant. There were no labels for the different areas or zones in the

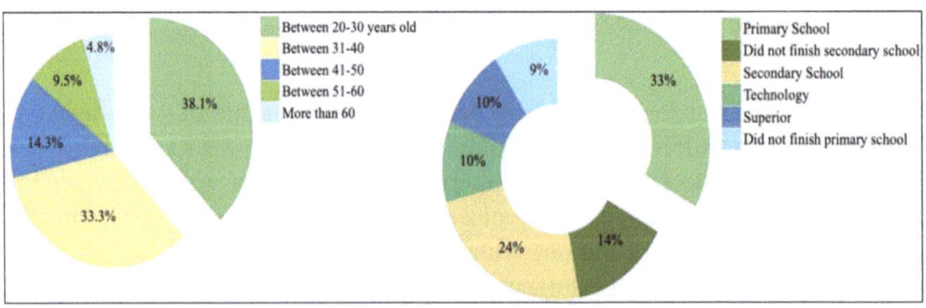

Fig. 1 Average age and level of education of AV. CORP.' Workers. *Source* Own Elaboration

Table 2 Detected problems in AV. CORP plant through a kaizen event

Aspect	Author	Criteria	Ideas	Clasif	Responsible
Slow flow of materials (space)	Juan Carlos (Project Leader) Modesto (Yard chief)	The materials enter in large quantities and remained stored for days, weeks and even months Economic resources are limited, it is difficult to make significant investments such as building a new warehouse	To open new points of sales in Quito (1 per month), out of the city and export materials or new products to the United States	C	Top Management Juan Carlos Carolina
	Jorge (Operator) Modesto (Yard chief)	There is a lack of advertising and promotion of the portfolio of products offered Competition in the market has lower prices	Hire expert sales advisors To eliminate competition from third parties (when possible)	B	Top Management Juan Carlos
	Germán (Zone chief)	No scheduled meetings between team leaders to plan the way they are going to work and fix the organization of personnel	Scheduling and coordinating weekly meetings to plan the work of every week	A	Modesto Santiago Germán
	Juan Carlos (Project Leader) Santiago (Zone chief)	Lack of documentation of materials in stock. It is not fully known what the best-selling product is, and the amount of it that is needed	Keep a record of the amount of material entering and leaving the plant Quantify the volume of different types of material occupying transfer/storage zone. Reduce the volume of materials in large quantities	A	Modesto Carolina Consulter (Estefanía)

(continued)

Table 2 (continued)

Aspect	Author	Criteria	Ideas	Clasif	Responsible
People	Germán (Zone chief) Vicente (Operator) Jorge (Operator)	The majority of workers only work the day shift and are not concern about productivity	Record individual productivity and exposing it in white boards inside the plant. Set targets for production. Give enticements to best workers	A	Carolina
	Modesto (Yard chief)	Uniforms for workers are not frequently changed. Personnel feel undermotivated because of that	Make an analysis/study of the durability of uniforms and make a new design of them	C	Human Resources
Clasification	Carolina (Administrative)	Materials in the transfer zone are not organized by types. There is a lack of order in the management of materials along the different stages	To stablish rules to prohibit the mixing of materials, apply 5S's in the transfer zone	A	Modesto Santiago
	Santiago (Zone chief) Juan (Forklift driver)	Personnel in the plant is specialized. Workers are good and skilled in their segment	Training in the field, multidisciplinary workers. Then, assign tasks and activities to people who best fit for them	B	Juan Carlos Germán Santiago

enterprise. Another obvious problem detected was the lack of yard space in every zone in the plant: storage zone, classification zone, disassemble zone, and so on. Figure 2 shows a scheme of the layout.

4.2 Measure

First of all, a value stream mapping was mapped to have a big picture of the current situation; this was done considering the five main activities in the operative processes. As we can see in Fig. 3, two important improvement opportunities are found, the first in transport and the second one in disassembly. Consecutively, it was decided to study

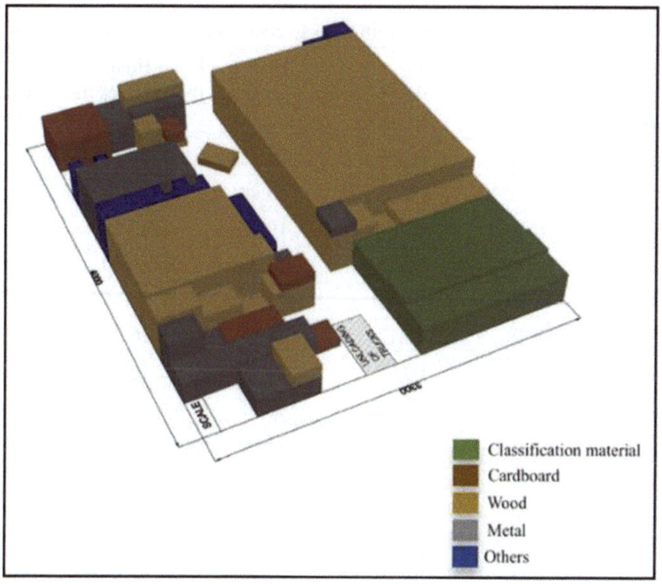

Fig. 2 Initial layout of materials in the transfer zone. *Source* Own elaboration—AutoCAD

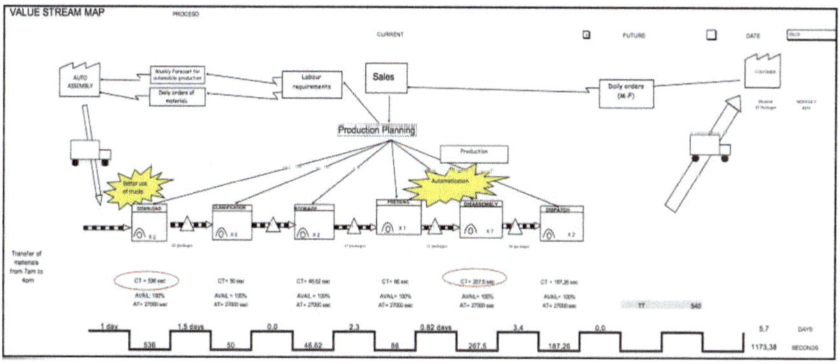

Fig. 3 Value stream mapping of AV. CORP. processes. *Source* Ow elaboration

and analyze three key performance indicators: utilized capacity in the transfer zone as "percentage of used space in the transfer zone," performance of workers as "time taken to disassembly of wooden board pallets," and finally the use of trucks as "time of trucks that aggregate value." All of these CTQs will be considered throughout the case study.

Regarding the transport, it was measured the time it takes to download materials from trucks versus the time they are within the plant. It is possible to notice that there is huge difference between both of them, which represents waste of time. The information collected is showed in Table 3. It was taken sixteen times of downloading

Table 3 Time of downloading materials from trucks versus time trucks is within the plant

Date	Hour parking to download	Hour end of download	T downloading (mins)	Hour truck arrives at the plant	Hour truck leaves the plant	T truck is within the plant (mins)
Jun-08	09:33	09:42	00:09	09:25	09:45	00:20
Jun-08	09:46	09:51	00:05	09:40	10:30	00:50
Jun-08	10:43	10:49	00:06	10:20	10:55	00:35
Jun-14	11:28	11:34	00:06	11:25	11:40	00:15
Jun-14	11:46	11:59	00:13	11:40	12:15	00:35
Jun-14	15:21	15:38	00:17	15:15	16:05	00:50
Jun-18	09:45	09:52	00:07	09:40	10:25	00:45
Jun-18	09:58	10:07	00:09	09:55	10:40	00:45
Jun-18	11:16	11:26	00:10	11:05	12:50	01:45
Jun-18	11:30	11:42	00:12	11:20	11:50	00:30
Jul-06	10:47	10:57	00:10	10:40	11:05	00:25
Jul-06	11:20	11:31	00:11	11:15	12:00	00:45
Jul-13	10:28	10:39	00:11	09:30	10:42	01:12
Jul-13	11:34	11:42	00:08	11:20	12:00	00:40
Jul-18	09:49	10:03	00:14	09:40	10:25	00:45
Jul-30	14:39	14:47	00:08	14:35	15:10	00:35

Source Own elaboration

and sixteen times of trucks within the plant. The initial plan was to take thirty dates of each one, but due to a change in the enterprise's reality it could not be possible. And, the study related to transportation/trucks had to be stopped.

In respect of the transfer zone, it was measured the total area used as well as the total volume used by different materials while they are classified. This was carried out in order to identify what was the "time of trucks that aggregate value." In Fig. 2, it is possible to see the different materials that are allocated in different spots in the transfer zone. Additionally, it is crucial to know the capacity of the transfer zone; it is shown in Table 4.

Table 4 Maximum capacity of transfer zone

Description	Dimension
Width (m)	33
Length (m)	40
Height (m)	4.3
Maximum capacity (area) (m^2)	1320
Maximum capacity (volume) (m^3)	5676

Source Own elaboration

From this stock taking of classified materials, it is observed that 1037 m^2 are occupied in terms of area and in terms of volume 3865 m^3. This suggests that 78.57% of available area is occupied and 68.09% of available volume.

4.3 Analyze

At this phase and using the data gathered in the previous stages, an in-depth analysis is carried out to identify the root causes of the problems identified. Other important tools that were used to analyze data are cause-and-effect diagram—which allows to determine the causes that are causing those problems related to 6 M's: Methods, Machinery, Materials, Manpower, Mother Nature, and Measurement—(see Fig. 4). Additionally, a Pareto diagram was used to visualize the importance of different materials taking into consideration the space that they occupy (see Fig. 5). This Pareto shows that about 70% of the occupied volume in the transfer zone is filled with wood—calculations for Pareto diagram for volume occupied by different materials are presented in Table 5. On the basis of this, it is on "wood" where it is necessary to optimize processes. This analysis is strongly correlated to two mentioned CTQs: "percentage of used space in the transfer zone" and "time taken to disassembly of wooden board pallets."

In respect of the times shown in Table 3, they were analyzed to prove if they were statistically different; a paired T analysis was conducted because the entity of study is in this case a truck. And, the same truck is measured in both conditions, so a paired T analysis is adequate (Fig. 6).

From these Minitab's results in regard of "time of trucks that aggregate value" CTQ, it is concluded that there is a statistically difference between download times and the time of stay of trucks. This shows that effectively there is waste in terms of time, which represents money for the enterprise. This conclusion results from the comparison between *p* value (0.000) and alpha level (0.05), given that p value is less than the established alpha.

In respect of the disassembly of wooden boards pallets, the operation of the activity was analyzed. The flow process chart presented in Fig. 7 shows the overall sequence of the operation considering the movements of the operators as well as the flow of the materials.

It can be noticed, in respect of "time taken to disassembly of wooden board pallets" CTQ, that many movements are part of the operation as well as the delays caused by the activity of dismounting wooden pallets or boards with nails. It is also self-evident that movements represent a third of the whole activity, so an improvement opportunity appears here.

Additionally, it was analyzed the time it takes to pull out all the nails and screws from the board, the final product of this activity is a clean wooden board without any nail or screw. It is a completely manual process, where there are no machines taking part of it. Is it important to mention that the pallets analyzed were the ones of 1.48 m × 1.14 m shown in Fig. 8; they were chosen because they are the dimensions in larger

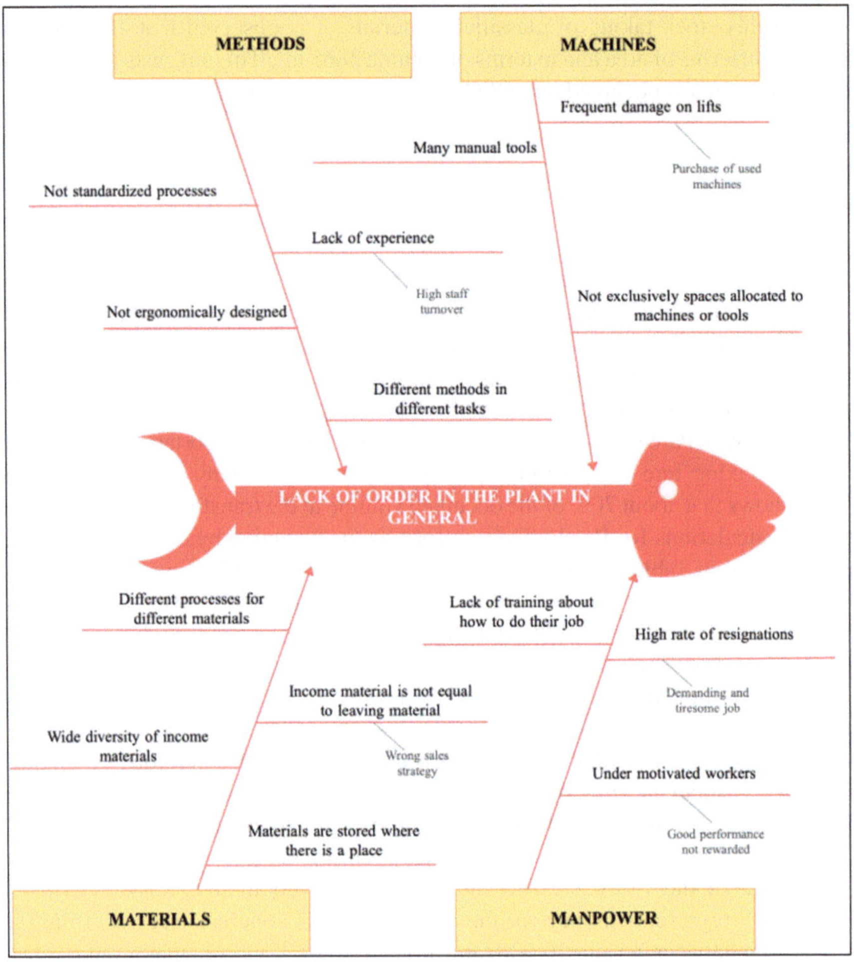

Fig. 4 Cause-and-effect diagram for the big problem "Lack of order in the plant in General." *Source* Own elaboration

quantities in the plant—there are approximately 17,000 pallets of these dimensions in total in AV. CORP's plant. In Fig. 8, a scheme of the pallet analyzed is shown.

Thirty times were collected due to the availability of the investigator and appealing to the central limit theorem which approves the validity of the sample which states that as the sample size increases, the shape for the means" distribution takes the shape of a normal distribution. This is shown in Table 6.

From Fig. 9, which show the times of disassembly a wooden pallet manually, it can be seen that, in average, the time it takes to disassembly one pallet of 1.14 m × 1.48 m is 238.98 s with a standard deviation of 86.06. As noted earlier in Fig. 7, this activity involves many movements and these times are reflecting that. So, in the future, these times will be reduced.

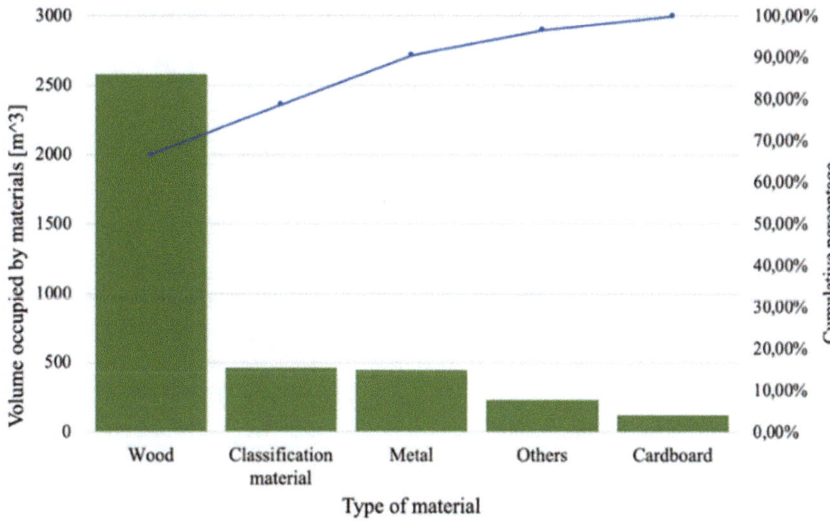

Fig. 5 Pareto diagram for volume occupied by different materials in the transfer zone. *Source* Own elaboration

Table 5 Calculations for Pareto diagram for volume occupied by different materials in the transfer zone

Materials	Occupied volume	Percentage (%)	Cumulative percentage (%)
Wood	2577.5475	66.69	66.69
Classification material	466.9	12.08	78.78
Metal	452.96465	11.72	90.50
Others	238.00132	6.16	96.65
Cardboard	129.309	3.35	100.00
TOTAL	3864.72247	100.00	

```
Paired T for t descarga - t parqueo

              N     Mean   StDev  SE Mean
t descarga   16     9,75    3,21     0,80
t parqueo    16    35,75   12,23     3,06
Difference   16   -26,00   12,14     3,03


95% CI for mean difference: (-32,47; -19,53)
T-Test of mean difference = 0 (vs ≠ 0): T-Value = -8,57  P-Value = 0,000
```

Fig. 6 Results for paired T (Minitab Software)

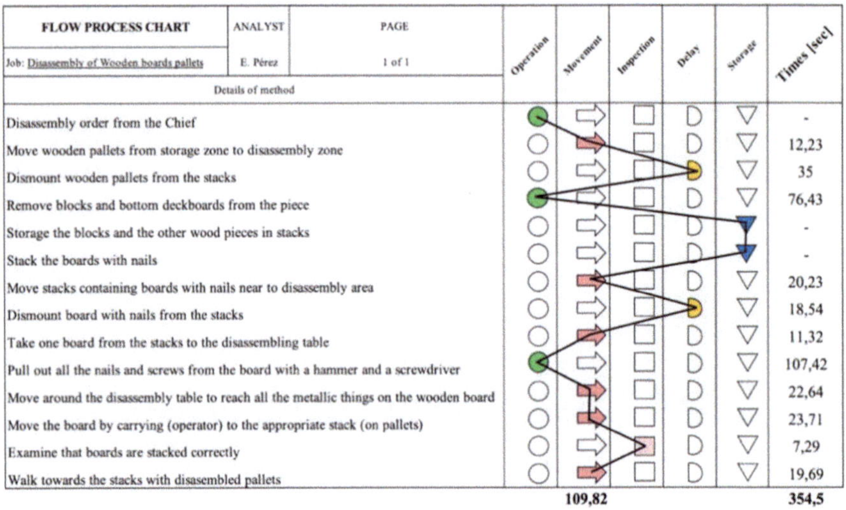

Fig. 7 Flow process chart of disassembling wooden board pallets. *Source* Own elaboration

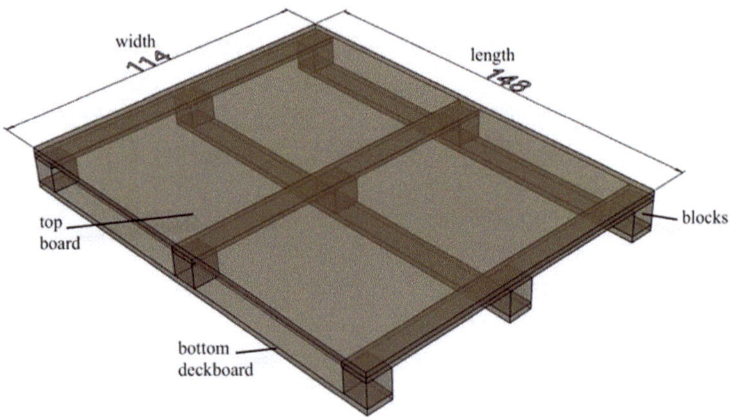

Fig. 8 Wooden boards pallets for disassembling. *Source* Own elaboration—AutoCAD

Table 6 Time taken to disassembly one pallet of 1.14 m × 1.48 m manually (s)

265.28	148.95	151.22	370	281.6
344.71	401.15	145.76	224.8	222.53
310.44	281.5	195.64	234.39	246.87
119.58	221.41	85.39	282.58	154.32
131.85	438.66	124.79	247.18	148.36
204.74	257.32	202.17	259.68	270.49

Source Own elaboration

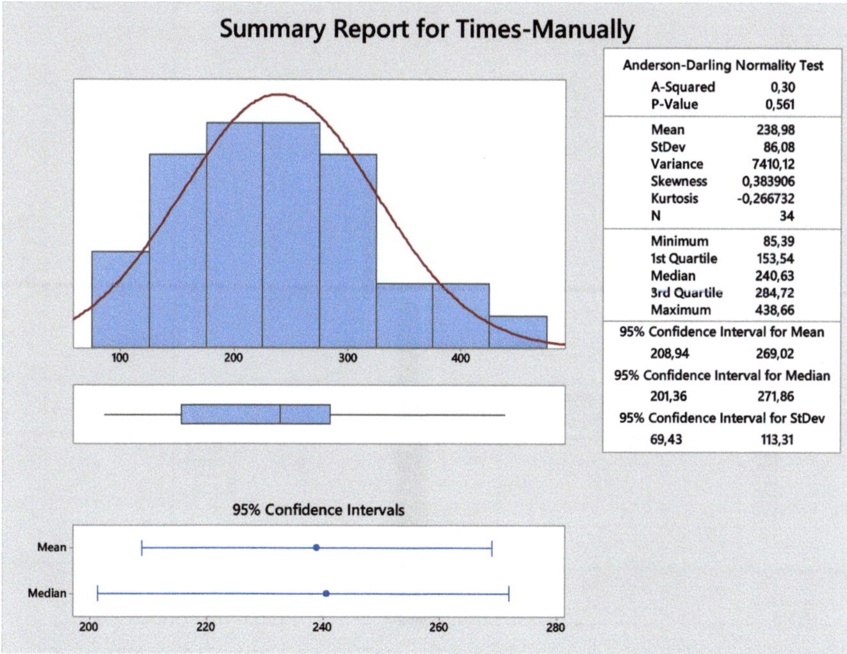

Fig. 9 Statistical summary of times for dissembling manually. *Source* Minitab Software

4.4 Improve

At this stage, a proposal to signalize the central courtyard was designed, making it easier to people at the time they choose the route they take to go somewhere. Through participant observation, it was noticed the lack of order and organization in the plant. This included the overall picture that customers perceive at the moment they arrived at the enterprise. Considering that the impression of clients in respect of the plant is important, it was observed a great opportunity to improve the physical appearance of the routes in the plant. Below there is a scheme where the tool poka-yoke helps to easily indicate where is the way to reach the different locations within the company.

The poka-yoke shown in Fig. 10 was not exactly implemented like in the proposal, there was a little variation. Below it is possible to see a before and an after the implementation, with an investment of $285 USD (see Table 7) (Fig. 11).

Regarding to the process of disassembling, it was semi-automatized trough the lean tool "Jidoka." This was implemented with the purpose to reduce the time spent in disassembling, that is to say, "time taken to disassembly of wooden board pallets" CTQ. Given that the process was completely manual, arose the need to implement some tool to facilitate the task. So, the enterprise acquired an air gun which helps to remove nails from the wooden board in less time than making it manually. The gun works with a pressure of 90 PSI. The tool is presented in the image (Fig. 12).

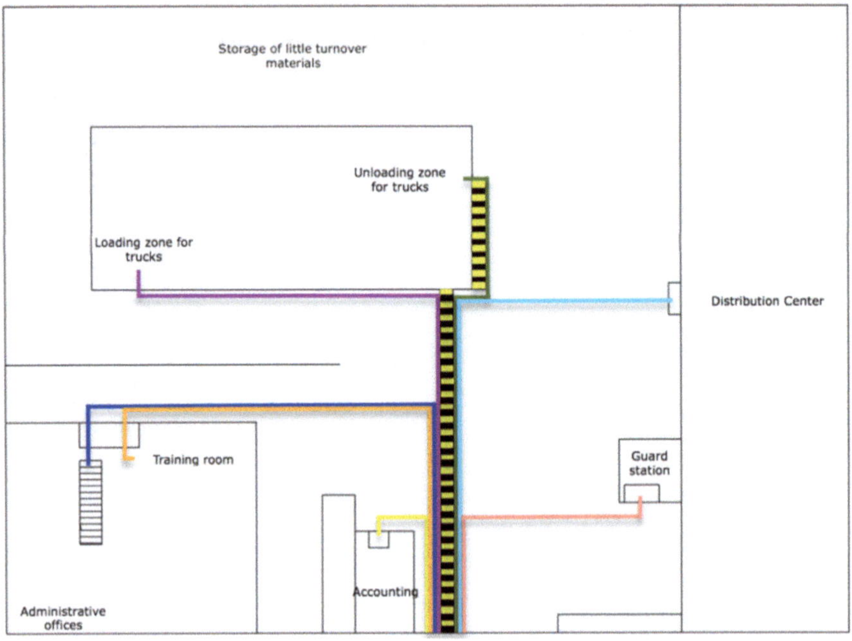

Fig. 10 Proposed poka-yoke for pedestrian path in AV. CORP

Table 7 Detailed economic analysis for poka-yoke implementation

Aspect	Invested money USd
Labor's cost ($)	225
Material's cost ($)	60
Total cost ($)	285

Fig. 11 Routes in AV. CORP before and after poka-yoke implementation

Fig. 12 Air gun implemented in the process of disassembling

Times were taking to compare and stablish the difference and the improvement from a manual process to a semi-automatized process (see Table 8). The calculated sample size was of 96 times; however, thirty times were taken due to the availability of the investigator and appealing to the central limit theorem which approves the validity of that sample size. Figure 13 shows the statistical summary:

In Fig. 13, it is possible to notice that the mean for disassembling one pallet of 1.14 m × 1.48 m is shorter than in the manual process. Another very important fact is that the process variability has been reduced. Now, there is a standard deviation of 44.07 s, with the manual process it was 86.08 s. A box plot for before and after of disassembly of pallets is shown in Fig. 14. In the same field, a proposal to facilitate the operators" work was done. In Fig. 15, it is depicted a movable table for the activity of disassembling with an approximate cost of $190 USD; and a mobile rack, which helps to move pallets in more efficiently way, with an approximate cost of $132 USD.

Table 8 Time taken to disassembly one pallet of 1.14 m × 1.48 m with an air gun (s)

207.98	161.33	211.82	225.82	172.45
211.6	247.54	207.32	234.81	137.2
367.38	206.58	197.22	144.94	170.91
158.83	200.25	212.65	199.65	194.54
189.8	190.31	153.37	156.33	234.28
205.15	171.77	235.64	194.45	265.91

Source Own elaboration

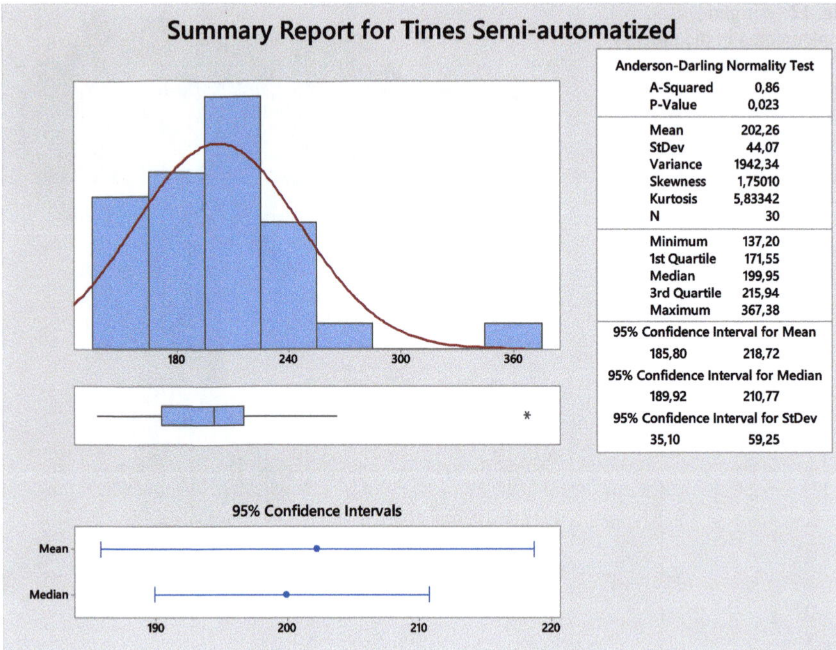

Fig. 13 Statistical summary of times for dissembling semi-automatized. *Source* Minitab Software

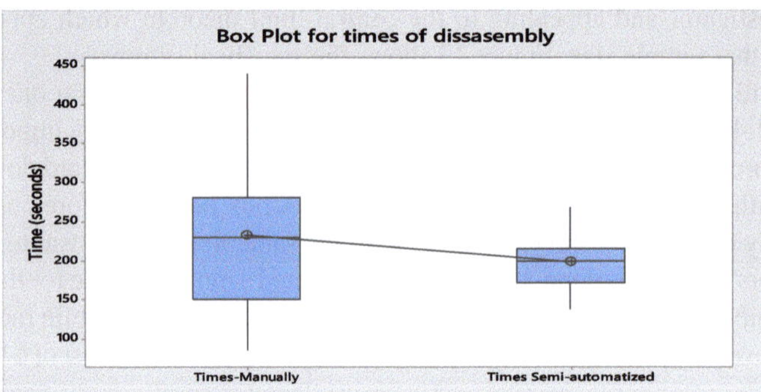

Fig. 14 Box plot for the before and after of disassembly of pallets

Additionally, after applying 5S's in the transfer zone, it is clearly visible that materials stored temporarily are more organized, in respect of "percentage of used space in the transfer zone" CTQ. And as shown in Fig. 16, materials are separately located depending on the type. Furthermore, the decision of disassembling wooden board pallets helped notably to reduce quantities of wood in the transfer zone as can be seen in the Fig. 16.

Fig. 15 Movable table to disassembly of wooden board pallets and mobile rack to transport them. *Source* Own elaboration

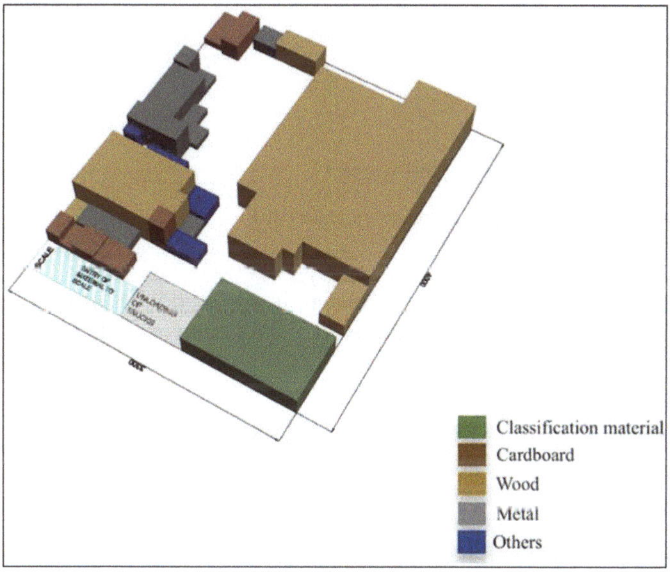

Fig. 16 Layout of transfer zone—materials are storage in an organized way. *Source* Own elaboration

After applying lean manufacturing in the plant, the space occupied by classified materials is considerably less than in the past. In terms of volume, the transfer zone has a volumetric capacity of 5676 m^3 and all the materials occupy 2886 m^3, which represents the 50.85% of the space. On the other hand, the area's capacity of this zone is 1320 m^2 and materials take up 817 m^2, which is 61.87% of the available area. This implementation represented an investment of $1650 USD (see Table 9);

Table 9 Investment made in the Jidoka implementation

Investment	US dollars [$]
Air gun	250
Air system	200
Compressor	1200
TOTAL	1650
Payment diferrals: 1 year	137.5

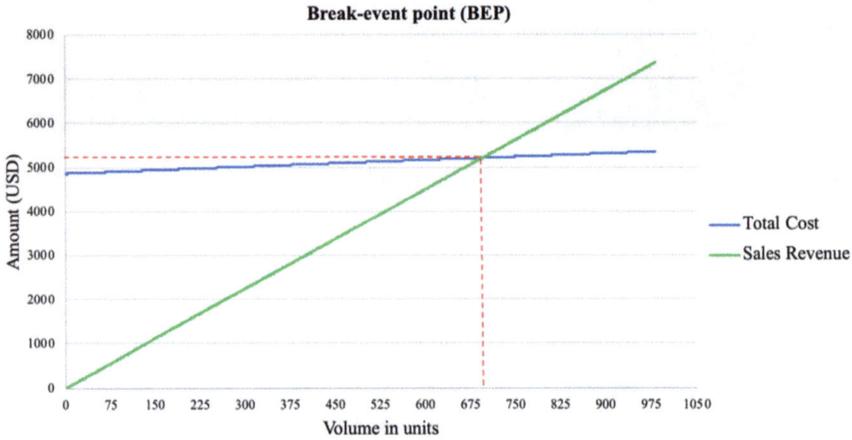

Fig. 17 Cost–volume relationship—breakeven point (BEP)

additionally, breakeven point was calculated, it was found that it was necessary to sell at least 695 units per month to recovery the investment made for one year (see Fig. 17).

4.5 *Control*

The scope of this study will be dealt with completely until the last stage of DMAIC cycle, but workers and leaders from AV. CORP. will be trained and prepared to manage the control phase with control charts and kaizen events with a structured manual format.

5 Discussion

In this enterprise, the problem identified was a lack of standardization in their processes and waste in terms of inventory and movements. The case study revealed that it is possible and effective the use of lean tools in a company of the recycling industry. The DMAIC methodology allowed to maintain an organized improvement process, which finished with very good results. The payoff was the space optimization with a relatively low cost of investment.

Through the implementation of lean tools, the results achieved in this case study showed that it was reduced the occupied space in transfer zone in 17.24% in terms of volume and in 16.7% in terms of area, with an economic benefit of $5789.39 USD. Tables 7 and 9 provide details of the expenses incurred to implement lean tools.

6 Conclusions

This case study contributes with a vision not only to identify, but also to eliminate different type of waste related to the recycling activities and points out the significance of the implementation of lean culture in an enterprise of the recycling industry. It was proved the effectiveness of it, even when the majority of the work force has a low level of education. All of these was done through user-friendly tools for workers in the enterprise and easy to implement. It is quite clear that the support of top management in this journey is a key element to achieve tangible improvements. It should be emphasized how challenging is the implementation in a company, where sometimes workers do not trust in the consultant. Therefore, it is necessary to have time with them and also perseverance to change their minds.

This study provides valuable insights to enterprises with similar problems of inventory, management of recycled materials, and manual activities in their processes. Additionally, it is a guide of a success implementation with use of DMAIC methodology for continuous improvement. Throughout the case study, different lean tools were applied and performed: value stream map (VSM), Jidok, poka-yoke, cause-and-effect diagram, kaizen event. And, it has also been proved in the reality of an enterprise that straightforward tools such as all mentioned above can generate great impact in the whole organization.

The main objective of this project was to analyze the applicability and effectiveness of lean tools to the recycling industry, and then measure the improvement as a result of the implementation of these tools in different processes. The CTQs considered were the percentage of space used in the transfer zone and time duration to disassemble one wooden board pallet. It has been observed that the processes variability has been reduced regarding to the disassembly of wood board pallets thanks to the semi-automatization of the process. Finally, it is important to mention that the "economic benefit" does not mean savings for the company, but rather an opportunity to give that space a better use to enhance other projects.

References

Abdul Rahman NA, Sharif SM, Esa MM (2013) Lean manufacturing case study with kanban system implementation. In: International conference on economics and business research 2013 (ICEBR 2013), vol 7, pp 174–180. https://doi.org/10.1016/S2212-5671(13)00232-3

Anshula G, Kalhan R, Richa S (2019) Reducing soldering defects in mobile phone manufacturing company: a DMAIC approach. In: Paper presented at the international conference on mechanical and energy technologies (ICMET 2019), Greater Noida, India

Banks J, Nelson BL, Carson JLSI (2005) Discrete-event system simulation, 4th edn. Pearson Prentice Hall

Beck C (2016) Your guide to 5S. Manufacturing management. Retrieved from https://www.manufacturingmanagement.co.uk/features/your-guide-to-5s

Bradley R, Jawahir IS, Badurdeen F, Rouch K (2018) A total life cycle cost model (TLCCM) for the circular economy and its application to post-recovery resource allocation. Resources. Conserv Recycl 1–9. https://doi.org/10.1016/j.resconrec.2018.01.017

Butzer S, Kemp D, Steinhilper R, Schötz S (2016) Identification of approaches for remanufacturing 4.0. Eur Technol Eng Manag Summit (E-TEMS) 1–6

Cabrita MR, Domingues JP, Requeijo J (2016) Application of lean six-sigma methodology to reducing production costs: case study of a Portuguese bolts manufacturer. Int J Manag Sci Eng Manag 11(4):222–230. https://doi.org/10.1080/17509653.2015.1094755

Dal Forno AJ, Pereira FA, Forcellini FA, Kipper LM (2014) Value Stream Mapping: a study about the problems and challenges found in the literature from the past 15 years about application of Lean tools. Int J Adv Manuf Technol 72(5–8):777–790. https://doi.org/10.1007/s00170-014-5712-z

Fercoq A, Lamouri S, Carbone V, Lelièvre A, Lemieux A (2013) Combining lean and green in manufacturing: a model of waste management. In: Paper presented at the 7th IFAC conference on manufacturing modelling, management, and control, Saint Petersburg, Russia.

Ferreira C, Sáa JC, Ferreira LP, Lopes MP, Pereira T, Ferreira LP (2019) iLeanDMAIC—a methodology for implementing the lean tools. In: Paper presented at the 8th manufacturing engineering society international conference, Madrid, España

Gupta A, Sharma P, Malik SC, Agarwal N, Jha PC (2016) Productivity improvement in the Chassis preparation stage of the amplifier production process: a DMAIC six sigma methodology. Int J Reliab Qual Saf Eng 23(6):13

Järvenpää E, Lanz M (2019) Lean manufacturing and sustainable development. In: Leal W, Azeiteiro U, Azul A, Brandli L, Özuyar P, Wall T (eds) Responsible consumption and production. Switzerland: Springer, Cham

Kumaravadivel A, Natarajan U (2013) Application of six-sigma DMAIC methodology to sand-casting process with response surface methodolog. Int J Manuf Technol 69:1403–1420. https://doi.org/10.1007/s00170-013-5119-2

Lewis J (2005) Identifying seven types of waste. Upholstery Manuf 18(10):20–24. (199360886)

McCandless K (2017) Squaring the circular economy. Autom Manuf Solutions 42–44

Mohammed AA, Mohammed A, Abdelhakim A, Ahmad M (2013) A proposed approach for setup time reduction through integrating conventional SMED method with multiple criteria decision-making techniques. Comput Ind Eng 66(2):461–469. https://doi.org/10.1016/j.cie.2013.07.011

Monteiro C, Ferreira LP, Fernandes NO, Sáa JC, Ribeiro MT, Silva FJG (2019) Improving the machining process of the metalworking industry using the lean tool SMED. In: Paper presented at the 8th manufacturing engineering society international conference, Madrid, España

Mudhafar A, Konstantinos S, Yuchun X (2017) The role of leadership in implementing lean manufacturing. Procedia CIRP 63:756–761. https://doi.org/10.1016/j.procir.2017.03.169

Nelson J (2016) Chapter 3—pull versus push: lessons from lean manufacturing author links open overlay panel, J. Nelson edn. Becoming a lean library: Elsevier Ltd.

Pardo Martínez CI, Piña WA (2017) Solid waste management in Bogotá: the role of recycling associations as investigated through SWOT analysis. Environ Dev Sustain 19(3):1067–1086. https://doi.org/10.1007/s10668-016-9782-y

Samant S, Prakash R (2020) Achieving lean and improving sustainability through value stream mapping for complex manufacturing. In: Paper presented at the international conference on mechanical and energy technologies (ICMET 2019), Greater Noida, India

Showkat N, Parveen H (2017) Non-probability and probability sampling. In: Gandhinagar, Gujarat, India, e-PG Pathshala, pp 1–9

Smith M, Paton S, MacBryde J (2018) Lean implementation in a service factory: views from the front-line. Prod Plann Control Manag Oper 29(4):280–288. https://doi.org/10.1080/09537287.2017.1418455

Srinivasan K, Muthu S, Devadasan SR, Sugumaran C (2014) Enhancing effectiveness of shell and tube heat exchanger through six sigma DMAIC phases. Procedia Eng 97:2064–2091. https://doi.org/10.1016/j.proeng.2014.12.449

Sutharsan SM, Mohan Prasad M, Vijay S (2020) Productivity enhancement and waste management through lean philosophy in Indian manufacturing industry. Mater Today: Proc. https://doi.org/10.1016/j.matpr.2020.02.976

Thürer M, Pan YH, Qu T, Luo H, Li CD, Huang GQ (2016) Internet of Things (IoT) driven kanban system for reverse logistics: solid waste collection. J Intell Manuf 1–10. https://doi.org/10.1007/s10845-016-1278-y

Tsao L, Patrick Rau PL, Ma L (2015) Development of a quick instrument measuring Kaizen culture (for Chinese). Procedia Manuf 3:4708–4715

Villareal B, García D, Rosas I (2009) Eliminating Transportation waste in food distribution: a case study. Transp J (Am Soc Transp Logistics Inc) 48(4):72–77

Vinodh S, Arvind KR, Somanaathan M (2011) Tools and techniques for enabling sustainability through lean initiatives. Clean Technol Environ Policy 13(3):469–479. https://doi.org/10.1007/s10098-010-0329-x

Case Study of Lean Manufacturing Application in a New Process Introduction into a Rail Company

Rosa Hilda Félix Jácquez, María Merced Cruz Rentería, Ma. Dolores Delgado Celis, Alfonso Alejandro Lara Negrete, Daniel Torres Ramírez, and Claudia Leticia Dávila Martínez

Abstract The main purpose of this chapter is to demonstrate the use of lean manufacturing tools to solve different problems appearing in the practice, into the manufacturing projects. The methodology focused on one case study where applied lean manufacturing techniques, based on searching the continuous improvement and better performance to meet customer demand. For the analysis, financial justification, and the introduction of the process, the lean philosophy using Define, Measure, Analyze, Improve, Control (DMAIC) methodology in the back shop area where the project carried out. It will have significant business impact as the annual increase of 18,772 man-hours for the workshop as well as $6.0 million USD of savings for the company by performing the process internally with long-term contracts.

Keywords Continuous improvement · Lean manufacturing · DMAIC

1 Introduction

A rail company in Mexico found a huge opportunity area for the repair of locomotives air compressors at one of its sites.

Currently, the economic situation of the repair shop at a rail company is critical, as the man-hours of yearly work have been reduced drastically in recent years due to the lack of New Product Introductions. Less working hours available mean a higher cost of labor due to all business expenses divide between the amounts of person-hours obtaining the labor cost rate annually. The increase of labor cost has made it harder to justify investment projects that are less than a year Return of Investment (ROI) as the corporation needs.

R. H. Félix Jácquez (✉) · M. M. Cruz Rentería · A. A. Lara Negrete · D. Torres Ramírez · C. L. Dávila Martínez
Department of Industrial Engineering, Instituto Tecnológico de San Luis Potosí, Soledad de Graciano Sánchez, México
e-mail: rosa.fj@slp.tecnm.mx

Ma. D. Delgado Celis
Department of CEA, Tecnológico Nacional de México, Soledad de Graciano Sánchez, México

J. L. García Alcaraz et al. (eds.), *Techniques, Tools and Methodologies Applied to Quality Assurance in Manufacturing*,
https://doi.org/10.1007/978-3-030-69314-5_18

That is why the company has the objective to gain more man-hours for the shop, become more productive, and reduce the labor cost. Due to this, the company aims to get more man-hours for the shop becoming more productive and reduced labor cost, all this through the introduction of new products. Besides, implementing better practices such as lean manufacturing to eliminate waste, standardize operations, improve flow and reduce the inventory in the new compressor repair line in the business unit.

This case study aims to achieve the introduction of a new process in the remanufacturing of compressors to increase the working time in the company through the documentation, analysis, and design of the process, applying lean manufacturing tools combined with DMAIC methodology therefore to ensure the quality levels required by the customer.

2 Literature Review

2.1 Continuous Improvement

Actually most companies live in a constant changing where customers demand better products, higher quality, and shorter delivery times. To achieve these customer requirements, companies rely on continuous improvement to reach their goals (Cardona 2014; Oakland 2014).

According to Japan Management Association (1989), now the costumers are the one that establish the prices, breaking the paradigm where the manufacturer imposes conditions of pricing. Moreover, because they have multiple options, they are demanding quality, faster response time and better production flexibility. Therefore, the only way to increase the profits is to reduce the hidden cost as poor quality, downtime, low efficiency, and scrap, among others, called waste. Then, the organizations rely on the continuous improvement to eliminate these wastes (Suárez-Barraza and Miguel-Davila 2009).

Continuous improvement traces its origins to two major historical trends, both dating from about 1950. The first occurred at Toyota, where Taiichi Ohno and Shigeo Shingo conceived the just-in-time (JIT) production system. The second trend underpinning continuous improvement is the quality movement and statistical reasoning, conceived in the 1920s by Shewhart (Zangwill and Kantor 1998).

Imai (1986) defines the continuous improvement as the organizational activities with the involvement of all people in the company.

The improvement can be defined as kaizen and innovation, where a kaizen strategy maintains and improves the standard of work through small and gradual improvements, and innovation produces redial improvements because of large investments in technology and equipment (Imai 1986).

The term kaizen is derived from two Japanese elements that means Kai = change and Zen = good (to improve) (Newitt 1996) Imai (1997) states kaizen as continuous improvement in all the company areas also refers to the creation of a process where most of the time is value added.

Zangwill and Kantor (1998) define the continuous improvement as a set of powerful techniques that has produced substantial improvements in numerous companies and organizations.

On other side, Suárez-Barraza (2007) defines it as a management philosophy that generates changes or small incremental improvements in the work processes that allows to reduce waste and as consequence improve work performance, taking to the organization into an incremental innovation spiral.

Several tools can support the continuous improvement. Bhuiyan and Baghel (2005) affirm that lean manufacturing, six sigma, and balance scorecard are three of the most common tools used.

Zangwill and Kantor (1998) propose the tools: total quality management (TQM), kaizen, Deming cycle or Plan, Do, Check, Act (PDCA), and just in time (JIT)_ to use in continuous improvement process. While Adams et al. (1999) mention lean manufacturing, kaizen, and Toyota Production System (TPS) as the tools can be used in continuous improvement process.

To Wei and Ling (2006), the most important tools used to continuous improvement are six sigma, 8D, TQM, TRIZ inventive solving problem theory, failure mode effect analysis (FMEA), and process flow.

2.2 *Lean Manufacturing*

Lean manufacturing was born of Toyota Company like a form to produce, with which it sought to have a smaller amount of waste and equal to the competitiveness of the US automotive companies (Villaseñor and Galindo 2008).

Lean manufacturing is a production practice that considers the expenditure of resources for any goal other than the creation of value for the end customer to be wasteful and thus a target for elimination (Čiarniene and Vienazindiene 2012).

Lean is as much about establishing a "lean corporate culture" as it is about lean manufacturing and production processes. A lean culture seeks to motivate workers to participate and even start lean initiatives to improve the corporate bottom line as stated Čiarniene and Vienazindiene (2012).

Working from the perspective of the customer who consumes a product or service, "value" defines as any action or process that a customer would be willing to pay for. Also known as the flexible mass production, the Toyota Production System (TPS) has two pillar concepts: just in time (JIT) or "flow," and "automation" (smart automation) (Barney 2004; Shingo 1989). The purpose is to move one piece at a time in each station and do not have work in process (WIP) (Villaseñor and Galindo 2008).

The critical starting point for lean thinking is value. The ultimate customer can only define value because it is the first person that the enterprises have to be in mind (Womack and Jones 2003). In addition, it is only meaningful when expressed in terms of a specific product, which meets the customer needs at specific price at a specific time.

Talking about value, we can mention value stream is the set of all the specific actions required to bring a specific product through the three critical management tasks of any business: the problem-solving task, information management task, and physical information task (Womack and Jones 2003).

Over the last years, much has written, taught, and discussed about "lean" and "six sigma," and many industries generally recognize that both have contributed substantially to bottomline results, enhanced competitiveness, and improved levels of customer service (Adams et al. 2003; Pavnaskar et al. 2003).

2.3 Define, Measure, Analize, Improve, Control (DMAIC) Methodology

One of the six sigma distinctive approaches to process and quality improvement is the DMAIC methodology. The structure DMAIC has five interconnected phases, i.e., Define, Measure, Analyze, Improve, and Control that systematically help organizations to solve problems and improve their processes (Rahman et al. 2017; Sharma and Rao 2014).

According to Zaman and Zerin (2017), the five phases include several activities as are described below.

- *Define*: problem selection and benefit analysis. Identifying and mapping relevant processes, identifying stakeholders, prioritizing customer needs and making a business case for the project.
- *Measure*: translation of the problem into a measurable form and measurement of the current situation.
- *Analyze*: identification of influence factors and causes, identifying potential influence factors and selecting the vital few influence factors.
- *Improve*: design and implementation of adjustments to the process to improve the performance and conduct pilot test of improvement actions.
- *Control*: empirical verification of the project's results and adjustment of the process management and control system in order that improvements are sustainable, the new process capability and implement control plans.

3 Case Study of Lean Manufacturing Application with DMAIC Methodology

For the application of the lean manufacturing, philosophy in the next case study, the DMAIC methodology was used because it provides a simple structure, and each step allows actions that aim to reduce waste and everything that does not add value to the product. Next, the phases of the DMAIC methodology and the lean manufacturing tools that were applied in the project are described.

3.1 Phase Define

The air compressor is one of the main components for the locomotive operation as it supplies air to the locomotive braking system. It primarily consists of a high-pressure head, two low pressure heads, an after cooler, two intercoolers, crankcase, electric motor, and the airflow lines. The first phase was an action workout to elaborate a process for the remanufacturing of the air compressor. The result of this event left various opportunities since repair activities carried out in different areas of the plant. According to the customer needs, the process must accomplish in a cycle time of 20 h. The results did not make it feasible to achieve the expected cost by the primary customers, nor the benefit projected by the company.

Among the opportunity areas identified stand out the absence of documentation of process sheets, standardization of operations, quality plan, Job Security Analysis (JSA's) and process Failure Mode and Effect Analysis (FMEA). The workstations and workflows were not well designed. And under these conditions, the operators could not be adequately trained to repair the air compressor.

The most critical activities in the project were the process sheets, quality plan, and FMEA because these activities took a lot of time to complete the documents to repair an air compressor.

3.2 Phase Measure

Benefits Array

The next phase of the project was to elaborate a benefits analysis to identify the area where the compressor line will install taking into consideration the needs for the process and the investment needed for the facility prep. This analysis contains the features that the teamwork considered most important as the equipment availability, washer machine repair, crane maintenance and availability, the process flow, distance to warehouse, and investment required.

The teamwork evaluated three different criteria to choose the best option, 1: poor 3: regular, and 9: good. The evaluation criteria shown in Table 1 indicates that the first area got 59 points and the second 91 points, and then, with this result, the optimal work area for the process installation was handpicked.

Table 1 Evaluation criteria to identify the best area of compressor line

Evaluation criteria	Option 1	Option 2	Pros and cons	
			Option 1	Option 2
Availability of the traveling crane	1	9	Traveling cranes must be shared with the electrical equipment area	Traveling cranes are exclusive to the compressor line
Functionality of the traveling crane	3	1	Traveling crane does not work	Three traveling cranes do not work
Hoist availability	9	9	There are two hoists for use	There are two hoists for use
Hoist functionality	3	1	One works and the other needs to be repaired	Both require repair and maintenance
Availability of PROCECO washing machine	1	9	It must be shared with the electrical equipment area	Exclusive use for the compressor line
PROCECO washing machine functionality	9	1	Ready to use	Requires repair or installation
KARCHER washing machine availability	9	9	Exclusive use for the compressor line	Exclusive use for the compressor line
KARCHER washing machine functionality	1	1	Requires be purchasing and installing	Requires be purchasing and installing
Paint cabin availability	1	9	It must be shared with various areas	Exclusive use for the compressor line
Paint cabin functionality	9	1	Ready to use	Requires be purchasing and installing
Area maintenance	3	1	Floor restoration	Floor, ceiling and wall restoration
Installation	3	1	KARCHER washing machine installation	Installation of KARCHER and PROCECO washing machine and paint cabin
Distance to warehouse	1	9	187 m of area	25 m of work area
Emergency exits	1	9	1	2
One-piece flow	1	9	Not met by sharing washing machine and paint cabin	100%
Available area	3	9	Area occupied by equipment and machinery of other lines	Free area to start with the installation

(continued)

Table 1 (continued)

Evaluation criteria	Option 1	Option 2	Pros and cons	
			Option 1	Option 2
5 S's	1	3	The work area must be released and organized	The area requires basic cleaning

Value Stream Mapping (VSM)

According to Rother and Shook (2009), value stream mapping is a tool that helps to see and understand the flow of material and information as a product makes its way through the value stream. What we mean by value stream mapping is simple: Follow a product´s production path from customer to supplier and carefully draw a visual representation every process in the material and information flow. Then, ask a set of key questions and draw a "future state" map of how value should flow.

There are two kinds of work scopes for the air compressor; one of them is the light version, where the high-pressure head, low-pressure heads crankcase, after cooler, and intercoolers test, and all the gaskets and seals change. The other one is the heavy repair that consists of the full disassembly, qualification of principal parts, non-destructive testing (Magnaflux), and the complete set of seals in gaskets.

The work scope for the air compressor repair used in this project is the light version.

By using VSM tool, the complete process of value-added and non-value-added operations is laid. This tool allowed to start identifying the main waste sources and opportunity areas, including the lack of exclusive space for the installation of the line, unfamiliarity with the necessary tools, the operator did not know the sequence of activities required by the absence of process sheets, among other things.

All this caused that the percentage of non-value added measured in the VSM represents 45.95% of the 20 h corresponding to the cycle time as represented in Fig. 1.

To reduce or eliminate non-value-added-time it was designed the future value stream mapping. To reduce cycle time by 25% and achieve 15 h were implemented several strategies.

To reduce the cycle time, the teamwork focused mainly on the elaboration of the process sheets, the identification of the necessary tools, layout design, process flow, and component transportation. See Fig. 2.

Spaghetti Diagram

Environment, health and safety (EHS) risks, non-conforming products, and long cycle times are the leading waste sources created by the transportation and material handling in the companies. A spaghetti diagram is so helpful to measure the total distance the component travels from a workstation to another. By using this tool, the

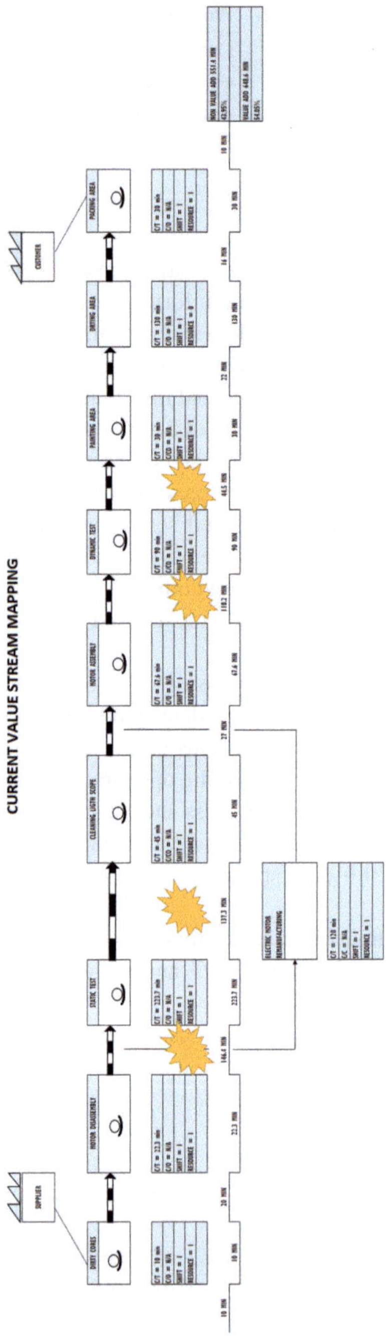

Fig. 1 Current value stream mapping light work scope

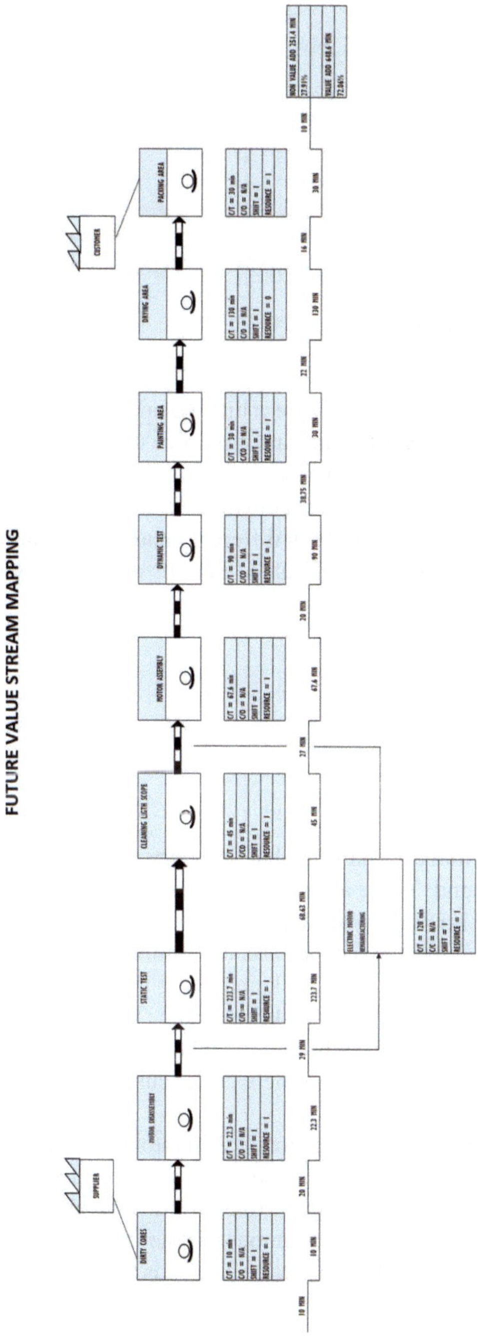

Fig. 2 Future value stream mapping light work scope

Fig. 3 Initial spaghetti diagram, Kaizen 1

materials flow and interference in the process line was identified, leading to redesign the process flow to avoid EHS problems and minimize traveling distance.

With the elaboration of spaghetti diagram, as that illustrated in Fig. 3, it allowed measuring the total number of meters that both the operator and the compressor travel to perform its remanufacturing process in different areas of the plant. As a result, it has been obtained 411 m, and this highlighted that the distances were too long due to the lack of a defined production line.

3.3 *Phase Analyze*

Takt Time

The next step was to define the takt time that in theory considers the total available time and customer demand. There are two scenarios, the best and the worst. In the first case, the cycle time for every workstation is calculated by the average number of fails during a static test, and in the second case, it is estimated considering the highest time of static analysis causing a bottleneck and being the most critical workstation in the line, not letting accomplish the weekly demand. The demand required by the customers was 25 compressors weekly, and the takt time indicates every few hours a compressor must exit the line ready to deliver the customer and achieve the goal of five compressors a day.

Once the layout designed and the compressor/operator route analyzed, the number of meters measured decreased to 40.1 m as represented in Fig. 4.

The takt time is obtained considering the total time available, subtracting the time for lunch, cleaning the work area, autonomous maintenance, and the meeting with the supervisor. The static test is the process that exceeds the takt time, as represented in Fig. 5.

A detailed analysis of each necessary activity for the remanufactured of the air compressor performed was assigned according to its criticality, considering the

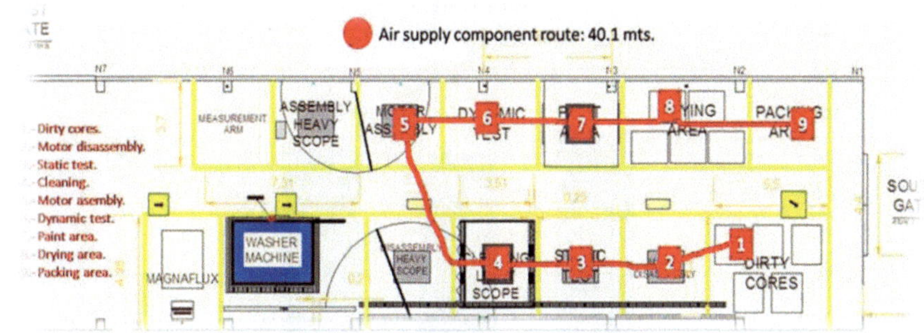

Fig. 4 Final layout and route compressor/operator, Kaizen 2

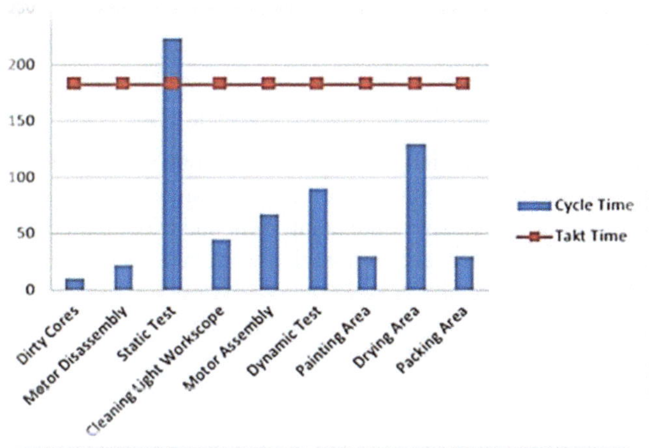

Fig. 5 Takt time versus cycle time

information described above. Additionally, it was taking into account the distances traveled by the operator along the line. The above includes EHS factors.

3.4 *Phase Improve*

Layout

The layout is used to design the spaces for a new line, redesign, and modify the existent process or demand increase. Due to this, before starting the design, a physical survey from the area intended for this project is performed, letting know the right dimensions. The layout design is elaborated on the AutoCAD software (Fig. 6).

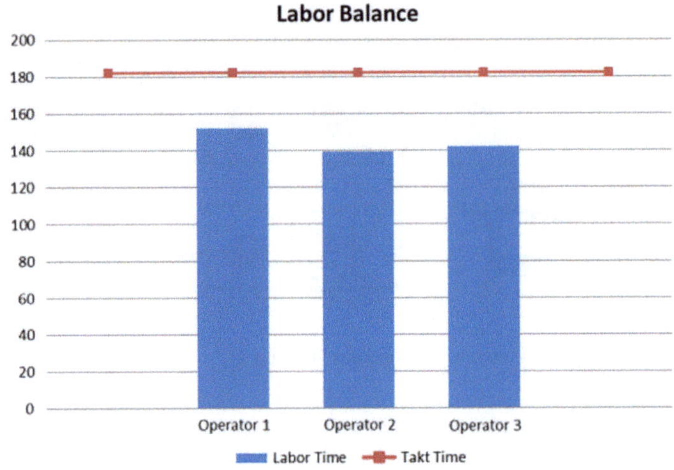

Fig. 6 Labor balance, Kaizen 3

Once that we completed the initial design of the layout, we started adding the lean manufacturing principles. The production line was designed in "U" shape to reduce the transportation and movement of the component and avoid material crossings by using production "pull" system and minimizing work in process (WIP) as illustrated in Fig. 7.

With the design of the layout, the total distance decreased 370.9 m; this has an impact on safety risks reducing 90.24% the probability of an accident while the compressor transports.

The validation of the dimensions of the layout allowed the physical delimitation of work areas in order to corroborate these measures and make changes if they were required.

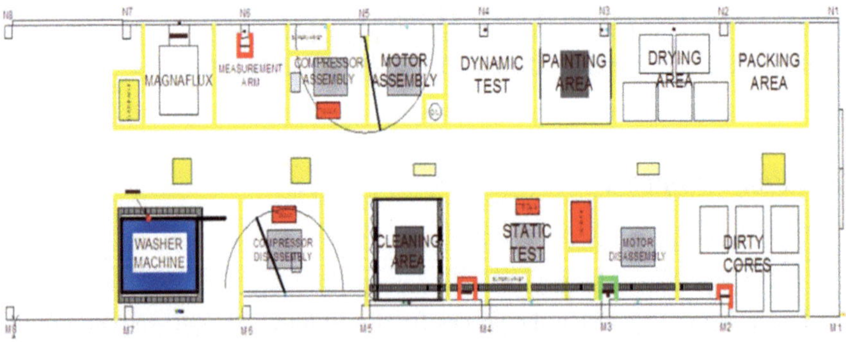

Fig. 7 Layout based on lean manufacturing principles, Kaizen 4

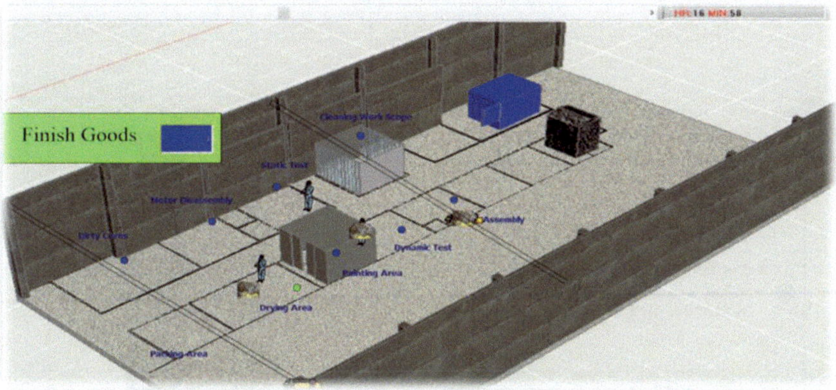

Fig. 8 Simulation model. Process simulation created on ProModel software for validation and verification of the proposal, Kaizen 5

Simulation Model

ProModel software is a simulator program that helps to display the behavior of a system using machinery, operators, and conveyors, among others. This is useful to make better and quick decisions for improving a process, and this is a very powerful tool for industrial engineers as in the industry, where it is critical to upgrade every day. Once the simulation model is made, it is possible to see the statistics and important information about the process like bottlenecks, the production in a determinate period of time, cycle time, and the number of operators required, and view opportunity areas to improve (Harrell and Tumay 1995).

For the air compressor remanufacturing process, two simulation models are elaborated for the air compressor remanufacturing process as depicted in Fig. 8, which emulated the compressor flow through the workstations. The first model was developed on a two-dimensional space allowing demonstrating that there was an existing bottleneck in the static tests workstation. This fact was blocking the fulfillment of the weekly demand because this station cycle time was over takt time due to the variability of compressor failures.

Regarding the second model, using a computer-aided design (CAD), it designed a three-dimensional layout, where it used different simulation elements such as cranes, operators, and status indicators. Once done the balance, the workload of the operators is distributed in a better way; the simulation model is again defined to meet the customer's demand.

Process Sheets

Since it did not have the methods and documentation of the line, the team has made the process sheets for the standardization of working methods.

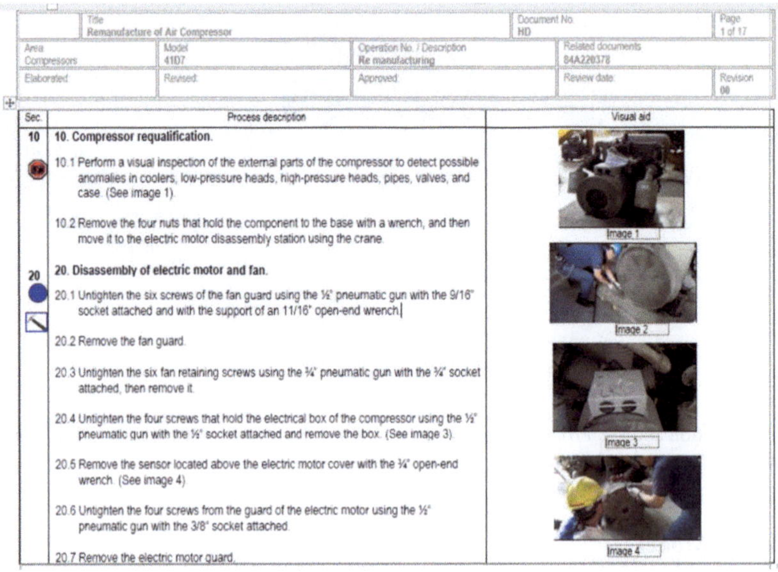

Title: Remanufacture of Air Compressor		Document No: HD	Page: 1 of 17
Area: Compressors	Model: 41D7	Operation No. / Description: Re manufacturing	Related documents: 84A220378
Elaborated:	Revised:	Approved:	Review date: / Revision: 00

Sec.	Process description	Visual aid
10	**10. Compressor requalification.** 10.1 Perform a visual inspection of the external parts of the compressor to detect possible anomalies in coolers, low-pressure heads, high-pressure heads, pipes, valves, and case. (See image 1). 10.2 Remove the four nuts that hold the component to the base with a wrench, and then move it to the electric motor disassembly station using the crane.	Image 1
20	**20. Disassembly of electric motor and fan.** 20.1 Untighten the six screws of the fan guard using the ½" pneumatic gun with the 9/16" socket attached and with the support of an 11/16" open-end wrench. 20.2 Remove the fan guard. 20.3 Untighten the six fan retaining screws using the ¾" pneumatic gun with the ¾" socket attached, then remove it. 20.4 Untighten the four screws that hold the electrical box of the compressor using the ½" pneumatic gun with the ½" socket attached and remove the box. (See image 3). 20.5 Remove the sensor located above the electric motor cover with the ¾" open-end wrench. (See image 4). 20.6 Untighten the four screws from the guard of the electric motor using the ½" pneumatic gun with the 3/8" socket attached. 20.7 Remove the electric motor guard.	Image 2 Image 3 Image 4

Fig. 9 Example of process sheet, Kaizen 6

The process sheets are information that helps the operators to do the activities with the correct tools and specifications in the adequate sequence.

One of the most important purposes of the process sheets is the standardization of the work methods to develop the air compressor remanufacturing. The process sheets are information that helps the operators to do the activities with the correct tools and specifications in the adequate sequence.

Process sheets are very helpful for the operators when the activities are too repetitive or when along the process the operator has to check any specification such as pressure, distances, among others that they could forget. Another important purpose is that process sheets are useful to train new operators for the air compressor line. The team completed the information with the videos, photographs, drawings, manuals, and bill of material (BOM) that had been collected during the action workout previously mentioned as represented in Fig. 9.

Quality Plan

Regarding the quality plan, this elaborated for the assurance of compliance and record of critical to quality that are the primary customer requirements that measured during the remanufacturing process as illustrated in Fig. 10.

The quality plan that the team developed includes different types of information, as they are critical to quality (CTQ), measuring gauges, tolerances, completion date, operator name, and shift to control the air compressor information. The people in the

	QUALITY PLAN	Rev. 16
	Area : Compressors	March 2017
	Operation: Air compressor remanufacturing	Page 1 of 10

Work order	Number serial	Customer	Date

PN:	Description:	Reference:	Remarks:
41D728786AFG5R	Air compressor remanufacturing	84A220378 HDP09CA001	Verify all measuring devices are within calibration period. NOTE. All calibrated gages must be tagged with ID sticker and dates.

Characteristic to be controlled		Specifications	Instrument	Reading	Operator	Date	Remarks
Air compressor visual inspection	⊖	Free of damage, inspect coolers, heads, piping, crankcase.	Visual	Fill in table 1			
Electric motor removal inspection.		No major defects in bearings, connecting rods and lubrication system.	Visual				
Air pressure test in housing at 12 - 15 psi. (Perform this test a total of 3 times.)		Inspect rotor and stator clearance for drag damage or burned motor	Visual				
Crankcase internal inspection		No major damage on bearings rods and lub system, crankshaft, look for abnormal wear	Visual	Fill in table 2			
Inspect crankshaft runout to the motor mounting face	⊖	Less than 0.008"	Magnetic base and dial indicator				Note: Perform compressor heavy work scope if not within spec.
Crankshaft perpendicularity to the motor mounting face	⊖	Less than 0.009"	Magnetic base and dial indicator				Note: Perform compressor heavy work scope if not within spec.

PIP10CA001

Fig. 10 Quality plan. Document elaborated for record and control of CTQ's, Kaizen 7

facility calls quality plan as "traveler sheet" because this document goes on with the component during its remanufacturing process in each workstation (Soin 1992).

Standard Work Combination Sheets (SWCS)

One of the essential principles of the lean manufacturing philosophy is the standardization of work that is why the elaboration of the standard work combination sheets (SWCS) helps to accomplish this goal.

The SWCS shows every operation per workstation and the interaction with the operator during the process. The standardization of work is crucial to provide a proper flow to the line, and it is possible to view the production sequence graphically and change it to improve capacity. In addition, this tool is useful to balance the labor for each operator and outlines the tasks, machinery, waiting time, transportation, and walking time (Meyers 2000).

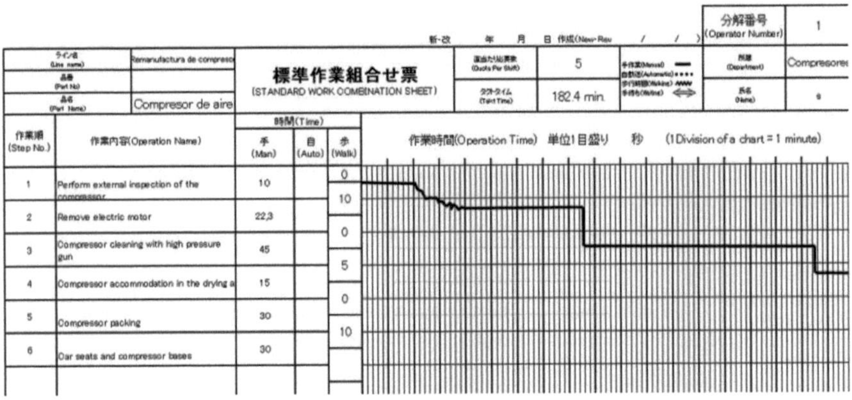

Fig. 11 Sample SWCS operator 1, Kaizen 8

One of the most important principles of the lean manufacturing philosophy is the standardization of labor, which is why the elaboration of the standard work combination sheets (SWCS) helps to accomplish this goal.

Through the elaboration of this document, each of the activities that the operators have to do gets to be standard and allows to have continuous flow in the line. Figure 11 illustrates SWCS of operator 1.

Job Safety Analysis (JSA)

The job safety analysis (JSAs) identifies the risks that expose the operator while working along the line. Besides, it indicates some operations dangerous, and preventive actions must avoid accidents (Wadden and Scheff 1987). Moreover, it identifies the ergonomic analysis for the correct positions.

This document determines the needs in each workstation and analyzes the personal protective equipment (PPE) required as illustrated in Fig. 12, and it contains general information, required training, workstation name, emergency phone numbers, and unsafe activities the operator must know before starting work. The JSAs are located near each workstation for operator query in case of any incident.

Failure Modes and Effect Analysis (FMEA)

The team has elaborated the failure mode and effect analysis (FMEA) to detect the possible failures and their potential effects on the process. This document is a tool used to identify and eliminate a potential product or process failures or defects. The FMEA is discussed with a multidisciplinary team including quality, engineering, materials, logistics, projects, EHS, and manufacturing. One of the main tasks of an

ELECTRIC MOTOR DISASSEMBLY				
GENERAL INFORMATION	SCOPE OF THE ANALYSIS	TRAINING REQUIRED	PERSONAL PROTECTION EQUIPMENT	IN CASE OF EMERGENCY
Discussed: Fernando C. **Authorized**: M.D. **Supervisors**: A.M. **Date**: Dec 2, 2018 **AST** No. 02 **Revision**: 01 **Air Supply complement**	ELECTRIC MOTOR DISASSEMBLY	❖ EYE PROTECTION ❖ HANDS PROTECTION ❖ TOOLS HANDLING ❖ CRANE HANDLING ❖ HAZARDOUS WASTE ❖ AST'S	**REQUIRED:** **NOTE:** GLOVES EARPLUGS TYVEK SUIT	In case of accident or emergency call EHS immediately Extensions: **2990** Mayra D. **2842** Leticia C. **2316** Dr. Tenoch C. **2840** L.E. Minerva P. Or via radio on **Channel 6.** Indicate the place of the incident and wait for the Emergency Response Brigade, providing all the requested information.

Fig. 12 Partial view JSA disassembly of electric motor, Kaizen 9

interdisciplinary team is to recommend actions if necessary and assign the owner to implement each one.

With FMEA, the teamwork analyzes every step in the process and prioritizes each one with a standard rating according to the risk for both, the product and the operator safety while he works in the line. Each possible failure has a score that indicates how many times it could happen in the process (occurrence), a score that indicates if the fault is difficult or easy to found (detectability), and finally a score that shows how dangerous the fail is for the product and the operator safety (severity). Because of the FMEA, the risk priority number (RPN) helps to determine the most dangerous steps and required action for their control as represented in Fig. 13.

One of the most important objectives of this meeting was to identify and analyze the different failures that the process could have along the value stream and determinate recommended actions if necessary. Additionally, it was determined the name of the person who has to supervise the accomplishment of each action.

In the case of the air compressor line, the most critical is the possibility that the electric motor does not be removed and the action to take is to create a special fixture for this operation.

Failure Modes and Effects Analysis
Process FMEA

ITEM A S Component   Process Responsibility   FMEA Number 1

Assembly ____ Compressor   Prepared By: Page 1 of 3

CORE Team ___.   FMEA DATE 02 Dec 0

Process Function Requirements	Potential Failure Modes	Potential Effect(s) of Failure	Sever	Potential Cause(s) Mechanism's of Failure	Occur	Current Process Controls	Detec	R. P. N.	Recommended Action	Responsibility & Target Completion Date	Action Results: Action Taken	Sev	Occ	Det	R. P. N.
Cores Reception	Wrong identification from customer	Spend more $ than standard to rebuild compressor with wrong components/damaged main components	3	Human mistake	2	Material planner to solve mistake	1	6	Include adders process	Master Scheduler					
	Core without paperwork	No able to start repair process, can't process work order	3	Human mistake	1	Material planner to solve mistake	1	3	Include adders process	Master Scheduler					
	Wrong PTN Numbers	No able to start repair process, can't process work order	2	Human mistake	6	Material planner to solve mistake	1	12							
Visual Inspection	Missing parts	Incomplete assemblies	3	Common issue	3	Add or complete missing parts at static testing	1	9	Included check list in quality plan						
	Wrong recording of information	Delay in process due to the missing component when needed	2	Human mistake	3	Add or complete missing parts at static testing	2	12	Include adders process	Master Scheduler					
Remove the electrical motor	Damaging the motor while removing	Scrap motor, or crankshaft major repair to motor	4	No right disassembly process in place	6	None	2	48	Fabriquate fixture table to hold motor	Fernando Cervantes					
		Spend more $ than standard													

Fig. 13 Partial view of FMEA, Kaizen 10

Fig. 14 Visual management implementation, Kaizen 11

Visual Management

Within the premises of a new production line is fundamental to implement a visual management system that allows identifying work areas, risks, and mandatory protective equipment. Among the main signs deployed are an emergency exit, workstations identification, classification of air, water and steam lines, fire extinguisher location, leak-proof kit and containers for scrap and garbage as depicted in Fig. 14.

3.5 *Phase Control*

Control Plan

Finally, it was necessary to create a control plan for project sustainability. This document includes actions, planned dates, owner, completion date, and frequency.The new workstation was involved in the quality audits annual program to review process compliance, record proper quality plans, and adjust measure gages. Also, the control

Table 2 Control plan Kaizen 12

Item to control	Responsible	Register	Date
Action			
Check 5'S tool cart and shadow board			
1. Tool box 1	Fernando C	Check list	Each month
2. Tool box 2	Fernando C	Check list	Each month
Housekeeping tours (extinguisher and EPP)	EHS Department	Lay out	Weekly
Check area conditions (machines, tools cart and fixtures)	José P	Lay out	Each month
Review and update next documentation			
Quality plan	Carlos R	Documentation	Bimonthly
FMEA	Fernando C	Documentation	Bimonthly
Process sheets	Fernando C	Documentation	Bimonthly
JSA´s	Fernando C	Documentation	Bimonthly
Standard work combination sheets	Fernando C	Documentation	Bimonthly
Control project plan			
Train operators in assembly methods	Angel M	Meeting	Every 6 months
Review lay out, new tools, equipment and others	Fernando C	Plan or scheme	Bimonthly
Design and approve fixtures	Fernando C. or Jorge H	Plan or scheme	Every 6 months

plan considered safety tours to verify the line conditions, 5S's, and ergonomics, as well as the preventive maintenance plan, as summarized in Table 2.

This document considers the dates of review as well as the person responsible for such activities, with the purpose that the project could have sustainability over the time.

The new production line got up in the annual plan for quality audits, where periodically the fulfillment of process is checked, the correct quality plan filling, and calibration of the measuring equipment.

4 Conclusion and Results

After the implementation of *lean manufacturing* philosophy and DMAIC methodology throughout the project of introducing the process of remanufacturing for air compressor, the team did twelve kaizen and achieved great impact benefits for the company, increasing in this way the number of man-hours in the repair shop with which they improved their economic situation. The financial benefit, as summarized in Table 3, was the labor cost decrease from 52.5 USD to 47.5 USD representing 9.5% cost reduction, efficiency increase 21%, distance reduction of 380 m, and annual savings 384 K USD.

The results obtained from the implementation and development of this project led to the approval decision of the new project.

The development of this remanufacturing line is the baseline for the management and implementation of future projects at the company, as it has its foundation on the *lean manufacturing* philosophy and represents a model to follow for the other rail sites.

The implementation of the lean manufacturing philosophy in conjunction with the DMAIC methodology generated positive results in the reduction of waste in cycle time, the distances traveled between operations, the cost of labor, among others; likewise, it allowed the achievement of the objectives set by the company and those proposed in the project.

In the introduction of new processes to ensure operational quality, there were identified opportunities for improvement that can be summarized in the following aspects: training, leadership, and control.

The participation of the personnel involved is a crucial factor in maintaining a quality work system and culture, so it is convenient for the company to establish a permanent training and education programme.

Additionally, the pledge of senior management and their participation in continuous improvement projects strengthen leadership, an essential element to achieving planned changes in the organization, and reaches the commitment of all involved.

Regarding control and monitoring, it is necessary to have a strategy so that the work system and the quality assurance system achieve the planned results. Part of the activities and results of this project were the control plan, which establishes some items identified as critical for carrying out audits or supervision regularly by those responsible, to verify their suitability or take action in the event of deviation. To complete this part, the organization must establish and monitor the quality indicators that impact the final product that are critical to end-user acceptance.

In the specific case of the introduction of new projects of this railway company, the use of continuous improvement methodologies implies the standardization of processes and their documentation to be monitored and thus ensure that the specifications oriented meet customer requirements.

Finally, in conclusion, the application of lean manufacturing tools provides abundant benefits to productivity and quality.

Table 3 Company benefits

Item	Increase/decrease	Total
Efficiency (cycle time)	I	21%
Savings	I	$384 K USD
Labor cost	D	$5.00 USD
Distance traveled	D	380 m

Acknowledgements Authors thankfully acknowledgments the opportunity to participate in this project, technical expertise, mentoring, and assistance provided by managers of Rail Company in Mexico.

References

Adams M, Componation P, Czarnecki H, Schroer B (1999) Simulation as tool for continuous process imptovement. In: Proceedings of the 1999 winter simulation conference. University of Alabama in Huntsville. https://doi.org/10.1109/WSC.1999.823256

Adams M, Kiemele M, Pollock L, Quan T (2003) Lean six sigma: a tools guide, 2nd edn. Academy Associates, Air Colorado

Barney H (2004) Toyota production system/lean manufacturing. In: Kirby S, Stecher B (eds) Organizational improvement and accountability: lessons for education from other sectors. RAND Corporation, Santa Monica, CA; Arlington, VA; Pittsburgh, PA, pp 35–50. Retrieved February 18, 2020, from www.jstor.org/stable/10.7249/mg136wfhf.9

Bhuiyan N, Baghel A (2005) An overview of continuous improvement: form the past to the present. Manage Decis 43(5):761–771. https://doi.org/10.1108/00251740510597761

Cardona JN (2014) Contonuous improvement strategy. Euro Sci J 10(34):117–121

Čiarniene R, Vienazindiene M (2012) Lean manufacturing: theory and practice. Econ Manage 17(2):726–732. https://doi.org/10.5755/j01.em.17.2.2205

Harrell Ch, Tumay K (1995) Simulation made easy: a manager's guide, 1st edn. Engineering & Management Press, USA

Imai M (1986) Kaizen. the Key to Japan's competitive succes, 1st edn. Irwin Professional, USA

Imai M (1997) Gemba Kaizen: a commonsense, low-cost approach to management, 1st edn. McGraw Hill, New York

Japan Management Association (1989) Kanban, just in time at Toyota: management begins at the workplace. Productivity Press, New York

Meyers F (2000) Estudio de tiempos y movimientos, 2nd edn. Pearson Educación, Mexico

Newitt D (1996) Beyond BPR & TQM—managing through processes: is Kaizen enough? Institution of Electric Engineers, Industrial Engineering, London, U.K. https://doi.org/10.1049/ic:19960785

Oakland J (2014) Total quality management and operational excellence: text with cases, 4th edn. Routledge, USA

Pavnaskar S, Gershenson J, Jambekar A (2003) Classification scheme for lean manufacturing tools. Int J Prod Res 41(13):3075–3090

Rahman A, Uddin S, Sarkar S, Hashem M, Hasan S, Mandal R, Islam U (2017) A case study of six sigma define-measure-analyze-improve-control (DMAIC) methodology in garment sector. Independent J Manage Prod 8(4):1309–1323. https://doi.org/10.14807/ijmp.v8i4.650

Rother M, Shook J (2009) Learning to see: value stream mapping to create value and eliminate Muda. Lean Enterprise Institute, Cambridge, MA

Shingo S (1989) A study of the Toyota production system from an industrial engineering viewpoint, 3rd edn. Productivity Press, Cambridge Massachusetts

Soin S (1992) Total quality control essential: key elements, methodologies, and managing for success, 1st edn. Mc Graw Hill, New York

Sharma G, Rao P (2014) A DMAIC approach for process capability improvement an engine crankshaft manufacturing process. J Ind Eng Int 10(65). https://doi.org/10.1007/s40092-014-0065-7

Suárez-Barraza M (2007) El Kaizen: la filosofía de mejora continua e innovación incremental de-trás de la administración por calidad total, 1st edn. Panorama, México

Suárez-Barraza M, Miguel-Davila J (2009) Encontrando al Kaizen: Un análisis teórico de la mejora continua. Pecvnia 7:285–331

Villaseñor A, Galindo E (2008) Conceptos y reglas de lean manufacturing, 2ª Ed. Limusa, México

Wadden R, Scheff P (1987) Engineering design for the control of workplace hazards, 2nd edn. McGraw Hill, USA

Womack J, Jones D (2003) Lean thinking, 1st edn. Free Press, New York

Wei C, Ling C (2006) An integrated structural model toward successful continuous improvement activity. Technovation 26:697–707

Zaman D, Zerin N (2017) Applying DMAIC methodology to reduce defects of sewing section in RMG: a case study. Am J Ind Bus Manage 7:1320–1329. https://doi.org/10.4236/ajibm.2017.712093

Zangwill W, Kantor P (1998) Toward a theory of continuous improvement and the learning curve. Manage Sci 44(7):910–920. https://doi.org/10.1287/mnsc.44.7.910

Personnel Training as a Tool for Quality Assurance: Case of Study at Plastic Injection Enterprise

Abril Salaices Medrano and G. Angelina López Pérez

Abstract New personnel, who is well trained, can feel confident about them daily work activities, therefore them reaction time to any problem may be reduced and decision-making capacity improves, making all personal more efficient. An automotive business dedicated to injection of plastics was often affected by delays in the planned production dates. Using some quality tools to identify the main causes, it was found that these delays were provoked by errors made by the quality auditors' team, damage in molds caused by machine operator neglect and high staff turnover. A 5 why methodology made possible to identify a common factor for all these incidents: the lack of a proper training for new personnel. This company managed to create a new training protocol which finally reduced the production delays.

Keywords Training · Defects · Staff · Tool · Plastics · Automotive

1 Introduction

Before starting with this article, we remember that according to (Guilló 2000) the direction of quality is a journey to continuous improvement. This remarks the historical importance of quality and its different tools for the enterprise's development.

However, the time organizations have witnessed the development of quality as a field and its effort to be recognized as a guideline for all enterprises in order to get success. This may be widely seen into the Toyota's competitive advantage and its philosophy based on the comprehension and motivation of its employees, betting for leadership and continuous learning (Jeffrey 2004).

The Toyota production system (TPS) is an example of the importance of cultivating knowledge and propitiate a good work environment for all personnel in order to allow them to grow professionally. This effort will be directly seen in the improvement of products and process quality. The total quality systems stand on people participation;

A. Salaices Medrano · G. Angelina López Pérez (✉)
Ingeniería en Tecnologías de Manufactura, Universidad Politécnica de Durango, Durango, México
e-mail: angelina.lopez@unipolidgo.edu.mx

J. L. García Alcaraz et al. (eds.), *Techniques, Tools and Methodologies Applied to Quality Assurance in Manufacturing*,
https://doi.org/10.1007/978-3-030-69314-5_19

likewise, any enterprise which decide to implement a total quality philosophy should make a special emphasis on human resource training (Danvila del Valle 2007; Guilló 2000). How the personnel training can help us in the way to get a philosophy of total quality on our companies? Personnel training is a tool which allow us to reduce the frequency of defects related to the lack of knowledge of the process by its own operators.

When personnel is well trained can feel confident about them daily work activities, therefore them reaction time to any problem may be reduced and decision-making capacity improves, making all personal more efficient.

The present case of study shows the problematic of an automotive enterprise which works with plastic injection. Due to privacy politics, the name of the company, which is placed on Queretaro city, will not be mentioned. This company frequently presents delays on production deadlines. The goal of this study is to identify and repair the main causes of this delays.

2 Methodology

2.1 *Identification of the Causes*

To achieve the objective, it is necessary to identify all the problems presented during the production time. This identification is developed using a Paretto diagram which concentrates the information of the reports made by the middle and high management of the production plant and quality area during the April to October period.

The 80% of the problems shown by the Paretto diagram will be analyzed alike using a 5 why methodology in order to establish an action plan to solve this problematic.

2.2 *Staff Turnover*

A previous red spot indicated by the personnel is that both, the production area and the quality area of the company, had a high staff turnover. Reported cases declare that some workers only attended for a few days and then left the company.

According to the book, Personnel Administration: An Approach to Quality, we can understand that the turnover rate is determined by the number of workers who join and leave in relation to the average total number of personnel in the organization, in a period of time (Castillo Aponte 2006). The personal turnover index is expressed in percentage terms using the following mathematical formula:

$$\mathrm{IRP} = \frac{\frac{A+D}{2} * 100}{\frac{F1+F2}{2}} \tag{1}$$

where

- A: The number of people hired during the considered period.
- D: Detached persons during the same period.
- $F1$: The number of workers at the beginning of the period considered
- $F2$: The number of workers at the end of the period.

The reasons why the worker had made the decision to leave the company is registered through a survey. Ex-workers report their reasons to human resources department when submitting their resignation by answering questions like: What are your reasons for leaving the company? Do you think you have received adequate training to carry out the activities required in your work position? and among others.

2.3 *Training Program*

The training program is defined as the detailed description of a set of instruction-learning activities structured in such a way that they lead to achieving a series of previously determined objectives (Alarcón 2020).

Conducting longer, well-structured training ensures that workers 100% understand the activities they must perform at their jobs (Danvila del Valle 2007). Long training should not be seen as a waste of time or resources because in the long term the benefits it provides will avoid a significant percentage of the defects or delays caused by human capital.

Also should be taken into account that by having full certainty of how to carry out the activities that correspond to their job position, the worker will be able to make decisions quickly, cutting the action time necessary to correct a defect and reduce the amount of scrap which is generated by the machine the operator is working on.

3 Results

3.1 *Analysis of Main Causes of Production Delay*

The Paretto diagram in Fig. 1 shown the main causes of production delay or breach of deadlines. The 80% of the problems were concentrated on: mistakes made by the quality auditor team, damage of the plastic injection mold caused by carelessness of the operator and a very high personnel rotation.

Quality team mistakes (quality auditors at production area) failure at detecting defects at pieces. This may be caused by slowness during the identification of a mistake or ignorance of the features and tolerances which pieces may have.

Damage of the molds which is caused by neglect of the operator, may produce non-programmed stops of production in order to repair the damaged mold.

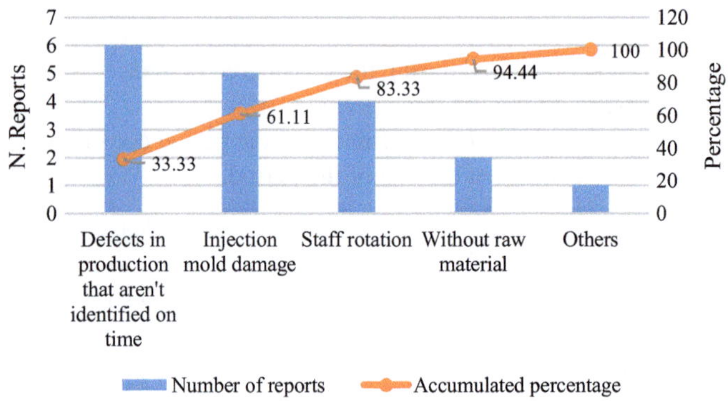

Fig. 1 Main causes of the production delays

These three causes of production delay are analyzed below to identify alike their own causes.

Staff Turnover

According to Eq. 1, the company had a staff turnover rate of 37.93%. The identified causes of the staff turnover are shown in Fig. 2. Although staff turnover is caused by various factors, including the search for better benefits or increase the salary perception, lack of training is an important factor when a worker decides to leave the company. As we can see, 27% of the interviewed former workers mentioned the unconformity with the training received as a reason to quit. This cause is only surpassed by the pursuit of better benefits.

Believing that providing workers with long and well-structured training is a waste of time could be a huge mistake if we have high rates of staff turnover. This provoke that workers do not feel satisfied and will not be able to perform properly in their activities.

Fig. 2 Job abandonment causes

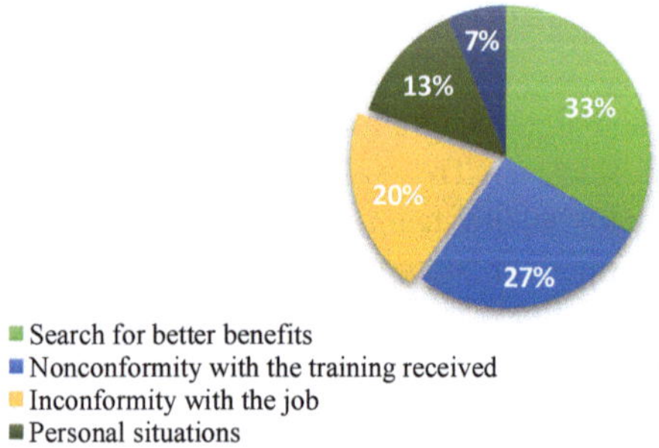

Production Defects Which Are not Identified on Time

When a quality auditor is not well trained, he makes more mistakes. These errors are usually presented by differences in the evaluation criteria of the pieces; this means, a quality auditor who has not been trained in a timely manner may believe that despite a no-go (NG) condition or a defect; it is not necessary to stop the process and report the situation to the production supervisor to correct it. Is convenient to remember that a small defect may be within the acceptance criteria, although not reporting it could turn this into an NG condition in most cases. As well, when the production continues running, the defect could appear in a greater quantity of pieces (Eiroá 2020). As an example of this situation, we may describe a case where the type of defect in a plastic part was a perforated injection point, which is the point where hot plastic material is injected to the mold in order to create the new plastic piece (Delgado 1975). The quality auditor did not notice this situation. Even though the defect was corrected by itself at the end of the turn, it reappeared during the next turn and was also not detected in a timely manner.

Finally, the defect was detected at the end of the next operation turn. As a result, the defect run over 20 h and affected to 1900 pieces with a NG condition. This is translated to economic loss of ten thousand Mexican pesos of raw material and production time. The amount of NG pieces produced can be seen at Fig. 3.

A 5-why technique was applied to know the reasons why this defect appeared and why it was not reported in a timely manner. Figure 4 shows what was found.

When interviewing the quality auditors to find out why they had not detected the defect in a timely manner, they first argued that the NG condition was not known. Both auditors, night work shift and morning work shift, gave similar arguments. Consequently, the inspection procedure had not been carried out correctly.

Why were the quality auditors unaware of the defect? The conclusion of 5 why's tool, shown in Fig. 5, pointed to them training, which consisted of being for a week accompanied by a quality auditor, who had more time working in the company. New workers helped the auditor with his work, and he instead explained the activities to

Fig. 3 Produced scrap by a wrong defect identification on production pieces

WHY'S
The found defect was a punched injection point on the piece
1. WHY?
The quality auditor did not notice the situation
2. WHY?
The inspection procedure was not performed correctly
3 WHY?
No such NG condition was known at that part number
4 WHY?
During their training week the part number mentioned above was not produced so they were unaware of the specific flaws that could occur in that part number
5 WHY?
There was no training plan that included all the part numbers produced by the company and their specific characteristics

Fig. 4 Five why's method

Fig. 5 Why does the operator makes this type of error?

be carried out during the shift. All the explanation process was only verbally. During the week of training, the part number mentioned above was not produced so they were unaware of the specific flaws that could occur in that part number.

Damage to Molds Caused by Operator Carelessness on the Machine

The next cause of the production delay is equipment damaged by mishandling. Information obtained from the April to October period regarding the damage caused to the molds by the production operators says that in most of the cases a common denominator is the fact that the operator started the injection process without having removed the previous piece or casting from the mold, which caused a double injection that caused internal damage to the different molds This may be seen at the next graph which includes a new 5 whys tool.

The machines of the company work in semi-automatic mode. The new workers were not properly trained about the correct order in which they had to carry out their work. The flow process should be:

Open the machine door.

Take out the hot runner channel scrap and put it on a special container.

Remove the injected part from the mold.

Close the machine door and start the cycle again.

Inspect the part 100% for defects.

Deposit the part in the box for the WIP warehouse.

This may cause confusion to the operator who frequently chose to skip steps, forgetting to remove the previous injected part or hot runner scrap from the mold before starting the cycle again.

Since April production time was increased to 168 h a week, the company was forced to hire more staff and create an additional working group to cover 24/7 work roll; this situation overwhelmed the training capabilities which the company had previously been managing. Production supervisors were forced to constantly leave operators unsupervised even knowing that they had not received adequate training.

Other causes of damaged molds were machine failures, deactivated sensors or lack of periodical preventive maintenance to the mold before being mounted on the injection machine. The amount of monetary loss due to this causes can be seen in Table 1. However, damage caused to the mold by operators actions was the main concern of the company due to the economic impact that this represented as can be seen in Fig. 6, which is why it was decided to mainly ensure adequate training for work personnel. The damage produced to the molds is shown in Fig. 7.

3.2 *Training Program*

In this case, the plant management worked with a new training plan developed jointly by the human resources department, the production and quality managers, production

Table 1 Monetary losses recorded due to damage to the molds in the April–October period

Mold Number	Part number	Date	Cost MNX	Cause
552	6924–1585	15–04-2019	$2000	Damage to the mold caused by carelessness of the operator in the machine
580	6952–1061	17–05-2019	$1000	Machine failure
580	6952–1061	04–06-2019	$2000	Damage to the mold caused by carelessness of the operator in the machine
584	6920–3245	10–07-2019	$10,000	Damage to the mold caused by carelessness of the operator in the machine
563	6924–1709	17–08-2019	$300	Preventive maintenance was not performed
580	6952–1061	14–09-2019	$2000	Damage to the mold caused by carelessness of the operator in the machine
584	6920–3245	16–10-2019	$3000	Deactivated sensors
585	6920–3246	22–10-2019	$2000	Machine failure
552	6924–1585	26–10-2019	$5000	Damage to the mold caused by carelessness of the operator in the machine

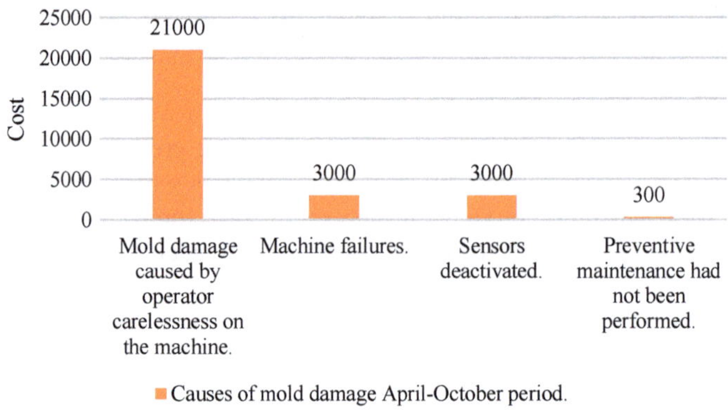

Fig. 6 Causes of mold damage in the April–October period

supervisors and the supervisor of quality auditors. Figure 8 shows personnel attending to the training plan. According to (Alarcón 2020), a training program should be developed due to benefits as shown in Fig. 9.

The new training plan created by the company for quality auditors provides a detailed description of the specific properties of each of the 30 part numbers that are produced, as well as a description of the areas where possible defects may occur

Fig. 7 Internal damage found after the operator closed the mold with a part already injected inside, the repair of this mold had a total cost of 12,000 Mexican pesos

Fig. 8 Training plan must include theoretical and practical content

on the pieces and which are the most common defects in each one. An example of the development of the new training plan is next shown following the detailed observations by the guide for the development of training programs issued by the Secretary of Labor and Social Welfare of Mexico.

In order to develop the training program, different authors were consulted (Alarcón 2020; Castillo Aponte 2006; Danvila del Valle 2007; Parra-Penagos and Rodriguez-Fonseca 2016). As defined in the guide for the development of a training program, it must have previously established objectives. What were the objectives to be pursued? Supporting ourselves with the interviews carried out together with the human resources department, three main objectives can be identified.

Show the worker which are the general characteristics of each part number, making them aware of the special inspection criteria for some part numbers, as well as their special characteristics.

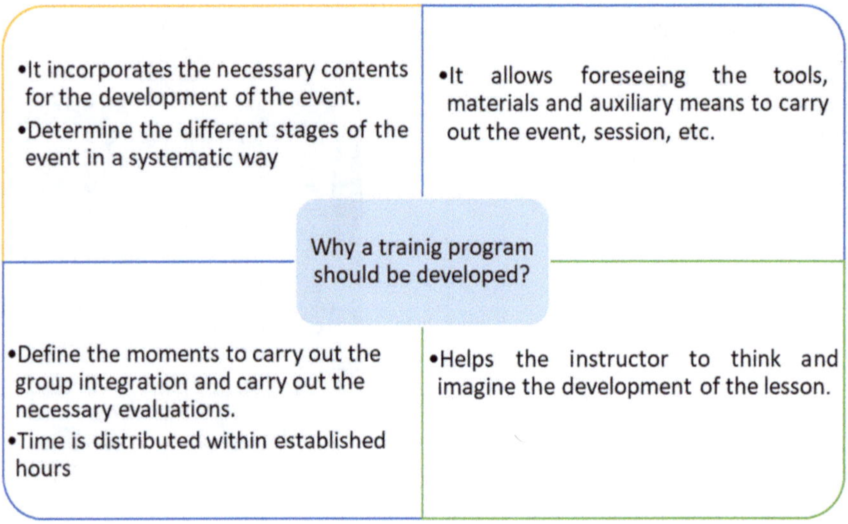

Fig. 9 Why should a training program be developed? *Source* Alarcón (2020)

Train the operator and make him aware of the correct procedure to carry out his work.

Inform the operator about the procedure to stop his machine and call his supervisor in case of identifying any abnormal situation.

The new training plan included:

A theoretical part with a duration of 2 days in which the mission, vision of the company will be disclosed to the worker. During these days, the worker will also know the regulations for absences, permits and other issues related to the department of human resources.

Another theoretical–practical part of 5 days in which the worker will receive the information about the characteristics of each part number, the possible defects of each part and the location of the same supported by inspection guides and visual aids, also to the worker he was asked to identify defective parts in a batch of 50 containing OK parts and others with NG conditions. This visual aids is shown in Figs. 10 and 11.

A practical part lasting 15 days, during this period, the personnel will know the procedure corresponding exclusively to their work area, as well as their responsibilities and tasks to be carried out; during the last 5 days, the worker would be fully integrated on your work team and schedule assigned by the production manager.

In the quality area, the procedure for the quality auditors changed so that an inspection of the parts produced had to be carried out every hour.

A set of parts produced every hour was put by the machine operator in the containers assigned for it in the production area, thus facilitating the inspection work for the quality auditors.

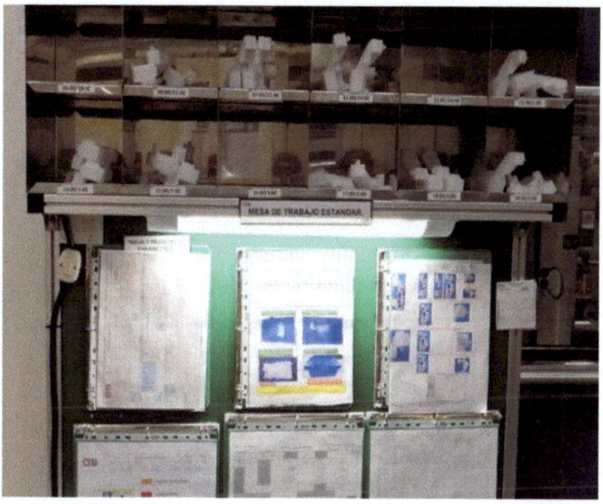

Fig. 10 Special containers for hour-by-hour inspection

Fig. 11 Special containers for hour-by-hour inspection

During the months of November, December and January, the new training plan for production operators and quality auditors was tested. This area of work also changed the way in which it carried out the activities to review the production process.

Example of a Training Program Made for Quality Auditors in the Production Process

Table 2 shows the main features of the final training program, as the main objectives of the program, the duration and the activities to be developed by the new personnel. Figure 12 shows all team members who attended to training program.

Table 2 Training program

Program	Training program for quality auditors	Duration	264 h
General objective	Identify the general and particular characteristics of the part numbers	Content	1. Part number 2. Characteristics 3. Main defects and its location
Aimed to	Quality auditors of production area		
Topic II	Characteristics of the part numbers produced in the company		
Theory hours	30	Practice hours	30
Particular objectives	The personnel must identify the different part numbers with their code, the personnel must differentiate between an ok condition and an ng condition in the parts		
Specific objectives	The personnel must identify the different part numbers with their code with 100% success in the theoretical-practical evaluation, the personnel must differentiate between an ok condition and a ng condition in the parts with a 100% successful events in the -practice		
Activities	The personnel will be provided information about the part numbers: features, codes, project to which they belong through the use of slides and physical examples of the parts The personnel will be provided information about the defects found in the parts, their location and their solution through the use of inspection guides, visual aids and physical examples of the parts		

Fig. 12 Personnel at training

Training is one of the best investments in human resources and one of the main sources of well-being for personnel and the organization (Parra-Penagos 2016).

4 Conclusion

In order to develop a training program, it is necessary to fully immerse in the problems experienced by the company, and as the project moves forward in the interviews with the personnel and in the approach of the challenges and objectives, it makes clear how deeply a bad training of the personnel affects a company.

After the implementation of the new training programs for quality auditors and production operators, a favorable increase in the commitment of employees toward the objectives pursued by the company and a positive and optimistic attitude toward production goals could be observed as it was expected according to Parra-Penagos and Rodriguez-Fonseca (2016).

There is now the ability to prevent a defect from continuing to appear in production for a long period of time. This is possible through a joint effort between quality auditors who now work inspecting batches of parts produced every hour and injection machine operators who can recognize an abnormal situation or defect in parts and notify your supervisor or quality auditor.

In Fig. 13, a significant decrease could be observed in the cases molds damage caused by carelessness of the operator. In the period from November 2019 to January 2020, only one case was presented. A comparison can be seen at the next graph. It is expected that by the end of 2020, an agreement can be reached with the client about reduce or eliminates the inspection from 200% of the product.

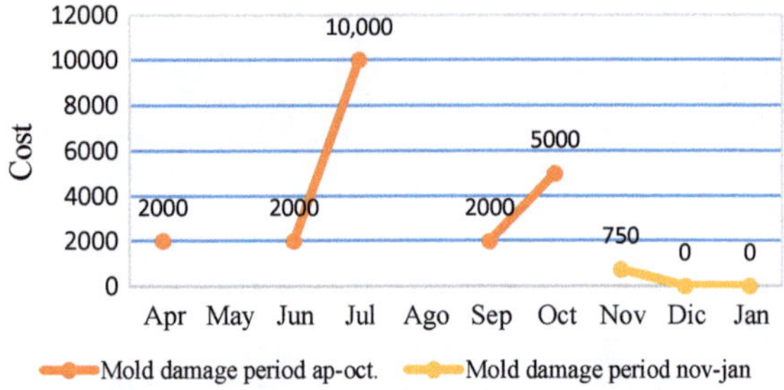

Fig. 13 Comparison of periods April–October and November–January

References

Alarcón LJL (2020) Elaboracion de programas de capacitacion, Anexo. [En línea]. Available at https://www.gob.mx/cms/uploads/attachment/file/160973/Elaboracion_de_programas_de_capacitaci_n_Anexo_1_250_1.pdf

Aponte JC (2006) El programa del personal. En: Administración de personal: un enfoque hacia la calidad. ECOE, Bogotá, p 68

Danvila del Valle SC (2007) El papel de la formación de personal en el proceso de implantacion de un sistema de calidad total. Contaduria y administracion, p 222

Delgado RA (1975) Moldeo por inyeccion.Blume, Madrid

Eiroá PL (2020) Inyección eficiente de plásticos. [En línea]. Available at https://inyeccion-eficiente-de-plasticos.blogspot.com/2013/01/defectos-en-piezas-inyectadas-y-su.html

Guilló JJT (2000) Calidad total fuente de ventaja competitiva. Publicaciones Universidad de Alicante, Alicante

Jeffrey IL (2004) The Toyota Way. McGray-Hill, Nueva York

Parra-Penagos C, Rodriguez-Fonseca F (2016) La capacitacion y su efecto en la calidad dentro de las empresas. Revista de investigacion desarrollo e innovación, pp 131–143

Operational Risk Management in the Supply Chain of Blood Products

Nataly Agudelo-Ibarguen, Mayerli Daniela Naranjo-Sánchez, and Juan Carlos Osorio-Gómez

Abstract Blood must be seen as a scarce resource in the world. Blood transfusions help to save and improve the quality of life of thousands of people around the world all the time. There is no alternative to meeting the demand for blood in medical procedures, and there is no substitute for human blood, which is voluntarily supplied by donor individuals. The rate of donation is currently considered low in Colombia. However, this is not the only problem for institutions responsible for the reception, storage and distribution of blood products. Other factors within the biological and logistical control of the chain directly affect the safety and availability of blood, making the functioning of the supply chain an interesting study problem to be addressed. This supply is exposed to the possibility of operational risks affecting the results of the whole blood supply chain. That is why it is important to manage risks from their identification and prioritization so that organizations can direct their efforts to mitigate or eliminate them. Considering all these aspects, a proposal is then presented for the management of operational risks in the transport and storage of blood products in Colombia. The proposal includes the identification, prioritization and definition of actions aimed at mitigating or eliminating the main risks in the chain.

Keywords Operational risk management · Blood products · Risk identification · Risk prioritization · Fuzzy QFD

1 Introduction

According to the World Health Organization, blood donations contribute to saving lives and improving people's health (Organización Mundial de la salud 2016). Types of patients requiring blood transfusions include: women with obstetric complications, children with severe anemia, people with severe trauma from natural disasters and

N. Agudelo-Ibarguen · M. D. Naranjo-Sánchez · J. C. Osorio-Gómez (✉)
Escuela de Ingeniería Industrial, Universidad del Valle, Cl. 13 #100-00. Cali, Valle del Cauca, Colombia
e-mail: juan.osorio@correounivalle.edu.co

J. L. García Alcaraz et al. (eds.), *Techniques, Tools and Methodologies Applied to Quality Assurance in Manufacturing*,
https://doi.org/10.1007/978-3-030-69314-5_20

those caused by man, cancer patients and many other patients who require surgery (Organización Mundial de la salud 2016). That is why the different authorities in this area feel the need to launch awareness campaigns on the importance of donating blood. A sample of the work carried out by these entities must be donated worldwide about 112.5 million blood units per year, according to data collected and published by the WHO in June 2017. Moreover, in Colombia, between 2014 and 2015 alone, according to figures revealed by the National Institute of Health, it was possible to collect around 792,000 blood units per year (Vengoechea 2016).

However, the blood that is donated is not sufficient to supply the demand for this supply. In Colombia, the rate of donations is 15.4 per 1000 inhabitants when the ideal would be for there to be between 30 and 40 donors for every 1000 people in the country, says Gabriel Cubillos, member of the board of directors of the Colombian Association of Blood Banks and Transfusion Medicine ACOBASMET (VIDA/SALUD 2016).

The shortage of blood, coupled with factors such as the risk of transmission of infections through unsafe blood use, among others, has resulted in WHO as the authority and coordinator of health action within the United Nations system; generate strategies to increase the safety and availability of blood globally. In Colombia, the National Institute for the Monitoring of Medicines and Food—INVIMA—was entrusted with the supervision, health control and monitoring of the functioning of blood banks in the country.

At the national level, blood banks have to follow the basic health regulations issued by the Ministry of Health and Social Protection, which are available from different sources, are publicly available and are responsible for managing the blood supply chain in the country.

Within these guidelines established by the country's health authorities, the management of the blood supply chain must include, inter alia, the implementation of proper cold chain management, considering good practices in the different technical and logistical aspects involved in the operation. The cold chain is the access to and use of appropriate technology for the safe storage and transport of blood from donation to transfusion in order to ensure the quality of blood.

Although the introduction of policies and regulations on the subject has made it possible to improve the availability and safety of blood in the country, organizations such as blood banks are not immune to the likelihood of unexpected events occurring that may negatively affect the performance of the blood supply chain, which is known as operational risk.

In the annual report of the Blood Bank Network for 2017, it is stated that:

Whereas on average in Colombia a unit of red blood cells has an economic value for the system of $291,733 (Instituto Nacional de Salud 2018) and that from blood banks more than 30,000 units of red blood cells were discarded due to controllable causes. It is estimated that the Health System lost in processing costs about eight thousand seven hundred million pesos ($8'751.900.000), which can result in a detriment to blood banks, and failures related to the satisfaction of demand, that for this component was 88.8 and 90% for platelets.

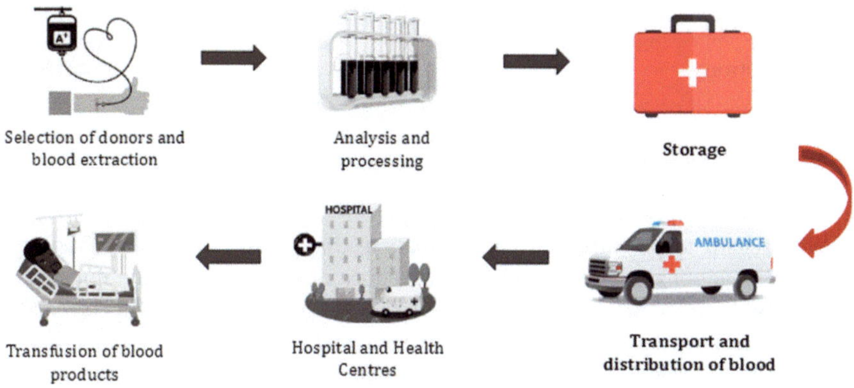

Fig. 1 Blood supply chain (Agudelo Ibarguen and Naranjo Sánchez 2019)

Poor risk management in this process not only results in a huge loss of economic resources for the country and the health sector, but also directly affects the patient's health and quality of life (Organización Mundial de la Salud 2004) the latter being the most important impact. This is why risk management is important, seeking to reduce, eliminate or avoid disruptions in the physical and information flows that affect the interaction of the different links in the supply chain. The string can be plotted graphically in Fig. 1.

2 Literature Review

Generally, blood management is a complex system. There are many factors involved in the system. Some previous research considered health-related risk, supply availability, demand fluctuation, short blood lifetime, to the cost incurred during the process (Mansur et al. 2018).

Researchers have tried to develop blood management improvement in every aspect, including procurement, production, storage/inventory and delivery. Actually, every solution offered is not comprehensive solution and sometimes it cannot apply directly. Therefore, there are still many problems which must be solved on the upstream nor the downstream of blood supply chain (Mansur et al. 2018).

One aspect to be considered in this direction is related to risk management, particularly operational risks present in supply chain management activities. According to Manotas et al. (2016), risk management consists of four phases which are seen in Fig. 2.

The identification phase is essential because risks that are not identified will be outside all other analyses and therefore the supply chain company(s) will not be able to develop management-oriented actions.

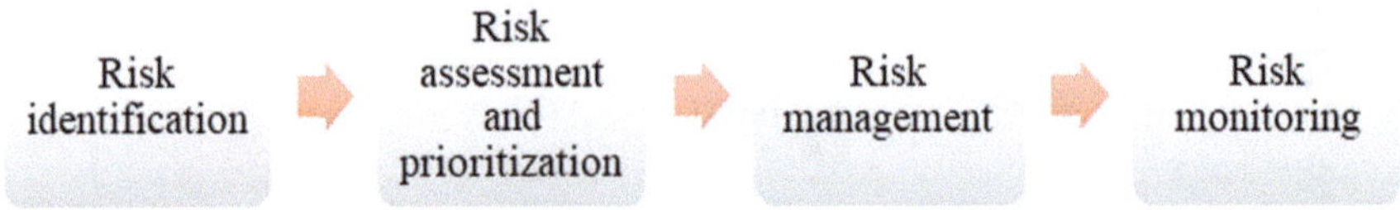

Fig. 2 Risk management system (adapted from Manotas et al. 2016)

With regard to prioritization and evaluation, this phase provides the basis for establishing actions that seek to eliminate, reduce or simply ignore the impacts of previously identified risks. The information of impact and probability of occurrence is obtained qualitatively through panels of experts, and in this sense, these qualitative data lead to the search for tools to analyze them. Table 1 illustrates prioritization techniques according to the literature review.

On the other hand, fuzzy logic allows and offers mathematical options to model the preferences that have been defined by experts of a characteristic process (Wang et al. 2012). However, the use of diffuse theory FST (fuzzy set theory) along with the deployment of quality function, QFD (quality function deployment), they form the basis for a strategic tool that seeks to respond to customer needs and translates them into engineering features (Osorio-Gomez et al. 2018).

Table 1 Prioritization techniques according to the literature (adapted from Manotas et al. 2016)

Techniques	Authors
HP, ANP, fuzzy AHP, fuzzy ANP	Borghesi and Gaudenzi (2013), Hanning et al. (2007), Sun et al. (2012), Zhang et al. (2013), Badea et al. (2014), Levary (2007, 2008), Wu et al. (2006), Li and Li (2010), Wang et al. (2012), Guan et al. (2011)
Discreet simulation, dynamic simulation, petri networks	Vilko et al. (2015), Schmitt and Singh (2012), Feng et al. (2010), Lee et al. (2011)
Fuzzy inference system, fuzzy multicriteria, fuzzy DEMATEL	Aqlan and Lam (2015), Behret et al. (2011), Wen and Xi (2007), Ya-feng and Qi-hua (2009)
Cvar and Var	Soleimani and Govindan (2014), Zhang et al. (2013), Mitra et al. (2015)
AMFE, FMECA	Lee et al. (2011), Tuncel and Alpan (2010), Lavastre et al. (2012)
Delphi method	Markmann et al. (2013), Hanning et al. (2007), Badenhorst-Weiss and Waugh (2014)
Others: stochastic programming, probability–impact matrix, DEA, multi-objective programming, multivariate statistical analysis, constraint programming, control graph P, uncertainty index, fault trees, social systems theory, scenario analysis, probabilistic risk analysis	Soleimani and Govindan (2014), Goh et al. (2007), Oke and Gopalakrishnan (2009), Cheng et al. (2012), Sun et al. (2012), Mojtahedi et al. (2010), Ritchie and Brindley (2007), Han and Chen (2007), Nan et al. (2009), Ouabouch and Amri (2013), Abolghasemi et al. (2015)

3 Methodological Approach

The methodology presented in Fig. 3 is proposed. It identifies three phases associated with the risk management system, which are: prioritization of risks and risk management when defining actions aimed at the elimination or mitigation of risks.

3.1 *Identification of Operational Risks Associated with the Transport and Storage of Blood Products*

From the literature and through the application of questionnaires to experts, the main risks associated with the transport and storage of blood products are identified. The application of the questionnaire was carried out on an individual basis and allowed the experts to rate the risk, in both probability and impact, using the linguistic scale illustrated in Table 2.

After the application of the questionnaire, responses are consolidated and processed according to the scale. Equations 1 and 2 are applied to obtain the percentages of application of risk and the weighted averages of probability of occurrence and magnitude of impact.

Equation 1. Weighted average risk magnitude i

$$\overline{X}_i = \frac{\sum_{j=1}^{n} B_{i,j} M_{i,j}}{n} \tag{1}$$

and

Equation 2. Weighted average probability of risk i

$$\overline{Y}_i = \frac{\sum_{j=1}^{n} B_{i,j} P_{i,j}}{n}; \quad \forall i \tag{2}$$

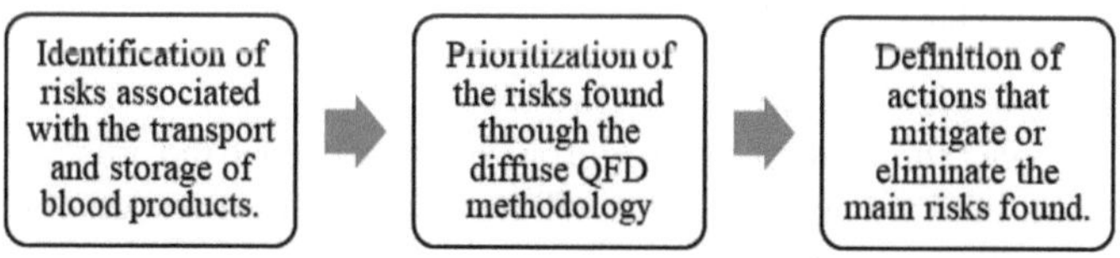

Fig. 3 Methodological design (Agudelo Ibarguen and Naranjo Sánchez 2019)

Table 2 Linguistic scale (Agudelo Ibarguen and Naranjo Sánchez 2019)

Very high	High	Medium	Low	Very low
VH	H	M	L	VL
5	4	3	2	1

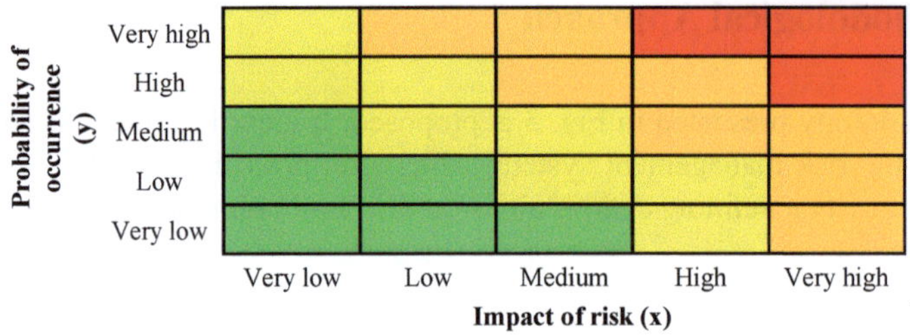

Fig. 4 Matrix impact–probability (adapted from Osorio et al. 2017)

Obtained this data, the matrix of impact–probability is established (see Fig. 4). The matrix is a tool that allows you to visually organize risks according to their impact level. The risks located in the green section are insignificant risks; the risks in the yellow section are risks with a moderate probability of occurrence, while the risks in the orange and red sections are critical risks.

3.2 Prioritization of Operational Risks Through Fuzzy QFD

The following methodology is based on the work of Osorio Gómez (2011), and also the specific methodological proposal for prioritization described in Osorio-Gomez et al (2018) in which the following phases are determined:

- **Identify internal variables—What's**
  In this phase, the panel of experts establishes a set of basic attributes that are expected to be achieved through the process to which operational risks will be prioritized. These variables are understood as the goals of the process.
- **Determine the relative importance of the "What's"**
- The panel of experts determines the importance of the What's, evaluating each attribute with a linguistic qualification according to the following scale used by the authors Bevilacqua et al (2006).

- **Identify Strategic Objectives or "How's"**
  The expert panel should define strategic objectives related to the process of storage and transport of blood products. This will result in the prioritization of risks being consistent with supply chain performance indicators.
- **Determine the correlation between the "What's" and "How's"**
  The same linguistic scale is applied in Table 3 where the expert panel rates the relationship between What's and How's.
- **Determining the Weight of the How's**
  Applying diffuse mathematics and taking the correlation between What's and How's can determine the weights of How's. The result is obtained by multiplying the ratings of each expert with respect to What–How by the weight of the What's obtained in Phase Two.

Table 3 Fuzzy rating scale adapted from Bevilacqua et al. (2006)

Linguistic scale		Triangular diffuse rating (*a*, *b*, *c*)
VL	Very low	(0, 1, 2)
L	Low	(2, 3, 4)
M	Medium	(4, 5, 6)
H	High	(6, 7, 8)
VH	Very high	(8, 9, 10%)

- **Determining the impact of risk on strategic objectives**
  At this stage, the relationship between each of the risks identified by the How's should be established; this relationship will be defined by the experience of the decision-making group.
- **Prioritizing the risks involved**
  To obtain the ranking of risks, the values obtained from the relationship between risks and their impact on the strategic objectives, obtained in the previous point, must be multiplied, for the relative importance of How's.

3.3 Strategies or Actions to Mitigate Operational Risks

Based on previous ratings, strategies must be defined to mitigate or eliminate the risks present in the process and thus improve the process (Osorio-Gomez et al. 2018). As a final step, it is important to emphasize the implementation of actions aimed at transferring, eliminating and/or reducing the risks of the process and applying strategies focused on the individual or associated machinery (Lavastre et al. 2012).

4 Results

The results are presented below using the methodology set out in the previous chapter for the identification and prioritization of operational risks associated with the transport and storage of blood products in Colombia, with the aim of developing actions aimed at the mitigation and reduction of risks.

4.1 Identification of Operational Risks

For the identification of operational risks, an analysis of the literature was carried out to define the most common risks in the processes of transport and storage of blood products. Once these risks, shown in Table 4, were identified, the questionnaire

Table 4 Definition of operational risks

R_i	Description of the risk
R_1	Equipment that makes up the cold chain without complying with minimum specifications and international standards (WHO)
R_2	Equipment used in nonstandard transport and storage
R_3	Lack of availability of technical support, spare parts and maintenance services for cold chain equipment
R_4	Power cuts to prevent continuous operation of refrigeration equipment
R_5	Use of picnic fridges due to shortage of portable refrigerators suitable for blood preservation
R_6	Equipment without sufficient technology for temperature control (devices with built-in temperature control failures and alarms exceeding preservation thresholds)
R_7	Inability of the equipment to maintain a stable temperature in extreme ambient temperature and humidity conditions (from +2 to +6 °C, being the operating temperature of the equipment +4 °C)
R_8	Lack of or inadequate advice from the operator/transporter to the sender on the technical specifications for the transport of total blood, blood components and samples
R_9	Lack of professionals specialized in transfusion medicine
R_{10}	Absence/error in labeling of transport units
R_{11}	Use of FIFO policies (first processed components, first sent for use)
R_{12}	Failure of the absorbent material located between the primary container (total blood unit or hemocomponent) and the secondary container (hermetically sealed plastic bag) in order to prevent leaks affecting the outer container (refrigerator where the blood is transported)
R_{13}	Documentation relating to the dispatch and transport of incomplete or poorly recorded total blood
R_{14}	No standardized or digital registration system
R_{15}	Traffic accident
R_{16}	Inefficient route for the transport of total blood, blood components and samples
R_{17}	Technical faults in the vehicle

designed for the panel of experts was applied, which applied a qualitative scale, and its quantitative equivalent was used for the analysis of the data as shown in Table 5.

Table 6 presents the risks identified and the results of the implementation of the questionnaire.

Table 5 Scales used for data analysis

Qualitative scale	Quantitative scale
Very high	5
High	4
Medium	3
Low	2

Table 6 Validation and weighted averages of operational risks

Risk description	Id	Probability of occurrence	Impact
Equipment that makes up the cold chain without meeting minimum specifications and international standards (WHO)	R_1	2,74	4,05
Lack of availability of technical support, spare parts and maintenance services for cold chain equipment	R_3	2,53	3,79
Cuts in the power supply that prevent the continuous operation of cooling equipment	R_4	2,05	4,11
Use of picnic fridges due to the shortage of portable refrigerators suitable for blood conservation	R_5	2,63	2,63
Equipment without sufficient technology for temperature control (devices with built-in temperature control failures and maintenance threshold overruns alarms)	R_6	2,95	4,05
Equipment's inability to maintain a stable temperature under extreme ambient temperature and humidity conditions (from + 2 to +6 °C, the operating temperature of the equipment being +4 °C)	R_7	2,53	3,95
Lack of or inadequate advice from the operator/transporter to the sender on the technical specifications for the transport of the total blood, hemocomponents and samples	R_8	2,47	4,00
Lack of professionals specialized in transfusion medicine	R_9	2,79	3,37
Absence/error in the labeling of transport units	R_{10}	2,26	4,05
Use of FIFO policies (first components processed, first to be sent for use)	R_{11}	0,89	1,42
Failure in the absorbent material between the primary container (total blood unit or hemocomponent) and the secondary container (hermetically sealed plastic bag) in order to prevent leakage affecting the outer container (refrigerator where blood is transported)	R_{12}	1,74	2,89
Documentation relating to the dispatch and transport of incomplete or poorly registered total blood	R_{13}	2,00	2,89
No standardized or digital recording system	R_{14}	2,32	2,68
It is a traffic accident	R_{15}	2,26	3,68
Inefficient route for the transport of total blood, blood cells and samples	R_{16}	1,79	2,53
Technical failures in the vehicle	R_{17}	1,84	3,00

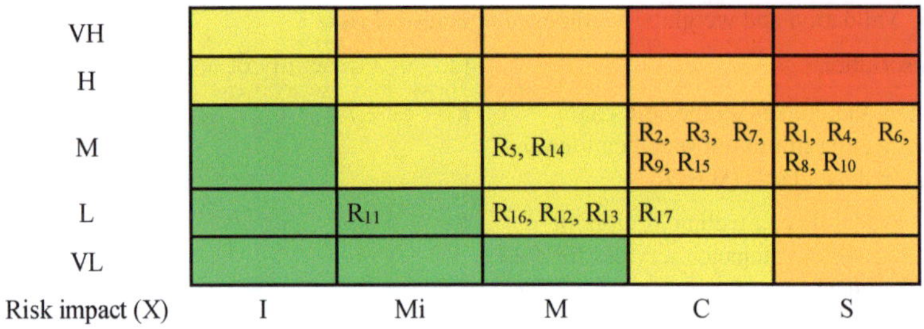

Fig. 5 Risk probability–impact matrix

Once the risks are identified with their probability of occurrence and the impact they generate, these results are placed in a probability–impact matrix, taking into account the rating each expert considers for each risk. This is shown in Fig. 5.

4.2 Prioritization of Operational Risks

The result of applying the above methodology is as follows.

Identify Internal Variables "What's"

The internal variables defined by the expert panel are expressed as follows:

Maintenance of the cold chain throughout the process—*Q*1.

Not to be contaminated at any time during the process—*Q*2.

That everything complies with the processes of cleaning and disinfection—*Q*3.

At the end of the process, you have a safe and quality product—*Q*4.

Determine the Relative Importance of the "What's"

According to the proposed methodology, a relative importance or weight should be assigned to each of the What's by making use of the linguistic scale and the weighting of the responses of each of the experts. The results are illustrated in Table 7.

Identify Strategic Objectives—Hows's How

The panel of experts established 7 strategic objectives as listed below:

- Ensuring quality in the process [C1]
- Have the least possible discard [C2]
- Ensuring the cold chain throughout the transport and storage process [C3]

Table 7 Relative importance and weighting of What's

	What's	*E*1	*E*2	*E*3	*E*4	Weighting w_i		
*Q*1	Maintenance of the cold chain throughout the process	VH	VH	VH	*M*	7.0	8.0	9.0
*Q*2	Never be contaminated during the process	VH	VH	VH	VH	8.0	9.0	10.0
*Q*3	Make sure everything complies with the cleaning and disinfection processes	*H*	VH	VH	*L*	6.0	7.0	8.0
*Q*4	At the end of the process, you have a safe and quality product	VH	VH	VH	*H*	7.5	8.5	9.5

	C1				C2				C3				C4				C5				C6				C7			
N°	*E1*	*E2*	*E3*	*E4*	*E1*	*E2*	*E3*	*E4*	*E1*	*E2*	*E3*	*E4*	*E1*	*E2*	*E3*	*E4*	*E1*	*E2*	*E3*	*E4*	*E1*	*E2*	*E3*	*E4*	*E1*	*E2*	*E3*	*E4*
Q1	VH	VH	--	H	H	VH	L	M	VH	VH	--	VH	VH	VH	--	H	VH	VH	VH	VH	L	L	VH	L	H	--	--	L
Q2	VH	VH	--	H	VH	VH	L	M	VH	VH	--	VH	VH	VH	--	M	VH	VH	VH	VH	M	H	L	L	H	--	--	M
Q3	VH	VH	--	H	VH	VH	L	M	H	VH	--	VH	VH	VH	--	M	VH	VH	VH	VH	H	M	VH	L	VH	--	--	M
Q4	VH	VH	--	M	VH	VH	L	H	VH	VH	--	H	VH	VH	--	H	VH	VH	VH	VH	H	M	VH	M	VH	--	--	L

Fig. 6 Correlation between What's and How's

- Constantly standardize and control the process [C4]
- Ensuring patient safety [C5]
- Ensuring the safety of all personnel involved in the process [C6]
- Ensuring the supply of units [C7].

Determine the Correlation Between What's and How's

At this stage, the panel of experts qualifies the relationship of the wishes with the strategic objectives defined, in a linguistic way, as shown in Fig. 6.

Determining the Weight of the How's

According to the ratings obtained in the determination of correlations, expressed in Table 8 and applying the formulas of diffuse mathematics explained in the methodology, the following triangular numbers are obtained for each strategic objective, which are shown in the table.

These values are obtained from the average of the multiplication of the weight of the "What's" and the valuation given by the experts to the relationship between said What and the strategic objective (How).

Table 8 How's weight

Strategic objectives		*A*	*B*	*C*
W_1	Ensure quality in the process	51.0	66.3	83.6
W_2	Have as little discard as possible	39.3	52.9	68.5
W_3	Ensure the cold chain throughout the transport and storage process	54.8	70.5	88.3
W_4	Constantly standardize and control the process	49.9	65.0	82.2
W_5	To ensure patient safety	57.0	73.1	91.3
W_6	Ensure the safety of all personnel involved in the process	30.9	43.4	57.9
W_7	Ensuring the supply of units	40.1	53.9	69.7

Determining the Impact of Risk on Strategic Objectives

The risks considered as critical located in the red and orange zones of the probability–impact matrix were evaluated according to the relationship between each of them and the How to's, and the following were defined:

R_1 Equipment that make up the cold chain without complying with minimum specifications and international standards (WHO).
R_2 Equipment used in nonstandard transport and storage.
R_3 Lack of availability of technical support, spare parts and maintenance services for cold chain equipment.
R_6 Equipment without sufficient technology for temperature control (devices with built-in temperature control failures and alarms exceeding preservation thresholds).
R_7 Inability of the equipment to maintain a stable temperature in extreme ambient temperature and humidity conditions (from +2 °C to +6, being the operating temperature of the equipment +4 °C).
R_8 Lack of or inadequate advice from the operator/transporter to the sender on the technical specifications for the transport of total blood, blood components and samples.
R_9 Lack of professionals specialized in transfusion medicine.
R_{10} Absence/error in labeling of transport units.

The impact ratings assigned by the panel of experts are shown in Fig. 7.

Prioritizing the Risks Involved

The results of the prioritization are reported in Table 9, where the risks have been sorted downward according to their final Risk Priority Index (IPRF). According to this prioritization, action plans are being drawn up to help reduce or mitigate the

	C1				C2				C3				C4				C5				C6				C7			
	E1	E2	E3	E4	E1	E2	E3	E4	E1	E2	E3	E4	E1	E2	E3	E4	E1	E2	E3	E4	E1	E2	E3	E4	E1	E2	E3	E4
R1	H	H	--	H	H	H	H	M	H	H	--	H	H	H	--	M	H	H	VH	VH	H	H	L	L	H	--	--	H
R2	VH	VH	--	H	VH	VH	H	M	VH	VH	--	H	VH	VH	--	M	VH	VH	VH	VH	L	L	L	L	VH	--	--	H
R3	H	H	--	M	H	H	H	M	H	H	--	H	H	H	--	H	H	H	VH	VH	L	L	L	ML	H	--	--	M
R6	VH	VH	--	H	VH	VH	H	H	VH	VH	--	VH	VH	VH	--	VH	VH	VH	VH	VH	L	L	L	L	H	--	--	H
R7	VH	H	--	H	VH	H	H	M	VH	H	--	H	VH	H	--	H	VH	VH	VH	VH	L	L	L	L	H	--	--	M
R8	H	VH	--	M	H	VH	H	M	H	VH	--	H	VH	VH	--	M	H	H	VH	M	M	M	VH	L	H	--	--	M
R9	VH	H	--	L	VH	H	H	L	H	H	--	L	H	H	--	L	VH	VH	VH	L	L	L	VH	L	H	--	--	L
R10	VH	VH	--	M	VH	VH	H	H	VH	M	--	H	VH	VH	--	M	VH	VH	VH	H	L	L	L	L	VH	--	--	M

Fig. 7 Impact of critical risks on How's to

Table 9 Results of prioritization

N	Type of the risk	Description of the risk	IPRF
Very high risk			552.0
6	Risks of equipment	Equipment without sufficient technology for temperature control (devices with built-in temperature control failures and alarms exceeding preservation thresholds)	486.4
2	Risks of equipment	Equipment used in nonstandard transport and storage	470.3
10	Personnel risks	Absence/error in labeling of transport units	442.0
7	Risks of equipment	Inability of the equipment to maintain a stable temperature in extreme ambient temperature and humidity conditions (from +2 to +6 °C, being the operating temperature of the equipment +4 °C)	439.4
High risk			431.1
8	Personnel risks	Lack of or inadequate advice from the operator or transporter to the sender on the technical specifications for the transport of total blood, blood components and samples	427.7
1	Risks of equipment	Equipment that makes up the cold chain without complying with minimum specifications and international standards (WHO)	419.0
3	Risks of equipment	Lack of availability of technical support, spare parts and maintenance services for cold chain equipment	396.2
9	Personnel risks	Lack of professionals specialized in transfusion medicine	370.4

impact of these plans on the strategic objectives of the process, for which the eight risks defined are classified into two categories, taking into account that they are risks arising from persons and equipment.

5 Actions Aimed at Mitigation

Once the risk prioritization phase was completed, it was decided to group risks according to their origin, resulting in two groups: risks associated with equipment and risks associated with staff. For each grouping, the cause-effect tool is used to investigate the causes related to each type of risk. After the identification of causes, interviews are conducted with experts in order to determine actions to mitigate or eliminate the prioritized risks.

5.1 Risks Associated with Equipment

Of the prioritized risks, the risks associated with the equipment are:

R_6 Equipment without sufficient technology for temperature control (devices with built-in temperature control failures and alarms exceeding preservation thresholds).

R_2 Equipment used in nonstandard transport and storage.

R_7 Inability of the equipment to maintain a stable temperature in extreme ambient temperature and humidity conditions (from +2 to +6 , being the operating temperature of the equipment +4).

R_1 Equipment that makes up the cold chain without complying with minimum specifications and international standards (WHO).

R_3 Lack of availability of technical support, spare parts and maintenance services for cold chain equipment.

The causes found after applying the cause-effect methodology for this risk group are illustrated in Fig. 8.

After proper validation with the panel of experts, the following actions were identified:

Actions aimed at risk mitigation:

- Suitable packaging must be provided to maintain the temperature over long distances. These packagings are: cardboard box, isothermal box and isothermal refrigerator. In addition, shock absorbers such as absorbent paper or bubble paper, among others, can be used to protect the blood components inside the refrigerator (Instituto Nacional de Salud 2019).
- A preventive review program is recommended for all equipment that makes up the chain, taking into account the specialty of each equipment and the time required in case of complete or partial replacement and the availability of its spare parts.
- It must have a program of review and calibration according to international standards and minimum specifications for the use of cold chain equipment for the process.

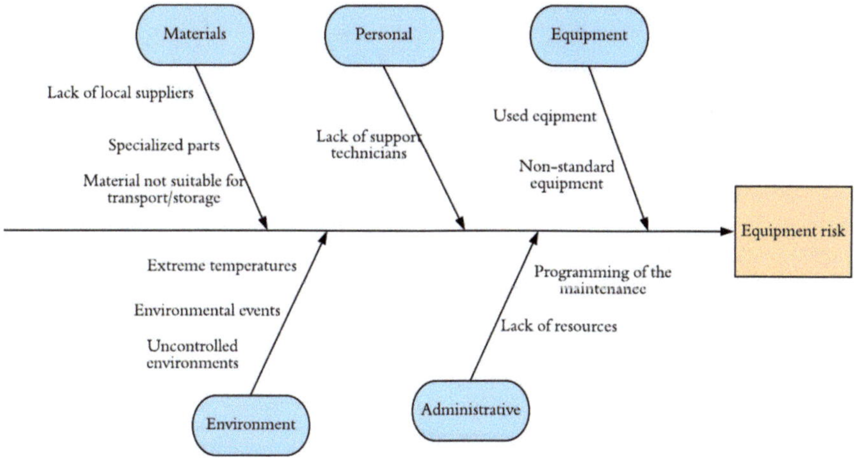

Fig. 8 Cause-effect equipment risk diagram

5.2 *Risks Associated with Staff*

Of the prioritized risks, the risks associated with staff are:

R_{10} Absence/error in labeling of transport units.
R_8 Lack of or inadequate advice from the operator/transporter to the sender on the technical specifications for the transport of total blood, blood components and samples.
R_9 Lack of professionals specialized in transfusion medicine.

Similarly, for this risk group, the cause-effect methodology is applied, the result of which is shown in Fig. 9.

Because of a literature review and validation with the expert group, the following actions were identified:

Actions aimed at risk mitigation:

- Human resource needs should be identified and ensured, including operational personnel and personnel for general storage and transport activities, who should be competent, technically up to date and qualified for the position held (Pardo and Cendales 2015).
- Staff should have knowledge of existing standards and working arrangements, continuous training in the quality system and sufficient time to do the work and for activities such as inspection and verification (Pardo and Cendales 2015).
- It should be borne in mind that individual roles and responsibilities are clearly defined, documented and disseminated to avoid gaps and overlaps. The responsibilities entrusted to each individual will not be so numerous as to constitute a risk to product quality (Pardo and Cendales 2015).

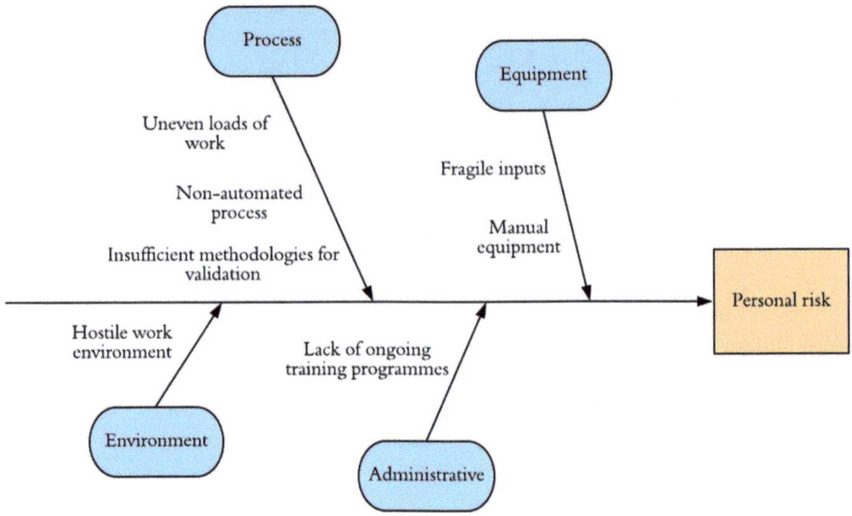

Fig. 9 Cause-effect staff risk diagram

- Written procedures on personnel selection, training and training processes are recommended (Pardo and Cendales 2015).

According to this prioritization, action plans are being drawn up to help reduce or mitigate the impact of these plans on the strategic objectives of the process, for which the eight risks defined are classified into two categories, taking into account that they are risks arising from persons and equipment.

6 Conclusions

The results of the prioritization assumed that the critical risks of the process target the teams and the people involved throughout the chain. It is important for the country and public health that the state invests in blood banks using prioritization techniques such as the one developed in this work to channel and target the best investment opportunities.

For future work, continuing research into risk management around the blood supply chain, the issue of risk quantification could also be addressed by considering the impact of the social component of blood supply.

References

Abolghasemi M, Khodakarami V, Tehranifard H (2015) A new approach for supply chain risk management: mapping SCOR into Bayesian network 8(1):280–302

Agudelo Ibarguen N, Naranjo Sánchez MD (2019) Gestión Del Riesgo Operacional En El Almacenamiento Y Transporte Terrestre De Productos Sanguíneos En Colombia Universidad Del Valle. https://doi.org/10.5281/zenodo.1477753

Aqlan F, Lam SS (2015) A fuzzy-based integrated framework for supply chain risk assessment. Int J Prod Econ 161:54–63. https://doi.org/10.1016/j.ijpe.2014.11.013

Badea A, Prostean G, Goncalves G, Allaoui H (2014) Assessing risk factors in collaborative supply chain with the analytic hierarchy process (AHP). Proc Soc Behav Sci 124:114–123. https://doi.org/10.1016/j.sbspro.2014.02.467

Badenhorst-Weiss JA, Waugh BJ (2014) Business environmental factors affecting South Africa's supply chains and economic growth and development. Probl Perspect Manage 12(4):238–291

Behret H, Öztayşi B, Kahraman C (2011) A fuzzy inference system for supply chain risk management. In: Wang Y, Li T (eds) Practical applications of intelligent systems. Advances in intelligent and soft computing, 124. Berlin, Heidelberg: Springer, pp 429–438. https://doi.org/10.1007/978-3-642-25658-5_52

Bevilacqua M, Ciarapica FE, Giacchetta G (2006) A fuzzy-QFD approach to supplier selection. J Purchasing Supply Manage 12(1):14–27. https://doi.org/10.1016/j.pursup.2006.02.001

Borghesi A, Gaudenzi B (2013) Risk Identification. In Borghesi A, Gaudenzi B (eds) Risk management. Springer, Verlag Italia, pp 43–52. https://doi.org/10.1007/978-88-470-2531-8

Cheng TCE, Yip FK, Yeung ACL (2012) Supply risk management via guanxi in the Chinese business context: the buyers perspective. Int J Prod Econ 139(1):3–13. https://doi.org/10.1016/j.ijpe.2011.03.017

Feng LFL, Jun-qi HJH, Dao-ming XDX (2010) Managing disruption risks in supply chain. In: 2010 IEEE international conference on emergency management and management sciences (ICEMMS), pp 434–438. https://doi.org/10.1109/ICEMMS.2010.5563408

Goh M, Lim JYS, Meng F (2007) A stochastic model for risk management in global supply chain networks. Eur J Oper Res 182:164–173. https://doi.org/10.1016/j.ejor.2006.08.028

Guan G, Dong Q, Li C (2011) Risk identification and evaluation research on F-AHP evaluation based supply chain. In: 2011 IEEE 18th international conference on industrial engineering and engineering management, part 3, pp 1513–1517. https://doi.org/10.1109/ICIEEM.2011.6035447

Han M, Chen J (2007) Managing operational risk in supply chain. In: 2007 international conference on wireless communications, networking and mobile computing, WiCOM 2007, pp 4914–4917. https://doi.org/10.1109/WICOM.2007.1205

Hanning Y, Jiangwei Y, Kang P (2007) Risk identification of electronic commerce project based on cloud-control. In: 2007 international conference on wireless communications, networking and mobile computing, WiCOM 2007, vol 2005047, pp 3648–3651. https://doi.org/10.1109/WICOM.2007.903

Instituto Nacional de Salud (2018) Informe anual red nacional de bancos de Sangre y servicios de transfusión, Colombia 2017. Recovered from https://www.ins.gov.co/Direcciones/RedesSaludPublica/DonacionSangre/AreasEstrategicas/Informe%20anual%20Red%20Sangre%202017%20v2.pdf

Instituto Nacional de Salud (2019) Lineamiento sobre especificaciones para el embalaje terciario en transporte de hemocomponentes distribuidos a nivel nacional Versión 03. Colombia. Recovered from https://www.ins.gov.co/Direcciones/RedesSaludPublica/DonacionSangre/Publicaciones/Boletin%20t%C3%A9cnico%20embalaje%20terciario.pdf

Lavastre O, Gunasekaran A, Spalanzani A (2012) Supply chain risk management in French companies. Decis Support Syst 52(4):828–838. https://doi.org/10.1016/j.dss.2011.11.017

Lee CKM, Yeung YC, Hong Z (2011) Managing the risks of outsourcing in supply chain networks. In: 1st international technology management conference (figure 1), pp 488–494. Retrieved from https://ieeexplore.ieee.org/xpls/abs_all.jsp?arnumber=5996017

Li Y, Li X (2010) Study on ANP cluster supply chain risk evaluation. In: 2010 international conference on logistics systems and intelligent management, ICLSIM 2010, vol 3, pp 1296–1299. https://doi.org/10.1109/ICLSIM.2010.5461173

Manotas DF, Osorio JC, Rivera L (2016) Operational risk management in third party logistics (3PL). In: Alor-Hernández G, Sánchez-Ramírez C, García-Alcaraz JL (eds) Handbook of research on managerial strategies for achieving optimal performance in industrial processes, IGI Global, pp 218–239. https://doi.org/10.4018/978-1-5225-0130-5

Mansur A, Vanany I, Indah Arvitrida N (2018) Challenge and opportunity research in blood supply chain management: a literature review. MATEC Web Conf 154:1–6. https://doi.org/10.1051/matecconf/201815401092

Markmann C, Darkow IL, von der Gracht H (2013) A Delphi-based risk analysis—identifying and assessing future challenges for supply chain security in a multi-stakeholder environment. Technol Forecast Soc Chang 80(9):1815–1833. https://doi.org/10.1016/j.techfore.2012.10.019

Mitra S, Karathanasopoulos A, Sermpinis G, Dunis C, Hood J (2015) Operational risk: Emerging markets, sectors and measurement. Eur J Oper Res 241(1):122–132. https://doi.org/10.1016/j.ejor.2014.08.021

Mojtahedi SMH, Mousavi SM, Makui A (2010) Project risk identification and assessment simultaneously using multi-attribute group decision making technique. Saf Sci 48(4):499–507. https://doi.org/10.1016/j.ssci.2009.12.016

Nan J, Huo JZ Liu HH (2009) Supply chain purchasing risk evaluation of manufacturing enterprise based on Fuzzy-AHP method. In: 2009 2nd international conference on intelligent computing technology and automation, ICICTA 2009, vol 3, no 70772077, pp 1001–1005. https://doi.org/10.1109/ICICTA.2009.707

Oke A, Gopalakrishnan M (2009) Managing disruptions in supply chains: a case study of a retail supply chain. Int J Prod Econ 118:168–174. https://doi.org/10.1016/j.ijpe.2008.08.045

Organización Mundial de la Salud (2004) Departamento de Tecnologías Sanitarias Esenciales. La cadena de frio de la sangre. Guía para la adquisición de equipos y accesorios

Organización Mundial de la salud (2016) OMS | ¿Por qué es importante donar sangre? Recovered from https://www.who.int/features/qa/61/es/Señalarlafuentedeestareferencia

Osorio Gómez JC (2011) Fuzzy QFD for multicriteria decision making—application example. Prospectiva 9(2):22–29

Osorio-Gomez JC, Manotas-Duque DF, Rivera L, Canales I (2018) Operational risk prioritization in supply chain with 3PL using Fuzzy-QFD. In: García-Alcaraz JL, Alor- Hernández G, Rivera Cadavid L, Canales Valdivieso, I (Coords) New perspectives on applied industrial tools and techniques, management an industrial engineering. Springer, pp 91–109. https://doi.org/10.1007/978-3-319-56871-3

Osorio JC, Manotas DF, Rivera L (2017) Priorización de Riesgos Operacionales para un Proveedor de Tercera Parte Logística - 3PL. Inf Tecnol 28(4):135–144. https://doi.org/10.4067/S0718-07642017000400016

Ouabouch L, Amri M (2013) Analysing supply chain risk factors: a probability-impact matrix applied to pharmaceutical industry. J Logist Manage 2(2):35–40. https://doi.org/10.5923/j.logistics.20130202.01

Pardo C, Cendales R. Incidencia, mortalidad y prevalencia de cáncer en Colombia, 2007–2011. Primera edición. Bogotá. D.C. Instituto Nacional de Cancerología, vol 1, p 148

Ritchie B, Brindley C (2007) Supply chain risk management and performance: a guiding framework for future development. Int J Oper Prod Manage 27(3):303–322. https://doi.org/10.1108/01443570710725563

Schmitt AJ, Singh M (2012) A quantitative analysis of disruption risk in a multi-echelon supply chain. Int J Prod Econ 139(1):22–32. https://doi.org/10.1016/j.ijpe.2012.01.004

Soleimani H, Govindan K (2014) Reverse logistics network design and planning utilizing conditional value at risk. Eur J Oper Res 237(2):487–497. https://doi.org/10.1016/j.ejor.2014.02.030

Sun J, Matsui M, Yin Y (2012) Supplier risk management: an economic model of P-chart considered due-date and quality risks. Int J Prod Econ 139(1):58–64. https://doi.org/10.1016/j.ijpe.2012.03.004

Tuncel G, Alpan G (2010) Risk assessment and management for supply chain networks: a case study. Comput Ind 61:250–259. https://doi.org/10.1016/j.compind.2009.09.008

Vengoechea M (2016) Bancos de sangre en Colombia, en saldo rojo. Noticias, canal RCN.com. Colombia

VIDA/SALUD (2016) Donación de sangre en Colombia_ bancos presentan saldos en rojo - Archivo Digital de Noticias de Colombia y el Mundo. Periódico el Tiempo. Colombia

Vilko J, Ritala P, Hallikas J (2015) Risk management abilities in multimodal maritime supply chains: visibility and control perspectives. Accid Anal Prev 123:469–481. https://doi.org/10.1016/j.aap.2016.11.010

Wang L, Juan YK, Wang J, Li KM, Ong C (2012) Fuzzy-QFD approach based decision support model for licensor selection. Expert Syst Appl 39(1):1484–1491. https://doi.org/10.1016/j.eswa.2011.08.037

Wen L Xi Z (2007) Supply chain risk evaluation based on fuzzy multi-criteria lattice-order decision-making. In: 2007 IEEE international conference on automation and logistics, Jinan, pp 1442–1445. https://doi.org/10.1109/ICAL.2007.4338797

Wu T, Blackhurst J, Chidambaram V (2006) A model for inbound supply risk analysis. Comput Ind 57:350–365. https://doi.org/10.1016/j.compind.2005.11.001

Ya-feng LYL, Qi-hua XQX (2009) A method of Identifying supply chain risk factors. WRI World Congr Softw Eng 4:369–373. https://doi.org/10.1109/WCSE.2009.275

Zhang AN, Goh, M, Terhorst, M, Lee, AJL, Pham MT (2013) An interactive decision support method for measuring risk in a complex supply chain under uncertainty. In: Proceedings—2013 IEEE international conference on systems, man, and cybernetics, SMC 2013, pp 633–638. https://doi.org/10.1109/SMC.2013.113

Taguchi's Loss Function in the Weight Quality of Products: Case Study of Cheese Making

Nicolás Francisco Mateo-Díaz, Lidilia Cruz-Rivero, Luis Adolfo Meraz-Rivera, and Jesús Ortiz-Martínez

Abstract The application of tools, techniques, and methodologies for the improvement of the quality of products or services creates opportunities to increase competitiveness in the market and significantly reduces production costs. One method that quantifies the poor quality of a product that is already on the market and therefore causes loss to society is that created by the quality guru Genichi Taguchi, called the quality loss function. This methodology was applied in a cheese company that belongs to the dairy industry to quantify the financial loss for both the consumer and the company by deviating a quality characteristic from its nominal value (product weight), determining the economic loss for five presentations, of product, and later establishing a mathematical equation and then applying an experimental design to improve the quality of the product in five different presentations.

Keywords Taguchi's loss function · Quality cost · MSD · Nominal is better

1 Introduction

Variation is inevitable in any manufacturing process. The nature of design always requires variation on either side of the target. The allowable amount of variation that will not affect the functional requirements of the component is called tolerance (Badu and Asha 2015). The loss function developed by Genichi Taguchi in 1950 is a key concept and tool to establish a financial measure of the negative impact on society (consumer, producer, etc.) of a product's performance when it deviates from a target

N. F. Mateo-Díaz (✉)
Departamento de Ciencias Básicas E Ingenierías- Ingeniería Industrial, Universidad del Caribe, Cancún, Quintana Roo, México
e-mail: nmateo@ucaribe.edu.mx

L. Cruz-Rivero · L. A. Meraz-Rivera
Subdirección de Posgrado e Investigación, Tecnológico Nacional de México, Campus Tantoyuca, Tantoyuca, Veracruz, México

J. Ortiz-Martínez
Dirección de posgrados, Universidad Euro-Hispanoamericana, Xalapa, Veracruz, México

J. L. García Alcaraz et al. (eds.), *Techniques, Tools and Methodologies Applied to Quality Assurance in Manufacturing*,
https://doi.org/10.1007/978-3-030-69314-5_21

value u target (t = target). This implies that the quality characteristic of a product, "y," must be closer and closer to its ideal value, "t," and everything that deviates from the ideal is considered a loss for society (Gutiérrez and De la Vara 2012).

According to Taguchi, there are three types of scenarios to calculate the loss function of a product; the first scenario is nominal and is better, this type of method is used when the characteristics of the product indicate that it must be an objective value, that is, not more, not less, an exact value. The second scenario is the smaller the better, they are variables or quality characteristics whose only requirement is that they do not exceed a certain minimum tolerated value or higher specification, and the smaller their value is the better. The third scenario is the bigger the better, they are variables or quality characteristics that are required to be greater than a minimum value or a certain lower specification, and the larger the value of the variable, the better. For this study, the scene-river to be used is nominal, since the weight of each type of presentation of the cheese must be exact and without any variation.

2 Literature Review

The Taguchi loss function has been used in various types of processes and has been integrated with other methodologies to obtain optimal results in solving specific problems. For example, Badu and Asha (2015) concluded that the Taguchi loss function focuses on a single sought value; but in final assembly processes, a range of values would have to be found, since the components of a product depend on tolerances. For this, they propose a method that they call interval-based symmetric Taguchi loss function (SIT) and complement the study with an algorithm of improved inheritance of the flock of sheep, thus obtaining a minimum loss value in the final assembly. Another integration with other methodologies was proposed by Chiu (2014), where they create a statistical economic design of the EWMA graph with variable time control limits incorporating the Taguchi loss function, thereby seeking to find the optimal solution in minimizing the cost of quality total. Zhang et al. (2019), discovered that the Taguchi loss function is not appropriate in the tolerance design of hierarchical products and proposed an extension to the model, which after being applied to fiberglass products, the results demonstrated good operability in the hierarchical product tolerance design.

Pasha et al. (2018) propose an integrated model of the Taguchi loss function and the economic statistical design of the $\bar{x}$ control charts, considering both normal and non-normal quality data. Their proposal has been achieved by modifying the Banerjee and Rahim cost model using a non-uniform sampling scheme for systems with increasing failure rate and the Weibull shock model. Then, they study and compare the effects of three most popular quality characteristic distributions (Normal, Burr, and Johnson), and the optimal parameters of the economic and economic statistical designs reveal that there is an insignificant difference between the Normal and Burr distributions, while it can be observed a relative difference in the case of the Johnson distribution.

Other research indicates that the Taguchi parametric design and the quality loss function model were successfully applied to improve the thickness of the solder paste in a component; since if the quality characteristic deviated from the nominal value, it had consequences that affected quality, thereby achieving stability in the manufacturing process (Huang 2018). In other works, it has been applied to evaluate the loss of quality during the spline carving operation of a shaft manufacturing process of five industries that manufacture the same type of shaft. Then, industries that have a comparative loss of quality are identified greater, and the necessary actions have been suggested to improve the quality of their products (Shilpa and Naidu 2014).

Even the application of the Taguchi loss function has been extended to the service sector, particularly in logistics, for example, to measure the arrival of buses and determine the loss when there are delays (Budaj and Hrnčiar 2016). In other cases, it has been used to make decisions evaluating the performance of suppliers; as in the study by Sharma and Kumar (2015), who created a model to solve the problem of strategic third-party logistics selection or logistics outsourcing which integrates the QFD technique and the Taguchi loss function. The QFD allowed to speed up the selection process while the loss function measured the performance of suppliers.

In educational institutions, it has been applied to assess the skills that a graduate of technical and engineering institutions contributes to the companies where they integrate to work (Upadhayay and Vrat 2016).

3 Study Area

The study was carried out in a cheese manufacturing company that belongs to the dairy agro-industrial sector, located in the municipality of Tempoal, Veracruz, Mexico, whose geographical coordinates are 21° 31′ north latitude and 98° 23′ west longitude at a height of 50 m above sea level. This municipality stands out since its main economic activity is livestock, with 19 economic units dedicated to the transformation of milk to produce different types of cheeses (DENUE 2019).

The company has 25 years of experience in the production of dairy products and derivatives (cream, cottage cheese and ground cheese in five presentations: 250 gr, 350 gr, 450 gr, 600 gr, and 900 gr, Asadero cheese, grain and panela cheese).

By 2020, the company has an average production capacity of 2000 kg of cheese daily in its various presentations and works throughout the year with 14 people in the production area, and the months of greatest demand are December, January, and a period of May through June with frequent orders, and the main source of demand for the company is toward the state of Hidalgo, Mexico.

4 Application: Cheese Making Case Study

To find the loss function of dairy products, specifically fresh cheese in its various presentations at the cheese manufacturing company, we proceeded to obtain fixed costs and monthly production costs; calculating first the monthly investment in liters of milk made by the company, for this, Table 1 shows the cost description.

As observed based on the calculations made, the monthly investment in liters of milk is $615,090.00, likewise, $30,000.00 is invested in milk powder, so the total monthly investment is $645,090.00. Subsequently, the fixed costs and production costs provided by the company's management are recorded on a monthly basis in Table 2.

Table 1 Description of costs

Monthly milk investment register	Quantity	Calculation methodology
Amount of milk (Lt.) recorded in one day	6658	Observation in company
Amount of powdered milk (Lt.) recorded in one day	2200	6.25 kg. Of milk powder yield 550 L of water, (there are 4 tubs with a capacity of 1100 Lt.)
Total milk (Lt)	8858	Total natural milk and total milk powder daily
Milk quantity destined for other processes (Lt)	3000	It is used to make Oaxaca type cheese and cream
Amount of milk used to make hoop cheese (Lt)	5858	Total daily milk (Lt) less total milk destined for other processes
Quantity of milk used to make 1 kg of cheese	7	3.5 Lt. of natural milk and 3.5 Lt. of powdered milk
Amount of milk used per month (specifically for hoop cheese) (Lt)	175,740	Sum of natural and powdered milk daily, multiplied by 30 days
Amount of cheese obtained per month (kg)	25,105.71	Amount of milk used per month (Lt)/amount of milk to make 1 kg of cheese
Cost per Lt. of milk	$3.50	Data provided by company management
Cost per kg of milk powder	$40.00	A 25 kg package has a cost of $1000.00 therefore the cost of a package is divided by the kg it contains
Monthly investment (natural milk)	$615,090.00	The product of multiplying the Lt. of milk per month by the cost of one Lt. of milk for 30 days
Monthly investment (Powdered milk)	$30,000.00	The product of multiplying the cost of a 25 kg package of powdered milk for 30 days
Total monthly investment in Lt. of milk	$645,090.00	Sum of total monthly investment of natural milk and milk powder

Table 2 Fixed and monthly production costs

Fixed cost		Production cost	
Description	Cost	Description	Cost
Water	$9000.00	Milk	$645,090.00
Electric light	$13,000.00	Packaging (Rings and others)	$3000.00
Workers salary	$63,000.00	Rennet and chemicals	$7000.00
Total	$85,000.00	Total	$655,090.00

Table 3 Cost of producing a kg of cheese

Concept	Quantity	Calculation methodology
Daily production	836.8571	kg of cheese obtained per month/30 days
Monthly production	25,105.71	Milk used per month (Lt)/amount of milk to make 1 kg of cheese
Cost of producing a kg of cheese	$29.48	Sum of fixed costs/monthly production in kg of cheese
Cost of producing one gram of cheese	$0.02948	Cost of producing a kg of cheese/1000 g

Based on the determination of the costs, the cost of producing a kg of cheese is calculated (Table 3), considering the sum of the fixed and production costs, as well as the daily and monthly production of kg of cheese.

As it is observed in Table 3, the cost of producing a kg of cheese is $29.48, in addition, the calculation was made to obtain the cost of a gram of cheese, and this data is used to know the loss function in the subsequent procedures.

4.1 *Statistical Sampling*

A sampling was carried out for each of the five cheese presentations (250 gr, 350 gr, 450 gr, 600 gr, and 900 gr). To observe the deviations that appear and to avoid setbacks and delays in production, a sample size of n = 12 cheeses was taken for each type of presentation, and the results on some descriptive statistics of interest are shown in Table 4. Figure 1: a, b, c, d, and e shows the deviation of the measured characteristic (cheese weight) for each product presentation with respect to its target value.

According to the calculated averages, the weight of the presentations of the product of 350 gr and 900 gr is below the nominal value, indicating that the product causes loss to society (Fig. 1b, e). While the averages of the presentations of 250, 450, and 600 are above the nominal value, which represents an economic loss for the company (Fig. 1a, c, d). Figure 2 shows the box diagram for the five presentations of the product, in which individual values of the product weights are observed; outliers are not shown.

Table 4 Some descriptive statistics by type of product presentation

Product presentation (gr)	250	350	400	600	900
Mean deviated weight (gr)	11.25	34.58	9.16	34.16	17.25
Total lost for the company (gr)	0	0	130	410	10
Total lost to society (gr)	135	415	20	0	197
Average product weight (gr)	261.25	315.41	459.16	634.16	882.75
Variance	175.52	122.74	203.47	203.47	167.35
Total cheeses with defect (gr)	10	12	12	12	12

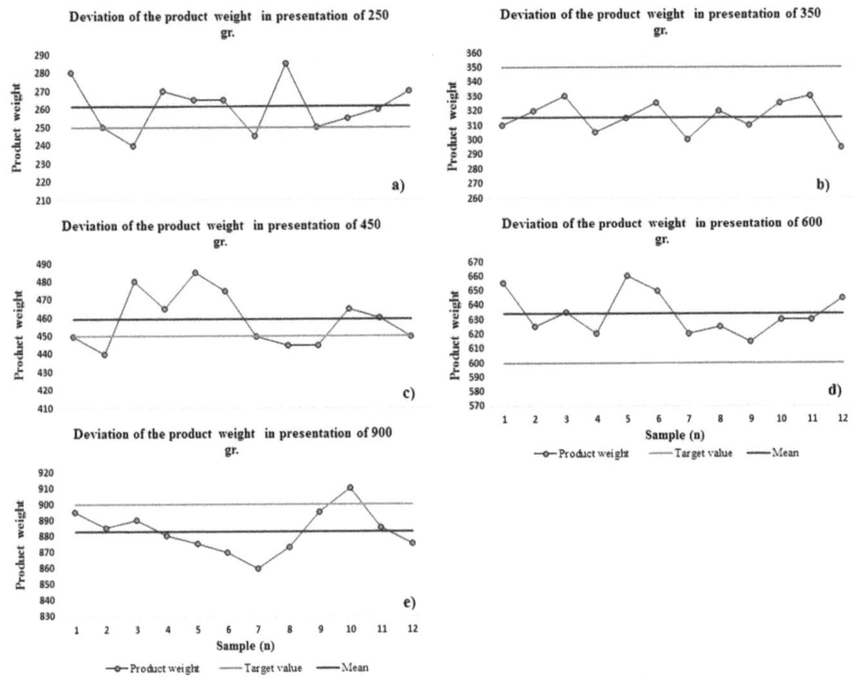

Fig. 1 Averages of the product weights in relation to the nominal value

4.2 Production Loss Calculation

30.6 kg of cheese were analyzed through the sampling carried out, and it was detected by the sum of the deviations that the lost grams are 2390 g (2.390 kg), and the daily losses in cheese production were calculated, as shown in Table 5.

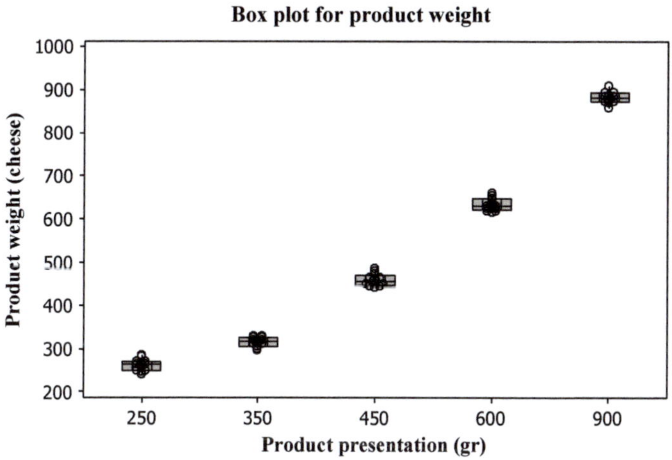

Fig. 2 Box plot for product weight

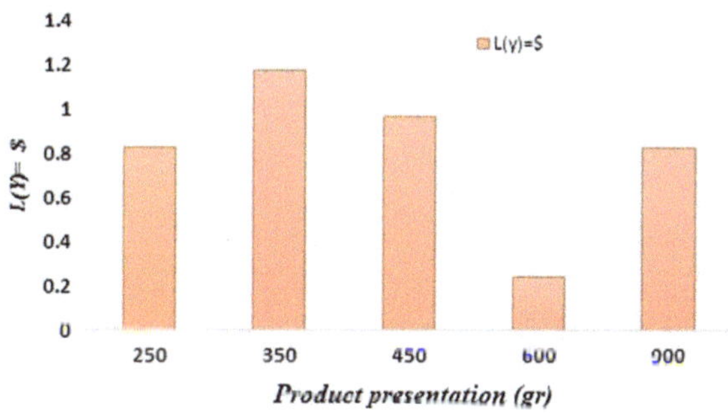

Fig. 3 Cost ($) of each product to deviate from the nominal value

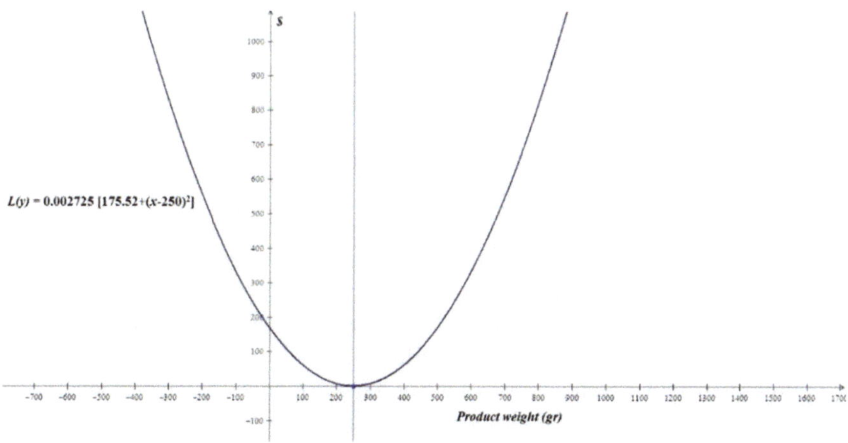

Fig. 4 Loss function for presentation 250 gr. $\mathrm{L}(y) = 0.0027\ [175.52 + (261.25–250)^2]$

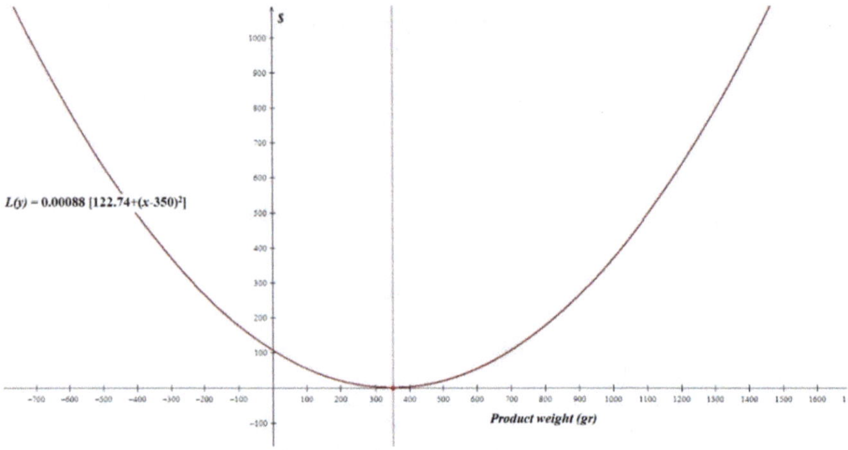

Fig. 5 Loss function for presentation 350 gr. L (y) = 0.00088 [122.74 + (315.41–350)^2]

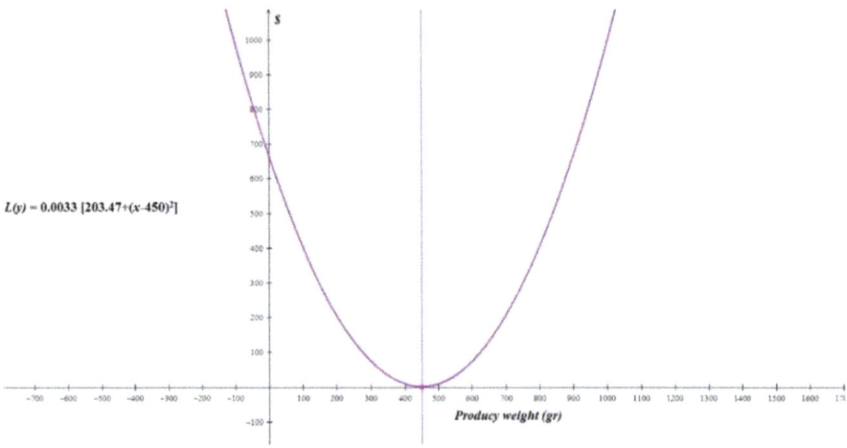

Fig. 6 Loss function for presentation 450 gr. L (y) = 0.0033 [203.47 + (459.16–450)^2]

4.3 Data for the Loss Function

To establish the loss function of the products of the company under study, it is important to define all the costs that directly influence the product; the costs involved for said calculation are the production costs, however, there is another cost that must be considered, which is the cost of transporting the products to the point of sale. For this purpose, in Tables 6 and 7, the results of the transport costs are presented.

Based on the sample of 60 cheeses in its various presentations, being 30.6 kg analyzed, the product deviations due to excess or lack in each presentation are shown in Table 8.

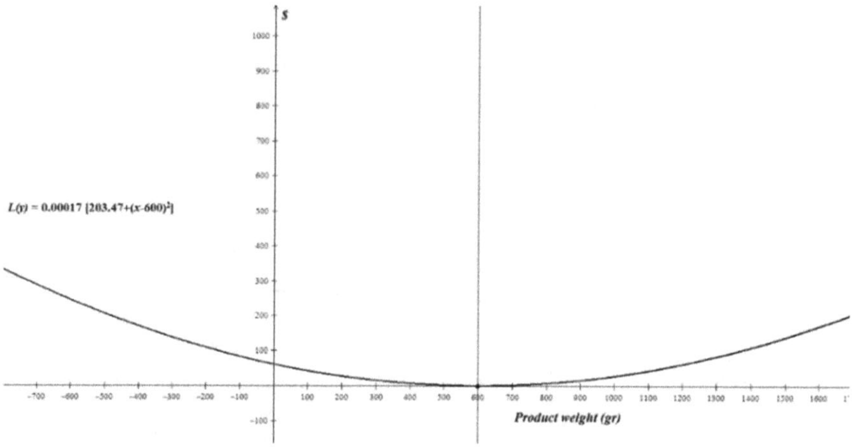

Fig. 7 Loss function for presentation 600 gr. L (y) = 0.00017 [203.47 + (634.16–600)^2]

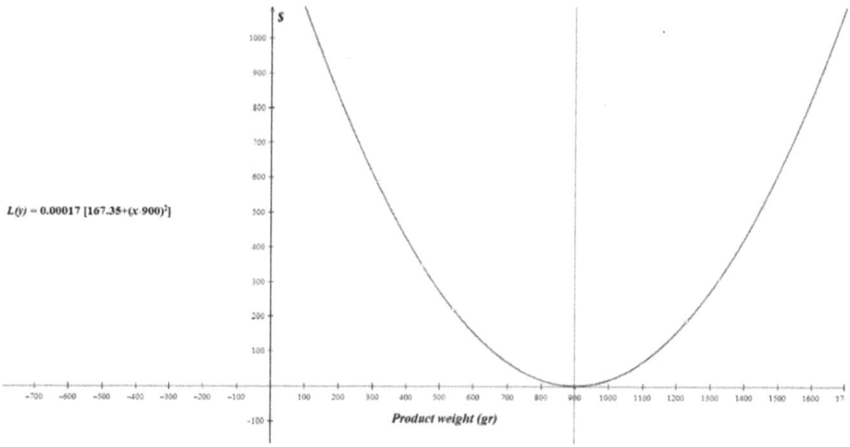

Fig. 8 Loss function for presentation 900 gr. L (y) = 0.0017 [167.35 + (882.75–900)^2]

5 Results

5.1 Calculation of the Loss Function for Nominal is Better

In column four of Table 8, the listed deviations correspond to the sample analysis carried out, and the values in this column represent a constant of proportionality (Δ_0) whose data allows us to calculate the loss function. The cost to transfer the excess quantity results from the multiplication of the cost to transfer a gr of cheese by the unit deviation in gr with respect to the nominal value of a cheese. The cost per surplus depending on the presentation of the cheese is obtained from the multiplication of

Table 5 Cost for daily waste

Concept	Quantity	Calculation methodology
Amount of product sampled (kg)	30.6	12 pieces were sampled for each type of presentation of fresh cheese (250 gr; 350 gr; 450 gr; 600gr; 900 gr.)
Loss for the company in gr for every 30,600 kg	2390	Cheese sampling
Equivalent in kg	2.39	Cheese sampling
Cost per 2390 gr wasted	$70.45	Losses in gr for the company multiplied by the cost of producing one gr of cheese
kg lost in a month	1960.871148	Monthly production of kg of cheese multiplied by the loss in kg of cheese for the company / the number of kg Sampled
Monthly loss cost	$57,804.42	Cost of producing a kg of cheese multiplied by the kg lost in a month
Annual loss cost	$693,652.98	Monthly loss cost multiplied by twelve (months of the year)
Daily waste (Kg.)	65.36237162	kg lost in a month/30 (days of a month)
Daily waste cost	$1926.81	Cost per monthly loss/30 (days of a month)

Table 6 Data for the loss function (daily)

Description	Quantity
Daily cheese production (kg)	836.86
Daily cheese production (gr)	836,857.14
The route of transfer of the produced product, has a capacity of 2000 kg of cheese	
Monthly fuel	$30,000.00

Table 7 Cost to transfer 1 kg of cheese

Description	Quantity
monthly fuel/Monthly cheese production (kg)	$1.19
Cost of producing a kg of cheese	$29.48
Sum of the above	$30.67
Cost to transfer 1 gr of cheese	
Monthly gasoline/Monthly production in gr	$0.001195
Cost of producing a kg of cheese/1000	$0.0294789
Sum of the above	$0.0306739

Table 8 Costs for weight deviations in each type of product presentation

Product presentation (gr.)	Equivalence of 2000 kg in different cheese presentations	Cheeses analyzed	Unit deviation ± (gr) with respect to the nominal value of a cheese (Δ_0)	Cost to transfer the excess quantity (gr)	Cost per excess depending on the type of cheese	Cost to deviate the weight of each cheese (A_0)
250	3347.4285	12	11.25	\$0.01	\$0.33	\$0.3450813
350	2391	12	34.58	\$0.04	\$1.02	\$1.0608055
450	1860	12	9.16	\$0.01	\$0.27	\$0.2811774
600	1395	12	175	\$0.21	\$5.16	\$5.3679313
900	930	12	17.25	\$0.02	\$0.51	\$0.5291247

the unit deviation of the nominal value of the cheese for the cost of producing one gr of cheese. The cost to deviate the weight of each cheese is the result of the sum of the cost to transfer the surplus quantity and the cost per surplus depending on the type of cheese, and these values represent the value of A_0, which is a data to be able to calculate the loss function.

Equation (1) is used to calculate the quality loss function for more than one piece studied.

$$(y) = k[\sigma^2 + (\overline{y} - m)^2] \tag{1}$$

where k – constant depending on tolerances and product repair costs; σ^2 represents the variance of the quality characteristic; y = mean of the quality characteristic; m = target, target value, that is, 250, 350, 450 , 600 and 900 gr. The expression in square brackets represents the mean square deviation (MSD).

First, we proceed to calculate the constant of proportionality k, with Eq. (2)

$$k = \frac{A_0}{\Delta_0^2} \tag{2}$$

where the values $\boldsymbol{A_0}$ and Δ_0 signify the cost originated by this loss and the deviation of the quality characteristic, respectively. The loss function for each type of cheese presentation is shown in Table 9.

Figure 3 indicates that the presentation of product of 350 gr presents greater loss, which means that deviating from its nominal value the loss costs \$1.1696; whose cost is cause to society since the weight is less than specified. On the contrary, the presentation of 600 gr has a low loss of just \$0.2402 toward the factory, since the weight is above the nominal value. The Taguchi loss functions obtained for each product presentation, structured according to Eq. 1, are presented in Table 10 with their respective figures.

Table 9 Costs for weight deviations in each type of product presentation

Product presentation (gr)	250	350	450	600	900
A_0	0.3450	1.060	0.2811	5.3679	0.5291
Δ_0^2	126.56	1196.00	84.0277	30,625	297.56
k	0.0027	0.00088	0.0033	0.00017	0.0017
$\bar{y}$	261.25	315.41	459.16	634.16	882.75
m	250	350	450	600	900
σ^2	175.52	122.74	203.47	203.47	167.35
MSD	302.08	1318.75	287.50	1370.83	469.91
L(y): ($)	0.8236	1.1696	0.9620	0.2402	0.8267

Table 10 Costs for weight deviations in each type of product presentation

Product presentation (gr)	Loss function	Figure
250	$L(y) = 0.0027\ [175.52 + (261.25 - 250)^2]$	4
350	$L(y) = 0.00088\ [122.74 + (315.41 - 350)^2]$	5
450	$L(y) = 0.0033\ [203.47 + (459.16 - 450)^2]$	6
600	$L(y) = 0.00017\ [203.47 + (634.16 - 600)^2]$	7
900	$L(y) = 0.0017\ [167.35 + (882.75 - 900)^2]$	8

The third column of Table X refers to the number with which the function is represented in Figs. 4, 5, 6, 7 and 8; for example, for the presentation of the product of 250 gr, the loss function corresponds to $\boldsymbol{L(y)} = 0.0027\ [175.52 + (261.25–250)^2]$ and is represented in Fig. 4.

The loss function is a quadratic function, a parabola that opens up and whose vertex touches the x axis, with coordinates (x, y) which represents the position of the nominal value of the desired quality characteristic.

The value k of the function causes the parabola to be more open or more closed, and the lower this value, the parabola has a larger opening.

6 Discussion

The losses for the products with presentations of 250, 350, 450, 600, and 900 gr are 0.8236, 1.1696, 0.9620, 0.2402, and 0.8267 pesos, respectively. The results are summarized in Figs. 4, 5, 6, 7 and 8. The loss of $0.8236 for the 250-gr product means that a randomly selected product shipped from the cheese factory is, on average, presenting a loss of $0.8236, indicating that of 3347 pieces produced from that product presentation, $2752 is lost daily. Someone spends that amount: a customer, the company itself, or an indirect consumer.

The lower the MSD, the lower the average loss for society. So, an improvement project should focus on obtaining high quality at a low cost, reducing the MSD. This can be accomplished using parameter design and tolerance design. The savings would represent customer satisfaction, reduced warranty costs, future market share, etc.

According to Taguchi, the loss function offers a way to quantify the benefits achieved by reducing variability around the target; it certainly helps justify an investment decision to improve a process that is already capable of meeting specifications.

Results on the effectiveness of this method have been reported in several works; some with other integrated methods depend on the application focus and type of problem. A study at airports used the Taguchi loss function to help make better decisions by setting different target values and quantifying quality loss as a metric to assess airport performance (Shojaei et al. 2018). Kumar et al. (2019) used the method to select reliable suppliers, where variables such as quality, delivery, price, and service were taken into account, then through a sensitivity analysis the credibility of the model was demonstrated, and the function of loss helped quantify the loss due to non-compliance by providers.

In the supply chain, it has been used to optimize processes and improve quality, particularly quantifying the cost of real quality loss in the economic product quantity (POQ) model, in such a way that analysis is involved from the retailer and the supplier maximizing the total benefit of the supply chain (Chuang and Wu 2018).

Furthermore, the loss function has even been used to determine sample sizes that minimize the cost to control processes using exponentially weighted moving average control charts (Huang and Lu 2015).

7 Conclusion

The loss function of Taguchi allowed knowing which product presentation is the one that generates the greatest economic loss when it deviates from the nominal value. In order to formulate the function, it was necessary to calculate the costs involved for the manufacture of cheeses, such as fixed, production, and logistics, and subsequently, the constant of proportionality k was calculated, which depends on the cost of this loss and the deviation of the characteristic of the quality. This value multiplies the mean square deviation (MSD), and the loss values are shown in Table 9 while the loss function equation in Table 10; the figures shown from 4, 5, 6 and 8 show the behavior of the curve when the weight of the product is deviated for each presentation.

Acknowledgements To the owner of the cheese manufacturing company of Tempoal, Veracruz for having allowed the project to be developed as part of the continuous improvement in its processes. To the Consejo Nacional de Ciencia y Tecnología (CONACYT) for the support of the scholarship for postgraduate studies.

References

Babu JR, Asha A (2015) Modelling in selective assembly with symmetrical interval-based Taguchi loss function for minimising assembly loss and clearance variation. Int J Manuf Technol Manage 29(5–6):288–308. https://doi.org/10.1504/IJMTM.2015.071223

Budaj P, Hrnčiar M (2016). Specifying the use of Taguchi's loss function in manufacturing and service sectors. In: 34th International conference mathematical methods in economics,MME

Chiu WC (2015) Economic-statistical design of EWMA control charts based on Taguchi's loss function. Commun Stat Simul Comput 44(1):137–153. https://doi.org/10.1080/03610918.2013.773346

Chuang CJ, Wu CW (2018) Optimal process mean and quality improvement in a supply chain model with two-part trade credit based on Taguchi loss function. Int J Prod Res 56(15):5234–5248. https://doi.org/10.1080/00207543.2017.1394591

Directorio Estadístico Nacional de Unidades Económicas (2019) Instituto Nacional de Estadística y Geografía. INEGI. https://en.www.inegi.org.mx/app/mapa/denue/

Gutiérrez PH, De la Vara SR (2012) Análisis y diseño de experimentos. Editorial Mc Graw Hill. 3a Ed. México, D.F. 489 pp.

Huang CJ, Lu SL (2015) Considering Taguchi loss function on statistically constrained economic sum of squares exponentially weighted moving average charts. J Stat Comput Simul 85(3):572–586. https://doi.org/10.1080/00949655.2013.829059

Huang CY (2018) Applying the Taguchi parametric design to optimize the solder paste printing process and the quality loss function to define the specifications. Solder Surface Mount Technol. https://doi.org/10.1108/SSMT-03-2017-0010

Kumar R, Padhi SS, Sarkar A (2019) Supplier selection of an Indian heavy locomotive manufacturer: an integrated approach using Taguchi loss function, TOPSIS, and AHP. IIMB Manag Rev 31(1):78–90. https://doi.org/10.1016/j.iimb.2018.08.008

Pasha MA, Moghadam MB, Fani S, Khadem Y (2018) Effects of quality characteristic distributions on the integrated model of Taguchi's loss function and economic statistical design of-control charts by modifying the Banerjee and Rahim economic model. Commun Stat Theor Methods 47(8):1842–1855. https://doi.org/10.1080/03610926.2017.1328512

Sharma SK, Kumar V (2015) Optimal selection of third-party logistics service providers using quality function deployment and Taguchi loss function. Int J Benchmarking. https://doi.org/10.1108/BIJ-02-2014-0016

Shilpa M, Naidu NVR (2014) Quantitative evaluation of quality loss for fraction defective case using Taguchi's quality loss function. Int J Logist Syst Manage 18(1):126–138. https://doi.org/10.1504/IJLSM.2014.062124

Shojaei P, Haeri SAS, Mohammadi S (2018) Airports evaluation and ranking model using Taguchi loss function, best-worst method and VIKOR technique. J Air Transp Manag 68:4–13. https://doi.org/10.1016/j.jairtraman.2017.05.006

Upadhayay L, Vrat P (2016) An ANP based selective assembly approach incorporating Taguchi's quality loss function to improve quality of placements in technical institutions. TQM J https://doi.org/10.1108/TQM-06-2014-0054

Zhang Y, Li L, Song M, Yi R (2019) Optimal tolerance design of hierarchical products based on quality loss function. J Intell Manuf 30(1):185–192. https://doi.org/10.1007/s10845-016-1238-6

MIX
Papier aus verantwortungsvollen Quellen
Paper from responsible sources
FSC® C105338

If you have any concerns about our products,
you can contact us on
ProductSafety@springernature.com

In case Publisher is established outside the EU,
the EU authorized representative is:
Springer Nature Customer Service Center GmbH
Europaplatz 3, 69115 Heidelberg, Germany

Printed by Libri Plureos GmbH
in Hamburg, Germany